Lecture Notes in Artificial Intelligence   10377

Subseries of Lecture Notes in Computer Science

More information about this series at http://www.springer.com/series/1244

Marcello Balduccini · Tomi Janhunen (Eds.)

Logic Programming and Nonmonotonic Reasoning

14th International Conference, LPNMR 2017
Espoo, Finland, July 3–6, 2017
Proceedings

 Springer

Editors
Marcello Balduccini
Drexel University
Philadelphia, PA
USA

Tomi Janhunen
Aalto University
Espoo
Finland

ISSN 0302-9743   ISSN 1611-3349   (electronic)
Lecture Notes in Artificial Intelligence
ISBN 978-3-319-61659-9   ISBN 978-3-319-61660-5   (eBook)
DOI 10.1007/978-3-319-61660-5

Library of Congress Control Number: 2017944353

LNCS Sublibrary: SL7 – Artificial Intelligence

Printed on acid-free paper

This Springer imprint is published by Springer Nature
The registered company is Springer International Publishing AG
The registered company address is: Gewerbestrasse 11, 6330 Cham, Switzerland

Preface

This volume contains the papers presented at the 14th International Conference on Logic Programming and Nonmonotonic Reasoning (LPNMR 2017) that was held July 3–6, 2017, in Espoo, Finland. The newly renovated Hanasaari Cultural Center served as the idyllic and scenic venue of the conference.

The LPNMR conference is a forum for exchanging ideas on declarative logic programming, nonmonotonic reasoning, and knowledge representation. The aim of the LPNMR conference series is to facilitate interactions between researchers and practitioners interested in the design and implementation of logic-based programming languages and database systems, and researchers who work in the areas of knowledge representation and nonmonotonic reasoning. The conference strives to encompass theoretical and experimental studies that have led or will lead to advances in declarative programming and knowledge representation, as well as their deployment in practical applications. The past editions of LPNMR were held in Washington, D.C., USA (1991), Lisbon, Portugal (1993), Lexington, Kentucky, USA (1995), Dagstuhl, Germany (1997), El Paso, Texas, USA (1999), Vienna, Austria (2001), Fort Lauderdale, Florida, USA (2004), Diamante, Italy (2005), Tempe, Arizona, USA (2007), Potsdam, Germany (2009), Vancouver, Canada (2011), Coruña, Spain (2013), and Lexington, Kentucky, USA (2015).

The 2017 edition received 47 submissions in three categories: technical papers, system descriptions, and application descriptions. Submissions as both long and short papers were considered and each submission was evaluated and reviewed by at least three Program Committee members. The final list of 27 accepted papers consists of 16 long and 11 short contributions, further divided into 15 technical papers, seven system descriptions, and five application descriptions.

The scientific program featured invited talks by Jõao Leite, Tran Cao Son, and Francesca Toni, as well as the oral presentations of the technical papers mentioned above. In addition, the program included sessions dedicated to the 7th Answer Set Programming Competition, the Doctoral Consortium of the conference, and a panel on the past and future of LPNMR. The main conference was preceded by five workshops offering an inspiring start for the conference. The conference proceedings contain extended abstracts for the invited talks, the 27 technical papers, and the report on the 7th Answer Set Programming competition. The social program of the conference included an informal get-together at Hanasaari, the official reception of Espoo City at Karhusaari Art Center, and a conference dinner served at Haltia Nature Center in Nuuksio National Park.

Many people played an important role in the success of LPNMR 2017 and deserve our warmest thanks and acknowledgments as follows. The Program Committee and the additional reviewers worked hard for the fair and thorough evaluation of submissions. The technical program is an essential contribution of the invited speakers and the authors of the accepted papers. The organizers of the programming competition,

Martin Gebser, Marco Maratea, and Francesco Ricca, have dedicated themselves to a long-term effort pushing ahead the improvement of state-of-the-art LPNMR systems. Marina De Vos took care of the organization of an excellent Doctoral Consortium program, guiding young researchers to plan their research and careers. Joohyung Lee's contribution was invaluable in coordinating the selection and organization of workshops for the conference. Peter Schüller advertised the conference and its workshops through a number of channels. Finally, the members of the Organizing Committee and volunteers deserve our special thanks for the success of the event.

The LPNMR 2017 conference received generous support from several organizations. We gratefully acknowledge our sponsors, the Finnish Cultural Foundation, the Foundation for Knowledge Representation and Reasoning (KR Inc), the Association for Logic Programming (ALP), and the European Association for Artificial Intelligence (EurAI). The Association for the Advancement of Artificial Intelligence (AAAI) helped to advertise the event through its channels. The possibilities for fast-track journal publications in the *Artificial Intelligence Journal and Theory and Practice of Logic Programming*, as well as the best paper prize offered by Springer, brought additional value and motivation. The functions of the EasyChair system were an indispensable help in the management of submissions and of the reviewing process. The representatives of Espoo City and Hanasaari Cultural Center kindly advised us of different possibilities and attractions that Espoo, the hometown of Aalto University, can offer for the organization of a scientific conference. The hosting university helped with practical matters such as payment transactions and accounting.

May 2017   Marcello Balduccini
                                                                          Tomi Janhunen

Organization

Program Committee Chairs

Marcello Balduccini	Drexel University, USA
Tomi Janhunen	Aalto University, Finland

Workshops Chair

Joohyung Lee	Arizona State University, USA

Publicity Chair

Peter Schüller	Marmara University, Turkey

Doctoral Consortium Chair

Marina De Vos	University of Bath, UK

The 7th ASP Competition

Martin Gebser	University of Potsdam, Germany
Marco Maratea	University of Genoa, Italy
Francesco Ricca	University of Calabria, Italy

Local Organization at Aalto University

Jori Bomanson	Shahab Tasharrofi
Tomi Janhunen	Mary-Ann Wikström
Noora Suominen de Rios	

Program Committee

Jose Julio Alferes	Universidade Nova de Lisboa, Portugal
Marcello Balduccini	Drexel University, USA
Chitta Baral	Arizona State University, USA
Bart Bogaerts	KU Leuven, Belgium
Gerhard Brewka	Leipzig University, Germany
Pedro Cabalar	University of Coruña, Spain
Francesco Calimeri	University of Calabria, Italy

Stefania Costantini	University of L'Aquila, Italy
James Delgrande	Simon Fraser University, Canada
Marc Denecker	KU Leuven, Belgium
Marina De Vos	University of Bath, UK
Agostino Dovier	University of Udine, Italy
Thomas Eiter	Vienna University of Technology, Austria
Wolfgang Faber	University of Huddersfield, UK
Paul Fodor	Stony Brook University, USA
Alfredo Gabaldon	GE Global Research, USA
Martin Gebser	University of Potsdam, Germany
Michael Gelfond	Texas Tech University, USA
Giovanni Grasso	Oxford University, UK
Giovambattista Ianni	University of Calabria, Italy
Daniela Inclezan	Miami University, USA
Tomi Janhunen	Aalto University, Finland
Matthias Knorr	Universidade Nova de Lisboa, Portugal
Joohyung Lee	Arizona State University, USA
Yuliya Lierler	University of Nebraska at Omaha, USA
Vladimir Lifschitz	University of Texas at Austin, USA
Fangzhen Lin	Hong Kong University of Science and Technology, Hong Kong, SAR China
Marco Maratea	University of Genoa, Italy
Alessandra Mileo	National University of Ireland, Ireland
Emilia Oikarinen	Finnish Institute of Occupational Health, Finland
Mauricio Osorio	Universidad de las Americas Puebla, Mexico
Ravi Palla	GE Global Research, USA
David Pearce	Universidad Politécnica de Madrid, Spain
Axel Polleres	Vienna University of Economics and Business, Austria
Enrico Pontelli	New Mexico State University, USA
Christoph Redl	Vienna University of Technology, Austria
Alessandra Russo	Imperial College, UK
Orkunt Sabuncu	University of Potsdam, Germany
Chiaki Sakama	Wakayama University, Japan
Torsten Schaub	University of Potsdam, Germany
Peter Schüller	Marmara University, Turkey
Tran Cao Son	New Mexico State University, USA
Hannes Strass	Leipzig University, Germany
Theresa Swift	Universidade Nova de Lisboa, Portugal
Shahab Tasharrofi	Aalto University, Finland
Eugenia Ternovska	Simon Fraser University, Canada
Hans Tompits	Vienna University of Technology, Austria
Mirek Truszczynski	University of Kentucky, USA
Agustin Valverde	Universidad de Màlaga, Spain
Kewen Wang	Griffith University, Australia
Yisong Wang	Guizhou University, China
Stefan Woltran	Vienna University of Technology, Austria

Jia-Huai You   University of Alberta, Canada
Yi Zhou   University of Western Sydney, Australia

Sponsors and Collaborators

Aalto University, Finland
Artificial Intelligence Journal, Elsevier
Association for the Advancement of Artificial Intelligence (AAAI)
Association for Logic Programming (ALP)
Espoo Innovation Garden, Finland
European Association for Artificial Intelligence (EurAI)
Finnish Cultural Foundation, Finland
Foundation for Knowledge Representation and Reasoning (KR Inc)
Theory and Practice of Logic Programming, Cambridge University Press

Additional Reviewers

Borja, Veronica   Germano, Stefano   Oetsch, Johannes
Carballido, Jose Luis   Kaminski, Tobias   Romero, Javier
Chowdhury, Md Solimul   Karimi, Arash   Van den Eynde, Tim
Dal Palù, Alessandro   Kaufmann, Benjamin   van der Hallen, Matthias
Devriendt, Jo   Kiesl, Benjamin   Wang, Zhe
Dodaro, Carmine   Lackner, Martin   Zangari, Jessica
Fandinno, Jorge   Lapauw, Ruben   Zhuang, Zhiqiang
Formisano, Andrea   Marzullo, Aldo
Fuscà, Davide   Nieves, Juan Carlos

Doctoral Consortium Contributions

Flavio Everardo   Sampling and Search Space with Answer
                           Set Programming
Markus Hecher   Structure-Driven Answer-Set Solving - Extended
                           Abstract
Tobias Kaminski   Answer Set Programs with External Source Access:
                           Integrated Evaluation and New Applications
Patrick Lühne   Discovering and Proving Invariants in Answer
                           Set Programming and Planning
Max Ostrowski   Modern Constraint Answer Set Solving
Javier Romero   Extending Answer Set Programming with Declarative
                           Heuristics, Preferences, and Online Planning
Sebastian Schellhorn   Theory Reasoning with Answer Set Programming
Richard Taupe   Lazy Grounding and Heuristic Solving in Answer
                           Set Programming

Contents

Answer Set Programming

LPNMR Systems

LPNMR Applications

Invited Talks

The Design of the Seventh Answer Set Programming Competition

Martin Gebser[1]([⊠]), Marco Maratea[2], and Francesco Ricca[3]

[1] Institute for Computer Science, University of Potsdam, Potsdam, Germany
gebser@cs.uni-potsdam.de
[2] DIBRIS, Università di Genova, Genoa, Italy
[3] Dipartimento di Matematica e Informatica, Università della Calabria, Rende, Italy

Abstract. Answer Set Programming (ASP) is a prominent knowledge representation language with roots in logic programming and non-monotonic reasoning. Biennial competitions are organized in order to furnish challenging benchmark collections and assess the advancement of the state of the art in ASP solving. In this paper, we report about the design of the Seventh ASP Competition, which is jointly organized by the University of Calabria (Italy), the University of Genova (Italy), and the University of Potsdam (Germany), in affiliation with the 14th International Conference on Logic Programming and Non-Monotonic Reasoning (LPNMR 2017). A novel feature of this competition edition is the re-introduction of a Model&Solve track, complementing the usual System track with problem domains where participants need to provide dedicated encodings and solving means.

1 Introduction

Answer Set Programming (ASP) [8,14,20,27,34,38,41] is a prominent knowledge representation language with roots in logic programming and non-monotonic reasoning. The goal of the ASP Competition series is to promote advancements in ASP methods, collect challenging benchmarks, and assess the state of the art in ASP solving (see, e.g., [1,3,9,15,16,24,25,37,39] for recent ASP systems). In this paper, we report about the design of the Seventh ASP Competition,[1] which is jointly organized by the University of Calabria (Italy), the University of Genova (Italy), and the University of Potsdam (Germany), in affiliation with the 14th International Conference on Logic Programming and Non-Monotonic Reasoning (LPNMR 2017).[2]

The Seventh ASP Competition includes a System track, oriented at the design of previous competition editions [17,26]: (*i*) benchmarks adhere to the ASP-Core-2 standard modeling language,[3] (*ii*) sub-tracks are based on language features utilized in problem encodings (e.g., aggregates, choice or disjunctive

---

[1] http://aspcomp2017.dibris.unige.it.
[2] http://lpnmr2017.aalto.fi.
[3] https://www.mat.unical.it/aspcomp2013/ASPStandardization/.

© Springer International Publishing AG 2017
M. Balduccini and T. Janhunen (Eds.): LPNMR 2017, LNAI 10377, pp. 3–9, 2017.
DOI: 10.1007/978-3-319-61660-5_1

rules, queries, and weak constraints), (*iii*) problem instances are classified and selected according to their expected hardness, and (*iv*) the best-performing systems are given more solving time in a Marathon track. A novel feature of this competition edition is the re-introduction of a Model&Solve track, complementing the System track with problem domains where participants need to provide dedicated encodings and solving means. In contrast to earlier ASP competitions with a Model&Solve track, i.e., the 2009, 2011, and 2013 editions (cf. [17]), the problem domains are purposefully limited to showcases in which features going beyond ASP-Core-2 are of interest. Namely, the Model&Solve track of the Seventh ASP Competition aims at domains involving discrete as well as continuous dynamics [7], so that extensions like Constraint Answer Set Programming (CASP) [40] and incremental ASP solving [23], which are beyond the scope of the System track, may be exploited.

The rest of this paper focuses on the System track of the Seventh ASP Competition and is organized as follows. Section 2 presents new problem domains contributed to this competition edition, followed by a survey of participant systems in Sect. 3, and Sect. 4 concludes the paper.

2 Benchmark Suite

Eight new problem domains, which are further detailed below, have been kindly provided for the System track of the Seventh ASP Competition. In addition, we acknowledge the contribution of new instances, augmenting the collection of benchmarks from previous competition editions, to the *Graph Colouring* domain.

Bayesian Network Learning. Bayesian networks are directed acyclic graphs representing (in)dependence relations between variables in multivariate data analysis. Learning the structure of Bayesian networks, i.e., selecting edges such that the resulting graph fits given data best, is a combinatorial optimization problem amenable to constraint-based solving methods like the one proposed in [18]. In fact, data sets from the literature serve as instances in this domain, while a problem encoding in ASP-Core-2 expresses optimal Bayesian networks, given by directed acyclic graphs whose associated cost is minimal.

Crew Allocation. This scheduling problem, which has also been addressed by related constraint-based solving methods [28], deals with allocating crew members to flights such that the amount of personnel with certain capabilities (e.g., role on board and spoken language) as well as off-times between flights are sufficient. Instances with different numbers of flights and available personnel further restrict the amount of personnel that may be allocated to flights in a way that no schedule is feasible under these restrictions.

Markov Network Learning. As with Bayesian networks, the learning problem for Markov networks [31] aims at the optimization of graphs representing the dependence structure between variables in statistical inference. In this domain, the graphs of interest are undirected and required to be chordal, while associated scores express marginal likelihood w.r.t. given data. Problem instances of varying hardness are obtained by taking samples of different size and density from literature data.

Paracoherent ASP. Given an incoherent logic program P, a paracoherent (or semi-stable) answer set corresponds to a gap-minimal answer set of the epistemic transformation of P [30]. The instances in this domain, used in [5] to evaluate genuine implementations of paracoherent ASP, are obtained by grounding and transforming incoherent programs stemming from previous editions of the ASP Competition. In particular, weak constraints single out answer sets of a transformed program such that the associated gap is cardinality-minimal.

Random Disjunctive ASP. The disjunctive logic programs in this domain express random 2QBF formulas, given as conjunctions of terms in disjunctive normal form, by an extension of the Eiter-Gottlob encoding in [19]. Parameters controlling the random generation of 2QBF formulas (e.g., number of variables and number of conjunctions) are set such that instances lie close to the phase transition, while having an expected average solving time below the competition timeout of 20 min per run.

Resource Allocation. This scheduling problem deals with allocating the activities of business processes to resources such that role requirements and temporal relations between activities are met [29]. Moreover, the total makespan of schedules is subject to an upper bound as well as optimization. The hardness of instances in this domain varies w.r.t. the number of activities, temporal relations, available resources, and upper bounds.

Supertree Construction. The goal of the supertree construction problem [33] is to combine the leaves of several given phylogenetic subtrees into a single tree fitting the subtrees as closely as possible. That is, the structures of subtrees shall be preserved, yet tolerating the introduction of intermediate nodes between direct neighbors, while avoiding such intermediate nodes is an optimization target as well. Instances of varying hardness are obtained by mutating projections of binary trees with different numbers of leaves.

Traveling Salesperson. The well-known traveling salesperson problem [6] is to optimize the round trip through a (directed) graph in terms of the accumulated edge cost. Instances in this domain are twofold by stemming from the TSPLIB repository[4] or being randomly generated to increase the variety in the ASP Competition, respectively.

---

[4] http://elib.zib.de/pub/mp-testdata/tsp/tsplib/tsplib.html.

3   Participant Systems

Fifteen systems, registered by four teams, participate in the System track of the Seventh ASP Competition. The majority of systems runs in the single-processor category, while two (indicated by the suffix "-MT" below) exploit parallelism in the multi-processor category. In the following, we survey the registered teams and systems.

Aalto. The team from Aalto University registered nine systems that utilize normalization [11,12] and translation [10,13,22,32,35] means. Two systems, LP2SAT+LINGELING and LP2SAT+PLINGELING-MT, perform translation to SAT and use LINGELING or PLINGELING, respectively, as back-end solver. Similarly, LP2MIP and LP2MIP-MT rely on translation to Mixed Integer Programming along with a single- or multi-threaded variant of CPLEX for solving. The LP2ACYCASP, LP2ACYCPB, and LP2ACYCSAT systems incorporate translations based on acyclicity checking, supported by CLASP run as ASP, Pseudo-Boolean, or SAT solver as well as the GRAPHSAT solver in case of SAT with acyclicity checking. Moreover, LP2NORMAL+LP2STS takes advantage of the SAT-TO-SAT framework to decompose complex computations into several SAT solving tasks. Unlike that, LP2NORMAL confines preprocessing to the (selective) normalization of aggregates and weak constraints before running CLASP as ASP solver.

ME-ASP. The ME-ASP team from the University of Genova, the University of Sassari, and the University of Calabria registered the multi-engine ASP system ME-ASP2, which is an updated version of ME-ASP [36,37], the winner system in the Regular track of the Sixth ASP Competition. Like its predecessor version, ME-ASP2 investigates features of an input program to select its back ends from a pool of ASP grounders and solvers. As regards grounders, ME-ASP2 can pick either DLV or GRINGO, while the available solvers include a selection of those submitted to the Sixth ASP Competition as well as CLASP.

UNICAL. The team from the University of Calabria plans to submit four systems utilizing the recent I-DLV grounder [16], developed as a redesign of (the grounder component of) DLV going along with the addition of new features. Moreover, back ends for solving will be selected from the variety of existing ASP solvers.

WASPINO. The WASPINO team from the University of Calabria and the University of Genova registered the WASPINO system. In case an input program is tight [21], WASPINO uses MAXINO [4], a MaxSAT solver extended with cardinality constraints, and otherwise the ASP solver WASP [2,3], winner in the Marathon track of the Sixth ASP Competition.

4   Conclusion

We have presented the design of the Seventh ASP Competition, with particular focus on new problem domains and systems registered for the System track.

A novel feature of this competition edition is the re-introduction of a Model&Solve track, complementing the System track with problem domains where features going beyond the ASP-Core-2 standard modeling language are of interest.

At the time of writing, we are finalizing the collection of benchmarks for both tracks. This goes along with the classification of problem instances according to their expected hardness and the installation of participant systems on the competition platform. The results and winners of the Seventh ASP Competition will be announced at LPNMR 2017.

References

1. Alviano, M., Dodaro, C., Ricca, F.: JWASP: a new Java-based ASP solver. In: Bistarelli, S., Formisano, A., Maratea, M. (eds.) Proceedings of RCRA 2015, vol. 1451 of CEUR Workshop Proceedings, pp. 16–23. CEUR-WS.org (2015)
2. Alviano, M., Dodaro, C., Faber, W., Leone, N., Ricca, F.: WASP: a native ASP solver based on constraint learning. In: Cabalar, P., Son, T.C. (eds.) LPNMR 2013. LNCS (LNAI), vol. 8148, pp. 54–66. Springer, Heidelberg (2013). doi:10.1007/978-3-642-40564-8_6
3. Alviano, M., Dodaro, C., Leone, N., Ricca, F.: Advances in WASP. In: Calimeri, F., Ianni, G., Truszczynski, M. (eds.) LPNMR 2015. LNCS (LNAI), vol. 9345, pp. 40–54. Springer, Cham (2015). doi:10.1007/978-3-319-23264-5_5
4. Alviano, M., Dodaro, C., Ricca, F.: A MaxSAT algorithm using cardinality constraints of bounded size. In: Yang, Q., Wooldridge, M. (eds.) Proceedings of IJCAI 2015, pp. 2677–2683. AAAI Press (2015)
5. Amendola, G., Dodaro, C., Faber, W., Leone, N., Ricca, F.: On the computation of paracoherent answer sets. In: Singh, S., Markovitch, S. (eds.) Proceedings of AAAI 2017, pp. 1034–1040. AAAI Press (2017)
6. Applegate, D., Bixby, R., Chvátal, V., Cook, W.: The Traveling Salesman Problem: A Computational Study. Princeton University Press, Princeton (2007)
7. Balduccini, M., Magazzeni, D., Maratea, M.: PDDL+ planning via constraint answer set programming. In: Bogaerts, B., Harrison, A. (eds.) Proceedings of ASPOCP 2016, pp. 1–12 (2016)
8. Baral, C.: Knowledge Representation, Reasoning and Declarative Problem Solving. Cambridge University Press, New York (2003)
9. Béatrix, C., Lefèvre, C., Garcia, L., Stéphan, I.: Justifications and blocking sets in a rule-based answer set computation. In: Carro, M., King, A. (eds.) Technical Communications of ICLP 2016, vol. 52 of OASIcs, pp. 6:1–6:15. Schloss Dagstuhl (2016)
10. Bogaerts, B., Janhunen, T., Tasharrofi, S.: Stable-unstable semantics: beyond NP with normal logic programs. Theory Pract. Logic Program. **16**(5–6), 570–586 (2016)
11. Bomanson, J., Gebser, M., Janhunen, T.: Improving the normalization of weight rules in answer set programs. In: Fermé, E., Leite, J. (eds.) JELIA 2014. LNCS (LNAI), vol. 8761, pp. 166–180. Springer, Cham (2014). doi:10.1007/978-3-319-11558-0_12
12. Bomanson, J., Gebser, M., Janhunen, T.: Rewriting optimization statements in answer-set programs. In: Carro, M., King, A. (eds.) Technical Communications of ICLP 2016, vol. 52 of OASIcs, pp. 5:1–5:15. Schloss Dagstuhl (2016)

13. Bomanson, J., Gebser, M., Janhunen, T., Kaufmann, B., Schaub, T.: Answer set programming modulo acyclicity. Fundamenta Informaticae **147**(1), 63–91 (2016)
14. Brewka, G., Eiter, T., Truszczyński, M.: Answer set programming at a glance. Commun. ACM **54**(12), 92–103 (2011)
15. Bruynooghe, M., Blockeel, H., Bogaerts, B., De Cat, B., De Pooter, S., Jansen, J., Labarre, A., Ramon, J., Denecker, M., Verwer, S.: Predicate logic as a modeling language: modeling and solving some machine learning and data mining problems with IDP3. Theory Pract. Logic Program. **15**(6), 783–817 (2015)
16. Calimeri, F., Fuscà, D., Perri, S., Zangari, J.: $\mathcal{I}$-DLV: the new intelligent grounder of DLV. In: Adorni, G., Cagnoni, S., Gori, M., Maratea, M. (eds.) AI*IA 2016. LNCS, vol. 10037, pp. 192–207. Springer, Cham (2016). doi:10.1007/978-3-319-49130-1_15
17. Calimeri, F., Gebser, M., Maratea, M., Ricca, F.: Design and results of the fifth answer set programming competition. Artif. Intell. **231**, 151–181 (2016)
18. Cussens, J.: Bayesian network learning with cutting planes. In: Cozman, F., Pfeffer, A. (eds.) Proceedings of UAI 2011, pp. 153–160. AUAI Press (2011)
19. Eiter, T., Gottlob, G.: On the computational cost of disjunctive logic programming: propositional case. Ann. Math. Artif. Intell. **15**(3–4), 289–323 (1995)
20. Eiter, T., Ianni, G., Krennwallner, T.: Answer set programming: a primer. In: Tessaris, S., Franconi, E., Eiter, T., Gutierrez, C., Handschuh, S., Rousset, M.-C., Schmidt, R.A. (eds.) Reasoning Web 2009. LNCS, vol. 5689, pp. 40–110. Springer, Heidelberg (2009). doi:10.1007/978-3-642-03754-2_2
21. Fages, F.: Consistency of Clark's completion and the existence of stable models. J. Methods Logic Comput. Sci. **1**, 51–60 (1994)
22. Gebser, M., Janhunen, T., Rintanen, J.: Answer set programming as SAT modulo acyclicity. In: Schaub, T., Friedrich, G., O'Sullivan, B. (eds.) Proceedings of ECAI 2014, pp. 351–356. IOS Press (2014)
23. Gebser, M., Kaminski, R., Kaufmann, B., Ostrowski, M., Schaub, T., Thiele, S.: Engineering an incremental ASP solver. In: Garcia de la Banda, M., Pontelli, E. (eds.) ICLP 2008. LNCS, vol. 5366, pp. 190–205. Springer, Heidelberg (2008). doi:10.1007/978-3-540-89982-2_23
24. Gebser, M., Kaminski, R., Kaufmann, B., Romero, J., Schaub, T.: Progress in *clasp* series 3. In: Calimeri, F., Ianni, G., Truszczynski, M. (eds.) LPNMR 2015. LNCS (LNAI), vol. 9345, pp. 368–383. Springer, Cham (2015). doi:10.1007/978-3-319-23264-5_31
25. Gebser, M., Kaminski, R., Schaub, T.: Grounding recursive aggregates: preliminary report. In: Denecker, M., Janhunen, T. (eds.) Proceedings of GTTV 2015 (2015)
26. Gebser, M., Maratea, M., Ricca, F.: The design of the sixth answer set programming competition. In: Calimeri, F., Ianni, G., Truszczynski, M. (eds.) LPNMR 2015. LNCS (LNAI), vol. 9345, pp. 531–544. Springer, Cham (2015). doi:10.1007/978-3-319-23264-5_44
27. Gelfond, M., Leone, N.: Logic programming and knowledge representation - the A-Prolog perspective. Artif. Intell. **138**(1–2), 3–38 (2002)
28. Guerinik, N., Caneghem, M.: Solving crew scheduling problems by constraint programming. In: Montanari, U., Rossi, F. (eds.) CP 1995. LNCS, vol. 976, pp. 481–498. Springer, Heidelberg (1995). doi:10.1007/3-540-60299-2_29
29. Havur, G., Cabanillas, C., Mendling, J., Polleres, A.: Resource allocation with dependencies in business process management systems. In: La Rosa, M., Loos, P., Pastor, O. (eds.) BPM 2016. LNBIP, vol. 260, pp. 3–19. Springer, Cham (2016). doi:10.1007/978-3-319-45468-9_1
30. Inoue, K., Sakama, C.: A fixpoint characterization of abductive logic programs. J. Logic Program. **27**(2), 107–136 (1996)

31. Janhunen, T., Gebser, M., Rintanen, J., Nyman, H., Pensar, J., Corander, J.: Learning discrete decomposable graphical models via constraint optimization. Stat. Comput. **27**(1), 115–130 (2017)
32. Janhunen, T., Niemelä, I.: Compact translations of non-disjunctive answer set programs to propositional clauses. In: Balduccini, M., Son, T.C. (eds.) Logic Programming, Knowledge Representation, and Nonmonotonic Reasoning. LNCS (LNAI), vol. 6565, pp. 111–130. Springer, Heidelberg (2011). doi:10.1007/978-3-642-20832-4_8
33. Koponen, L., Oikarinen, E., Janhunen, T., Säilä, L.: Optimizing phylogenetic supertrees using answer set programming. Theory Pract. Logic Program. **15**(4–5), 604–619 (2015)
34. Lifschitz, V.: Answer set programming and plan generation. Artif. Intell. **138**(1–2), 39–54 (2002)
35. Liu, G., Janhunen, T., Niemelä, I.: Answer set programming via mixed integer programming. In: Brewka, G., Eiter, T., McIlraith, S. (eds.) Proceedings of KR 2012, pp. 32–42. AAAI Press (2012)
36. Maratea, M., Pulina, L., Ricca, F.: A multi-engine approach to answer-set programming. Theory Pract. Logic Program. **14**(6), 841–868 (2014)
37. Maratea, M., Pulina, L., Ricca, F.: Multi-level algorithm selection for ASP. In: Calimeri, F., Ianni, G., Truszczynski, M. (eds.) LPNMR 2015. LNCS (LNAI), vol. 9345, pp. 439–445. Springer, Cham (2015). doi:10.1007/978-3-319-23264-5_36
38. Marek, V., Truszczyński, M.: Stable models and an alternative logic programming paradigm. In: Apt, K., Marek, V., Truszczyński, M., Warren, D. (eds.) The Logic Programming Paradigm - A 25-Year Perspective, pp. 375–398. Springer, Heidelberg (1999)
39. Marple, K., Gupta, G.: Dynamic consistency checking in goal-directed answer set programming. Theory Pract. Logic Program. **14**(4–5), 415–427 (2014)
40. Mellarkod, V., Gelfond, M., Zhang, Y.: Integrating answer set programming and constraint logic programming. Ann. Math. Artif. Intell. **53**(1–4), 251–287 (2008)
41. Niemelä, I.: Logic programming with stable model semantics as a constraint programming paradigm. Ann. Math. Artif. Intell. **25**(3–4), 241–273 (1999)

A Bird's-Eye View of Forgetting in Answer-Set Programming

João Leite[(✉)]

NOVA LINCS & Departamento de Informática,
Universidade Nova de Lisboa, Lisbon, Portugal
jleite@fct.unl.pt

Abstract. Forgetting is an operation that allows the removal, from a knowledge base, of middle variables no longer deemed relevant, while preserving all relationships (direct and indirect) between the remaining variables. When investigated in the context of Answer-Set Programming, many different approaches to forgetting have been proposed, following different intuitions, and obeying different sets of properties.

This talk will present a bird's-eye view of the complex landscape composed of the properties and operators of forgetting defined over the years in the context of Answer-Set Programming, zooming in on recent findings triggered by the formulation of the so-called *strong persistence*, a property based on the strong equivalence between an answer-set program and the result of forgetting modulo the forgotten atoms, which seems to best encode the requirements of the forgetting operation.

Keywords: Forgetting · Answer-Set Programming · Variable elimination

1 Introduction

Whereas keeping memory of information and knowledge has always been at the heart of research in Knowledge Representation and Reasoning, with tight connections to broader areas such as Databases and Artificial Intelligence, we have recently observed a growing attention being devoted to the complementary problem of *forgetting*.

Forgetting – or variable elimination – is an operation that allows the removal of *middle* variables no longer deemed relevant. It is most useful when we wish to eliminate (temporary) variables introduced to represent auxiliary concepts, with the goal of restoring the declarative nature of some knowledge base, or just to simplify it. Furthermore, it is becoming increasingly necessary to properly deal with legal and privacy issues, including, for example, to enforce the new EU General Data Protection Regulation [3], which includes the *right to*

The author was partially supported by FCT UID/CEC/04516/2013. This paper is substantially based on [1,2].

M. Balduccini and T. Janhunen (Eds.): LPNMR 2017, LNAI 10377, pp. 10–22, 2017.
DOI: 10.1007/978-3-319-61660-5_2

be forgotten. Recent applications of forgetting to cognitive robotics [4–6], resolving conflicts [7–10], and ontology abstraction and comparison [11–14], further witness its importance.

With its early roots in Boolean Algebra [15], forgetting has been extensively studied in the context of classical logic [7,16–21] and, more recently, in the context of logic programming, notably of Answer Set Programming (ASP). The non-monotonic rule-based nature of ASP creates very unique challenges to the development of forgetting operators – just as it happened with other belief change operations such as revision and update, cf. [22–28] – making it a special endeavour with unique characteristics distinct from those for classical logic.

Over the years, many have proposed different approaches to forgetting in ASP, through the characterization of the result of forgetting a set of atoms from a given program up to some equivalence class, and/or through the definition of concrete operators that produce a specific program for each input program and atoms to be forgotten [8,9,29–34].

All these approaches were typically proposed to obey some specific set of properties that their authors deemed adequate, some adapted from the literature on *classical* forgetting [30,33,35], others specifically introduced for the case of ASP [9,29–32,34]. Examples of such properties include *strengthened consequence*, which requires that the answer sets of the result of forgetting be bound to the answer-sets of the original program modulo the forgotten atoms, or the so-called *existence*, which requires that the result of forgetting belongs to the same class of programs admitted by the forgetting operator, so that the same reasoners can be used and the operator be iterated, among many others.

All this resulted is a *complex* landscape filled with operators and properties, with very little effort put into drawing a map that could help to better understand the relationships between properties and operators. This was recently addressed in [1], through the presentation of a systematic study of *forgetting* in Answer Set Programming (ASP), thoroughly investigating the different approaches found in the literature, their properties and relationships.

In the first part of this invited talk, we will present a bird's-eye view of this complex landscape investigated in [1].

One of the main conclusions drawn from observing the landscape of existing operators and properties is that there cannot be a one-size-fits-all forgetting operator for ASP, but rather a family of operators, each obeying a specific set of properties. Furthermore, it is clear that not all properties bear the same relevance. Whereas some properties can be very important, such as *existence*, since it guarantees that we can use the same automated reasoners after forgetting, despite not being a property specific of forgetting operators, other properties are less important, sometimes perhaps even questionable, as discussed in [1].

There is nevertheless one property – *strong persistence* [32] – which seems to best capture the essence of forgetting in the context of ASP. The property of *strong persistence* essentially requires that all existing relations between the atoms not to be forgotten be preserved, captured by requiring that there be a correspondence between the answer sets of a program before and after forgetting a set of atoms, and that such correspondence be preserved in the presence of additional rules not containing the atoms to be forgotten. Referring to the notation introduced in the appendix, an operator f is said to obey strong persistence if, for any program P and any set of atoms to be forgotten V, it holds that $\mathcal{AS}(f(P, V) \cup R) = \mathcal{AS}(P \cup R)_{\|V}$, for all programs R not containing atoms in V, where $f(P, V)$ denotes the result of forgetting V from P, $\mathcal{AS}(P)$ the answer sets of P, and $\mathcal{AS}(P)_{\|V}$ their restriction to atoms not in V.

Whereas it seems rather undisputed that *strong persistence* is a desirable property, it was not clear to what extent one could define operators that satisfy it. Whereas in [32], the authors proposed an operator that obeys such property, it is only defined for a restricted class of programs and can only be applied to forget a single atom from a program in a very limited range of situations.

The limits of forgetting while obeying *strong persistence* were investigated in [2]. There, after showing that sometimes it is simply not possible to forget some set of atoms from a program, while maintaining the relevant relations between other atoms, since the atoms to be forgotten play a pivotal role, the following three fundamental questions addressed: (a) *When can't we forget some set of atoms from an ASP while obeying strong persistence?*, (b) *When (and how) can we forget some set of atoms from an ASP while obeying strong persistence?*, and (c) *What can we forget from a specific ASP while obeying strong persistence?*

In the second part of this invited talk, we will zoom in on the limits of *forgetting* under *strong persistence* investigated in [2], and point to the future.

2   Forgetting in Answer-Set Programming

Forgetting. The principal idea of forgetting in ASP is to remove or hide certain atoms from a given program, while preserving its semantics for the remaining atoms. As the result, rather often, a representative up to some notion of equivalence between programs is considered. In this sense, many notions of forgetting for logic programs are defined semantically, i.e., they introduce a class of operators that satisfy a certain semantic characterization. Each single operator in such a class is then a concrete function that, given a program P and a non-empty set of atoms V to be forgotten, returns a unique program, the result of forgetting about V from P. Given a class of logic programs[1] $\mathcal{C}$ over $\mathcal{A}$, a *forgetting operator* (over $\mathcal{C}$) is a partial function $f : \mathcal{C} \times 2^{\mathcal{A}} \to \mathcal{C}$ s.t. $f(P, V)$ is a program over $\mathcal{A}(P) \backslash V$, for each $P \in \mathcal{C}$ and $V \subseteq \mathcal{A}$. We call $f(P, V)$ the *result of forgetting about V from P*. Unless stated otherwise, we will be focusing on $\mathcal{C} = \mathcal{C}_e$, and we leave $\mathcal{C}$ implicit. Furthermore, f is called *closed* for $\mathcal{C}' \subseteq \mathcal{C}$ if, for every $P \in \mathcal{C}'$ and $V \subseteq \mathcal{A}$, we have $f(P, V) \in \mathcal{C}'$. A *class* F *of forgetting operators (over* $\mathcal{C}$) is a set of forgetting operators (over $\mathcal{C}'$) s.t. $\mathcal{C}' \subseteq \mathcal{C}$.

Properties. Over the years, many have introduced a variety of properties that forgetting operators *should* obey, which we now briefly discuss.

The first three properties were proposed by Eiter and Wang [9], though not formally introduced as such. The first two were in fact guiding principles for

---

[1] See Appendix for definitions and notation on Answer-Set Programming.

defining their notion of forgetting, while the third was later formalized by Wang et al. [31]:

– F^2 satisfies *strengthened Consequence* (**sC**) if, for each $\mathsf{f} \in \mathsf{F}$, $P \in \mathcal{C}$ and $V \subseteq \mathcal{A}$, we have $\mathcal{AS}(\mathsf{f}(P,V)) \subseteq \mathcal{AS}(P)_{\|V}$. Strengthened Consequence requires that the answer sets of the result of forgetting be answer sets of the original program, ignoring the atoms to be forgotten.

– F satisfies *weak Equivalence* (**wE**) if, for each $\mathsf{f} \in \mathsf{F}$, $P, P' \in \mathcal{C}$ and $V \subseteq \mathcal{A}$, we have $\mathcal{AS}(\mathsf{f}(P,V)) = \mathcal{AS}(\mathsf{f}(P',V))$ whenever $\mathcal{AS}(P) = \mathcal{AS}(P')$. Weak Equivalence requires that forgetting preserves equivalence of programs.

– F satisfies *Strong Equivalence* (**SE**) if, for each $\mathsf{f} \in \mathsf{F}$, $P, P' \in \mathcal{C}$ and $V \subseteq \mathcal{A}$: if $P \equiv P'$, then $\mathsf{f}(P,V) \equiv \mathsf{f}(P',V)$. Strong Equivalence requires that forgetting preserves strong equivalence of programs.

The next three properties were introduced by Zhang and Zhou [35] in the context of forgetting in modal logics, and later adopted by Wang et al. [30,33] for forgetting in ASP:

– F satisfies *Weakening* (**W**) if, for each $\mathsf{f} \in \mathsf{F}$, $P \in \mathcal{C}$ and $V \subseteq \mathcal{A}$, we have $P \models_{\mathsf{HT}} \mathsf{f}(P,V)$. Weakening requires that the HT-models of the original program also be HT-models of the result of forgetting, thus implying that the result of forgetting has at most the same consequences as the original program.

– F satisfies *Positive Persistence* (**PP**) if, for each $\mathsf{f} \in \mathsf{F}$, $P \in \mathcal{C}$ and $V \subseteq \mathcal{A}$: if $P \models_{\mathsf{HT}} P'$, with $P' \in \mathcal{C}$ and $\mathcal{A}(P') \subseteq \mathcal{A}\backslash V$, then $\mathsf{f}(P,V) \models_{\mathsf{HT}} P'$. Positive Persistence requires that the HT-consequences of the original program not containing atoms to be forgotten be preserved in the result of forgetting.

– F satisfies *Negative Persistence* (**NP**) if, for each $\mathsf{f} \in \mathsf{F}$, $P \in \mathcal{C}$ and $V \subseteq \mathcal{A}$: if $P \not\models_{\mathsf{HT}} P'$, with $P' \in \mathcal{C}$ and $\mathcal{A}(P') \subseteq \mathcal{A}\backslash V$, then $\mathsf{f}(P,V) \not\models_{\mathsf{HT}} P'$. Negative Persistence requires that a program not containing atoms to be forgotten not be a HT-consequence of the result of forgetting, unless it was already a HT-consequence of the original program.

The property *Strong (addition) Invariance* was introduced by Wong [29], and assigned this name in [1]:

– F satisfies *Strong (addition) Invariance* (**SI**) if, for each $\mathsf{f} \in \mathsf{F}$, $P \in \mathcal{C}$ and $V \subseteq \mathcal{A}$, we have $\mathsf{f}(P,V) \cup R \equiv \mathsf{f}(P \cup R, V)$ for all programs $R \in \mathcal{C}$ with $\mathcal{A}(R) \subseteq \mathcal{A}\backslash V$. Strong (addition) Invariance requires that it be (strongly) equivalent to add a program without the atoms to be forgotten before or after forgetting.

The property called *existence* was discussed by Wang et al. [30], formalized by Wang et al. [31], and refined by [1]. It requires that a result of forgetting for P in $\mathcal{C}$ exists in the class $\mathcal{C}$, important to iterate:

– F satisfies *Existence for $\mathcal{C}$* (**$E_{\mathcal{C}}$**), i.e., F is *closed for a class of programs $\mathcal{C}$* if there exists $\mathsf{f} \in \mathsf{F}$ s.t. f is closed for $\mathcal{C}$. Existence for class $\mathcal{C}$ requires that the a result of forgetting for a program in $\mathcal{C}$ exists in the class $\mathcal{C}$. Operators that satisfy Existence for class $\mathcal{C}$ are said to be closed for that class.

The property *Consequence Persistence* was introduced by Wang et al. [31] building on the ideas behind (**sC**) by Eiter and Wang [9]:

---

^{2} Unless stated otherwise, F is a class of forgetting operators, and $\mathcal{C}$ the class of programs over $\mathcal{A}$ of a given $\mathsf{f} \in \mathsf{F}$.

– F satisfies *Consequence Persistence* (**CP**) if, for each $f \in F$, $P \in C$ and $V \subseteq \mathcal{A}$, we have $\mathcal{AS}(f(P, V)) = \mathcal{AS}(P)_{\parallel V}$. Consequence persistence requires that the answer sets of the result of forgetting correspond exactly to the answer sets of the original program, ignoring the atoms to be forgotten.

The following property was introduced by Knorr and Alferes [32] with the aim of imposing the preservation of all dependencies contained in the original program:

– F satisfies *Strong Persistence* (**SP**) if, for each $f \in F$, $P \in C$ and $V \subseteq \mathcal{A}$, we have $\mathcal{AS}(f(P, V) \cup R) = \mathcal{AS}(P \cup R)_{\parallel V}$, for all programs $R \in C$ with $\mathcal{A}(R) \subseteq \mathcal{A} \backslash V$. Strong Persistence strengthens (**CP**) by imposing that the correspondence between answer-sets of the result of forgetting and those of the original program be preserved in the presence of any additional set of rules not containing the atoms to be forgotten.

The final property here[3] is due to Delgrande and Wang [34], although its name was assigned in [1]:

– F satisfies *weakened Consequence* (**wC**) if, for each $f \in F$, $P \in C$ and $V \subseteq \mathcal{A}$, we have $\mathcal{AS}(P)_{\parallel V} \subseteq \mathcal{AS}(f(P, V))$. Weakened Consequence requires that the answer sets of the original program be preserved while forgetting, ignoring the atoms to be forgotten.

These properties are not orthogonal to one another: (**CP**) is incompatible with (**W**) as well as with (**NP**) (for F closed for C, where C contains normal logic programs); (**W**) is equivalent to (**NP**); (**SP**) implies (**PP**); (**SP**) implies (**SE**); (**W**) and (**PP**) together imply (**SE**); (**CP**) and (**SI**) together are equivalent to (**SP**); (**sC**) and (**wC**) together are equivalent to (**CP**); (**CP**) implies (**wE**); (**SE**) and (**SI**) together imply (**PP**).

Operators. Over the years, many operators of forgetting have been introduced, implementing certain intuitions and obeying particular sets of properties.

Strong and Weak Forgetting. The first proposals are due to Zhang and Foo [8] introducing two syntactic operators for normal logic programs, termed Strong and Weak Forgetting. Both start with computing a reduction corresponding to the well-known weak partial evaluation (WGPPE) [38]. Then, the two operators differ on how they subsequently remove rules containing the atom to be forgotten. In Strong Forgetting, all rules containing the atom to be forgotten are simply removed. In Weak Forgetting, rules with negative occurrences of the atom to be forgotten in the body are kept, after such occurrences are removed. The motivation for this difference is whether such negative occurrences of the atom to be forgotten are seen as support for the rule head (Strong) or not (Weak). Both operators are closed for C_n.

Semantic Forgetting. Eiter and Wang [9] proposed Semantic Forgetting to improve on some of the shortcomings of the two purely syntax-based operators of Zhang and Foo [8]. The basic idea is to characterize a result of forgetting just by its answer sets, obtained by considering only the minimal sets among the answer sets of the initial program ignoring the atoms to be forgotten.

---

[3] An additional set of properties was introduced in [29]. The reader is referred to [36, 37] for a detailed discussion regarding these properties.

Three concrete algorithms are presented, two based on semantic considerations and one syntactic. Unlike the former, the latter is not closed for $\mathcal{C}_d^+$ and $\mathcal{C}_n^+$ ($^+$ denotes the restriction to consistent programs), since double negation is required in general.

Semantic Strong and Weak Forgetting. Wong [29] argued that semantic forgetting should not be focused on answer sets only, as these do not contain all the information present in a program. He defined two classes of forgetting operators for disjunctive programs, building on HT-models. The basic idea is to start with the set of rules HT-entailed by the original program without those with positive occurrences of the atoms to be forgotten, and after removing positive occurrences of the atoms to be forgotten from the head of rules, whenever their negation appears in the body, and then, in Semantic Strong Forgetting, rules containing the atoms to be forgotten are removed, while in Semantic Weak Forgetting rules with negative (resp. positive) occurrences of the atoms to be forgotten in the body (resp. head) are kept, after such occurrences are removed. Wong [29] defined one construction closed for $\mathcal{C}_d$.

HT-Forgetting. Wang et al. [30,33] introduced HT-Forgetting, building on properties introduced by Zhang and Zhou [35] in the context of modal logics, with the aim of overcoming problems with Wongs notions, namely that each of them did not satisfy one of the properties **(PP)** and **(W)**. HT-Forgetting is defined for extended programs, characterising the set of HT-models of the result of forgetting as being composed of the HT-models of the original program, modulo any occurrence of the atoms to be forgotten. A concrete operator is presented [33] that is shown to be closed for $\mathcal{C}_e$ and $\mathcal{C}_H$, and it is also shown that no HT-Forgetting operator exists that is closed for either $\mathcal{C}_d$ or $\mathcal{C}_n$.

SM-Forgetting. Wang et al. [31] defined a modification of HT-Forgetting, SM-Forgetting, for extended programs, with the objective of preserving the answer sets of the original program (modulo the forgotten atoms). As with HT-Forgetting, it is defined for extended programs though a characterisation of the HT-models of the result of forgetting, which are taken to be maximal subsets of the HT-models of the original program, modulo any occurrence of the atoms to be forgotten, such that the set of their answer-sets coincides with the set of answer-sets of the original program, modulo the forgotten atoms. A concrete operator is provided that is shown to be closed for $\mathcal{C}_e$ and $\mathcal{C}_H$. It is also shown that no SM-Forgetting operator exists that is closed for either $\mathcal{C}_d$ or $\mathcal{C}_n$.

Strong AS-Forgetting. Knorr and Alferes [32] introduced Strong AS-Forgetting with the aim of preserving both the answer sets of the original program, and also those of the original program augmented by any set of rules over the signature without the atoms to be forgotten. A concrete operator is defined for $\mathcal{C}_{nd}$, but not closed for $\mathcal{C}_n$ and only defined for certain programs with double negation.

SE-Forgetting. Delgrande and Wang [34] recently introduced SE-Forgetting based on the idea that forgetting an atom from a program is characterized by the set of those SE-consequences, i.e., HT-consequences, of the program that do not mention the atoms to be forgotten. The notion is defined for disjunctive programs building on an inference system by Wong [39] that preserves strong equivalence.

	sC	wE	SE	W	PP	NP	SI	CP	SP	wC	$E_{\mathcal{C}_H}$	$E_{\mathcal{C}_n}$	$E_{\mathcal{C}_d}$	$E_{\mathcal{C}_{nd}}$	$E_{\mathcal{C}_e}$
F_{strong}	×	×	×	√	×	√	√	×	×	×	√	√	-	-	-
F_{weak}	×	×	×	×	√	×	√	×	×	×	√	√	-	-	-
F_{sem}	√	√	×	×	×	×	×	×	×	×	√	√	√	-	-
F_S	×	×	√	√	√	√	×	×	×	×	√	×	√	-	-
F_W	√	√	√	×	√	×	√	×	×	×	√	√	√	-	-
F_{HT}	×	×	√	√	√	√	√	×	×	×	√	×	×	×	√
F_{SM}	√	√	√	×	√	×	×	√	×	√	√	×	×	×	√
F_{Sas}	√	√	√	×	√	×	√	√	√	√	√	×	×	×	×
F_{SE}	×	×	√	√	√	√	×	×	×	×	√	×	√	-	-

Fig. 1. Satisfaction of properties for known classes of forgetting operators. For class F and property **P**, '√' represents that F satisfies **P**, '×' that F does not satisfy **P**, and '-' that F is not defined for the class $\mathcal{C}$ in consideration.

An operator is provided, which is closed for $\mathcal{C}_d$. Gonçalves et al. [1] have shown SE-Forgetting to coincide with Semantic Strong Forgetting [29].

Figure 1 summarises the satisfaction of properties for known classes of forgetting operators.

3   Forgetting Under Strong Persistence

Among the desirable properties of classes of forgetting operators recalled in the previous section, *strong persistence* (**SP**) [32] is of particular interest, as it ensures that forgetting preserves all existing relations between all atoms occurring in the program, but the forgotten. In this sense, a class of operators satisfying (**SP**) removes the desired atoms, but has no negative semantical effects on the remainder. The importance of (**SP**) is also witnessed by the fact that a class of operators that satisfies (**SP**) also satisfies all the other previously mentioned properties with the exception of (**W**) and (**NP**), which happen to be equivalent and can hardly be considered desirable [1].

However, determining a forgetting operator that satisfies (**SP**) has been a difficult problem, since, for the verification whether a certain program P' should be the result of forgetting about V from P, none of the well-established equivalence relations can be used, i.e., neither equivalence nor strong equivalence hold in general between P and P', not even relativized equivalence [40], even though it is close in spirit to the ideas of (**SP**). Hence, maybe not surprisingly, there was no known general class of operators that satisfies (**SP**) and which is closed (for the considered class of logic programs).

And, until recently, the two known positive results concerning the satisfiability of (**SP**) were the existence of several known classes of operators that satisfy (**SP**) *when restricted to Horn programs* [1], which is probably of little relevance given the crucial role played by (default) negation in ASP, and the existence of one specific operator that permits forgetting about V from P while satisfying (**SP**) [32], but only in a very restricted range of situations based on a non-trivial syntactical criterion which excludes large classes of cases where forgetting about V from P is possible.

All this begged the question of whether there exists a forgetting operator, defined over a class of programs $\mathcal{C}$ beyond the class of Horn programs, that satisfies **(SP)**, which was given a negative answer in [2]: – it is not always possible to forget a set of atoms from a given logic program satisfying the property **(SP)**.

Whereas this negative result shows that in general it is not always possible to forget while satisfying **(SP)**, its proof presented in [2] provided some hints on why this is the case. Some atoms play an important role in the program, being pivotal in establishing the relations between the remaining atoms, making it simply not possible to forget them and expect that the relations between other atoms be preserved. That is precisely what happens with the pair of atoms p and q in the program

$$a \leftarrow p \qquad b \leftarrow q \qquad p \leftarrow not\, q \qquad q \leftarrow not\, p$$

It is simply not possible to forget them both and expect all the semantic relations between a and b to be kept. No program over atoms $\{a, b\}$ would have the same answer sets as those of the original program (modulo p and q), when both are extended with an arbitrary set of rules over $\{a, b\}$.

This observation lead to another central question: under what circumstances is it not possible to forget about a given set of atoms V from P while satisfying **(SP)**? In particular, given a concrete program, which sets of atoms play such a pivotal role that they cannot be jointly forgotten without affecting the semantic relations between the remaining atoms in the original program?

This question was answered in [2] through the introduction of a criterion (Ω) which characterizing the instances $\langle P, V \rangle$ for which we cannot expect forgetting operators to satisfy $\textbf{(SP)}_{\langle P,V \rangle}$.[4]

Definition 1 (Criterion Ω). *Let P be a program over $\mathcal{A}$ and $V \subseteq \mathcal{A}$. An instance $\langle P, V \rangle$ satisfies criterion Ω if there exists $Y \subseteq \mathcal{A} \setminus V$ such that the set of sets*

$$\mathcal{R}_{\langle P,V \rangle}^{Y} = \{R_{\langle P,V \rangle}^{Y,A} \mid A \in Rel_{\langle P,V \rangle}^{Y}\}$$

is non-empty and has no least element, where

$$R_{\langle P,V \rangle}^{Y,A} = \{X \setminus V \mid \langle X, Y \cup A \rangle \in \mathcal{HT}(P)\}$$
$$Rel_{\langle P,V \rangle}^{Y} = \{A \subseteq V \mid \langle Y \cup A, Y \cup A \rangle \in \mathcal{HT}(P) \text{ and }$$
$$\nexists A' \subset A \text{ such that } \langle Y \cup A', Y \cup A \rangle \in \mathcal{HT}(P)\}.$$

It turns out that Ω is a necessary and sufficient criterion to determine that some set of atoms V cannot be forgotten from a program P while satisfying *strong persistence*.

---

[4] $\textbf{(SP)}_{\langle P,V \rangle}$ is a restriction of property **(SP)** to specific forgetting instances. A forgetting operator f over $\mathcal{C}$ satisfies $\textbf{(SP)}_{\langle P,V \rangle}$ if $\mathcal{AS}(\text{f}(P, V) \cup R) = \mathcal{AS}(P \cup R)_{\parallel V}$, for all programs $R \in \mathcal{C}$ with $\mathcal{A}(R) \subseteq \mathcal{A} \setminus V$.

Whereas at a technical level, criterion Ω is closely tied to certain conditions on the HT-models of the program at hand, it seems that what cannot be forgotten from a program are atoms used in rules that are somehow equivalent to *choice rules* [41], and those atoms are pivotal in the sense that they play an active role in determining the truth of other atoms in some answer sets i.e., there are rules whose bodies mention these atoms and they are true at least in some answer sets.

Nevertheless, sometimes it is possible to forget while satisfying *strong persistence* and, in such cases, the following class $\mathsf{F_{SP}}$ of forgetting operators, dubbed *SP-Forgetting*, precisely characterises the desired result of forgetting:

$$\mathsf{F_{SP}} = \{\mathsf{f} \mid \mathcal{HT}(\mathsf{f}(P,V)) = \{\langle X, Y \rangle \mid Y \subseteq \mathcal{A}(P) \backslash V \wedge X \in \bigcap \mathcal{R}^Y_{\langle P, V \rangle}\}\}$$

Thus, given an instance $\langle P, V \rangle$, we can test whether Ω is not satisfied, i.e., whether we are allowed to forget V from P while preserving **(SP)**, in which case the HT-models that characterise a result can be obtained from $\mathsf{F_{SP}}$. It was further shown in [2] that $\mathsf{F_{SP}}$ is closed in the general case and for Horn programs, but not for disjunctive or normal programs.

If we restrict our attention to the cases where we can forget, i.e., where the considered instance does not satisfy Ω, then most of the properties mentioned before are satisfied. In particular, restricted to instances $\langle P, V \rangle$ that do not satisfy Ω, $\mathsf{F_{SP}}$ satisfies **(sC)**, **(wE)**, **(SE)**, **(PP)**, **(SI)**, **(CP)**, **(SP)** and **(wC)**. The properties which are not satisfied – **(W)** and **(NP)** – have been proved orthogonal to **(SP)** [1], hence of little relevance in our view.

4 Outlook

We began by presenting a bird's-eye view of *forgetting* in Answer Set Programming (ASP), covering the different approaches found in the literature, their properties and relationships. We then zoomed in on the important property of *strong persistence*, and reviewed the most relevant known results, including that it is not always possible to forget a set of atoms from a program while obeying this property, a precise characterisation of what can and cannot be forgotten from a program established through a necessary and sufficient criterion, and a characterisation of the class of forgetting operators that achieve the correct result whenever forgetting is possible.

But what happens if we *must* forget, but cannot do it without violating *strong persistence*? This may happen for legal and privacy issues, including, for example, the implementation of court orders to eliminate certain pieces of illegal information. Investigating weaker requirements, e.g. by imposing only a subset of the three properties – **(sC)**, **(wC)** and **(SI)** – that together compose **(SP)**, or by considering weaker notions of equivalence such as *uniform equivalence* [42,43], is the subject of ongoing work.

Other interesting avenues for future research include investigating different forms of forgetting which may be required in practice, such as those that preserve some aggregated meta-level information about the forgotten atoms, or even going beyond maintaining all relationships between non-forgotten atoms which may be required by certain legislation.

Acknowledgments. I would like to thank my close colleagues Ricardo Gonçalves and Matthias Knorr for all their dedication and contributions to our joint projects, turning it into a more fun and rewarding ride.

A   Answer-Set Programming

We assume a *propositional signature* $\mathcal{A}$, a finite set of propositional atoms[5]. An *(extended) logic program* P over $\mathcal{A}$ is a finite set of *(extended) rules* of the form

$$a_1 \vee \ldots \vee a_k \leftarrow b_1, ..., b_l, not\, c_1, ..., not\, c_m, not\, not\, d_1, ..., not\, not\, d_n \,, \qquad (1)$$

where all $a_1, \ldots, a_k, b_1, \ldots, b_l, c_1, \ldots, c_m$, and $d_1, \ldots, d_n$ are atoms of $\mathcal{A}$.[6] Such rules r are also commonly written in a more succinct way as

$$A \leftarrow B, not\, C, not\, not\, D \,, \qquad (2)$$

where we have $A = \{a_1, \ldots, a_k\}$, $B = \{b_1, \ldots, b_l\}$, $C = \{c_1, \ldots, c_m\}$, $D = \{d_1, \ldots, d_n\}$, and we will use both forms interchangeably. By $\mathcal{A}(P)$ we denote the set of atoms appearing in P. This class of logic programs, $\mathcal{C}_e$, includes a number of special kinds of rules r: if $n = 0$, then we call r *disjunctive*; if, in addition, $k \leq 1$, then r is *normal*; if on top of that $m = 0$, then we call r *Horn*, and *fact* if also $l = 0$. The classes of *disjunctive*, *normal* and *Horn* programs, $\mathcal{C}_d$, $\mathcal{C}_n$, and $\mathcal{C}_H$, are defined resp. as a finite set of disjunctive, normal, and Horn rules. We also call extended rules with $k \leq 1$ *non-disjunctive*, thus admitting a non-standard class $\mathcal{C}_{nd}$, called *non-disjunctive programs*, different from normal programs. Given a program P and a set I of atoms, the *reduct* P^I is defined as $P^I = \{A \leftarrow B : r$ of the form (2) in $P, C \cap I = \emptyset, D \subseteq I\}$.

An *HT-interpretation* is a pair $\langle X, Y \rangle$ s.t. $X \subseteq Y \subseteq \mathcal{A}$. Given a program P, an HT-interpretation $\langle X, Y \rangle$ is an *HT-model of* P if $Y \models P$ and $X \models P^Y$, where $\models$ denotes the standard consequence relation for classical logic. We admit that the set of HT-models of a program P are restricted to $\mathcal{A}(P)$ even if $\mathcal{A}(P) \subset \mathcal{A}$. We denote by $\mathcal{HT}(P)$ the set of *all HT-models of* P. A set of atoms Y is an *answer set* of P if $\langle Y, Y \rangle \in \mathcal{HT}(P)$, but there is no $X \subset Y$ such that $\langle X, Y \rangle \in \mathcal{HT}(P)$. The set of all answer sets of P is denoted by $\mathcal{AS}(P)$. We say that two programs P_1, P_2 are *equivalent* if $\mathcal{AS}(P_1) = \mathcal{AS}(P_2)$ and *strongly equivalent*, denoted by $P_1 \equiv P_2$, if $\mathcal{AS}(P_1 \cup R) = \mathcal{AS}(P_2 \cup R)$ for any $R \in \mathcal{C}_e$. It is well-known that $P_1 \equiv P_2$ exactly when $\mathcal{HT}(P_1) = \mathcal{HT}(P_2)$ [45]. We say that P' is an *HT-consequence* of P, denoted by $P \models_{HT} P'$, whenever $\mathcal{HT}(P) \subseteq \mathcal{HT}(P')$. The *$V$-exclusion* of a set of answer sets (a set of HT-interpretations) $\mathcal{M}$, denoted $\mathcal{M}_{\|V}$, is $\{X \backslash V \mid X \in \mathcal{M}\}$ $(\{\langle X \backslash V, Y \backslash V \rangle \mid \langle X, Y \rangle \in \mathcal{M}\})$. Finally, given two sets of atoms $X, X' \subseteq \mathcal{A}$, we write $X \sim_V X'$ whenever $X \backslash V = X' \backslash V$.

---

[5] Often, the term propositional variable is used synonymously.

[6] Extended logic programs [44] are actually more expressive, but this form is sufficient here.

References

1. Goncalves, R., Knorr, M., Leite, J.: The ultimate guide to forgetting in answer set programming. In: Baral, C., Delgrande, J., Wolter, F. (eds.) Proceedings of KR, pp. 135–144. AAAI Press (2016)
2. Gonçalves, R., Knorr, M., Leite, J.: You can't always forget what you want: on the limits of forgetting in answer set programming. In: Fox, M.S., Kaminka, G.A. (eds.) Proceedings of ECAI. IOS Press (2016)
3. European Parliament: General data protection regulation. Official Journal of the European Union L119/59, May 2016
4. Lin, F., Reiter, R.: How to progress a database. Artif. Intell. **92**(1–2), 131–167 (1997)
5. Liu, Y., Wen, X.: On the progression of knowledge in the situation calculus. In: Walsh, T. (ed.) Proceedings of IJCAI, IJCAI/AAAI, pp. 976–982 (2011)
6. Rajaratnam, D., Levesque, H.J., Pagnucco, M., Thielscher, M.: Forgetting in action. In: Baral, C., Giacomo, G.D., Eiter, T. (eds.) Proceedings of KR. AAAI Press (2014)
7. Lang, J., Liberatore, P., Marquis, P.: Propositional independence: formula-variable independence and forgetting. J. Artif. Intell. Res. (JAIR) **18**, 391–443 (2003)
8. Zhang, Y., Foo, N.Y.: Solving logic program conflict through strong and weak forgettings. Artif. Intell. **170**(8–9), 739–778 (2006)
9. Eiter, T., Wang, K.: Semantic forgetting in answer set programming. Artif. Intell. **172**(14), 1644–1672 (2008)
10. Lang, J., Marquis, P.: Reasoning under inconsistency: a forgetting-based approach. Artif. Intell. **174**(12–13), 799–823 (2010)
11. Wang, Z., Wang, K., Topor, R.W., Pan, J.Z.: Forgetting for knowledge bases in DL-Lite. Ann. Math. Artif. Intell. **58**(1–2), 117–151 (2010)
12. Kontchakov, R., Wolter, F., Zakharyaschev, M.: Logic-based ontology comparison and module extraction, with an application to dl-lite. Artif. Intell. **174**(15), 1093–1141 (2010)
13. Konev, B., Ludwig, M., Walther, D., Wolter, F.: The logical difference for the lightweight description logic EL. J. Artif. Intell. Res. (JAIR) **44**, 633–708 (2012)
14. Konev, B., Lutz, C., Walther, D., Wolter, F.: Model-theoretic inseparability and modularity of description logic ontologies. Artif. Intell. **203**, 66–103 (2013)
15. Lewis, C.I.: A survey of symbolic logic. University of California Press (1918). Republished by Dover (1960)
16. Bledsoe, W.W., Hines, L.M.: Variable elimination and chaining in a resolution-based prover for inequalities. In: Bibel, W., Kowalski, R. (eds.) CADE 1980. LNCS, vol. 87, pp. 70–87. Springer, Heidelberg (1980). doi:10.1007/3-540-10009-1_7
17. Larrosa, J.: Boosting search with variable elimination. In: Dechter, R. (ed.) CP 2000. LNCS, vol. 1894, pp. 291–305. Springer, Heidelberg (2000). doi:10.1007/3-540-45349-0_22
18. Larrosa, J., Morancho, E., Niso, D.: On the practical use of variable elimination in constraint optimization problems: 'still-life' as a case study. J. Artif. Intell. Res. (JAIR) **23**, 421–440 (2005)
19. Middeldorp, A., Okui, S., Ida, T.: Lazy narrowing: strong completeness and eager variable elimination. Theor. Comput. Sci. **167**(1&2), 95–130 (1996)
20. Moinard, Y.: Forgetting literals with varying propositional symbols. J. Log. Comput. **17**(5), 955–982 (2007)

21. Weber, A.: Updating propositional formulas. In: Expert Database Conference, pp. 487–500 (1986)
22. Alferes, J.J., Leite, J.A., Pereira, L.M., Przymusinska, H., Przymusinski, T.C.: Dynamic updates of non-monotonic knowledge bases. J. Logic Program. **45**(1–3), 43–70 (2000)
23. Eiter, T., Fink, M., Sabbatini, G., Tompits, H.: On properties of update sequences based on causal rejection. Theor. Pract. Logic Program. (TPLP) **2**(6), 721–777 (2002)
24. Sakama, C., Inoue, K.: An abductive framework for computing knowledge base updates. Theor. Pract. Logic Program. (TPLP) **3**(6), 671–713 (2003)
25. Slota, M., Leite, J.: Robust equivalence models for semantic updates of answer-set programs. In: Brewka, G., Eiter, T., McIlraith, S.A. (eds.) Proceeding of KR, pp. 158–168. AAAI Press (2012)
26. Slota, M., Leite, J.: A unifying perspective on knowledge updates. In: Cerro, L.F., Herzig, A., Mengin, J. (eds.) JELIA 2012. LNCS, vol. 7519, pp. 372–384. Springer, Heidelberg (2012). doi:10.1007/978-3-642-33353-8_29
27. Delgrande, J.P., Schaub, T., Tompits, H., Woltran, S.: A model-theoretic approach to belief change in answer set programming. ACM Trans. Comput. Log. **14**(2), 14 (2013)
28. Slota, M., Leite, J.: The rise and fall of semantic rule updates based on SE-models. TPLP **14**(6), 869–907 (2014)
29. Wong, K.S.: Forgetting in Logic Programs. Ph.D. thesis, The University of New South Wales (2009)
30. Wang, Y., Zhang, Y., Zhou, Y., Zhang, M.: Forgetting in logic programs under strong equivalence. In: Brewka, G., Eiter, T., McIlraith, S.A. (eds.) Proceedings of KR, pp. 643–647. AAAI Press (2012)
31. Wang, Y., Wang, K., Zhang, M.: Forgetting for answer set programs revisited. In: Rossi, F. (ed.) Proceedings of IJCAI, IJCAI/AAAI (2013)
32. Knorr, M., Alferes, J.J.: Preserving strong equivalence while forgetting. In: Fermé, E., Leite, J. (eds.) JELIA 2014. LNCS, vol. 8761, pp. 412–425. Springer, Cham (2014). doi:10.1007/978-3-319-11558-0_29
33. Wang, Y., Zhang, Y., Zhou, Y., Zhang, M.: Knowledge forgetting in answer set programming. J. Artif. Intell. Res. (JAIR) **50**, 31–70 (2014)
34. Delgrande, J.P., Wang, K.: A syntax-independent approach to forgetting in disjunctive logic programs. In: Bonet, B., Koenig, S. (eds.) Proceedings of AAAI, pp. 1482–1488. AAAI Press (2015)
35. Zhang, Y., Zhou, Y.: Knowledge forgetting: properties and applications. Artif. Intell. **173**(16–17), 1525–1537 (2009)
36. Gonçalves, R., Knorr, M., Leite, J.: Forgetting in ASP: the forgotten properties. In: Michael, L., Kakas, A. (eds.) JELIA 2016. LNCS, vol. 10021, pp. 543–550. Springer, Cham (2016). doi:10.1007/978-3-319-48758-8_37
37. Gonçalves, R., Knorr, M., Leite, J.: On some properties of forgetting in ASP. In: Booth, R., Casini, G., Klarman, S., Richard, G., Varzinczak, I.J. (eds.) Proceedings of the International Workshop on Defeasible and Ampliative Reasoning (DARe-16). CEUR Workshop Proceedings, vol. 1626. CEUR-WS.org (2016)
38. Brass, S., Dix, J.: Semantics of (disjunctive) logic programs based on partial evaluation. J. Log. Program. **40**(1), 1–46 (1999)
39. Wong, K.S.: Sound and complete inference rules for SE-consequence. J. Artif. Intell. Res. (JAIR) **31**, 205–216 (2008)
40. Eiter, T., Fink, M., Woltran, S.: Semantical characterizations and complexity of equivalences in answer set programming. ACM Trans. Comput. Log. **8**(3) (2007)

41. Lifschitz, V.: What is answer set programming? In: Fox, D., Gomes, C.P. (eds.) Proceedings of AAAI, pp. 1594–1597. AAAI Press (2008)
42. Sagiv, Y.: Optimizing datalog programs. In: Minker, J. (ed.) Foundations of Deductive Databases and Logic Programming, pp. 659–698. Morgan Kaufmann (1988)
43. Eiter, T., Fink, M.: Uniform equivalence of logic programs under the stable model semantics. In: Palamidessi, C. (ed.) ICLP 2003. LNCS, vol. 2916, pp. 224–238. Springer, Heidelberg (2003). doi:10.1007/978-3-540-24599-5_16
44. Lifschitz, V., Tang, L.R., Turner, H.: Nested expressions in logic programs. Ann. Math. Artif. Intell. **25**(3–4), 369–389 (1999)
45. Lifschitz, V., Pearce, D., Valverde, A.: Strongly equivalent logic programs. ACM Trans. Comput. Log. **2**(4), 526–541 (2001)

Answer Set Programming and Its Applications in Planning and Multi-agent Systems

Tran Cao Son[✉]

Department of Computer Science, New Mexico State University,
Las Cruces, NM 88003, USA
tson@cs.nmsu.edu

Abstract. The paper presents some applications in planning and multi-agent systems of answer set programming. It highlights the benefits of answer set programming based techniques in these applications. It also describes a class of multi-agent planning problems that is challenging to answer set programming.

Keywords: Answer set programming · Planning · Multi-agent system

1 Introduction

The invention of answer set programming (ASP) [18,20] and the development of efficient answer set solvers such as smodels [24], dlv [5], and clingo [8] enable the use of logic programming under answer set semantics in several practical applications [6]. The fundamental idea of ASP is to represent solutions to a problem by answer sets of a logic program. That is, to solve a problem, one first represents it as a logic program whose answer sets correspond one-to-one to its solutions; next, to find a solution, one computes an answer set of that program and extracts the solution from the answer set.

Formally, a logic program Π is a set of rules of the form

$$c_1 \mid \ldots \mid c_k \leftarrow a_1, \ldots, a_m, not\ a_{m+1}, \ldots, not\ a_n \qquad (1)$$

where $0 \leq m \leq n$, $0 \leq k$, each a_i or c_j is a literal of a propositional language[1] and *not* represents *default negation*. Both the head and the body can be empty. When the head is empty, the rule is called a *constraint*. When the body is empty, the rule is called a *fact*. The semantics of a program Π is defined by a set of *answer sets* [10]. An answer set is a distinguished model of Π that satisfies all the rules of Π and is minimal and well-supported.

To increase the expressiveness of logic programs and simplify its use in applications, the language has been extended with several features such as *weight constraints or choice atoms* [24], or *aggregates* [7,21,25]. Standard syntax for these extensions has been proposed and adopted in most state-of-the-art ASP-solvers such as clingo and dlv.

In recent years, attempts have been made to consider continuously changing logic programs or external atoms. For example, the system clingo enables the multi-shot model as oppose to the traditional single-shot model. In this model, ASP programs are

---

[1] Rules with variables are viewed as a shorthand for the set of their ground instances.

© Springer International Publishing AG 2017
M. Balduccini and T. Janhunen (Eds.): LPNMR 2017, LNAI 10377, pp. 23–35, 2017.
DOI: 10.1007/978-3-319-61660-5_3

extended with Python procedures that control the answer set solving process along with the evolving logic programs. This feature provides an effective way for the application of ASP in a number of applications that were difficult to deal with previously.

This paper describes the application of ASP in planning in the presence of incomplete information and sensing actions (Sect. 2), in goal recognition design (Sect. 3), and in various settings of multi-agent planning (Sect. 4). It highlights the advantage of ASP in these researches and, when possible, identifies the challenging issues faced by ASP.

2 Planning with Incomplete Information and Sensing Actions

Answer set planning was one of the earliest applications of answer set programming [3,16,32]. The logic program encoding proposed in these papers are suitable for classical planning problems with complete information about the initial state and deterministic actions. In a series of work, we applied ASP to *conformant planning* and *conditional planning* (e.g., [30,31,34,35]). The former refers to planning with incomplete information about the initial state whose solutions are action sequences that achieve the goal from any possible initial state (and hence, the terms *comformant planning*). The latter refers to planning with incomplete information and sensing actions whose solutions often contain *branches* in the form of conditional statements (e.g., **if-then-else** or **case-statement**) that leads to the terms *conditional planning*.

Conditional planning is computationally harder than conformant planning which, in turn, is computationally harder than classical planning. When actions are deterministic and the plan's length is polynomially bounded by the size of the problem, the complexity of conditional and comformant planning are PSPACE-complete and Σ_2^P-complete, respectively, [1]. As such, there are problems that has conditional plan as solution but does not have conformant plan as solution. The following example highlights this issue.

Example 1 (From [34]). Consider a security window with a lock that can be in one of the three states *opened, closed*[2] or *locked*[3]. When the window is closed or opened, pushing it *up* or *down* will *open* or *close* it respectively. When the window is closed or locked, flipping the lock will lock or close it respectively.

Suppose that a robot needs to make sure that the window is locked and initially, the robot knows that the window is not open (but whether it is locked or closed is unknown).

No conformant plan can achieve the goal. Instead, the robot needs a conditional plan consisting of the following steps: (1) checks the window to determine the window's status; if the window is closed, (2.a) locks the window; otherwise (i.e., the window is already locked), (2.b) does nothing.   □

The proposed ASP-based systems for conditional and conformant planning in [30,31,34,35] show that ASP-based planners performed well comparing to state-of-the-art planning systems of the same kind in several domains. Their performance can be attributed to the following key ideas:

---

[2] The window is closed and unlocked.
[3] The window is closed and locked.

- The use of an input language that allows for the representation and reasoning with static causal laws (a.k.a. axiom or domain constraints). It should be noted that the original specification of the Planning Domain Description Language (PDDL) – a language frequently used for the specification of planning problems by the planning community – includes axioms[4] which correspond to non-recursive static causal laws in our terminology [11]. However, the conformant planning benchmarks designed by the planning community do not use static causal laws.
- The employment of an approximation semantics that reduces the computational complexity of planning with incomplete information to **NP**-complete.

The next example highlights the advantage of directly dealing with static causal laws.

Example 2 (Dominos Domain [31]). Suppose that we have n dominos standing on a line in such a way that if one domino falls then the domino on its right also falls. There is also a ball hanging close to the leftmost domino. Swinging the ball will cause the leftmost domino to fall. Initially, the ball stays still and whether or not the dominos are standing is unknown. The goal is to have the rightmost domino to fall. Obviously, swinging the ball is the only plan to achieve this goal, no matter how big n is.

The problem can be easily expressed by a theory with a set of objects $1, \ldots, n$ denoting the dominos from left to right and a single action *swing* that causes $down_1$ (the leftmost domino falls) to be true, and $n - 1$ axioms (state constraints) $down_i \Rightarrow down_{i+1}$ representing the fact that $down_{i+1}$ is true if $down_i$ is true. The goal is to have $down_n$ become true.

State constraints are usually dealt with by compiling them away. According to the compilation suggested in [33], for each axiom $down_i \Rightarrow down_{i+1}$, we introduce a new action e_i whose effect is $down_{i+1}$ and whose precondition is $down_i$. Clearly, under this compilation, the plan to achieve the goal is the sequence of actions $[swing, e_1, \ldots, e_{n-1}]$.

The main problem with this compilation is that the plan length increases with the number of objects. Even when it is only linear to the size of the original problem, it proves to be challenging for planners following this approach. Most planners have problem when plan length is greater than 500 (i.e., more than 500 dominos).   □

The input language is the action language $\mathcal{A}^c$ (in [30,31,35]) and $\mathcal{A}^c_K$ (in [34]). Since $\mathcal{A}^c_K$ is an extension of $\mathcal{A}^c$ with sensing actions, we summarize the features of $\mathcal{A}^c_K$ below. An action theory in $\mathcal{A}^c_K$ is a collection of statements of the following forms:

$$\textbf{initially}(l) \qquad (2)$$

$$\textbf{executable\_if}(a, \psi) \qquad (3)$$

$$\textbf{causes}(a, l, \phi) \qquad (4)$$

$$\textbf{if}(l, \varphi) \qquad (5)$$

$$\textbf{determines}(a, \theta) \qquad (6)$$

where a is an action, l is a fluent literal, and $\psi, \phi, \varphi, \theta$ are sets of literals[5]. (2) says that l holds in the initial situation. (3) says that a is executable in any situation in which ψ holds (the precise meaning of *hold* will be given later). (4) represents a conditional effect of an action. It says that performing a in a situation in which ϕ holds causes l to hold in the successor situation. (5), called a *static causal law*, states that l holds in any situation in which φ holds. (6) states that the values of literals in θ, sometimes referred to as *sensed-literals*, will be known after a is executed.

---

[4] In our view, static causal laws can be used to represent relationships between fluents and thus could be considered as axioms in PDDL.

[5] A set of literals is interpreted as the conjunction of its members. $\emptyset$ denotes *true*.

The complete semantics of $\mathcal{A}_K^c$ can be found in [2]. It defines a transition function Φ over pairs of actions and sets of belief states. The approximation semantics employed in the systems in [30,31,34,35] defines a transition function Φ_a over pairs of actions and approximation states where an approximation state is a set of consistent literals satisfying the static causal laws. Φ_a can be defined in set theoretical terms [31,34] or by logic program rules [35]. Its precise definition can be found in the aforementioned papers. Φ_a can be used for conformant planning in ASP, in the same way that the transition function of the action language is used for planning as described in [16].

Given a planning problem instance $\mathcal{P} = (\mathcal{D}, \mathcal{I}, \mathcal{G})$, where $\mathcal{D}$ is a set of statements of the forms (3)–(6), $\mathcal{I}$ is a set of statements of the form (2), and $\mathcal{G}$ is a fluent formula. Let k be an integer denoting the maximal length of the desirable solutions. We solve $\mathcal{P}$ by translating it into a logic program $\pi_k(\mathcal{P})$ such that each answer set of $\pi_k(\mathcal{P})$ corresponds to a solution of at most k actions. Besides atoms defining the actions, fluents, literals, etc., $\pi_k(\mathcal{P})$ uses the following main predicates:

- $holds(L, T)$: literal L holds at step T.
- $poss(A, T)$: action A is executable at T.
- $occ(A, T)$: action A occurs at T.
- $pc(L, T)$: literal L may change at $T + 1$.
- $goal(T)$: the goal is satisfied at T.

The rules for encoding direct effects of actions in $\pi_k(\mathcal{P})$ are similar to the rules used for classical planning:

$$holds(L, T + 1) \leftarrow occ(A, T), causes(A, L, \varphi), holds(\varphi, T).$$

The difference with classical planning lies in the rules defining $pc(L, T)$ and the rule encoding of the inertial axiom:

$$holds(L, T + 1) \leftarrow holds(L, T), not\ pc(\neg L, T + 1).$$

Rules in $\pi_k(\mathcal{P})$ ensure that, for an answer set S of $\pi_k(\mathcal{P})$, if $\delta = \{l \mid holds(l, t) \in S\}$, $\delta' = \{l \mid holds(l, t + 1) \in S\}$, and $occ(a, t) \in S$ then (a) a is executable in δ; and (b) $\delta' = \Phi_a(a, \delta)$.

As shown in Example 1, conformant plans are insufficient when sensing actions are needed. In this situation, conditional plans are required. Formally, a conditional plan is (i) [] is a conditional plan, denoting the empty plan, i.e., the plan containing no action; (ii) if a is a non-sensing action and p is a conditional plan then $[a; p]$ is a conditional plan; (iii) if a is a sensing action with proposition (6), where $\theta = \{g_1, \dots, g_n\}$, and p_j's are conditional plans then $[a; \mathbf{cases}(\{g_j \rightarrow p_j\}_{j=1}^n)]$ is a conditional plan; and (iv) Nothing else is a conditional plan.

To encode a conditional planning problem in ASP, we need to accommodate possible cases of a conditional plan. Let us observe that each conditional plan p corresponds to a labeled plan tree T_p defined as below.

- If $p = []$ then T_p is a tree with a single node.
- If $p = [a]$, where a is a non-sensing action, then T_p is a tree with a single node and this node is labeled with a.
- If $p = [a; q]$, where a is a non-sensing action and q is a non-empty plan, then T_p is a tree whose root is labeled with a and has only one subtree which is T_q. Furthermore, the link between a and T_q's root is labeled with an empty string.

– If $p = [a; \textbf{cases}(\{g_j \rightarrow p_j\}_{j=1}^n)]$, where a is a sensing action that determines g_j's, then T_p is a tree whose root is labeled with a and has n subtrees $\{T_{p_j} \mid j \in \{1, \ldots, n\}\}$. For each j, the link from a to the root of T_{p_j} is labeled with g_j.

For instance, Fig. 1 shows the trees for the following four conditional plans in the domain of Example 1:

(i) $p_1 = [push\_down; flip\_lock]$;

(ii) $p_2 = check; \textbf{cases} \begin{pmatrix} open & \rightarrow & [] \\ closed & \rightarrow & [flip\_lock] \\ locked & \rightarrow & [] \end{pmatrix}$;

(iii) $p_3 = check; \textbf{cases} \begin{pmatrix} open & \rightarrow & [push\_down; flip\_lock] \\ closed & \rightarrow & [flip\_lock; flip\_lock; flip\_lock] \\ locked & \rightarrow & [] \end{pmatrix}$; and

(iv) $p_4 = check; \textbf{cases} \begin{pmatrix} open & \rightarrow & [] \\ closed & \rightarrow & p_2 \\ locked & \rightarrow & [] \end{pmatrix}$

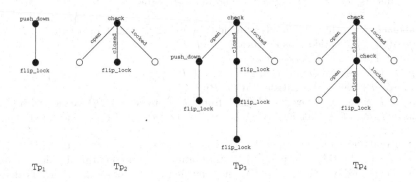

Fig. 1. Sample plan trees

Observe that each trajectory of the plan p corresponds to a path from the root to a leave of T_p. Furthermore, let α (or the *width* of T_p) be the number of leaves of T_p and β (or the *height* of T_p) be the number of nodes along the longest path from the root to the leaves of T_p. Let w and h be two integers such that $\alpha \leq w$ and $\beta \leq h$ and the leaves of T_p be $x_1, \ldots, x_\alpha$. We map each node y of T_p to a pair of integers $n_y = (t_y, p_y)$, where t_y is the number of nodes along the path from the root to y, and p_y is defined in the following way.

– For each leaf x_i of T_p, p_{x_i} is an arbitrary integer between 1 and w such that (*i*) there exists a leaf x with $p_x = 1$, and (*ii*) $i \neq j$ implies $p_{x_i} \neq p_{x_j}$.
– For each interior node y of T_p with children $y_1, \ldots, y_r$, $p_y = \min\{p_{y_1}, \ldots, p_{y_r}\}$.

Figure 2 shows some possible mappings for the four trees in Fig. 1. It is easy to see that if $\alpha \leq w$ and $\beta \leq h$ then such a mapping always exists and $(1, 1)$ is always assigned to the root. Furthermore, given a labeled tree T_p whose nodes are numbered according to the about rules, the plan p can easily be reconstructed. This means that

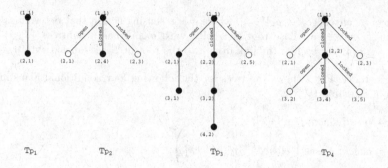

Fig. 2. Possible mappings for T_{p_i} $(i = 1, 2, 3, 4)$ with $w = 5$ and $h = 4$

computing a solution p of a planing problem $\mathcal{P} = (\mathcal{D}, \mathcal{I}, \mathcal{G})$ is equivalent to identifying its tree T_p. This property allows for the encoding of $\mathcal{P}$ as a logic program $\pi_{h,w}(\mathcal{P})$ that generates labeled trees corresponding to solutions of $\mathcal{P}$ whose width and height are bounded by w and h, respectively. In addition to the usual predicates defining the actions, fluents, etc. $\pi_{h,w}(\mathcal{P})$ uses the following predicates:

- $holds(L, T, P)$: literal L holds at node (T, P) (i.e., at step T of path P).
- $poss(A, T, P)$: action A is executable at (T, P).
- $occ(A, T, P)$: action A occurs at (T, P).
- $pc(L, T, P)$: literal L may change at $(T + 1, P)$.
- $goal(T, P)$: the goal is satisfied at (T, P).
- $br(G, T, P, P_1)$: there exists a branch from (T, P) to $(T + 1, P_1)$ labeled with G.
- $used(T, P)$: (T, P) belongs to some extended trajectory of the constructed plan.

Observe that most of the predicates used in $\pi_{h,w}(\mathcal{P})$ are similar to those in $\pi_k(\mathcal{P})$ extended with the third parameter encoding branches of a conditional plan, the last two predicates are specific to $\pi_{h,w}(\mathcal{P})$. They encode the cases of the solution. The detail encoding of $\pi_{h,w}(\mathcal{P})$ and its soundness and completeness can be found in [34].

One disadvantage of the proposed approach is the incompleteness of the ASP based planners. To address this issue, we identified completeness condition of the approximation [35]. Saturation and meta-programming techniques (see Sect. 3) could be used for a complete ASP-based planner.

3 Answer Set Programming in Goal Recognition Design

Goal recognition, a special form of plan recognition, deals with online problems aiming at identifying the goal of an agent as quickly as possible given its behavior [9,23]. For example, Fig. 3(a) shows an example gridworld application, where the agent starts at cell $E3$ and can move in any of the four cardinal directions. Its goal is one of three possible ones G1, G2, and G3. The traditional approach has been to find efficient algorithms that observe the trajectory of the agent and predict its actual goal [9,23].

Goal recognition design (GRD) [12] aims at identifying possible changes to the environment in which the agents operate, typically by making a subset of feasible actions infeasible, so that agents are forced to reveal their goals as early as possible. For example, under the assumption that agents follow optimal plans to reach their

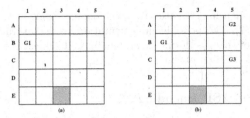

Fig. 3. Example Problem

goal, by making the action that moves the agent from cells $E3$ to $D3$ infeasible, the agent is forced to either move left to $E2$, which would immediately reveal that its goal is G1, or move right to $E4$, revealing that it is either G2 or G3. In [12], the authors introduced the notion of *worst-case distinctiveness* (*wcd*), as a goodness measure that assesses the ease of performing goal recognition within an environment. The *wcd* of a problem is the longest sequence of actions an agent can take without revealing its goal. The objective in GRD is then to find a subset of feasible actions to make infeasible such that the resulting *wcd* is minimized. We will next present two ASP-based solutions of the GRD problem. Abusing the notation, we represent a GRD problem $\mathcal{P}$ by the triple $(\mathcal{D}, \mathcal{I}, \mathcal{G})$ with the understanding that $\mathcal{G}$ is the set of possible goals of the agent. By wcd($\mathcal{P}$), we denote the *wcd* of $\mathcal{P}$.

3.1   A Saturation-Based Meta Encoding

The first encoding of the GRD problem in ASP utilizes meta-programming and saturation techniques. The saturation technique is an advanced guess and check methodology used in disjunctive ASP to check whether *all* possible guesses in a problem domain satisfy a certain property [4]. It can be used to encode Σ_2^P-complete problems such as the satisfiability problem for $\exists\forall$-QBF. For instance, in a typical encoding for satisfiability of a $\exists\forall$-QBF the *guess part* uses disjunction to generate all possible truth values for the propositional atoms that are quantified by $\forall$ ($\forall$-atoms) and the *check part* checks the satisfiability of the formula for all valuations of the $\forall$-atoms (i.e., it checks whether the resulting formula after applying choices made for $\exists$-atoms is a tautology or not). To achieve this, the fact that answer sets are minimal w.r.t. the atoms defined by disjunctive rules is utilized. To this end, the *saturation part* of the program derives (saturates) all atoms defined in the guess part for generating the search space. It should be noted that the saturation technique puts syntactical restrictions on the program parts by forbidding the use of saturated atoms as default negation literals in a rule or as positive literals in a constraint [4,15].

As it turns out, the *wcd* of a problem can be formulated as a $\exists\forall$-QBF formula as follows. Let $g \in \mathcal{G}$ and π_g^* denote the minimal cost plan achieving g. Let $vl(x, y, c)$ denote that c is the common prefix of minimal cost plans of π_x^* and π_y^*. The *wcd* definition of $\mathcal{P}$ can be encoded by the following $\exists\forall$-QBF:

$$\exists x, y, c[vl(x, y, c) \wedge [\forall x', y', c'[vl(x', y', c') \rightarrow |c| \geq |c'|]] \tag{7}$$

where, for the sake of simplicity, we omit some details such as $x, y, x', y' \in \mathcal{G}$, and that c and c' correspond to sequences of actions that are the common prefix of cost-optimal plans π_x^* and π_y^*, $\pi_{x'}^*$ and $\pi_{y'}^*$, respectively.

To compute the *wcd* using the saturation technique, we only need to encode the satisfiability of formula (7). Two possible implementations of the saturation techniques are detailed in [29]; one of them performs exceptionally well against the system developed by the inventor of the GRD problem.

3.2 A Multi-shot ASP Encoding

The second encoding of the GRD problem employs a hybrid implementation made possible by multi-shot ASP. Given a GRD $\mathcal{P} = (\mathcal{D}, \mathcal{I}, \mathcal{G})$, an integer k denoting the maximal number of actions that can be blocked for reducing the wcd, and an integer max be denoting the maximal length of plans in $\mathcal{P}$. We develop a multi-shot ASP program $\Pi(\mathcal{P})'$ for computing (i) wcd$(\mathcal{P})$; and (ii) a solution of $\mathcal{P}$ wrt. k (a set of actions that should be blocked) to achieve wcd$(\mathcal{P})$. Specifically, $\Pi(\mathcal{P})$ implements Algorithm 1 in multi-shot ASP and consists of a logic program $\pi(\mathcal{P})$ and a Python program $GRD(\mathcal{P}, k, max)$.

Algorithm 1. $GRD(\mathcal{P}, k, max)$

1: **Input**: a GRD problem $\mathcal{P} = (\mathcal{D}, \mathcal{I}, \mathcal{G})$ & integers k, max.
2: **Output**: wcd$(\mathcal{P})$, and a solution R of $\mathcal{P}$ w.r.t. k or unsolvable if some goal is not achievable.
3: **for** each goal g in $\mathcal{G}$ **do**
4:     compute the length of minimal plan for g
5:     **if** plan of length $i \leq max$ exists **then** set $m_g = i$
6:     **else return** *unsolvable*
7: **end for**
8: let $\pi_1 = \pi^*(\mathcal{P}) \cup \{min\_goal(g, m_g), activate(g) \mid g \in \mathcal{G}\}$
9: set $len = \max\{m_g \mid g \in \mathcal{G}\}$ in π_1
10: add the optimization module of $\pi(\mathcal{P})$ to π_1
11: compute an answer set Y of π_1
12: let wcd$(\mathcal{P}) = d$ where $wcd(d) \in Y$ % Note: π_1 defines the atom $wcd(d)$
13: compute a set S of actions that can potentially change wcd$(\mathcal{P})$ when they are removed
14: set $w = $ wcd$(\mathcal{P})$ and $R = \emptyset$
15: **for** each set X of at most k actions in S **do**
16:     let $\pi_2 = \pi_1 \cup \{blocked(a) \mid a \in X\} \cup$ the blocking module of $\pi(\mathcal{P})$
17:     compute an answer set Z of π_2
18:     **if** $wcd(d') \in Z$ & $d' < w$ **then** set $w = d'$ and $R = X$
19: **end for**
20: **return** $\langle w, R \rangle$

The program $\pi(\mathcal{P})$ consists of the following modules:

- *Planning*: A program encoding the domain information D of $\mathcal{P}$ and the rules for generating optimal plan for each $g \in \mathcal{G}$. This module is similar to the standard encoding in ASP planning [16] with an extension to allow for the generation of multiple plans for multiple goals at the same time (i.e., similar to that used in conditional planning in Sect. 2).

- *Optimization*: A set of rules for determining the longest prefix between two plans of two goals g_I and g_J on trajectories $I \neq J$ given a set of plans for the goals in $\mathcal{G}$. It also contains the optimization statement for selecting answer sets containing $\text{wcd}(\mathcal{P})$.
- *Blocking*: A set of rules that interact with the Python program to block actions from the original problem.

The multi-shot ASP implementations of the GRD problem performs reasonably well against the system developed by the inventor of the GRD problem [29].

4   ASP in Multi-agent System

4.1   ASP and Distributed Constraint Optimization Problems

A *distributed constraint optimization problem* (DCOP) is defined by $\langle \mathcal{X}, \mathcal{D}, \mathcal{F}, \mathcal{A}, \alpha \rangle$, where: $\mathcal{X} = \{x_1, \dots, x_n\}$ is a set of *variables*; $\mathcal{D} = \{D_1, \dots, D_n\}$ is a set of finite *domains*, where D_i is the domain of variable x_i; $\mathcal{F} = \{f_1, \dots, f_m\}$ is a set of *constraints*, where each k_i-ary constraint $f_i : D_{i_1} \times D_{i_2} \times \dots \times D_{i_{k_i}} \mapsto \mathbb{N} \cup \{-\infty, 0\}$ specifies the utility of each combination of values of the variables in its *scope*, $scope(f_i) = \{x_{i_1}, \dots, x_{i_{k_i}}\}$; $\mathcal{A} = \{a_1, \dots, a_p\}$ is a set of *agents*; and $\alpha : \mathcal{X} \to \mathcal{A}$ maps each variable to one agent.

Figure 4(a,b) shows an example of a DCOP with three agents (see [14]), where each agent a_i controls variable x_i with domain $\{0, 1\}$. Figure 4(a) shows its constraint graph and Fig. 4(b) shows the utility functions, assuming that all of the three constraints have the same function.

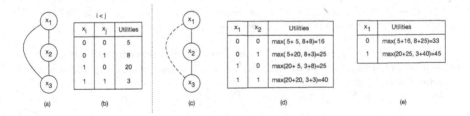

Fig. 4. DCOP graph (a), utility table (b); DPOP graph (c), UTIL-Phase Computation Table (d, e)

A *solution* is a value assignment for all variables and its corresponding utility is the evaluation of all utility functions on such solution. The goal is to find a utility-maximal solution. Solutions of a DCOP can be computed in three steps [22]: (*i*) constructing a pseudo-tree from the constraint graph (Fig. 4(c) for the example); (*ii*) UTIL-computation phase: each agent, starting from the leafs of the pseudo-tree, (*ii.x*) computes the optimal sum of utilities in its subtree for each value combination of variables in the set of variables owned by ancestor agents that are constrained with variables owned by the agents in the subtree (Fig. 4(d) shows the UTIL-computation of the agent a_3) and (*ii.xx*) sends the maximal value to its parent; and (*iii*) VALUE-propagation phase: each agent, starting from the root of the pseudo-tree, determines the optimal value for its variables upon receiving the VALUE message from its parent and sends to its children (Fig. 4(e) shows the VALUE-propagation of the agent a_1) in a VALUE message.

In [14], we presented an ASP-based system, ASP-DPOP, for computing solutions of DCOP. In this system, each agent consists of two modules, an ASP module and a controller. The ASP module is responsible for computing the UTIL and VALUE messages when the agent needs to do so. The controller, written in SICStus© Prolog, is responsible for all communications between the agent and other agents. When an agent receives all UTIL messages from its children, the ASP module computes its UTIL-message and the controller sends the message to its parent. When an agent receives its parent's VALUE message, the ASP module computes its own VALUE message and the controller sends the message to its children. The flexibility and expressiveness of ASP allows ASP-DPOP to work with agents who control multiple variables while state-of-the-art DCOP solvers assume that each agent controls only one variable. ASP-DPOP performs well against state-of-the-art DCOP solvers in several domains and has better performance, both in scalability and efficiency, in domains with hard constraints. The approach has been extended to deal with uncertainty in constraint utilities [13].

4.2   Multi-agent Planning

Multi-agent planning (MAP) is the problem of planning for multiple agents. The presence of multiple agents that can change the environment simultaneously brings about a number of issues:

- can the planning process be done centralized or must it be done distributed?
- what is the protocol for agents to communicate with each other?
- what types of actions are available for the agents (e.g., whether group actions are available? whether knowledge and/or belief changing actions are involved? etc.)?
- what are the representation languages used by individual agents?

For simplicity of the presentation, let us assume that all agents use the same representation language. The answer to the other questions depends on the degree of cooperativeness between agents.

Generally, a MAP for the agents $\{1, \ldots, n\}$ can be represented by a tuple $(\mathcal{P}_1, \ldots, \mathcal{P}_n)$ where $\mathcal{P}_i$ is a planning problem for agent i extended with information about other agents who can affect the view of the environment locally to i.

- When agents are fully cooperative and planning can be done by one single agent, the encoding for single-agent planning (e.g., in [16]) can be extended to deal with MAP by
  - creating the program $\pi_k(\mathcal{P}_i)$ for $\mathcal{P}_i$; and
  - adding constraints to eliminate conflicts that arise due to the (potentially) parallel execution of actions among agents

A prototype of an ASP based MAP system was proposed in [28]. Recently, we extend this prototype to deal with an interesting application the *Multi-Agent Path Finding* (MAPF) problem that deals with teams of agents that need to find collision-free paths from their respective starting locations to their respective goal locations on a graph. This model has attracted a lot of attention due to the success of the autonomous warehouse systems [36]. In these systems (illustrated by Fig. 5), robots (in orange) navigate around a warehouse to pick up inventory pods

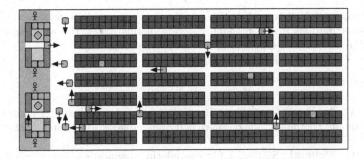

Fig. 5. Layout of an autonomous warehouse system [Wurman *et al.*, 2008] (Color figure online)

from their storage locations (in green) and drop them off at designated inventory stations (in purple) in the warehouse.

As it turns out, the ASP based system does not perform very well in this application comparing to state-of-the-art (e.g., [17]). The interesting part of this problem is that the basic ASP encoding is fairly simple. Yet, the problem quickly becomes *unsolvable* when its size increases.

By adding domain-knowledge to the encoding and decomposing the problem into smaller sub-problems, the scalability of the system improves significantly [19]. For example, it is easy to see that by adding some designated locations to the map, a path can be seen as multiple segments among the designated locations. As such, a path can be generated in multiple steps. In the first step, segments of a path are generated using a simplified map. The final path is then obtained by generating the concrete path for each segment.

– For self-interested agents, solving an MAP requires that agents negotiate with each other and thus an integration of a negotiation framework with MAP will be necessary. As shown in [27], ASP can also be used effectively for the development of negotiation systems. It is worth noticing that any negotiation framework used for this purpose must consider the dynamic of the environment. In [26], we developed an ASP based prototype for planning with negotiation in a dynamic environment. While the underlying encoding for planning does not change, special attentions need to be made to deal with the "effects" of negotiations. We envision that this approach will be necessary for some future extensions of the MAPF problem that might require stronger interactions between agents. For example, an agent might request help from another agent to continue its job if it realizes that its battery will run out before it can complete its job.

5   Conclusions

In this paper, we describe the application of answer set programming in planning with incomplete information and sensing actions, goal recognition design, distributed constraint optimization problem, and various settings of multi-agent planning. We discuss the key techniques that contribute to the good performance of ASP based solutions and present a challenging application for ASP.

Acknowledgement. The author wishes to thank his many collaborators and students for their contributions in the research reported in this paper. He would also like to acknowledge the partial support from various NSF grants.

References

1. Baral, C., Kreinovich, V., Trejo, R.: Computational complexity of planning and approximate planning in the presence of incompleteness. Artif. Intell. **122**, 241–267 (2000)
2. Baral, C., McIlraith, S., Son, T.C.: Formulating diagnostic problem solving using an action language with narratives and sensing. In: KR, pp. 311–322 (2000)
3. Dimopoulos, Y., Nebel, B., Koehler, J.: Encoding planning problems in non-monotonic logic programs. In: ECP, pp. 169–181 (1997)
4. Eiter, T., Ianni, G., Krennwallner, T.: Answer set programming: a primer. In: Tessaris, S., Franconi, E., Eiter, T., Gutierrez, C., Handschuh, S., Rousset, M.-C., Schmidt, R.A. (eds.) Reasoning Web 2009. LNCS, vol. 5689, pp. 40–110. Springer, Heidelberg (2009). doi:10.1007/978-3-642-03754-2_2
5. Eiter, T., Leone, N., Mateis, C., Pfeifer, G., Scarcello, F.: The KR system dlv: progress report, comparisons, and benchmarks. In: KR 1998, pp. 406–417 (1998)
6. Erdem, E., Gelfond, M., Leone, N.: Applications of answer set programming. AI Mag. **37**(3), 53–68 (2016)
7. Faber, W., Leone, N., Pfeifer, G.: Recursive aggregates in disjunctive logic programs: semantics and complexity. In: Alferes, J.J., Leite, J. (eds.) JELIA 2004. LNCS, vol. 3229, pp. 200–212. Springer, Heidelberg (2004). doi:10.1007/978-3-540-30227-8_19
8. Gebser, M., Kaufmann, B., Neumann, A., Schaub, T.: *clasp*: a conflict-driven answer set solver. In: Baral, C., Brewka, G., Schlipf, J. (eds.) LPNMR 2007. LNCS (LNAI), vol. 4483, pp. 260–265. Springer, Heidelberg (2007). doi:10.1007/978-3-540-72200-7_23
9. Geffner, H., Bonet, B.: A Concise Introduction to Models and Methods for Automated Planning. Morgan & Claypool Publishers (2013)
10. Gelfond, M., Lifschitz, V.: Logic programs with classical negation. In: ICLP, pp. 579–597 (1990)
11. Ghallab, M., Howe, A., Knoblock, C., McDermott, D., Ram, A., Veloso, M., Weld, D., Wilkins, D.: PDDL – the planning domain definition language, version 1.2. Technical report, CVC TR98003/DCS TR1165, Yale Center for Comp, Vis and Ctrl (1998)
12. Keren, S., Gal, A., Karpas, E.: Goal recognition design. In: ICAPS (2014)
13. Le, T., Fioretto, F., Yeoh, W., Son, T.C., Pontelli, E.: ER-DCOPs: a framework for distributed constraint optimization with uncertainty in constraint utilities. In: AAMAS, pp. 606–614. ACM (2016)
14. Le, T., Son, T.C., Pontelli, E., Yeoh, W.: Solving distributed constraint optimization problems using logic programming. In: AAAI, pp. 1174–1181. AAAI Press (2015)
15. Leone, N., Rosati, R., Scarcello, F.: Enhancing answer set planning. In: IJCA Workshop on Planning under Uncertainty and Incomplete Information (2001)
16. Lifschitz, V.: Answer set programming and plan generation. Artif. Intell. **138**(1–2), 39–54 (2002)
17. Ma, H., Koenig, S.: Optimal target assignment and path finding for teams of agents, pp. 1144–1152 (2016)

18. Marek, V., Truszczyński, M.: Stable models and an alternative logic programming paradigm. In: The Logic Programming Paradigm: a 25-Year Perspective, pp. 375–398 (1999)
19. Nguyen, V.D., Obermeier, P., Son, T.C., Schaub, T., Yeoh, W.: Generalized target assignment and path finding using answer set programming. Technical report, NMSU (2017)
20. Niemelä, I.: Logic programming with stable model semantics as a constraint programming paradigm. Ann. Math. Artif. Intell. **25**(3,4), 241–273 (1999)
21. Pelov, N., Denecker, M., Bruynooghe, M.: Partial stable models for logic programs with aggregates. In: Lifschitz, V., Niemelä, I. (eds.) LPNMR 2004. LNCS (LNAI), vol. 2923, pp. 207–219. Springer, Heidelberg (2003). doi:10.1007/978-3-540-24609-1_19
22. Petcu, A., Faltings, B.: A scalable method for multiagent constraint optimization. In: IJCAI, pp. 1413–1420 (2005)
23. Ramírez, M., Geffner, H.: Goal recognition over pomdps: inferring the intention of a POMDP agent. In: IJCAI, pp. 2009–2014 (2011)
24. Simons, P., Niemelä, N., Soininen, T.: Extending and implementing the stable model semantics. Artif. Intell. **138**(1–2), 181–234 (2002)
25. Son, T.C., Pontelli, E.: A constructive semantic characterization of aggregates in answer set programming. Theory Pract. Logic Program. **7**(03), 355–375 (2007)
26. Son, T.C., Pontelli, E., Sakama, C.: Logic programming for multiagent planning with negotiation. In: Hill, P.M., Warren, D.S. (eds.) ICLP 2009. LNCS, vol. 5649, pp. 99–114. Springer, Heidelberg (2009). doi:10.1007/978-3-642-02846-5_13
27. Son, T.C., Pontelli, E., Nguyen, N., Sakama, C.: Formalizing negotiations using logic programming. ACM Trans. Comput. Log. **15**(2), 12 (2014)
28. Son, T.C., Pontelli, E., Nguyen, N.-H.: Planning for multiagent using ASP-prolog. In: Dix, J., Fisher, M., Novák, P. (eds.) CLIMA 2009. LNCS (LNAI), vol. 6214, pp. 1–21. Springer, Heidelberg (2010). doi:10.1007/978-3-642-16867-3_1
29. Son, T.C., Sabuncu, O., Schulz-Hanke, C., Schaub, T., Yeoh, W.: Solving goal recognition design using ASP. In: AAAI (2016)
30. Son, T.C., Tu, P.H., Gelfond, M., Morales, R.: An approximation of action theories of $\mathcal{AL}$ and its application to conformant planning. In: LPNMR, pp. 172–184 (2005)
31. Son, T.C., Tu, P.H., Gelfond, M., Morales, R.: Conformant planning for domains with constraints – a new approach. In: AAAI, pp. 1211–1216 (2005)
32. Subrahmanian, V., Zaniolo, C.: Relating stable models and AI planning domains. In: ICLP, pp. 233–247 (1995)
33. Thiebaux, S., Hoffmann, J., Nebel, B.: In defense of PDDL axioms. In: Proceedings of the 18th International Joint Conference on Artificial Intelligence (IJCAI 2003) (2003)
34. Tu, P., Son, T.C., Baral, C.: Reasoning and planning with sensing actions, incomplete information, and static causal laws using logic programming. Theory Pract. Logic Program. **7**, 1–74 (2006)
35. Tu, P., Son, T.C., Gelfond, M., Morales, R.: Approximation of action theories and its application to conformant planning. Artif. Intell. J. **175**(1), 79–119 (2011)
36. Wurman, P., D'Andrea, R., Mountz, M.: Coordinating hundreds of cooperative, autonomous vehicles in warehouses. AI Mag. **29**(1), 9–20 (2008)

From Logic Programming and Non-monotonic Reasoning to Computational Argumentation and Beyond

Francesca Toni[✉]

Imperial College London, London, UK
ft@imperial.ac.uk

Abstract. Argumentation has gained popularity in AI in recent years to support several activities and forms of reasoning. This talk will trace back the logic programming and non-monotonic reasoning origins of two well-known argumentation formalisms in AI (namely abstract argumentation and assumption-based argumentation). Finally, the talk will discuss recent developments in AI making use of computational argumentation, in particular to support collaborative decision making.

1 Introduction

Computational Argumentation (CA, aka 'Argumentation in AI') amounts to the definition of formalisms, semantics, algorithms and systems to support reasoning with conflicting and incomplete information, as well as, in many instances, explaining the outcomes of this reasoning. Abstract argumentation (AA) [Dun95] and Assumption-based Argumentation (ABA) [BTK93,BDKT97, DKT09,Ton14] are two well-known CA formalisms, equipped with a variety of semantics, algorithms and systems, and deployed to support a number of applications. AA frameworks can be simply thought of as directed graphs whose nodes are arguments and whose edges represent conflicts (where an edge from A to B represents an *attack* from A to B). Whereas in AA frameworks arguments and attacks are primitive notions, in ABA they are defined in terms of other, primitive notions, and have, as a result, an internal structure. Thus, ABA is a form *structured CA* [BH14]. In the case of ABA the primitive notions based on which arguments and attacks are obtained are those of *rules* in an underlying *deductive system, assumptions* and their *contraries*: arguments are supported by rules and assumptions and attacks are directed against (assumptions deducible from) assumptions supporting arguments, by building arguments for the contrary of these assumptions. Semantics of AA are characterised in terms of sets of arguments (or *extensions*) [Dun95,DMT07] and semantics of ABA frameworks in terms of sets of assumptions or arguments (or *extensions*, again) [BTK93,BDKT97,DMT07] meeting desirable requirements, including, but not limited to, the core requirement of *conflict-freeness* (where an extension is *conflict-free* iff none of its elements attack any of its elements).

M. Balduccini and T. Janhunen (Eds.): LPNMR 2017, LNAI 10377, pp. 36–39, 2017.
DOI: 10.1007/978-3-319-61660-5_4

2 From Logic Programming and Non-monotonic Reasoning to AA/ABA

The AA/ABA semantics of admissible, preferred, complete, grounded, stable and ideal extensions [Dun95,BDKT97,DMT07] differ in which additional desirable requirements they impose upon extensions, but can all be seen as providing argumentative counterparts of semantics that had previously been defined for logic programming and, in the case of stable extensions, other non-monotonic reasoning frameworks, by appropriately instantiating AA/ABA frameworks [BTK93,Dun95,BDKT97] to "match" the original logic programming and non-monotonic reasoning frameworks.

AA/ABA are equipped with a range of computational tools, in the form of algorithms and/or systems. Some of these are top-down, query-oriented, based on *dispute trees*, as defined in [DKT06,DMT07], and amount to *dispute derivations* of various kinds for different semantics [DKT06,DMT07,TDH09,Ton13]. These dispute derivations generalise in turn existing SLD-based procedures for logic programming [Ton13]. Other computational tools are bottom-up, based on the computation of extensions, and are based on mappings of CA frameworks onto Answer Set Programming (ASP) and the use of ASP solvers [EGW10] or onto constraint problems and the use of constraint solvers [BS11].

3 Applications of CA

Computational tools for AA/ABA based on dispute trees have been used to support explanations of reasoning outputs, in various settings and senses, e.g. to explain (non-)membership in answer sets of logic programs [ST16], to explain "goodness" of decisions [FT14,FCS+13,ZFTL14] and, more generically, to explain admissibility of sentences in ABA [FT15], and to explain predictions of recommendations in case-based reasoning [CST16]. Moreover, they can be used to support collaborative decision-making in multi-agent systems, e.g. to speed up the agents' individual learning, as in [GT14], or to allow agents to converge to socially optimal but privacy preserving solutions, as in [GTWX16].

References

[BDKT97] Bondarenko, A., Dung, P.M., Kowalski, R.A., Toni, F.: An abstract, argumentation-theoretic approach to default reasoning. Artif. Intell. **93**(1–2), 63–101 (1997)

[BH14] Besnard, P., Hunter, A.: Constructing argument graphs with deductive arguments: a tutorial. Argument Comput. **5**(1), 5–30 (2014)

[BS11] Bistarelli, S., Santini, F.: Conarg: a constraint-based computational framework for argumentation systems. In: IEEE 23rd International Conference on Tools with Artificial Intelligence, ICTAI, pp. 605–612. IEEE Computer Society (2011)

[BTK93] Bondarenko, A., Toni, F., Kowalski, R.A.: An assumption-based framework for non-monotonic reasoning. In: Pereira, L.M., Nerode, A. (eds.) Proceedings of the 2nd International Workshop on Logic Programming and Non-monotonic Reasoning (LPNMR 1993), pp. 171–189, Lisbon, Portugal. MIT Press, June 1993

[CST16] Cyras, K., Satoh, K., Toni, F.: Abstract argumentation for case-based reasoning. In: Baral, C., Delgrande, J.P., Wolter, F. (eds.) Proceedings of the Fifteenth International Conference on Principles of Knowledge Representation and Reasoning, KR 2016, Cape Town, South Africa, 25–29 April 2016, pp. 549–552. AAAI Press (2016)

[DKT06] Dung, P.M., Kowalski, R.A., Toni, F.: Dialectic proof procedures for assumption-based, admissible argumentation. Artif. Intell. **170**, 114–159 (2006)

[DKT09] Dung, P.M., Kowalski, R.A., Toni, F.: Assumption-based argumentation. In: Rahwan, I., Simari, G.R. (eds.) Argumentation in AI, pp. 25–44. Springer, Heidelberg (2009)

[DMT07] Dung, P.M., Mancarella, P., Toni, F.: Computing ideal sceptical argumentation. Artif. Intell. **171**(10–15), 642–674 (2007)

[Dun95] Dung, P.M.: On the acceptability of arguments and its fundamental role in non-monotonic reasoning, logic programming and n-person games. Artif. Intell. **77**, 321–357 (1995)

[EGW10] Egly, U., Gaggl, S.A., Woltran, S.: Answer-set programming encodings for argumentation frameworks. Argument Comput. **1**(2), 147–177 (2010)

[FCS+13] Fan, X., Craven, R., Singer, R., Toni, F., Williams, M.: Assumption-based argumentation for decision-making with preferences: a medical case study. In: Leite, J., Son, T.C., Torroni, P., Torre, L., Woltran, S. (eds.) CLIMA 2013. LNCS, vol. 8143, pp. 374–390. Springer, Heidelberg (2013). doi:10.1007/978-3-642-40624-9_23

[FT14] Fan, X., Toni, F.: Decision making with assumption-based argumentation. In: Black, E., Modgil, S., Oren, N. (eds.) TAFA 2013. LNCS, vol. 8306, pp. 127–142. Springer, Heidelberg (2014). doi:10.1007/978-3-642-54373-9_9

[FT15] Fan, X., Toni, F.: On computing explanations in argumentation. In: Bonet, B., Koenig, S. (eds.) Proceedings of the Twenty-Ninth AAAI Conference on Artificial Intelligence, pp. 1496–1502. AAAI Press (2015)

[GT14] Gao, Y., Toni, F.: Argumentation accelerated reinforcement learning for cooperative multi-agent systems. In: Schaub, T., Friedrich, G., O'Sullivan, B. (eds.) 21st European Conference on Artificial Intelligence, ECAI 2014, vol. 263 of Frontiers in Artificial Intelligence and Applications, pp. 333–338. IOS Press (2014)

[GTWX16] Gao, Y., Toni, F., Wang, H., Xu, F.: Argumentation-based multi-agent decision making with privacy preserved. In: Jonker, C.M., Marsella, S., Thangarajah, J., Tuyls, K. (eds.) Proceedings of the 2016 International Conference on Autonomous Agents & Multiagent Systems (AAMAS 2016), pp. 1153–1161. ACM (2016)

[ST16] Schulz, C., Toni, F.: Justifying answer sets using argumentation. Theory Pract. Logic Program. **16**(1), 59–110 (2016)

[TDH09] Thang, P.M., Dung, P.M., Hung, N.D.: Towards a common framework for dialectical proof procedures in abstract argumentation. J. Logic Comput. **19**(6), 1071–1109 (2009)

[Ton13] Toni, F.: A generalised framework for dispute derivations in assumption-based argumentation. Artif. Intell. **195**, 1–43 (2013)

[Ton14] Toni, F.: A tutorial on assumption-based argumentation. Argument Comput. Spec. Issue Tutorials Struct. Argumentation **5**(1), 89–117 (2014)

[ZFTL14] Zhong, Q., Fan, X., Toni, F., Luo, X.: Explaining best decisions via argumentation. In: Proceedings of the European Conference on Social Intelligence (ECSI-2014), Barcelona, Spain, 3–5 November 2014, pp. 224–237 (2014)

Nonmonotonic Reasoning

Modular Construction of Minimal Models

Rachel Ben-Eliyahu-Zohary[1]([✉]), Fabrizio Angiulli[2], Fabio Fassetti[2], and Luigi Palopoli[2]

[1] Azrieli College of Engineering, Jerusalem, Israel
rbz@jce.ac.il
[2] DIMES, University of Calabria, Rende, Italy
{f.angiulli,fassetti,palopoli}@dimes.unical.it

Abstract. We show that minimal models of positive propositional theories can be decomposed based on the structure of the dependency graph of the theories. This observation can be useful for many applications involving computation with minimal models. As an example of such benefits, we introduce new algorithms for minimal model finding and checking that are based on model decomposition. The algorithms' temporal worst-case complexity is exponential in the size s of the largest connected component of the dependency graph, but their actual cost depends on the size of the largest source actually encountered, which can be far smaller than s, and on the class of theories to which sources belong. Indeed, if all sources reduce to an HCF or HEF theory, the algorithms are polynomial in the size of the theory.

1 Introduction

The tasks of minimal model finding and checking are central in Artificial Intelligence (AI). These computational tasks are at the heart of several knowledge representation systems[1].

Reasoning with minimal models has been the subject of several studies in the AI community. Given a theory T, the *Minimal Model Finding* task consists of computing a minimal model of T, whereas the *Minimal Model Checking* task is concerned with the problem of checking whether a given set of atoms is indeed a minimal model of T. Both tasks have been proven to be intractable even if only positive theories are considered [3,4]. Therefore, we deem it relevant and interesting to single out classes of theories for which these problems can be solved efficiently.

This work looks for methods to decompose a theory into disjoint subsets of clauses, such that the formidable task of minimal model computation is split between subsets of the original theory. We do so by investigating the relationship between a propositional theory and its super-dependency graph. We show that a minimal model of a theory can be generated by first computing, separately and in parallel, the minimal models of the theories corresponding to sources of the graph and then by computing the minimal models of the rest of the theory, after

---

[1] This is a short paper, all missing references can be found in [2].

© Springer International Publishing AG 2017
M. Balduccini and T. Janhunen (Eds.): LPNMR 2017, LNAI 10377, pp. 43–48, 2017.
DOI: 10.1007/978-3-319-61660-5_5

propagating the assignment to variables by the minimal models computed at the sources. Regarding the opposite direction, we show that given a minimal model, if its projection on a source is a minimal model of the theory corresponding to the source, then the rest of the model is a minimal model of the theory updated by the content of the minimal model computed at the sources.

To demonstrate the merits of theory decomposition, we present two new algorithms- one for minimal model generation and one for minimal model checking. The basic idea of the model generation algorithm is to compute the minimal models bottom to up while traversing the graph source following source. Intuitively, the algorithm starts with an empty model and iteratively adds to it "necessary" atoms. When a source in the graph is encountered during the computation, first the algorithm calls an external procedure like, for example, IGEA, to compute a minimal model of the sub-theory induced by that source. In many cases, this external computation will successfully terminate in polynomial time. Clearly enough, any algorithm possibly proposed in the future might be plugged into the algorithmic schema to ameliorate its performance. The model checking algorithm works in a way opposite to the model finding algorithm. It starts with a model, and it decomposes the model and the theory until both become empty, which means the model is, indeed, a minimal model of the given theory.

Noteworthy, almost all the studies indicate that the source of intractability in minimal model finding stems from the presence of head-loops in the dependency graphs of the theories. In fact, in HCF theories no such a loop occurs, whereas in HEF theories only specific kinds of loops are allowed. Starting from this, the work reported in this manuscript presents an algorithm that finds a minimal model of any positive theory in time exponential in the size of the largest head-loop that induces a sub-theory on which the incomplete algorithm of [1] fails. In particular, when run on HEF theories, our algorithm is guaranteed to find a minimal model in polynomial time.

2   Preliminaries

We focus on propositional theories. We will refer to a theory as a set of clauses of the form

$$a_1 \wedge a_2 \wedge \ldots \wedge a_m \supset c_1 \vee c_2 \vee \ldots \vee c_n \tag{1}$$

where all the a's and the c's are atoms[2]. We assume that all the c's are different. The expression to the left of $\supset$ is called the *body* of the clause, while the expression to the right of $\supset$ is called the *head* of the clause. We will sometimes denote a clause by $B \supset H$, where B is the set of atoms in the body of the clause and H the set of atoms in its head. A clause is disjunctive if $n > 1$. A theory is called *positive* if, for every clause, $n > 0$. From now on, when we refer to a theory it is a positive theory.

---

[2] Note that the syntax of (1) is a bit unusual for a clause; usually, the equivalent notation $\neg a_1 \vee \neg a_2 \vee \ldots \vee \neg a_m \vee c_1 \vee c_2 \vee \ldots \vee c_n$ is employed.

Let X be a set of atoms. X *satisfies the body of a clause* if and only if all the atoms in the body of the clause belong to X. X *violates a clause* if and only if X satisfies the body of the clause, but none of the atoms in the head of the clause belongs to X. X is a *model* of a theory if none of its clauses is violated by X. A model X of a theory T is *minimal* if there is no $Y \subset X$, which is also a model of T. Note that positive theories always have at least one minimal model.

With every theory T we associate a directed graph, called the *dependency graph* of T, in which (a) each atom and each clause in T is a node, and (b) there is an arc directed from a node a to a clause δ if and only if a is in the body of δ. There is an arc directed from δ to a if a is in the head of δ[3].

A *super-dependency graph* SG is an acyclic graph built from a dependency graph G as follows: for each strongly connected component c in G, there is a node in SG, and for each arc in G from a node in a strongly connected component c_1 to a node in a strongly connected component c_2 there is an arc in SG from the node associated with c_1 to the node associated with c_2. A theory T is Head-Cycle-Free (HCF) if there are no two atoms in the head of some clause in T that belong to the same component in the super-dependency graph of T.

A *source* in a directed graph is a node with no incoming edges. By abuse of terminology, we will sometimes use the term "source" as the set of atoms in the source. *A source in a propositional theory* will serve as a shorthand for "a source in the super dependency graph of the theory." A source is called *empty* if the set of atoms in it is empty. Given a source S of a theory T, T_S denotes the set of clauses in T that uses only atoms from S.

Our algorithms use function $\text{Reduce}(T, X, Y)$ which resembles many reasoning methods in knowledge representation, like, for example, unit propagation in DPLL and other constraint satisfaction algorithms. Reduce returns the theory obtained from T where all atoms in X are set to true and all atoms in Y are set to false. More specifically, Reduce returns the theory obtained by first removing all clauses that contain atoms in X in the head and atoms in Y in the body, and second removing all remaining atoms in $X \cup Y$ from T.

3   Minimal Model Finding

The algorithm for minimal model finding exploits the graph-based decompositions presented in the following theorems.

Theorem 1 (Theory decomposition). *Let T be a theory, let G be the SG of T. For any source S in G, let X be a minimal model of T_S. Moreover, let $T' = Reduce(T, X, S - X)$. Then, for any minimal model M' of T', $M' \cup X$ is a minimal model of T.*

Theorem 2 (Minimal model decomposition). *Let T be a positive theory, let G be the SG of T, and let M be a minimal model of T. Moreover, assume*

---

[3] Clause nodes in the dependency graph are mandatory to achieve a graph which is linear in the size of the theory.

Algorithm 1. Algorithm ModuMin

Input: A positive theory T
Output: A minimal model for T

1 $M := \emptyset$;
2 **while** $T \neq \emptyset$ **do**
3   **if** *There is a clause δ in T violated by M such that* $|\text{head}(\delta)| = 1$ **then**
4   let $X := \text{head}(\delta)$; $M := M \cup X$;
5          $T := \text{Reduce}(T, X, \emptyset)$;
6   **else**
7   let G be the super-dependency graph of T;
8   Iteratively delete from G all the empty sources ;
9   let S be the set of atoms in a source of G ;
10   let T_S be the subset of T containing all the clauses from T having only atoms from S;
11   let X be a minimal model of T_S;
12         $M := M \cup X$;
13         $T := T - T_S$; $T := \text{Reduce}(T, X, S - X)$;

14 **return** M

---

there is a source S in G such that $X = M \cap S$ is a minimal model of T_S, and let $T' = Reduce(T, X, S - X)$. Then $M - X$ is a minimal model of T'.

Algorithm ModuMin uses the function *head*. Given a clause δ, *head* returns the set of all atoms belonging to the head of δ. The algorithm works on the super-dependency graph of the theory, from bottom to up. It starts with the empty set as a minimal model and adds to it atoms only when proved to be necessary to build a model.

In [2] we provide a proof that Algorithm ModuMin is correct: it outputs a minimal model of the input theory.

Let n be the size of a theory T, s the size of the largest connected component, and k be the number of connected components in the dependency graph of T. The cost of ModuMin is upper bounded by $t_{\text{ModuMin}}^{ub}(n) = O(n + k \cdot 2^s)$.

The ideas of algorithm ModuMin can be adopted to solving the minimal model checking problem. The minimal model checking problem is defined as follows: *Given a theory T and a model M, check whether M is a minimal model of T.*

In [2], we present Algorithm CheckMin that can be used to check whether a model M of a theory T is a minimal model. It works through the super dependency graph of T, and it recursively deletes from M sets of atoms that are minimal models of the sources of T. T is reduced after each such deletion, to reflect the minimal models found for the sources. This process goes on until T shrinks to the empty set. When this happens, we check if M has shrunk to be the empty set as well. If this is the case, we conclude that M is indeed a minimal model of T.

4 Completeness

While ModuMin is guaranteed to return a minimal model, for some theories there are minimal models that will never be generated by ModuMin. On the other hand, there are clearly theories for which ModuMin is complete. The question remains if we can find cases in which the algorithms will be complete. We provide a partial answer here, and leave the rest for further investigations.

We first define recursively a property called the *Modular property*.

Definition 1. *1. A minimal model M of a positive theory T has the Modular property with respect to T, if the SG of T has only one component.*
2. A minimal model M of a positive theory T has the Modular property with respect to T, if there is a source S in T such that $X = M \cap S$ is a minimal model of T_S, and $M - X$, which is a minimal model of $T' = Reduce(T, X, S - X)$ according to Theorem 2, has the Modular property with respect to T'.

The following theorems hold:

Theorem 3. *Let T be the theory which is input into the algorithm ModuMin. If every minimal model of T has the modular property w.r.t. T, then ModuMin is complete for T.*

Theorem 4. *Assume the theory T and a minimal model M of T are given as input to the algorithm CheckMin. If M has the modular property w.r.t. T, then CheckMin will return* **true**.

Theorems 3 and 4 give us a useful analysis of the cases in which the algorithms presented in this manuscript are complete. They guide us to look for subclasses of theories with respect to which any minimal model has the modular property. One example is theories that have the *OSH Property*, defined next.

Definition 2 (one-source-head (OSH) Property). *A theory T has the one-source-head (OSH) Property if there is a source S in T such that for every atom $P \in S$, if P is in the head of some clause δ in T, then all the other atoms in the head of δ are also in S.*

Theories having the OSH property are useful for completeness:

Theorem 5. *If a theory T has the OSH property, then for every minimal model M of T there is a source S in T such that $X = M \cap S$ is a minimal model of T_S.*

Corollary 3. *Assume T has the OSH property, let M be a minimal model of T, let S be a source such that $X = M \cap S$ is a minimal model of T_S (note that by Theorem 5 there is such S), and let $T' = Reduce(T, X, S - X)$. If $M - X$ (which is a minimal model of T' according to Theorem 2) has the modular property w.r.t. T', then M can be generated by ModuMin.*

Corollary 4. *Assume T has the OSH property, let M be a minimal model of T, let S be a source such that $X = M \cap S$ is a minimal model of T_S (note that by Theorem 5 there is such S), and let $T' = Reduce(T, X, S - X)$. If $M - X$ (which is a minimal model of T' according to Theorem 2) has the modular property w.r.t. T', then* CheckMin *will return* **true** *when given T and M as input.*

The notion of OSH property has practical implications. If T and all the smaller and smaller theories generated by algorithm CheckMin while working on a the input theory T and a candidate minimal model M has the OSH property, then it can be certain that *CheckMin* will return **true** if and only if M is a minimal model of T. Since the OSH property can be checked in linear time, we can easily check whether it holds for the theories generated during the execution of CheckMin.

5   Conclusions

It has long been realized that the source of complexity in computing minimal models of theories is the loops between atoms that lie in the heads of disjunctive clauses. Algorithm ModuMin presented in this paper enables us to compute minimal models in time complexity that is directly dependent on the size of the disjunctive head loops. Past algorithms for computing minimal model did make efforts to exploit the structure of the dependency graph of the theory, but they did not manage to decompose the theory to totally independent sub-theories that can be computed in parallel as we do here. Hence past algorithms did not achieve the complexity analysis that we provide here, which shows that the complexity of model finding is exponential in the size of the largest strongly connected component of the dependency graph of the theory. A detailed discussion of relevant work can be found in [2].

References

1. Angiulli, F., Ben-Eliyahu-Zohary, R., Fassetti, F., Palopoli, L.: On the tractability of minimal model computation for some CNF theories. Artif. Intell. **210**, 56–77 (2014). doi:10.1016/j.artint.2014.02.003
2. Ben-Eliyahu-Zohary, R., Angiulli, F., Fassetti, F., Palopoli, L.: Modular construction of minimal models. A full version of this paper, can be obtained from the authors
3. Cadoli, M.: The complexity of model checking for circumscriptive formulae. Inf. Process. Lett. **44**(3), 113–118 (1992)
4. Cadoli, M.: On the complexity of model finding for nonmonotonic propositional logics. In: Proceedings of the 4th Italian Conference on Theoretical Computer Science, pp. 125–139. World Scientific Publishing Co., October 1992

A Hasse Diagram for Weighted Sceptical Semantics with a Unique-Status Grounded Semantics

Stefano Bistarelli and Francesco Santini$^{(\boxtimes)}$

Dipartimento di Matematica e Informatica, Università di Perugia, Perugia, Italy
{bista,francesco.santini}@dmi.unipg.it

Abstract. We provide an initial study on the Hasse diagram that represents the partial order -w.r.t. set inclusion- among weighted sceptical semantics in Argumentation: grounded, ideal, and eager. Being our framework based on a parametric structure of weights, we can directly compare weighted and classical approaches. We define a unique-status weighted grounded semantics, and we prove that the lattice of strongly-admissible extensions becomes a semi-lattice.

1 Introduction

An *Abstract Argumentation Framework (AAF)* [9] is essentially a pair $\langle \mathscr{A}_{rgs}, R \rangle$ consisting of a set of arguments and a binary oriented relation of attack defined among them (*e.g.*, $a, b \in \mathscr{A}_{rgs}$, and $R(a,b)$). The key idea behind *extension-based* semantics is to identify some subsets of arguments (called *extensions*) that survive the conflict "together". For example, the arguments in an *admissible* extension [9] $\mathscr{B}$ are not in conflict and they counter-attack attacked arguments in $\mathscr{B}$, i.e., arguments in $\mathscr{B}$ are *defended*.

Several notions of weighted defence have been defined in the literature [4,5,8,11,12]. Attacks are associated with a weight indicating a "strength" value, thus generalising the notion of AAF into *Weighted AAF (WAAF)* [4,5,11]. In [3] we provide a new definition of defence for WAAFs, called *w-defence*, and we use this to redefine classical semantics [9] to their weighted counterpart, that is *w-semantics* [3] (e.g., *w-admissible*).

In formal Abstract Argumentation, as well as in non-monotonic inference in general, it is possible for a semantics to yield more than one extension: in a framework with two arguments a and b, if $R(a,b)$ and $R(b,a)$ then both $\{a\}$ and $\{b\}$ are admissible. Often, this is dealt with by using a sceptical approach: hence, it is desirable to also have sceptical semantics that always yields exactly one extension. The most well-known example of such unique-status semantics is the *grounded* semantics [9]; however, in the literature it is sometimes supposed to be too sceptical [7]. The *ideal* [10] and *eager* [7] semantics try to be less sceptical, i.e., grounded $\subseteq$ ideal $\subseteq$ eager [7].[1]

---

[1] The eager is a unique-status semantics only for finite AAFs [1], which we study in this paper.

© Springer International Publishing AG 2017
M. Balduccini and T. Janhunen (Eds.): LPNMR 2017, LNAI 10377, pp. 49–56, 2017.
DOI: 10.1007/978-3-319-61660-5_6

In the paper we provide an initial study on the Hasse diagram that represents the partially order set (or *poset*) -w.r.t. set inclusion- among w-extensions; in particular, we curb to sceptical semantics. In a Hasse diagram, each vertex corresponds to an extension $\mathscr{B}$, and there is an edge between extensions $\mathscr{B}$ and $\mathscr{C}$ whenever $\mathscr{B} \subseteq \mathscr{C}$ and there is no extension $\mathscr{D}$ such that $\mathscr{B} \subseteq \mathscr{D} \subseteq \mathscr{C}$. We will build this study on a parametric framework based on an algebraic structure, which can represent [3] the frameworks in [8,9,12]. Differently from [8,11,12], our solution always provides a single grounded extension.

2   Background

C-semirings are commutative semirings where $\otimes$ is used to compose values, while an idempotent $\oplus$ is used to represent a partial order among them.

Definition 1 (C-semirings [2]). *A c-semiring is a five-tuple* $\mathbb{S} = \langle S, \oplus, \otimes, \perp, \top \rangle$ *such that S is a set, $\top, \perp \in S$, and $\oplus, \otimes : S \times S \to S$ are binary operators making the triples $\langle S, \oplus, \perp \rangle$ and $\langle S, \otimes, \top \rangle$ commutative monoids, satisfying,* (i) *distributivity* $\forall a, b, c \in S.a \otimes (b \oplus c) = (a \otimes b) \oplus (a \otimes c)$, (ii) *annihilator* $\forall a \in A.a \otimes \perp = \perp$, *and* (iii) *absorptivity* $\forall a, b \in S.a \oplus (a \otimes b) = a$.

The idempotency of $\oplus$, which derives from absorptivity, leads to the definition of a partial order $\leq_{\mathbb{S}}$ over S: $a \leq_{\mathbb{S}} b$ iff $a \oplus b = b$, which means that b is "better" than a. $\oplus$ is the *least upper bound* of the lattice $\langle S, \leq_{\mathbb{S}} \rangle$. Some c-semiring instances are: *Boolean* $\langle \{F, T\}, \vee, \wedge, F, T \rangle$, *Fuzzy* $\langle [0, 1], \max, \min, 0, 1 \rangle$, and *Weighted* $\langle \mathbb{R}^+ \cup \{+\infty\}, \min, +, +\infty, 0 \rangle$. Thus, the definition of WAAFs can represent different problems.

Definition 2 (c-semiring-based WAAF [5]). *A semiring-based Argumentation Framework (WAAF$_{\mathbb{S}}$) is a quadruple* $\langle \mathscr{A}_{rgs}, R, W, \mathbb{S} \rangle$, *where $\mathbb{S}$ is a semiring $\langle S, \oplus, \otimes, \perp, \top \rangle$, $\mathscr{A}_{rgs}$ is a set of arguments, R the attack binary-relation on $\mathscr{A}_{rgs}$, and $W : \mathscr{A}_{rgs} \times \mathscr{A}_{rgs} \to S$ is a binary function. Given $a, b \in \mathscr{A}_{rgs}$, $\forall (a, b) \in R$, $W(a, b) = s$ means that a attacks b with a weight $s \in S$. Moreover, we require that $R(a, b)$ iff $W(a, b) <_{\mathbb{S}} \top$.*

In [3] we define w-defence: a set $\mathscr{B}$ defends an argument b from a if the set-wise $\otimes$ of the attacks from all $c \in \mathscr{B}$ that defend b, i.e., $W(\mathscr{B}, a) = \bigotimes_{c \in \mathscr{B}} W(c, a)$, is worse than (i.e., stronger) or equal to the attacks to b and all the arguments in $\mathscr{B}$, i.e., $W(a, \mathscr{B} \cup b)$.

Definition 3 (w-defence [3]). *Given* $WF = \langle \mathscr{A}_{rgs}, R, W, S \rangle$, $\mathscr{B} \subseteq \mathscr{A}_{rgs}$ w-defends $b \in \mathscr{A}_{rgs}$ from $a \in \mathscr{A}_{rgs}$ s.t. $R(a, b)$, iff $W(a, \mathscr{B} \cup \{b\}) \geq_{\mathbb{S}} W(\mathscr{B}, a)$; $\mathscr{B}$ w-defends b iff it defends b from any a s.t. $R(a, b)$.

Note that (the weights of) counter-attacks of b can be exploited in the defence offered by $\mathscr{B}$, as in [9] happens for self-defence: this is why we consider all the attacks from $\mathscr{B} \cup \{b\}$ to a, and vice-versa. From [3] we report the definitions of w-semantics.

Definition 4 (*w*-semantics [3]). *Given* $WF = \langle \mathscr{A}_{rgs}, R, W, \mathbb{S} \rangle$, $\mathscr{B}$ *is a conflict-free set* [9] *iff* $W(\mathscr{B}, \mathscr{B}) = \top$ *(where* $W(\mathscr{B}, \mathscr{D}) = \bigotimes_{b \in \mathscr{B}, d \in \mathscr{D}} W(b, d)$*).* $\mathscr{B}$ *can be:*

- *a* w-*admissible (*wadm*) extension iff all the arguments in* $\mathscr{B}$ *are* w-*defended by* $\mathscr{B}$;
- *a* w-*complete (*wcom*) extension iff each argument* $b \in \mathscr{A}_{rgs}$ *s.t.* $\mathscr{B} \cup \{b\}$ *is* w-*admissible belongs to* $\mathscr{B}$;
- *a* w-*preferred (*wprf*) extension iff it is a maximal (w.r.t. set inclusion)* w-*admissible subset of* $\mathscr{A}_{rgs}$;
- w-*semi-stable (*wsst*) iff, given the range of* $\mathscr{B}$ *defined as* $\mathscr{B} \cup \mathscr{B}^{+}$, *where* $\mathscr{B}^{+} = \{a \in \mathscr{A}_{rgs} : W(\mathscr{B}, a) <_{\mathbb{S}} \top\}$, $\mathscr{B}$ *is a* w-*complete extension with maximal (w.r.t. set inclusion) range.*[2]
- *a* w-*stable extension (*wstb*) iff* $\forall a \notin \mathscr{B}, \exists b \in \mathscr{B}.W(b, a) <_{\mathbb{S}} \top$.

If we use the *Boolean* c-semiring and consider $W(a, b) = false$ whenever $R(a, b)$, for each semantics in Definition 4 we exactly obtain the corresponding original Dung's one [3]: i.e., respectively admissible (*adm*), complete (*com*), preferred (*prf*), semi-stable (*sst*), and stable (*stb*) [9,13]. In this case, the notion of w-defence collapses to classical defence [9]: $\mathscr{B}$ w-defends a iff $\mathscr{B}$ defends a.[3]

We conclude the background by recalling in Definition 5 the definitions of the classical sceptical semantics, which we will later weigh in Sect. 3.

Definition 5 (Sceptical semantics). *Given a framework* $F = \langle \mathscr{A}_{rgs}, R \rangle$:

(a) $\mathscr{B} \subseteq \mathscr{A}_{rgs}$ *is grounded (*grd*) iff* $\mathscr{B}$ *is complete and* $\forall \mathscr{B}' \in com(F)$, $\mathscr{B} \subseteq \mathscr{B}'$. *The grounded extension is the minimal (w.r.t. set inclusion) complete set* [9].

(b) $\mathscr{B} \subseteq \mathscr{A}_{rgs}$ *is ideal (*ide*) iff* $\mathscr{B}$ *is admissible and* $\forall \mathscr{B}' \in prf(F)$, $\mathscr{B} \subseteq \mathscr{B}'$. *The ideal extension is the maximal (w.r.t. set inclusion) ideal set* [10].

(c) $\mathscr{B} \subseteq \mathscr{A}_{rgs}$ *is eager (*eag*) iff* $\mathscr{B}$ *is admissible and* $\forall \mathscr{B}' \in sst(F)$, $\mathscr{B} \subseteq \mathscr{B}'$. *The eager extension is the maximal (w.r.t. set inclusion) eager set* [7].

In the following, we use $WF_{\Downarrow}$ to denote the classical framework $F = \langle \mathscr{A}_{rgs}, R \rangle$ of Dung [9] that can be obtained by just lifting W and $\mathbb{S}$ from $WF = \langle \mathscr{A}_{rgs}, R, W, \mathbb{S} \rangle$. In practice, $WF_{\Downarrow}$ drops the weighted system in WF.

3   Definitions and Formal Results

We focus on weighted sceptical semantics, starting from the grounded one. If we directly extend Definition 5a to WAAFs, there are frameworks where the set of complete extensions has more than one minimal element: hence, there is no unique least-set as it happens in [9], and as we also desire for WAAFs.

---

[2] Even if new and not in [3], we introduce this semantics in Definition 4 for the sake of presentation.

[3] Since Dung's definitions of semantics are directly encompassed by our framework (just by using the *Boolean* semiring), we do not introduce them in this paper for the sake of brevity.

For example, we consider a WAAF with arguments $\mathscr{A}_{rgs} = \{a, b, c, d\}$, and $R(a, b)$, $R(b, c)$, $R(b, d)$, all with a weight of 1 (using the *Weighted* semiring): $W(a, b) = 1, W(b, c) = 1, W(b, d) = 1$. The set of w-complete extensions is $\{\{a, c\}, \{a, d\}\}$, and then there is no single least element. According to [9] instead, the grounded extension in $WF_\Downarrow$ is $\{a, c, d\}$: the least element exists.

Since our goal is to preserve its uniqueness in WAAFs, we improve Definition 5 in order to always have one single solution. We follow the same approach used in [1] to define the ideal and eager semantics.

Definition 6 (w-grounded). *Given $WF = \langle \mathscr{A}_{rgs}, R, W, \mathbb{S} \rangle$, an extension $\mathscr{B} \in wgrd(F)$, iff $\mathscr{B} \in wadm(WF)$, and $\mathscr{B} \subseteq \bigcap wcom(WF)$, and $\nexists \mathscr{B}' \in wadm(WF)$ satisfying $\mathscr{B}' \subseteq \bigcap wcom(WF)$ s.t. $\mathscr{B} \subsetneq \mathscr{B}'$.*

In words, the w-grounded extension is any maximal (w.r.t. set inclusion) w-admissible extension included in the intersection of w-complete extensions. We now relate the w-grounded semantics in Definition 6 to the classical one by Dung [9].

Proposition 1. *In $WF = \langle \mathscr{A}_{rgs}, R, W, \mathbb{S} \rangle$, if $\mathbb{S}$ is Boolean then the w-grounded extension is equivalent to the classical grounded extension on $WF_\Downarrow$.*

Moreover, from Definition 6 we can derive some noticeable properties in the following.

Proposition 2. *The w-grounded extension always exists and is unique.*

Proposition 3. *The w-grounded extension corresponds to the set of sceptically accepted arguments in $wcom(WF)$: $grd(WF) = \{a \in \mathscr{A}_{rgs} \mid \forall \mathscr{B} \in wcom(WF), a \in \mathscr{B}\}$.*

According to Proposition 2, the w-grounded extension is a subset of $\bigcap wcom(WF)$, which always exists, is unique and w-admissible. This uniqueness is novel w.r.t. [8, 11, 12], where the described frameworks offer several grounded scenarios. Furthermore, when there is only one minimal w-complete extension, it corresponds to the w-grounded one.

Theorem 1. *Given $WF = \langle \mathscr{A}_{rgs}, R, W, \mathbb{S} \rangle$, and $\mathbb{S}$ any semiring, if $\forall \mathscr{B}' \in wcom(F)$, $\mathscr{B} \in wcom(F)$ s.t. $\mathscr{B} \subseteq \mathscr{B}'$, then $\mathscr{B} = wgrd(WF)$. Consequently, $\mathscr{B}$ is w-complete.*

Theorem 1 is always satisfied when the *Boolean* semiring is used. Moreover, we have that the w-grounded extension is a subset of each minimal w-complete extension.

Proposition 4. *Given $WF = \langle \mathscr{A}_{rgs}, R, W, \mathbb{S} \rangle$, $\mathbb{S}$ any semiring, and $wcom_\subseteq (WF) = \{\mathscr{B} \in wcom(WF) \mid \nexists \mathscr{B}' \in wcom(WF).\mathscr{B}' \subset \mathscr{B}'\}$, then $\forall \mathscr{B} \in wcom_\subseteq(WF).wgrd(WF) \subseteq \mathscr{B}$.*

In addition, each minimal w.r.t. set inclusion of the w-complete extensions is always a subset of the classical grounded extension in Definition 5a.

Proposition 5. *Given $WF = \langle \mathscr{A}_{rgs}, R, W, \mathbb{S} \rangle$, $\mathbb{S}$ any semiring, and $wcom_{\subseteq}$ $(WF) = \{\mathscr{B} \in wcom(WF) \mid \nexists \mathscr{B}' \in wcom(WF).\mathscr{B}' \subset \mathscr{B}'\}$, then $\forall \mathscr{B} \in wcom_{\subseteq}(WF).\mathscr{B} \subseteq grd(WF_{\Downarrow})$.*

We now introduce the w-strongly-admissible semantics, from which Proposition 6 follows, relating it with the w-grounded one.

Definition 7 (w-strongly-admissible). *Given $WF = \langle \mathscr{A}_{rgs}, R, W, \mathbb{S} \rangle$, $\mathscr{B} \subseteq \mathscr{A}_{rgs}$ is w-strongly-admissible iff every $b \in \mathscr{B}$ is w-defended by some $\mathscr{B}' \subseteq \mathscr{B} \setminus \{b\}$.*

Proposition 6. *The w-grounded extension is w-strongly-admissible.*

In Fig. 1 we present a Hasse diagram summarising some of the formal results above. The reference WAAF is $\mathscr{A}_{rgs} = \{a, b, c, d\}$, and $R(a, b)$, $R(b, c)$, $R(b, d)$. The w-grounded extension is $\{a\}$, the minimal w-complete ones are $\{a, c\}$ and $\{a, d\}$, and the grounded extension (Definition 5) is $\{a, c, d\}$. Hence, $\{a\}$ corresponds to the set of sceptically accepted arguments in $wcom(WF)$ (Proposition 3), $\{a\} \subseteq \{a, c\}$ and $\{a\} \subseteq \{a, d\}$ (Proposition 4), and both $\{a, c\}$ and $\{a, d\}$ are a subset of $\{a, c, d\}$ (Proposition 5). In the following we extend the other two well-known sceptical semantics in order to make them consider weights:

Definition 8 (w-ideal). *Given $WF = \langle \mathscr{A}_{rgs}, R, W, \mathbb{S} \rangle$, $\mathscr{B} = w\text{-}ideal(F)$ (wide), iff $\mathscr{B}$ is w-admissible and $\forall \mathscr{B}' \in wprf(WF)$, $\mathscr{B} \subseteq \mathscr{B}'$. The w-ideal extension is the maximal (w.r.t. set inclusion) w-ideal set.*

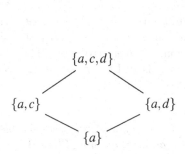

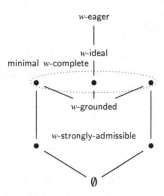

Fig. 1. Given the WAAF with $\mathscr{A}_{rgs} = \{a, b, c, d\}$, and $R(a, b)$, $R(b, c)$, $R(b, d)$, the w-grounded extension is $\{a\}$, the minimal w-complete (they are w-admissible and include all w-defended arguments) ones are $\{a, c\}$ and $\{a, d\}$, and the grounded extension [9] is $\{a, c, d\}$.

Fig. 2. The partial ordered for sceptical w-semantics.

Definition 9 (*w-eager*). *Given* $WF = \langle \mathscr{A}_{rgs}, R, W, \mathbb{S} \rangle$, $\mathscr{B} = w\text{-}eager(F)$ *(weag),* *iff* $\mathscr{B}$ *is admissible and* $\forall \mathscr{B}' \in wsst(F)$, $\mathscr{B} \subseteq \mathscr{B}'$. *The w-eager extension is the maximal (w.r.t. set inclusion) w-eager set.*

As done in Proposition 1 for the w-grounded semantics, we relate the w-ideal and w-eager semantics to their classical counterparts:

Proposition 7. *In* $WF = \langle \mathscr{A}_{rgs}, R, W, \mathbb{S} \rangle$, *if* $\mathbb{S}$ *is Boolean then the w-ideal and w-eager extensions are equivalent to the classical ideal and eager extension, i.e.,* $wide(WF) = ide(WF_{\downarrow})$ *and* $weag(WF) = eag(WF_{\downarrow})$.

We prove the semantics in Definitions 8 and 9 are satisfied by only one extension:

Proposition 8. *The w-ideal and the w-eager are unique-status semantics.*

From Definitions 6, 8 and 9 we obtain the same inclusion-result as for their corresponding unweighted semantics [7]: $grd(F) \subseteq ide(F) \subseteq eag(F)$.

Theorem 2. *Given* $WF = \langle \mathscr{A}_{rgs}, R, W, \mathbb{S} \rangle$, *then* $wgrd(WF) \subseteq wide(WF) \subseteq weag(WF)$.

Then, we can relate sceptical w-semantics to their unweighted counterpart given in Definition 5: each of such extensions results to be a subset of the corresponding original one:

Theorem 3. *Given* $WF = \langle \mathscr{A}_{rgs}, R, W, \mathbb{S} \rangle$ *and* $F = \langle \mathscr{A}_{rgs}, R \rangle$ *(same* $\mathscr{A}_{rgs}$ *and* R), *then* $wgrd(WF) \subseteq grd(WF_{\downarrow})$, $wide(WF) \subseteq ide(WF_{\downarrow})$, *and* $weag(WF) \subseteq eag(WF_{\downarrow})$.

Classical strongly-admissible extensions form a lattice w.r.t. $\subseteq$ [7], that is a partial order where all the subsets of elements have an infimum, in this case $\emptyset$, and a supremum, in this case the grounded extension. With w-strongly-admissible ones instead, only a semi-lattice can be obtained: $\emptyset$ is still the infimum, but no supremum in this case. The minimal w-complete extensions are the maximal elements of such a semi-lattice.

Theorem 4. *Given* $WF = \langle \mathscr{A}_{rgs}, R, W, \mathbb{S} \rangle$, *the set of w-strongly-admissible extensions forms a semi-lattice w.r.t.* $\subseteq$, *with* $\emptyset$ *as the infimum. With* $WF = \langle \mathscr{A}_{rgs}, R, W, Boolean \rangle$, *the lattice structure is preserved, with* $\emptyset$ *as infimum and* $grd(WF_{\downarrow})$ *as supremum.*

Figure 2 presents the Hasse diagram (w.r.t. $\subseteq$) for sceptical semantics, considering a generic semiring: in case of a *Boolean* semiring, we still have a complete lattice. Note that w-ideal and w-eager are not w-strongly-admissible, as for classical frameworks [7].

4   Conclusion

We have extended the weighted framework in [3] by proposing a unique-status grounded semantics, and a Hasse diagram that represents the partial order - w.r.t. set inclusion- among sceptical extensions and w-strongly admissible ones. By having a general framework based on semirings, it is easier to check which relations among semantics change when the defence considers weights. According to its sceptical nature, it is desirable to provide a single grounded extension, differently from the frameworks in [8,11,12].

In the future we will study the framework in [6], which partitions the arguments into sets satisfying the same semantics. In addition, we would like to define the upper part of the Hasse diagram: for instance, what the relation is between w-preferred and w-stable extensions, or the conditions when the preferred or stable extensions are unique.

References

1. Baumann, R., Spanring, C.: Infinite argumentation frameworks. In: Eiter, T., Strass, H., Truszczyński, M., Woltran, S. (eds.) Advances in Knowledge Representation, Logic Programming, and Abstract Argumentation. LNCS, vol. 9060, pp. 281–295. Springer, Cham (2015). doi:10.1007/978-3-319-14726-0_19
2. Bistarelli, S., Montanari, U., Rossi, F.: Semiring-based constraint satisfaction and optimization. J. ACM **44**(2), 201–236 (1997)
3. Bistarelli, S., Rossi, F., Santini, F.: A collective defence against grouped attacks for weighted abstract argumentation frameworks. In: Florida Artificial Intelligence Research Society Conference, Flairs, pp. 638–643. AAAI (2016)
4. Bistarelli, S., Rossi, F., Santini, F.: A relaxation of internal conflict and defence in weighted argumentation frameworks. In: Michael, L., Kakas, A. (eds.) JELIA 2016. LNCS (LNAI), vol. 10021, pp. 127–143. Springer, Cham (2016). doi:10.1007/978-3-319-48758-8_9
5. Bistarelli, S., Santini, F.: A common computational framework for semiring-based argumentation systems. In: ECAI - European Conference on Artificial Intelligence. FAIA, vol. 215, pp. 131–136. IOS (2010)
6. Bistarelli, S., Santini, F.: Coalitions of arguments: an approach with constraint programming. Fundam. Inform. **124**(4), 383–401 (2013)
7. Caminada, M.: Comparing two unique extension semantics for formal argumentation: ideal and eager. In: Belgian-Dutch Conference on Artificial Intelligence (BNAIC), pp. 81–87 (2007)
8. Coste-Marquis, S., Konieczny, S., Marquis, P., Ouali, M.A.: Weighted attacks in argumentation frameworks. In: Principles of Knowledge Representation and Reasoning (KR), pp. 593–597. AAAI (2012)
9. Dung, P.M.: On the acceptability of arguments and its fundamental role in non-monotonic reasoning, logic programming and n-person games. Artif. Intell. **77**(2), 321–357 (1995)
10. Dung, P.M., Mancarella, P., Toni, F.: A dialectic procedure for sceptical, assumption-based argumentation. In: Computational Models of Argument (COMMA). FAIA, vol. 144, pp. 145–156. IOS (2006)

11. Dunne, P.E., Hunter, A., McBurney, P., Parsons, S., Wooldridge, M.: Weighted argument systems: basic definitions, algorithms, and complexity results. Artif. Intell. **175**(2), 457–486 (2011)
12. Martínez, D.C., García, A.J., Simari, G.R.: An abstract argumentation framework with varied-strength attacks. In: Principles of Knowledge Representation and Reasoning (KR), pp. 135–144. AAAI (2008)
13. Verheij, B.: Two approaches to dialectical argumentation: admissible sets and argumentation stages. In: Dutch Conference on Artificial Intelligence, vol. 96, pp. 357–368 (1996)

Foundations for a Probabilistic Event Calculus

Fabio Aurelio D'Asaro[✉], Antonis Bikakis, Luke Dickens, and Rob Miller

Department of Information Studies, University College London, London, UK
{uczcfad,a.bikakis,l.dickens,r.s.miller}@ucl.ac.uk

Abstract. We present PEC, an Event Calculus (EC) style action language for reasoning about probabilistic causal and narrative information. It has an action language style syntax similar to that of the EC variant $\mathcal{M}$odular-$\mathcal{E}$. Its semantics is given in terms of *possible worlds* which constitute possible evolutions of the domain, and builds on that of Epistemic Functional EC (EFEC). We also describe an ASP implementation of PEC and show the sense in which this is sound and complete.

1 Introduction

The Event Calculus (EC) [6] is a well-known approach to reasoning about the effects of a narrative of action occurrences (events) along a time line. This paper briefly summarises [3], which describes PEC, an adaptation of EC able to reason with probabilistic causal knowledge. There are numerous applications for this kind of probabilistic reasoning, e.g. in modelling medical, environmental, legal and commonsense domains, and in complex activity recognition and security monitoring. Full technical details of PEC are in [3]. Its main characteristics are: (i) it supports EC-style narrative reasoning, (ii) it uses a tailored action language syntax and semantics, (iii) it uses a *possible worlds* semantics to naturally allow for *epistemic* extensions, (iv) for a wide subset of domains it has a sound and complete ASP implementation, and (v) its generality allows in principle for the use of other models of uncertainty, e.g. truth-functional belief or Dempster-Schafer theory. Although other formalisms exist for probabilistic reasoning about actions (see e.g. [1,2,5,11,12]), PEC is, to our knowledge, the only framework to combine together these features. As shown in [3] it can be used to model scenarios such as:

Scenario 1 (Coin Toss). *A coin initially (instant 0) shows Heads. A robot can attempt to toss the coin, but there is a small chance that it will fail to pick it up, leaving the coin unchanged. The robot attempts to toss the coin (instant 1).*

Scenario 2 (Antibiotic). *A patient has a rash often associated with a bacterial infection, and can take an antibiotic known to be reasonably effective. Treatment is not always successful, and if not may still clear the rash. Failed treatment leaves the bacteria resistant. The patient is treated twice (instants 1 and 3).*

Other than the primary author, authors are listed alphabetically.

© Springer International Publishing AG 2017
M. Balduccini and T. Janhunen (Eds.): LPNMR 2017, LNAI 10377, pp. 57–63, 2017.
DOI: 10.1007/978-3-319-61660-5_7

2   Overview of PEC's Syntax and Semantics

A PEC *domain language* consists of a a finite non-empty set $\mathcal{F}$ of *fluents*, a finite set $\mathcal{A}$ of *actions*, a finite non-empty set $\mathcal{V}$ of *values* such that $\{\top, \bot\} \subseteq \mathcal{V}$, a function $vals : \mathcal{F} \cup \mathcal{A} \to 2^{\mathcal{V}} \setminus \emptyset$, and a non-empty set $\mathcal{I}$ of *instants* with minimum element $\bar{0}$ w.r.t. total ordering $\leq$. For $A \in \mathcal{A}$ we impose $vals(A) = \{\top, \bot\}$. Our approach is to model a given domain with action-language-like propositions that specify (probabilistic) causal and narrative information. For example, Scenario 1 is modelled using the following domain description $\mathcal{D}_C$:

$$Coin \textbf{ takes-values } \{Heads,\ Tails\} \tag{C1}$$

$$\textbf{initially-one-of}\{(Coin{=}Heads, 1)\} \tag{C2}$$

$$Toss \textbf{ causes-one-of} \tag{C3}$$
$$\{(\{Coin{=}Heads\}, 0.49), (\{Coin{=}Tails\}, 0.49), (\emptyset, 0.02)\}$$

$$Toss \textbf{ performed-at } 1 \tag{C4}$$

More generally *v-propositions*, such as (C1), have the form

$$F \textbf{ takes-values } \{V_1, \ldots, V_m\} \tag{1}$$

for $F \in \mathcal{F}$, $m \geq 1$, $V_i \in \mathcal{V}$ for all $1 \leq i \leq m$, and $\{V_1, \ldots, V_m\} = vals(F)$. *c-propositions* such as (C3) modeling causal relationships are of the form

$$\theta \textbf{ causes-one-of} \{O_1, O_2, \ldots, O_m\} \tag{2}$$

where formula θ captures preconditions, and $O_1, \ldots, O_m$ are alternative *outcomes* – partial assignments of fluent values paired with probabilities that sum to 1. Initial conditions are declared via *i-propositions* of the form

$$\textbf{initially-one-of}\{O_1, O_2, \ldots, O_m\} \tag{3}$$

and action occurrences are identified through *p-propositions* of the form

$$A \textbf{ performed-at } I \textbf{ with-prob } P^+ \tag{4}$$

for $A \in \mathcal{A}$, $I \in \mathcal{I}$ and $P^+ \in (0, 1]$. When $P^+ = 1$ it is omitted as in (C4).

A *domain description* is a finite set $\mathcal{D}$ of v-, c-, p- and i-propositions such that: (i) for any two distinct c-propositions in $\mathcal{D}$ with bodies θ and θ', there is no state compatible[1] with both θ and θ', (ii) $\mathcal{D}$ contains exactly one i-proposition, and (iii) $\mathcal{D}$ contains exactly one v-proposition for each $F \in \mathcal{F}$.

PEC's semantics describes how domain descriptions entail *h-propositions* of the form

$$\varphi \textbf{ holds-with-prob } P \tag{5}$$

where $P \in [0, 1]$, and φ is an *i-formula* (time-stamped formula). For example, $\mathcal{D}_C$ entails '$[Coin{=}Heads]@2$ **holds-with-prob** 0.51'.

---

[1] i.e. Taking literals as propositions, there is no state that is a classical Herbrand model of both θ and θ'.

PEC has a *possible-worlds* semantics. A *world* is an evolution of the environment, i.e. a function $W : \mathcal{I} \rightarrow \mathcal{S}$, where $\mathcal{S}$ is the set of all *states* (complete assignments of values to fluents and actions). $\mathcal{W}$ denotes the set of all worlds. Worlds can be pictured as timelines with information about the current state attached at each instant. E.g. two worlds for Scenario 1 can be visualised as:

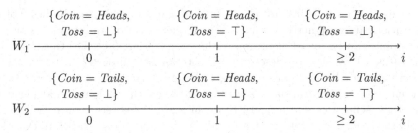

Intuitively, W_1 is consistent with domain description $\mathcal{D}_C$ as it represents a coherent history of what could have happened in Scenario 1, whereas W_2 does not (e.g., changes occur when no action is performed, an infinite number of actions are performed, etc...). For this reason, world W_1 said to be *well-behaved w.r.t.* $\mathcal{D}_C$, whereas W_2 is not. The semantics captures this notion with the concept of a *trace* – a chain of effects matching both a unique world W and the domain description $\mathcal{D}$, through consistency with propositions in $\mathcal{D}$ and a persistence condition. In other words, a world W represents an evolution of *state*, whereas a trace of W represents a legal causal history w.r.t. W and a corresponding domain description. For example, two traces for W_1 w.r.t. $\mathcal{D}_C$ are:

$$t_1 = \langle (\{Coin = Heads\}, 1)@\bar{X}, (\{Coin = Heads\}, 0.49)@1 \rangle$$

$$t_2 = \langle (\{Coin = Heads\}, 1)@\bar{X}, (\emptyset, 0.02)@1 \rangle$$

where the special symbol $\bar{X}$ is used to deal with the initial condition. The *evaluation* of a trace tr, written $\epsilon(tr)$, is the product of all real values appearing in it. In our example, $\epsilon(t_1) = 0.49$ and $\epsilon(t_2) = 0.02$. W_2 has no trace w.r.t. $\mathcal{D}_C$ and so is not well-behaved w.r.t. $\mathcal{D}_C$.

PEC's semantics defines a probability distribution over worlds. To show this, we first define a *[0,1]-interpretation* as a function from $\mathcal{W}$ to $[0, 1]$, and, given a domain description $\mathcal{D}$, single out a unique [0,1]-interpretation $M_{\mathcal{D}} : \mathcal{W} \mapsto [0, 1]$ called the *model* of $\mathcal{D}$. For world W, well-behaved w.r.t. $\mathcal{D}$, $M_{\mathcal{D}}(W)$ is the sum of values $\epsilon(tr)$ for all corresponding traces tr of W. If W is not well-behaved, then $M_{\mathcal{D}}(W) = 0$. $M_{\mathcal{D}}$ is extended to a function $M_{\mathcal{D}}^*$ over i-formulas as follows:

$$M_{\mathcal{D}}^*(\varphi) = \sum_{W \models \varphi} M_{\mathcal{D}}(W) \tag{6}$$

where $W \models \varphi$ indicates that φ is *satisfied* in W (in the obvious sense, see [3]). We say that 'φ **holds-with-prob** P' *is entailed by* $\mathcal{D}$ iff $M_{\mathcal{D}}^*(\varphi) = P$.

In [3], we prove that M^* is a *probability function* (see assumptions (P1) and (P2) from [9, Chap. 1]). Note that M^* could be alternatively defined to satisfy different axioms (see e.g. (DS1–3) from [9, Chap. 1] for Dempster-Schafer belief functions or [9, Chap. 5] for truth-functional belief functions).

3   ASP Implementation

We have implemented PEC for the class of domains in which the bodies θ of c-propositions are conjunctive formulas. The implementation and example domain descriptions can be found at https://github.com/dasaro/pec. For this, a translator turns a PEC domain description $\mathcal{D}$ into an ASP program using standard lexical analyser Flex and parser generator Bison. A grounder and solver, Clingo [4], then processes the translator's output together with the domain-independent part of the semantics and a query (both in ASP). This returns a collection of answer sets, each representing a trace and the corresponding well-behaved world. A standard text processing tool, AWK, then evaluates $\epsilon(tr)$ for each answer set trace tr and sums these to give a probability for the query using Eq. (6). The following diagram illustrates this procedure:

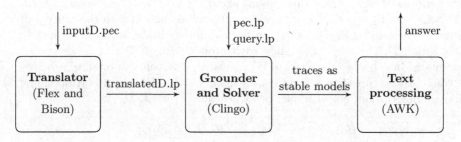

A general translation procedure from this class of PEC domain descriptions to ASP programs is given in [3]. To illustrate, $\mathcal{D}_C$ is translated to:

$fluent(coin).$ (TC0)
$action(toss).$
$instant(0..maxinst).$

$possVal(coin, heads).$ (TC1)
$possVal(coin, tails).$

$belongsTo((coin, heads), id_1^0).$ (TC2)
$initialCondition((id_1^0, 1)).$

$belongsTo((coin, heads), id_1^1).$ (TC3.1)
$causesOutcome((id_1^1, 49/100), I) \leftarrow holds(((toss, true), I)).$

$belongsTo((coin, tails), id_2^1).$ (TC3.2)
$causesOutcome((id_2^1, 49/100), I) \leftarrow holds(((toss, true), I)).$

$causesOutcome((id_3^1, 2/100), I) \leftarrow holds(((toss, true), I)).$ (TC3.3)

$performed(toss, 1).$ (TC4)

The domain-independent part of the implementation is as follows:

$possVal(A, true) \leftarrow action(A).$ (PEC1)
$possVal(A, false) \leftarrow action(A).$

$fluentOrAction(X) \leftarrow fluent(X); action(X).$ (PEC2)

$literal((X, V)) \leftarrow possVal(X, V).$ (PEC3)

$iLiteral((L, I)) \leftarrow literal(L), instant(I).$ (PEC4)

$1\{ holds(((X, V), I)) : iLiteral(((X, V), I)) \}1$ (PEC5)
$\leftarrow instant(I), fluentOrAction(X).$

$inOcc(I) \leftarrow instant(I), causesOutcome(O, I).$ (PEC6)

$1\{ effectChoice(O, I) : causesOutcome(O, I) \}1 \leftarrow inOcc(I).$ (PEC7)

$1\{ initialChoice(O) : initialCondition(O) \}1.$ (PEC8)

$\perp \leftarrow action(A), instant(I),$ (PEC9)
$holds(((A, true), I)), not\, performed(A, I).$

$\perp \leftarrow action(A), instant(I),$ (PEC10)
$not\, holds(((A, true), I)), performed(A, I).$

$\perp \leftarrow initialChoice((S, P)), literal(L),$ (PEC11)
$belongsTo(L, S), not\, holds((L, 0)).$

$\perp \leftarrow instant(I), effectChoice((X, P), I),$ (PEC12)
$fluent(F), belongsTo((F, V), X),$
$not\, holds(((F, V), I + 1)), I < maxinst.$

$\perp \leftarrow instant(I), fluent(F), not\, holds(((F, V), I)),$ (PEC13)
$effectChoice((X, P), I), not\, belongsTo((F, V), X),$
$holds(((F, V), I + 1)), I < maxinst.$

$\perp \leftarrow fluent(F), instant(I), holds(((F, V), I)), not\, inOcc(I),$ (PEC14)
$not\, holds(((F, V), I + 1)), I < maxinst.$

Correctness of the translation and implementation are guaranteed by the following proposition, more details of which (including a proof) are given in [3].

Proposition 1 (Soundness and Completeness). *Z is a stable model of the translated domain description $\mathcal{D}$ together with the domain-independent part of PEC iff Z represents a well-behaved world W and one of its traces w.r.t. $\mathcal{D}$.*

Intuitively, this proposition states that if we interpret the elements of a stable model Z of the translated domain description and the domain-independent part of PEC as their natural semantic counterpart (e.g., $holds(((F, V), I))$ is interpreted as $F = V \in W(I)$), then this interpretation is a trace together with its corresponding well-behaved world W w.r.t. $\mathcal{D}$. The trace and world are then said to be *represented* by Z. Conversely, for every well-behaved world W w.r.t. $\mathcal{D}$ and one of its traces tr there exists a stable model Z of the program such that

Z represents W and *tr*. It is in this sense that our implementation is *sound* and *complete*. Our proof in [3] relies on the Splitting Theorem [7].

4 Future Work

PEC semantics is defined in terms of (possible) *worlds* with a view to adding epistemic features in the future (see e.g. [10]). Our initial investigations in this respect focus on representing *imperfect sensing actions* and *actions conditioned on knowledge* acquired in previous instants (similar to the approach in the EFEC extension of FEC [8]). We envisage including *s-propositions* such as

$$See \textbf{ senses } Coin \textbf{ with-accuracies } \begin{pmatrix} 0.9 \ 0.1 \\ 0.3 \ 0.7 \end{pmatrix}$$

which represents that our coin-tossing robot can imperfectly sense the current face showing on the coin, and *conditional p-propositions* such as

$$Toss \textbf{ performed-at } 2 \textbf{ if-believes } (Coin = Tails, (0.65, 1])$$

which represents that the robot will toss again if it believes with a greater than 65% probability that the first toss resulted in *Tails*. Preliminary results indicate that our possible worlds semantics can be readily extended to cover these notions.

There are several other ways in which the present work can be continued. For instance, the problem of *elaboration tolerance*, which plays an important role in classical reasoning about actions, needs to be reviewed and solved in our setting. A related point is that of *underspecification*, i.e. what an agent can reasonably infer from a domain in which the initial conditions and the effects of actions are not entirely specified (even probabilistically). Finally, a crucial point is that of *computational efficiency*. Indeed, the intractability of several computational problems arising in this setting (such as temporal projection) suggests that techniques (e.g. Monte Carlo Markov Chain) are needed to efficiently approximate the correct answer to a given query with an appropriate degree of confidence.

Related Work: For a discussion of related work see [3].

References

1. Bacchus, F., Halpern, J.Y., Levesque, H.J.: Reasoning about noisy sensors and effectors in the situation calculus. Artif. Intell. **111**(1), 171–208 (1999)
2. Baral, C., Tran, N., Tuan, L.C.: Reasoning about actions in a probabilistic setting. In: AAAI/IAAI, pp. 507–512 (2002)
3. D'Asaro, F.A., Bikakis, A., Dickens, L., Miller, R.: Foundations for a probabilistic event calculus: Technical report (2017). http://arxiv.org/abs/1703.06815
4. Gebser, M., Kaminski, R., Kaufmann, B., Schaub, T.: Clingo = ASP + control: Preliminary report (2014). http://arxiv.org/abs/1405.3694
5. Iocchi, L., Lukasiewicz, T., Nardi, D., Rosati, R.: Reasoning about actions with sensing under qualitative and probabilistic uncertainty. TOCL **10**(1), 1–39 (2009)

6. Kowalski, R., Sergot, M.: A logic-based calculus of events. In: Schmidt, J.W., Thanos, C. (eds.) Foundations of Knowledge Base Management, pp. 23–55. Springer, Heidelberg (1989)
7. Lifschitz, V., Turner, H.: Splitting a logic program. In: ICLP, pp. 23–37 (1994)
8. Ma, J., Miller, R., Morgenstern, L., Patkos, T.: An epistemic event calculus for asp-based reasoning about knowledge of the past, present and future. In: LPAR-19, vol. 26, pp. 75–87 (2014)
9. Paris, J.B.: The Uncertain Reasoner's Companion: A Mathematical Perspective, vol. 39. Cambridge University Press, Cambridge (2006)
10. Scherl, R.B., Levesque, H.J.: Knowledge, action, and the frame problem. Artif. Intell. **144**(1–2), 1–39 (2003)
11. Skarlatidis, A., Artikis, A., Filippou, J., Paliouras, G.: A probabilistic logic programming event calculus. TPLP **15**, 213–245 (2015)
12. Skarlatidis, A., Paliouras, G., Artikis, A., Vouros, G.A.: Probabilistic event calculus for event recognition. ACM Trans. Comput. Logic **16**(2), 11:1–11:37 (2015)

Contextual Reasoning:
Usually Birds Can Abductively Fly

Emmanuelle-Anna Dietz Saldanha[1]([✉]), Steffen Hölldobler[1,2],
and Luís Moniz Pereira[3]

[1] International Center for Computational Logic,
TU Dresden, 01062 Dresden, Germany
{dietz,sh}@iccl.tu-dresden.de
[2] North-Caucasus Federal University, Stavropol, Russian Federation
[3] NOVA Laboratory for Computer Science and Informatics,
Departamento de Informática Faculdade de Ciências e Tecnologia,
Universidade Nova de Lisboa, 2829-516 Caparica, Portugal
lmp@fct.unl.pt

Abstract. We present a new logic programming approach to contextual reasoning, based on the Weak Completion Semantics (WCS), the latter of which has been successfully applied in the past to adequately model various human reasoning tasks. One of the properties of WCS is the open world assumption with respect to undefined atoms. This is a characteristic that is different to other common Logic Programming semantics, a property that seems suitable when modeling human reasoning. Notwithstanding, we have noticed that the famous Tweety default reasoning example, originally introduced by Reiter, cannot be modeled straightforwardly under WCS. Hence, to address the issue and taking Pereira and Pinto's inspection points as inspiration, we develop a notion of contextual reasoning for which we introduce contextual logic programs. We reconsider the formal properties of WCS with respect to these and verify whether they still hold. Finally, we set forth contextual abduction and show that not only the original Tweety example can be nicely modeled within the new approach, but more sophisticated examples as well, where context plays an important role.

1 Introduction

We shall develop a characterization of contextual reasoning, which has its origins in [21,22], and which in turn was inspired by [23]. This will be done in the context of logic programming (cf. [18]), weak completion [13], abduction [16], Stenning and van Lambalgen's representation of implications as well as their semantic operator Φ [25] and our use of three-valued Łukasiewicz logic [19], instead of their use of [10] which has been assembled together in [9,12–15]. This approach–which we call WCS for *Weak Completion Semantics*–has been applied to adequately model various human reasoning tasks [5,7,8] and to reasoning about conditionals [4,6].

L.M. Pereira—The authors are mentioned in alphabetical order.

© Springer International Publishing AG 2017
M. Balduccini and T. Janhunen (Eds.): LPNMR 2017, LNAI 10377, pp. 64–77, 2017.
DOI: 10.1007/978-3-319-61660-5_8

Consider the famous Tweety example from [24]: *Usually birds can fly. Tweety and Jerry are birds.* They can be encoded by the following (datalog) program, $\mathcal{P}_1$, where '*ab*' stands for 'abnormal':

$$can\_fly(X) \leftarrow bird(X) \wedge \neg ab(X). \qquad ab(X) \leftarrow \bot. \qquad bird(tweety) \leftarrow \top.$$
$$bird(jerry) \leftarrow \top.$$

We derive $can\_fly(tweety)$ and $can\_fly(jerry)$ as nothing is abnormal with respect to Tweety and Jerry. We modify the example by replacing the first statement with: *Usually birds can fly, but kiwis and penguins cannot.* This is encoded by $\mathcal{P}_2$:

$$can\_fly(X) \leftarrow bird(X) \wedge \neg ab(X). \quad ab(X) \leftarrow kiwi(X). \qquad bird(tweety) \leftarrow \top.$$
$$ab(X) \leftarrow penguin(X). \quad bird(jerry) \leftarrow \top.$$

The ground instances of the weak completion of this program are as follows:

$$can\_fly(tweety) \leftrightarrow bird(tweety) \wedge \neg ab(tweety). \quad bird(tweety) \leftrightarrow \top.$$
$$can\_fly(jerry) \leftrightarrow bird(jerry) \wedge \neg ab(jerry). \quad bird(jerry) \leftrightarrow \top.$$
$$ab(tweety) \leftrightarrow kiwi(tweety) \vee penguin(tweety).$$
$$ab(jerry) \leftrightarrow kiwi(jerry) \vee penguin(jerry).$$

Different than under Clark's completion semantics [3], the Stable Model Semantics [11] or the Well-Founded Semantics [26], $kiwi(tweety)$, $penguin(tweety)$, $kiwi(jerry)$ and $penguin(jerry)$ are not assumed to be false by the closed world assumption but stay unknown by WCS's open world assumption. In other words, they are neither true nor false.[1] This leads to the following consequence under WCS: As we don't know whether Tweety and Jerry are penguins or kiwis, we cannot derive that they can fly. Even if we model this case with the help of abduction, e.g. we observe that Jerry flies, $\mathcal{O} = \{can\_fly(jerry)\}$, we still need to state in its explanation that Jerry must be neither a penguin nor a kiwi.

We want to avoid explicitly stating that all exception cases are false such as the Completion Semantics, the Stable Model Semantics or Well-Founded Semantics will do. We don't think that humans actively apply the closed world assumption in reality, i.e. that they explicitly add negation to the cases they don't know anything about. Instead, we assume that humans, if they are not for some reason aware of exceptions, simply ignore these cases. In other words, they do not consciously become aware of all exceptions when they reason.[2] Accordingly, when modeling these cases with logic programs, we should leave the truth values of these exception cases unknown and find a mechanism that just ignores them. At the moment, we cannot express this syntactically in WCS programs.

After Sects. 2 and 3 have introduced preliminaires and abduction, we present contextual programs in Sect. 4, and verify whether the same properties of the Φ operator hold for contextual programs as for the programs we have considered so far in our modeling of applications. Section 5 presents contextual abductive reasoning and specifies the notion of contextual side-effects. We terminate with conclusions, including open questions.

---

[1] If those atoms were assumed to be false, then typical human reasoning tasks like the suppression task [2] could not be modeled adequately [7].

[2] Currently, we know of at least 40 species of birds that can't fly.

Table 1. T, ⊥, and U denote *true*, *false*, and *unknown*, respectively.

F	$\neg F$
T	⊥
⊥	T
U	U

$\wedge$	T	U	⊥
T	T	U	⊥
U	U	U	⊥
⊥	⊥	⊥	⊥

$\vee$	T	U	⊥
T	T	T	T
U	T	U	U
⊥	T	U	⊥

$\leftarrow_L$	T	U	⊥
T	T	T	T
U	U	T	T
⊥	⊥	U	T

$\leftrightarrow_L$	T	U	⊥
T	T	U	⊥
U	U	T	U
⊥	⊥	U	T

2   Preliminaries

(Program) clauses are expressions of the forms $A \leftarrow L_1 \wedge \ldots \wedge L_n$ (called *rules*), $A \leftarrow \top$ (called *facts*), and $A \leftarrow \bot$ (called *assumptions*), where $n \geq 1$, A is an atom, and each L_i, $1 \leq i \leq n$, is a literal. A is called *head* and $L_1 \wedge \ldots \wedge L_n$ as well as $\top$ and $\bot$, standing for *true* and *false*, respectively, are called *body* of the corresponding clauses. A *(logic) program* is a set of clauses. Throughout this paper, $\mathcal{P}$ denotes a program. We assume for each $\mathcal{P}$ that the alphabet consists precisely of the symbols occurring in $\mathcal{P}$ and that non-propositional programs contain at least one constant symbol.

$g\mathcal{P}$ denotes the set of all ground instances of clauses occurring in $\mathcal{P}$. Let $\mathcal{P}$ be a ground program and A a ground atom. An *exception clause* in $\mathcal{P}$ is a clause whose head is an abnormality predicate, i.e. it features predicate ab or ab_i, where $1 \leq i \leq n$. A is *defined in* $\mathcal{P}$ iff $\mathcal{P}$ contains a rule or a fact with head A. A is *undefined in* $\mathcal{P}$ iff A is not defined in $\mathcal{P}$. The *definition of A in $\mathcal{P}$* is as follows:

$$def(A, \mathcal{P}) = \{A \leftarrow Body \mid A \leftarrow Body \text{ is a rule or a fact occurring in } \mathcal{P}\}.$$

$\neg A$ is *assumed in* $\mathcal{P}$ iff $\mathcal{P}$ contains an assumption with head A and $def(A, \mathcal{P}) = \emptyset$.

Let $\mathcal{P}$ be a ground program. Consider the following transformation: (1) Replace all clauses with the same head $A \leftarrow Body_1$, $A \leftarrow Body_2$, ... by $A \leftarrow Body_1 \vee Body_2 \vee \ldots$. (2) Replace all occurrences of $\leftarrow$ by $\leftrightarrow$. The resulting set is called *weak completion* of $\mathcal{P}$ or $wc\mathcal{P}$. Note that undefined atoms are not identified with $\bot$ as in the completion of $\mathcal{P}$ [3].

We consider the three-valued Łukasiewicz (or Ł-) logic [19] (see Table 1) and represent each interpretation I by $\langle I^\top, I^\bot \rangle$, where $I^\top = \{A \mid I(A) = \top\}$, $I^\bot = \{A \mid I(A) = \bot\}$, $I^\top \cap I^\bot = \emptyset$, and each ground atom $A \notin I^\top \cup I^\bot$ is mapped to U (*unknown*). Let $\langle I^\top, I^\bot \rangle$ and $\langle J^\top, J^\bot \rangle$ be two interpretations. We define $\langle I^\top, I^\bot \rangle \subseteq \langle J^\top, J^\bot \rangle$ iff $I^\top \subseteq J^\top$ and $I^\bot \subseteq J^\bot$. Logic programs as well as their weak completions admit a least model under Ł-logic. The least Ł-model of $wc\mathcal{P}$ can be obtained as the least fixed point of the following operator, which is due to Stenning and van Lambalgen [25]: $\Phi_{\mathcal{P}}(\langle I^\top, I^\bot \rangle) = \langle J^\top, J^\bot \rangle$, where

$$J^\top = \{A \mid \text{there exists } A \leftarrow Body \in g\mathcal{P} \text{ with } I(Body) = \top\},$$
$$J^\bot = \{A \mid \text{there exists } A \leftarrow Body.\in g\mathcal{P}$$
$$\text{and for all } A \leftarrow Body \in g\mathcal{P} \text{ we find } I(Body) = \bot\}.$$

$(\mathcal{I}, \subseteq)$ is a complete partial order, where $\mathcal{I}$ is the set of all interpretations, the least one being $\langle \emptyset, \emptyset \rangle$. $\varPhi_\mathcal{P}$ is monotonic on $(\mathcal{I}, \subseteq)$, i.e., $I_1 \subseteq I_2$ implies $\varPhi_\mathcal{P}(I_1) \subseteq \varPhi_\mathcal{P}(I_2)$. By the Knaster-Tarski theorem, $\varPhi_\mathcal{P}$ has a least fixed point. Moreover, $\varPhi_\mathcal{P}$ is continuous for finite datalog programs. For details see [13,17].

Weak Completion Semantics (WCS) is the approach that considers weakly completed logic programs to reason with respect to the least L-models of these programs. We write $\mathcal{P} \models_{wcs} F$ iff formula F holds in the least L-model of $wc\mathcal{P}$. In the remainder of this paper, $\mathcal{M}_\mathcal{P}$ denotes the L-model of $wc\mathcal{P}$, which has been computed by the $\varPhi_\mathcal{P}$ operator.

A set of *integrity constraints* $\mathcal{IC}$ comprises clauses of the form $\mathrm{U} \leftarrow Body$, where $Body$ is a conjunction of literals. Given $\mathcal{P}$ and $\mathcal{IC}$, $\mathcal{P}$ *satisfies* $\mathcal{IC}$ iff for all $\mathrm{U} \leftarrow Body \in \mathcal{IC}$, we find that $\mathcal{M}_\mathcal{P}(Body) \neq \top$. This definition allows us to specify that literals can be either unknown or false. This understanding is similar to the definition of the integrity constraints for the Well-founded Semantics in [20]. Note that in Sect. 4, when we introduce contextual programs we can restrict $\mathcal{IC}$s to have $\perp$ in heads instead, because contextual literals, as we will see, can compensate for that in $\mathcal{IC}$ bodies, without loss of generality.

3 Abduction

Let $\mathcal{P}$ be a ground program. The *set of abducibles of* $\mathcal{P}$ is

$$\mathcal{A}_\mathcal{P} = \{A \leftarrow \top \mid A \text{ is undefined in } \mathcal{P} \text{ or } A \text{ is head of an exception clause in } \mathcal{P}\}$$
$$\cup \{A \leftarrow \perp \mid A \text{ is undefined in } \mathcal{P} \text{ and } \neg A \text{ is not assumed in } \mathcal{P}\}.$$

It consists of facts and assumptions for the undefined ground atoms occurring in $\mathcal{P}$ as well as of defeaters for assumptions and exception clauses occurring in $\mathcal{P}$. A fact of the form $A \leftarrow \top$ is called *defeater* for the assumption $A \leftarrow \perp$ or for an exception clause $ab(X) \leftarrow Body$, because the weak completion of the fact together with these clauses is semantically equivalent to $A \leftrightarrow \top$.[3]

An *abductive framework* consists of a logic program $\mathcal{P}$, a set of *abducibles* $\mathcal{A} \subseteq \mathcal{A}_\mathcal{P}$, a set of *integrity constraints* $\mathcal{IC}$ and the entailment relation $\models_{wcs}$. An abductive framework is denoted by $\langle \mathcal{P}, \mathcal{A}, \mathcal{IC}, \models_{wcs} \rangle$.

In the sequel, we consider datalog programs, i.e. abductive frameworks are defined with respect to the ground instances of the program. Consider again $\mathcal{P}_1$: Its least L-model, $\mathcal{M}_{\mathcal{P}_1} = \langle \mathcal{M}_{\mathcal{P}_1}^\top, \mathcal{M}_{\mathcal{P}_1}^\perp \rangle$, is

$$\mathcal{M}_{\mathcal{P}_1}^\top = \{bird(tweety), bird(jerry), can\_fly(tweety), can\_fly(jerry)\},$$
$$\mathcal{M}_{\mathcal{P}_1}^\perp = \{ab(tweety), ab(jerry)\}.$$

The set of abducibles of $\mathcal{P}_1$ is $\mathcal{A}_{\mathcal{P}_1} = \{ab(tweety) \leftarrow \top, \ ab(jerry) \leftarrow \top\}$.

One should observe that each program and, in particular, each finite set of facts and assumptions has an L-model. For the latter case, it can be obtained

---

[3] Defeaters were not part of the first specification of the set of abducibles under WCS [7,15], but without the defeaters the first example discussed in this section cannot be solved.

by mapping all heads occurring in the set to true. Thus, in the next definition, explanations as well as the union of a program and an explanation are satisfiable.

An *observation* $\mathcal{O}$ is a set of ground literals; it is *explainable* in the framework $\langle \mathcal{P}, \mathcal{A}, \mathcal{IC}, \models_{wcs} \rangle$ iff there exists an $\mathcal{E} \subseteq \mathcal{A}$ called *explanation* such that $\mathcal{M}_{\mathcal{P} \cup \mathcal{E}} \models_{wcs} L$ for all $L \in \mathcal{O}$ and $\mathcal{P} \cup \mathcal{E}$ satisfies $\mathcal{IC}$. Sometimes explanations are required to be *minimal* in that they cannot be subsumed by another explanation.

Consider the framework $\langle \mathcal{P}_1, \mathcal{A}_{\mathcal{P}_1}, \emptyset, \models_{wcs} \rangle$ and let $\mathcal{O} = \{\neg can\_fly(tweety)\}$. There are two explanations, viz. $\mathcal{E}_1 = \{ab(tweety) \leftarrow \top\}$ and $\mathcal{E}_2 = \{ab(tweety) \leftarrow \top, ab(jerry) \leftarrow \top\}$ with $\mathcal{E}_1$ being minimal. We obtain $\mathcal{M}_{\mathcal{P}_1 \cup \mathcal{E}_1}$ with

$$\mathcal{M}^{\top}_{\mathcal{P}_1 \cup \mathcal{E}_1} = \{bird(tweety), bird(jerry), can\_fly(jerry), ab(tweety)\},$$
$$\mathcal{M}^{\perp}_{\mathcal{P}_1 \cup \mathcal{E}_1} = \{ab(jerry), can\_fly(tweety)\}.$$

Consider again $\mathcal{P}_2$: We obtain $\mathcal{M}_{\mathcal{P}_2} = \langle \{bird(tweety), bird(jerry)\}, \emptyset \rangle$ and

$$\mathcal{A}_{\mathcal{P}_2} = \mathcal{A}_{\mathcal{P}_1} \cup \{ \begin{array}{ll} kiwi(tweety) \leftarrow \perp, & penguin(tweety) \leftarrow \perp, \\ kiwi(tweety) \leftarrow \top, & penguin(tweety) \leftarrow \top, \\ kiwi(jerry) \leftarrow \perp, & penguin(jerry) \leftarrow \perp, \\ kiwi(jerry) \leftarrow \top, & penguin(jerry) \leftarrow \top \end{array} \}$$

Considering $\langle \mathcal{P}_2, \mathcal{A}_{\mathcal{P}_2}, \emptyset, \models_{wcs} \rangle$ and $\mathcal{O} = \{can\_fly(jerry)\}$ we obtain the minimal explanation $\{kiwi(jerry) \leftarrow \perp, penguin(jerry) \leftarrow \perp\}$. In other words, in order to explain that *Jerry can fly* we have to assume that *Jerry* does not belong to any of the known exceptions. The question that arises already in [24] is whether and how we can avoid the explicit investigation into all known exceptions and conclude instead that *Jerry can fly by default*? Posed differently, does an ordinary human knowing the statements mentioned above and observing that *Jerry can fly* accept this observation as *default* or does he or she accept this observation only after explicitly reasoning just in case *Jerry is not a kiwi and not a penguin*?

4 Contextual Programs

In [23], Pereira and Pinto have introduced *inspection points* of the form $inspect(L)$, where L is a literal. Inspection points are treated as meta-predicates belonging to a special case of abducibles: $inspect(L)$ can only be abduced to explain some observation in case L is abduced to explain some given observation. More formally, $\mathcal{E}$ is an explanation if for each $inspect(L) \in \mathcal{E}$ we find that $L \in \mathcal{E}$ too. That is, $inspect(L)$ is only accepted in the context of L.

Table 2. Truth table for ctxt(L).

L	ctxt(L)
$\top$	$\top$
$\perp$	$\perp$
U	$\perp$

Inspired by the idea underlying inspection points, we introduce a new truth-functional operator ctxt (called *context*), whose meaning is specified in Table 2. With the help of ctxt, preferences on explanations, among other things, can be syntactically specified. These preferences are context-dependent.

The interpretation of ctxt can be understood as a mapping from three-valuedness to two-valuedness. It is one possible way to capture negation as failure (or negation by default) under WCS. The original idea of negation as failure [3] is to derive the negation of A in case we fail to derive A, where the meaning of derivation failure depends on the semantics. Negation as failure does not exist under WCS, quite the contrary is the case. Let $\mathcal{P}_3 = \{p \leftarrow q, \; p \leftarrow \bot\}$. Its weak completion is $wc\mathcal{P}_3 = \{p \leftrightarrow q \vee \bot\}$, which is semantically equivalent to $\{p \leftrightarrow q\}$. The least model of $wc\mathcal{P}_3$ is $\langle \emptyset, \emptyset \rangle$, so p and q are both unknown, whereas they would be false if negation as failure had been adopted under WCS. The assumption $p \leftarrow \bot$ has been overridden by the first clause of $\mathcal{P}_3$ and does not have any effect at all. On the other hand, $\mathsf{ctxt}(L) = \bot$ if L is unknown.

We extend the definition of programs by allowing expressions of the form $\mathsf{ctxt}(L)$ in the body of clauses. Formally, *contextual rules* are expressions of the form $A \leftarrow L_1 \wedge \ldots \wedge L_m \wedge \mathsf{ctxt}(L_{m+1}) \wedge \ldots \wedge \mathsf{ctxt}(L_{m+p})$, where $m, p \in \mathbb{N}$ such that $m + p \geq 1$. A *contextual (datalog) program* $\mathcal{P}$ is a finite set of contextual rules, facts and assumptions.

Consider the following two programs and their weak completions:

$$\begin{aligned} \mathcal{P}_4 &= \{p \leftarrow \neg q\}, & \mathcal{P}_5 &= \{p \leftarrow \mathsf{ctxt}(\neg q)\}, \\ wc\mathcal{P}_4 &= \{p \leftrightarrow \neg q\}, & wc\mathcal{P}_5 &= \{p \leftrightarrow \mathsf{ctxt}(\neg q)\}. \end{aligned}$$

Starting with $I_0 = \langle \emptyset, \emptyset \rangle$ we compute the corresponding least fixed points of $\Phi_{\mathcal{P}_4}$ and $\Phi_{\mathcal{P}_5}$ as $\Phi_{\mathcal{P}_4}(\langle \emptyset, \emptyset \rangle) = \langle \emptyset, \emptyset \rangle$ and $\Phi_{\mathcal{P}_5}(\langle \emptyset, \emptyset \rangle) = \langle \emptyset, \{p\} \rangle = \Phi_{\mathcal{P}_5}(\langle \emptyset, \{p\} \rangle)$. $\mathsf{ctxt}(\neg q)$ in $\mathcal{P}_5$ mimics negation as failure: Nothing is known about q, therefore we derive that p is false in $\mathcal{P}_5$.

By means of ctxt, the common syntactical form for integrity constraints can be re-established: $\mathcal{IC}$ comprises clauses of the form $\bot \leftarrow Body$, where $Body$ is a conjunction of $L_1 \wedge \ldots \wedge L_m \wedge \mathsf{ctxt}(L_{m+1}) \wedge \ldots \wedge \mathsf{ctxt}(L_{m+p})$, $m, p \in \mathbb{N}$ and $m + p \geq 1$. Given $\mathcal{P}$ and $\mathcal{IC}$, $\mathcal{P}$ *satisfies* $\mathcal{IC}$ iff for all $\bot \leftarrow Body \in \mathcal{IC}$, we find that $\mathcal{M}_\mathcal{P}(Body) = \bot$. Because ctxt is allowed in the body of these clauses, the same understanding as in Sect. 2 can be maintained: If a literal L should be either unknown or false, then we can simply write $\bot \leftarrow \mathsf{ctxt}(L)$.

In the remainder of this paper we assume that the underlying semantics of these contextual programs is the Ł-logic defined in Sect. 2 extended by the truth table for ctxt defined in Table 2. Although the Φ-operator admits a least fixed point for programs $\mathcal{P}_4$ and $\mathcal{P}_5$, we need to check whether this holds in general under the modified logic. As another example consider

$$\mathcal{P}_6 = \{p \leftarrow q \wedge \mathsf{ctxt}(r), \; r \leftarrow \neg s, \; s \leftarrow \mathsf{ctxt}(t)\}.$$

Applying Φ to $\mathcal{P}_6$ starting with the empty interpretation $I_0 = \langle \emptyset, \emptyset \rangle$ results in:

$$\begin{aligned} \Phi_{\mathcal{P}_6}(I_0) &= \langle \emptyset, \{p, s\} \rangle = I_1, & \Phi_{\mathcal{P}_6}(I_1) &= \langle \{r\}, \{p, s\} \rangle = I_2, \\ & & \Phi_{\mathcal{P}_6}(I_2) &= \langle \{r\}, \{s\} \rangle = \Phi_{\mathcal{P}_6}(\langle \{r\}, \{s\} \rangle). \end{aligned}$$

One should observe that $I_1 \subseteq I_2$, but $\Phi_{\mathcal{P}_6}(I_1) \subseteq \Phi_{\mathcal{P}_6}(I_2)$ does not hold. Hence, Φ is not monotonic anymore and, therefore, we cannot employ the Knaster-Tarski theorem to conclude that Φ has a least fixed point. In fact, it not always has a least fixed point, as the program $\mathcal{P}_7 = \{p \leftarrow \mathsf{ctxt}(\neg p)\}$ shows. Starting with $I_0 = \langle \emptyset, \emptyset \rangle$ we obtain:

$$\Phi_{\mathcal{P}_7}(I_0) = \langle \emptyset, \{p\} \rangle = I_1, \quad \Phi_{\mathcal{P}_7}(I_1) = \langle \{p\}, \emptyset \rangle = I_2, \quad \Phi_{\mathcal{P}_7}(I_2) = \langle \{\emptyset, \{p\} \rangle = I_1.$$

Is it possible to show that we can guarantee a least fixed point for a particular class of contextual programs? In this paper we follow an idea first developed by Fitting in [10] for programs under Kripke-Kleene logic. The idea is adapted to programs under Ł-logic in [12,17]. The Banach Contraction Theorem [1] states that every contraction mapping has a unique fixed point. Hence, the goal is to show that $\Phi_{\mathcal{P}}$ is a contraction on an appropriately defined metric space. Unfortunately, semantic operators are in general not contractions. This holds for Fitting's approach as well as for its adaptation discussed in [12,17], and program $\mathcal{P}_7$ shows that it also does not apply for the logic considered in this Section. But, as we will show, it applies to acyclic contextual programs. Before being able to do so, we need some definitions.

A *metric* on a space $\mathcal{M}$ is a mapping $d : \mathcal{M} \times \mathcal{M} \mapsto \mathbb{R}$, such that for all $x, y, z \in \mathcal{M}$ we have: $d(x, y) = 0$ iff $x = y$, $d(x, y) = d(y, x)$, and $d(x, y) \leq d(x, z) + d(z, y)$. A metric space $(\mathcal{M}, d)$ is *complete* if every Cauchy sequence converges. A sequence $s_1, s_2, s_3, \ldots$ is *Cauchy* if, for every $\epsilon > 0$ there is an integer N such that for all $n, m \geq N, d(s_n, s_m) \leq \epsilon$. The sequence *converges* if there is an s such that, for every $\epsilon > 0$, there is an integer N such that for all $n \geq N, d(s_n, s) \leq \epsilon$. Let $(\mathcal{M}, d)$ be a metric space: A mapping $f : \mathcal{M} \mapsto \mathcal{M}$ is a *contraction* if for all $x, y \in \mathcal{M}$ there exists a $k \in \mathbb{R}$ with $0 < k < 1$ such that $d(f(x), f(y)) \leq k \cdot d(x, y)$.

Theorem 1 [1]. *A contraction mapping f on a complete metric space has a unique fixed point. The sequence x, $f(x)$, $f(f(x))$, $\ldots$ converges to this fixed point for any x.*

A *level mapping* for a contextual program $\mathcal{P}$ is a function ℓ which assigns to each ground atom a natural number. It is extended to ground literals and expressions of the form $\mathsf{ctxt}(L)$ as follows, where L is a ground literal and A a ground atom: $\ell(\neg A) = \ell(A)$ and $\ell(\mathsf{ctxt}(L)) = \ell(L)$. A *contextual program* $\mathcal{P}$ is *acyclic with respect to a level mapping* ℓ iff for every rule $A \leftarrow L_1 \wedge \ldots \wedge L_m \wedge \mathsf{ctxt}(L_{m+1}) \wedge \ldots \wedge \mathsf{ctxt}(L_{m+p}) \in \mathcal{P}$ and for all $1 \leq i \leq m$, we find that $\ell(A) > \ell(L_i)$ and for all $m + 1 \leq j \leq m + p$, we find that $\ell(A) > \ell(\mathsf{ctxt}(L_j))$. A *contextual program* $\mathcal{P}$ is *acyclic* iff it is acyclic with respect to some level mapping ℓ.

Let ℓ be a level mapping and I and J be interpretations. The function $d_\ell : \mathcal{I} \times \mathcal{I} \mapsto \mathbb{R}$ is defined as

$$d_\ell(I, J) = \begin{cases} (\frac{1}{2})^n & I \neq J \text{ and } I(A) = J(A) \neq \mathsf{U} \text{ for all } A \text{ with } \ell(A) < n \text{ and,} \\ & \text{for some } A \text{ with } \ell(A) = n, I(A) \neq J(A) \text{ or } I(A) = J(A) = \mathsf{U}, \\ 0 & \text{otherwise.} \end{cases}$$

Proposition 2 [17]. *d_ℓ is a metric and $(\mathcal{I}, d_\ell)$ is a complete metric space.*

Theorem 3. *Let $\mathcal{P}$ be an acyclic contextual program with respect to the level mapping ℓ. Then $\Phi_{\mathcal{P}}$ is a contraction on $(\mathcal{I}, d_\ell)$.*

Proof. Given that $\mathcal{P}$ is a contextual program, we have to show that

$$d_\ell(\Phi_{\mathcal{P}}(I), \Phi_{\mathcal{P}}(J)) \leq \frac{1}{2} \cdot d_\ell(I, J). \tag{1}$$

If $I = J$, then $\Phi_{\mathcal{P}}(I) = \Phi_{\mathcal{P}}(J)$, $d_\ell(\Phi_{\mathcal{P}}(I), \Phi_{\mathcal{P}}(J)) = d_\ell(I, J) = 0$, and (1) holds. If $I \neq J$, then since ℓ is total, we obtain $d_\ell(I, J) = \frac{1}{2}^n$ for some $n \in \mathbb{N}$. We will show that $d_\ell(\Phi_{\mathcal{P}}(I), \Phi_{\mathcal{P}}(J)) \leq (\frac{1}{2})^{n+1}$, i.e. for all ground atoms $A \in \mathsf{g}\mathcal{P}$, with $\ell(A) \leq n$ we have that $\Phi_{\mathcal{P}}(I)(A) = \Phi_{\mathcal{P}}(J)(A)$.
Let us take some A with $\ell(A) \leq n$ and let $clauses(A, \mathcal{P})$ be the set of all clauses in $\mathsf{g}\mathcal{P}$ where A is the head of. As $\mathcal{P}$ is acyclic, for all rules

$$A \leftarrow L_1 \wedge \ldots \wedge L_m \wedge \mathsf{ctxt}(L_{m+1}) \wedge \ldots \wedge \mathsf{ctxt}(L_{m+p}) \in def(A, \mathcal{P})$$

for all $1 \leq i \leq m$ we obtain $\ell(L_i) < \ell(A) \leq n$, and for all $m + 1 \leq j \leq m + p$, we obtain $\ell(\mathsf{ctxt}(L_j)) < \ell(A) \leq n$. We know that $d_\ell(I, J) \leq (\frac{1}{2})^n$, so for all $1 \leq i \leq m$, $I(L_i) = J(L_i)$, and for all $m + 1 \leq j \leq m + p$, $I(\mathsf{ctxt}(L_j)) = J(\mathsf{ctxt}(L_j))$. Therefore, I and J interpret identically all bodies of clauses with A in the head. Consequently, $\Phi_{\mathcal{P}}(I)(A) = \Phi_{\mathcal{P}}(J)(A)$. ∎

Corollary 4. *Let $\mathcal{P}$ be an acyclic contextual program. $\Phi_{\mathcal{P}}$ has a unique fixed point. This fixed point can be reached in finite time by iterating $\Phi_{\mathcal{P}}$ starting from any interpretation.*

Proof. 1. By Proposition 2, $(\mathcal{I}, d_\ell)$ is a complete metric space.
2. By Theorem 3, $\Phi_{\mathcal{P}}$ is a contraction for acyclic contextual $\mathcal{P}$ on $\mathcal{I}$ using d_ℓ.
3. By 1. and 2. and the Banach Contraction Theorem, Theorem 1, for any acyclic contextual $\mathcal{P}$, $\Phi_{\mathcal{P}}$ has a unique fixed point. Furthermore, this fixed point can be reached starting from any interpretation.
4. By 3. and because $\mathcal{P}$ is finite, $\Phi_{\mathcal{P}}$ can be reached in a finite number of times starting from any interpretation. ∎

The following result can be shown analogously to the proof for $\mathcal{P}$ in [17].

Proposition 5. *Let $\mathcal{P}$ be an acyclic contextual program. Then, the least fixed point of $\Phi_{\mathcal{P}}$ is a model of $wc\mathcal{P}$.*

Proof. Assume that the least fixed point of $\Phi_{\mathcal{P}} = \langle I^\top, I^\perp \rangle$ and $A \leftrightarrow F \in wc\mathcal{P}$. We distinguish between 3 cases:

1. If $I(A) = \top$, then according to the definition of $\Phi_{\mathcal{P}}$, there exists a clause $A \leftarrow L_1 \wedge \cdots \wedge L_m \wedge \mathsf{ctxt}(L_{m+1}) \wedge \cdots \wedge \mathsf{ctxt}(L_{m+p})$, such that for all $1 \leq i \leq m+p$, $I(L_i) = \top$. As $L_1 \wedge \cdots \wedge L_m \wedge \mathsf{ctxt}(L_{m+1}) \wedge \cdots \wedge \mathsf{ctxt}(L_{m+p})$ is one of the disjuncts in F, $I(F) = \top$, and thus $I(A \leftrightarrow F) = \top$.

2. If $I(A) = \mathsf{U}$, then according to the definition of $\Phi_{\mathcal{P}}$, there is no clause $A \leftarrow L_1 \wedge \cdots \wedge L_m \wedge \mathsf{ctxt}(L_{m+1}) \wedge \cdots \wedge \mathsf{ctxt}(L_{m+p})$, such that for all $1 \leq i \leq m+p$, $I(L_i) = \top$ and there is at least one clause $A \leftarrow L_1 \wedge \cdots \wedge L_m \wedge \mathsf{ctxt}(L_{m+1}) \wedge \cdots \wedge \mathsf{ctxt}(L_{m+p})$, such that for all $1 \leq i \leq m+p$, $I(L_i) \neq \bot$ and there exists $1 \leq j \leq m$, $I(L_j) = \mathsf{U}$. As none of the disjuncts in F is true, and at least one is unknown, $I(F) = \mathsf{U}$ and thus $I(A \leftrightarrow F) = \top$.

3. If $I(A) = \bot$, then according to the definition of $\Phi_{\mathcal{P}}$, there exists a clause $A \leftarrow L_1 \wedge \cdots \wedge L_m \wedge \mathsf{ctxt}(L_{m+1}) \wedge \cdots \wedge \mathsf{ctxt}(L_{m+p})$ and for all clauses $A \leftarrow L_1 \wedge \cdots \wedge L_m \wedge \mathsf{ctxt}(L_{m+1}) \wedge \cdots \wedge \mathsf{ctxt}(L_{m+p})$, there exists $1 \leq i \leq m+p$ such that $I(L_i) = \bot$ or there exists $m+1 \leq j \leq m+p$, such that $I(L_j) \neq \top$. As all disjuncts in F are false, $I(F) = \bot$ and thus $I(A \leftrightarrow F) = \top$. ∎

As has been shown in [14], for non-contextual programs, the least fixed point of $\Phi_{\mathcal{P}}$ is identical to the least model of the weak completion of $\mathcal{P}$. As the following example shows this does not hold for contextual programs, as the weak completion of contextual programs might have more than one minimal model. Consider $\mathcal{P} = \{s \leftarrow \neg r, \; r \leftarrow \neg p \wedge q, \; q \leftarrow \mathsf{ctxt}(\neg p)\}$. Its weak completion is $wc\mathcal{P} = \{s \leftrightarrow \neg r, \; r \leftrightarrow \neg p \wedge q, \; q \leftrightarrow \mathsf{ctxt}(\neg p)\}$. The least fixed point of $\Phi_{\mathcal{P}}$ is $\langle\{s\}, \{q, r\}\rangle$, which is a minimal model of $wc\mathcal{P}$. However, yet another minimal model of $wc\mathcal{P}$ is $\langle\{q, r\}, \{p, s\}\rangle$. But this model is not supported in the sense that if we iterate $\Phi_{\mathcal{P}}$ starting with this model, then we will compute $\langle\{s\}, \{q, r\}\rangle$. As the fixpoint of $\Phi_{\mathcal{P}}$ is unique and the only supported minimal model of $wc\mathcal{P}$, we define $\mathcal{P} \models_{wcs} F$ if and only if F holds in the least fixed point of $\Phi_{\mathcal{P}}$.

5   Contextual Abduction

How can we prefer explanations that explain the normal cases to explanations that explain the exception cases? How can we express that some explanations have to be considered only if there is some evidence for considering the exception cases? We want to avoid having to consider all explanations if there is no evidence for considering exception cases. On the other hand, we don't want to state that all exception cases are false, as we must do for $\mathcal{O} = \{can\_fly(jerry)\}$ given $\mathcal{P}_1$.

Consider the following definition for strong dependency in contextual programs: Given a rule $A \leftarrow L_1 \wedge \ldots \wedge L_m \wedge \mathsf{ctxt}(L_{m+1}) \wedge \ldots \wedge \mathsf{ctxt}(L_{m+p})$ for all $1 \leq i \leq m$, A 'strongly depends on' L_i. The 'strongly depends on' relation is transitive. If A strongly depends on L_i, then $\neg A$ strongly depends on L_i. Furthermore, if $L_i = B$, then A strongly depends on $\neg B$ and if $L_i = \neg B$, then A strongly depends on B. Consider program $\mathcal{P} = \{p \leftarrow r, \; p \leftarrow \mathsf{ctxt}(q)\}$: p strongly depends on r, p strongly depends on $\neg r$, $\neg p$ strongly depends on r and $\neg p$ strongly depends on $\neg r$. However, p does not strongly depend on q, nor on $\mathsf{ctxt}(q)$.

A *contextual abductive framework* is a quadruple $\langle \mathcal{P}, \mathcal{A}, \mathcal{IC}, \models_{wcs}\rangle$, consisting of an acyclic contextual program $\mathcal{P}$, a set of abducibles $\mathcal{A} \subseteq \mathcal{A}_{\mathcal{P}}$, a set of integrity constraints $\mathcal{IC}$, and the entailment relation $\models_{wcs}$, where $\mathcal{A}_{\mathcal{P}}$ is defined as

$$\mathcal{A}_{\mathcal{P}} = \{A \leftarrow \top \mid A \text{ is undefined in } \mathcal{P} \text{ or } A \text{ is head of an exception clause in } \mathcal{P}\}$$
$$\cup \; \{A \leftarrow \bot \mid A \text{ is undefined in } \mathcal{P} \text{ and } \neg A \text{ is not assumed in } \mathcal{P}\}.$$

Let an *observation* $\mathcal{O}$ be a non-empty set of ground literals. Note that $\mathcal{O}$ does not contain formulas of the form $\mathsf{ctxt}(L)$.

Let $\langle \mathcal{P}, \mathcal{A}, \mathcal{IC}, \models_{wcs} \rangle$ be a contextual abductive framework where $\mathcal{P}$ satisfies $\mathcal{IC}$, $\mathcal{E} \subseteq \mathcal{A}$ and $\mathcal{O}$ is an observation. $\mathcal{O}$ is *contextually explained by $\mathcal{E}$ given $\mathcal{P}$ and $\mathcal{IC}$* iff $\mathcal{O}$ is explained by $\mathcal{E}$ given $\mathcal{P}$ and $\mathcal{IC}$ and for all $A \leftarrow \top \in \mathcal{E}$ and for all $A \leftarrow \bot \in \mathcal{E}$ there exists $L \in \mathcal{O}$, such that L strongly depends on A. $\mathcal{O}$ is *contextually explainable given $\mathcal{P}$ and $\mathcal{IC}$* iff there exists some $\mathcal{E}$ such that $\mathcal{O}$ is contextually explained by $\mathcal{E}$ given $\mathcal{P}$ and $\mathcal{IC}$.

Similar to explanations in abduction, we assume that contextual explanations are minimal, that is, there is no other contextual explanation $\mathcal{E}' \subset \mathcal{E}$ for $\mathcal{O}$ given $\mathcal{P}$ and $\mathcal{IC}$. Note that, compared to explanations, contextual explanations have an additional requirement: There has to be a literal in the observation, which strongly depends on some atom for which there exists a fact or assumption in $\mathcal{E}$. We distinguish between skeptical and credulous reasoning in the usual way:

F *contextually follows skeptically from $\mathcal{P}$, $\mathcal{IC}$ and $\mathcal{O}$* iff $\mathcal{O}$ can be contextually explained given $\mathcal{P}$ and $\mathcal{IC}$, and for all $\mathcal{E}$ for $\mathcal{O}$ it holds that $\mathcal{P} \cup \mathcal{E} \models_{wcs} F$.
F *contextually follows credulously from $\mathcal{P}$, $\mathcal{IC}$ and $\mathcal{O}$* iff there exists some $\mathcal{E}$ that contextually explains $\mathcal{O}$ and it holds that $\mathcal{P} \cup \mathcal{E} \models_{wcs} F$.

Let us adapt $\mathcal{P}_2$ from the introduction such that all exceptions, viz. X being a penguin or a kiwi, are evaluated with respect to their context, ctxt, instead:

$$\mathcal{P}_8 = \{ \; can\_fly(X) \leftarrow bird(X) \wedge \neg ab_1(X), \quad bird(tweety) \leftarrow \top,$$
$$ab_1(X) \leftarrow \mathsf{ctxt}(kiwi(X)), \qquad\qquad bird(jerry) \leftarrow \top,$$
$$ab_1(X) \leftarrow \mathsf{ctxt}(penguin(X)) \qquad\qquad\qquad\qquad\quad \}.$$

We obtain $\mathcal{M}_{\mathcal{P}_8}$ with

$$\mathcal{M}_{\mathcal{P}_8}^{\top} = \{ bird(tweety), bird(jerry), can\_fly(tweety), can\_fly(jerry) \},$$
$$\mathcal{M}_{\mathcal{P}_8}^{\bot} = \{ ab_1(tweety), ab_1(jerry) \}.$$

This model already entails the observation $\mathcal{O} = \{ can\_fly(jerry) \}$ and, thus, $\mathcal{O}$ does not need any explanation beyond the minimal empty one.

5.1   More About Tweety

Consider, the following extension of $\mathcal{P}_8$:

$$\mathcal{P}_9 = \mathcal{P}_8 \cup \{ \; kiwi(X) \leftarrow featherslikeHair(X) \wedge \neg ab_2(X),$$
$$penguin(X) \leftarrow blackAndWhite(X) \wedge \neg ab_3(X),$$
$$ab_2(X) \leftarrow \mathsf{ctxt}(inEurope(X)),$$
$$ab_3(X) \leftarrow \mathsf{ctxt}(inEurope(X)) \qquad\qquad\qquad \}.$$

We obtain $\mathcal{M}_{\mathcal{P}_9}$ with

$$\mathcal{M}_{\mathcal{P}_9}^{\top} = \{ bird(tweety), bird(jerry), can\_fly(jerry), can\_fly(tweety) \},$$
$$\mathcal{M}_{\mathcal{P}_9}^{\bot} = \{ ab_i(tweety) \mid 1 \leq i \leq 3 \} \cup \{ ab_i(jerry) \mid 1 \leq i \leq 3 \}.$$

We observe that Tweety does not fly and that Tweety has feathers like hair: $\mathcal{O} = \{\neg can\_fly(tweety),\ featherslikeHair(tweety)\}$. The set of abducibles is:

$$\mathcal{A}_{\mathcal{P}_9} = \mathcal{A}_{\mathcal{P}_2} \cup \{\ \begin{aligned}
&featherslikeHair(jerry) \leftarrow \top,\quad &&featherslikeHair(jerry) \leftarrow \bot,\\
&featherslikeHair(tweety) \leftarrow \top,\quad &&featherslikeHair(tweety) \leftarrow \bot,\\
&blackAndWhite(jerry) \leftarrow \top,\quad &&blackAndWhite(jerry) \leftarrow \bot,\\
&blackAndWhite(tweety) \leftarrow \top,\quad &&blackAndWhite(tweety) \leftarrow \bot,\\
&inEurope(jerry) \leftarrow \top,\quad &&inEurope(jerry) \leftarrow \bot,\\
&inEurope(tweety) \leftarrow \top,\quad &&inEurope(tweety) \leftarrow \bot\quad\ \}.
\end{aligned}$$

$\mathcal{E} = \{featherslikeHair(tweety) \leftarrow \top\}$ is the only (minimal) contextual explanation for $\mathcal{O}$. We obtain $\mathcal{M}_{\mathcal{P}_9 \cup \mathcal{E}}$ with

$$\mathcal{M}^{\top}_{\mathcal{P}_9 \cup \mathcal{E}} = \{\ bird(tweety), featherslikeHair(tweety), kiwi(tweety), ab_1(tweety),$$
$$bird(jerry), can\_fly(jerry)\ \},$$
$$\mathcal{M}^{\bot}_{\mathcal{P}_9 \cup \mathcal{E}} = \{\ can\_fly(tweety), ab_2(tweety), ab_3(tweety)\} \cup \{ab_i(jerry) \mid 1 \leq i \leq 3\ \}.$$

From this model we derive that Tweety is a kiwi, even though $inEurope(tweety)$ is unknown. This is exactly what we assume humans do while reasoning: They do not assume anything about Tweety living in Europe, and by using ctxt we can model their ignorance concerning it.

5.2   More About Jerry

Consider again $\mathcal{P}_9$ from the previous example together with the observation that Jerry flies and Jerry lives in Europe: $\mathcal{O} = \{can\_fly(jerry), inEurope(jerry)\}$. $\mathcal{A}_{\mathcal{P}_9}$ is defined in the previous example. The only contextual explanation for $\mathcal{O}$ is $\mathcal{E} = \{inEurope(jerry) \leftarrow \top\}$ and we obtain $\mathcal{M}_{\mathcal{P}_9 \cup \mathcal{E}}$ with

$$\mathcal{M}^{\top}_{\mathcal{P}_9 \cup \mathcal{E}} = \{\ inEurope(jerry), bird(jerry), can\_fly(jerry), ab_2(jerry), ab_3(jerry),$$
$$bird(tweety), can\_fly(tweety)\ \},$$
$$\mathcal{M}^{\bot}_{\mathcal{P}_9 \cup \mathcal{E}} = \{\ kiwi(jerry), penguin(jerry), ab_1(jerry)\} \cup \{ab_i(tweety) \mid 1 \leq i \leq 3\}.$$

From this model we can derive, from the contextual clauses, that Jerry is not a penguin and Jerry is not a kiwi!

5.3   Contextual Side-Effects

Consider the following definitions originally presented in [21,22], which captures the idea of contextual side-effects: Given a program $\mathcal{P}$ and a set of integrity constraints $\mathcal{IC}$, let $\mathcal{O}_1$ and $\mathcal{O}_2$ be two observations and $\mathcal{E}_1$ be a contextual explanation for $\mathcal{O}_1$. $\mathcal{O}_2$ is a *necessary contextual side-effect* of $\mathcal{O}_1$ given $\mathcal{P}$ and $\mathcal{IC}$ iff $\mathcal{O}_2$ cannot be contextually explained but $\mathcal{O}_1 \cup \mathcal{O}_2$ is contextually explained by $\mathcal{E}_1$. $\mathcal{O}_2$ is a *possible contextual side-effect* of $\mathcal{O}_1$ given $\mathcal{P}$ and $\mathcal{IC}$ iff $\mathcal{O}_2$ cannot be contextually explained by $\mathcal{E}_1$ but $\mathcal{O}_1 \cup \mathcal{O}_2$ is contextually explained by $\mathcal{E}_1$. The idea behind contextual side-effects is that every explanation $\mathcal{E}_1$ for $\mathcal{O}_1$ gives us an explanation for $\mathcal{O}_2$. Note that a necessary contextual side-effect is also a possible contextual side-effect.

Consider again the Tweety example of Sect. 5.1, where the observation is $\mathcal{O} = \{\neg can\_fly(tweety), featherslikeHair(tweety)\}$ and its only contextual explanation is $\mathcal{E} = \{featherslikeHair(tweety) \leftarrow \top\}$. $\mathcal{O}_2 = \{\neg can\_fly(tweety)\}$ cannot be contextually explained by $\mathcal{E}$. However, $\mathcal{O}$ can be contextually explained by $\mathcal{E}$ and $\mathcal{O}_2$ is a necessary contextual side-effect of $\mathcal{O} \setminus \mathcal{O}_2 = \{featherslikeHair(tweety)\}$.

On the other hand, consider again the Jerry example of Sect. 5.2, where the observation is $\mathcal{O} = \{can\_fly(jerry), inEurope(jerry)\}$, for which the only explanation is $\mathcal{E} = \{inEurope(jerry) \leftarrow \top\}$. As $\mathcal{O}_2 = \{can\_fly(jerry)\}$ already follows from the empty explanation, $\mathcal{E}_2 = \emptyset$, $\mathcal{O}_2$ cannot be considered a contextual side-effect of $\mathcal{O} \setminus \mathcal{O}_2 = \{inEurope(jerry)\}$.

6   Conclusion

Prompted by the famous Tweety example, we first show that the Weak Completion Semantics does not yield the desired results. We would like to avoid having to abductively consider all exception cases and to automatically prefer normal explanations to those explanations specifying such exception cases. To do so, we set forth contextual programs, for the purpose of which we introduce ctxt, a new truth-functional operator, which turns out to fit quite well with the interpretation of negation as failure under three-valued semantics. Unfortunately, the usual Φ operator is not monotonic with respect to these contextual programs anymore. Even worse, the Φ operator might not even have a least fixed point for some contextual programs. However, we can show that the Φ operator does always have a least fixed point if we restrict contextual programs to the class of acyclic ones. After that, we define contextual abduction and show that the Tweety example from the introduction leads to the desired results. Furthermore, we can now specify the relations between observations and explanations under contextual abduction, allowing us to define notions pertaining to contextual side-effects.

Some open questions are left to be investigated in the future. For instance, can the requirements for the classes of acyclic contextual programs be relaxed to those that are only acyclic with respect to the truth functional operator ctxt, so that the Φ operator is still guaranteed to yield a fixed point? Furthermore, as the Weak Completion Semantics seems to adequately model human reasoning, a natural question to ask is whether contextual reasoning can help us model pertinent psychological experiments. For this purpose, we are particularly interested in psychological findings that deal with context sensitive information.

Acknowledgements. LMP acknowledges support from FCT/MEC NOVA LINCS Pest UID/CEC/04516/2013. Many thanks to Tobias Philipp and Christoph Wernhard.

References

1. Banach, S.: Sur les opérations dans les ensembles abstraits et leur application aux équations intégrales. Fund. Math. **3**, 133–181 (1922)
2. Byrne, R.: Suppressing valid inferences with conditionals. Cognition **31**, 61–83 (1989)
3. Clark, K.: Negation as failure. In: Gallaire, H., Minker, J. (eds.) Logic and Databases, pp. 293–322. Plenum, New York (1978)
4. Dietz, E.-A., Hölldobler, S.: A new computational logic approach to reason with conditionals. In: Calimeri, F., Ianni, G., Truszczynski, M. (eds.) LPNMR 2015. LNCS (LNAI), vol. 9345, pp. 265–278. Springer, Cham (2015). doi:10.1007/978-3-319-23264-5_23
5. Dietz, E.-A., Hölldobler, S., Höps, R.: A computational logic approach to human spatial reasoning. In: IEEE Symposium Series on Computational Intelligence, pp. 1634–1637 (2015)
6. Dietz, E.-A., Hölldobler, S., Pereira, L.M.: On conditionals. In: Gottlob, G., Sutcliffe, G., Voronkov, A. (eds.), Global Conference on Artificial Intelligence. Epic Series in Computing, vol. 36, pp. 79–92. EasyChair (2015)
7. Dietz, E.-A., Hölldobler, S., Ragni, M.: A computational logic approach to the suppression task. In: Miyake, N., Peebles, D., Cooper, R.P. (eds.), Proceedings of the 34th Annual Conference of the Cognitive Science Society, pp. 1500–1505. Cognitive Science Society (2012)
8. Dietz, E.-A., Hölldobler, S., Ragni, M.: A computational logic approach to the abstract and the social case of the selection task. In: Proceedings of the Eleventh International Symposium on Logical Formalizations of Commonsense Reasoning (2013). commonsensereasoning.org/2013/proceedings.html
9. Dietz, E.-A., Hölldobler, S., Wernhard, C.: Modelling the suppression task under weak completion and well-founded semantics. J. Appl. Non-Classical Logics **24**, 61–85 (2014)
10. Fitting, M.: Metric methods - three examples and a theorem. J. Logic Program. **21**(3), 113–127 (1994)
11. Gelfond, M., Lifschitz, V.: The stable model semantics for logic programming. In: Kowalski, R., Bowen, K. (eds.), Proceedings of the International Joint Conference and Symposium on Logic Programming, pp. 1070–1080. MIT Press (1988)
12. Hölldobler, S., Kencana Ramli, C.D.P.: Contraction properties of a semantic operator for human reasoning. In: Li, L., Yen, K.K. (eds.), Proceedings of the Fifth International Conference on Information, pp. 228–231. International Information Institute (2009)
13. Hölldobler, S., Kencana Ramli, C.D.P.: Logic programs under three-valued Łukasiewicz semantics. In: Hill, P.M., Warren, D.S. (eds.) ICLP 2009. LNCS, vol. 5649, pp. 464–478. Springer, Heidelberg (2009). doi:10.1007/978-3-642-02846-5_37
14. Hölldobler, S., Kencana Ramli, C.D.P.: Logics and networks for human reasoning. In: Alippi, C., Polycarpou, M., Panayiotou, C., Ellinas, G. (eds.) ICANN 2009. LNCS, vol. 5769, pp. 85–94. Springer, Heidelberg (2009). doi:10.1007/978-3-642-04277-5_9
15. Hölldobler, S., Philipp, T., Wernhard, C.: An abductive model for human reasoning. In: Proceedings of the Tenth International Symposium on Logical Formalizations of Commonsense Reasoning (2011). commonsensereasoning.org/2011/proceedings.html

16. Kakas, A.C., Kowalski, R.A., Toni, F.: Abductive logic programming. J. Logic Comput. **2**(6), 719–770 (1993)
17. Kencana Ramli, C.D.P.: Logic programs and three-valued consequence operators. Master's thesis, International Center for Computational Logic, TU Dresden (2009)
18. Lloyd, J.W.: Foundations of Logic Programming. Springer, Heidelberg (1984)
19. Łukasiewicz, J.: O logice trójwartościowej. Ruch Filozoficzny **5**, 169–171 (1920). English translation: On Three-Valued Logic. In: Jan Łukasiewicz Selected Works. (L. Borkowski, ed.), North Holland, 87–88 (1990)
20. Pereira, L.M., Aparício, J.N., Alferes, J.: Hypothetical reasoning with well founded semantics. In: Mayoh, B. (ed.), Proceedings of the 3th Scandinavian Conference on AI, pp. 289–300. IOS Press, 1991
21. Pereira, L.M., Dietz, E.-A., Hölldobler, S.: An abductive reasoning approach to the belief-bias effect. In: Baral, C., Giacomo, G.D., Eiter, T. (eds.) Principles of Knowledge Representation and Reasoning: Proceedings of the 14th International Conference, pp, pp. 653–656. AAAI Press, Cambridge (2014)
22. Pereira, L.M., Dietz, E.-A., Hölldobler, S.: Contextual abductive reasoning with side-effects. In: Niemelä, I. (ed.), Theory and Practice of Logic Programming (TPLP), vol. 14, pp. 633–648. Cambridge University Press, Cambridge (2014)
23. Moniz Pereira, L., Pinto, A.M.: Inspecting side-effects of abduction in logic programs. In: Balduccini, M., Son, T.C. (eds.) Logic Programming, Knowledge Representation, and Nonmonotonic Reasoning. LNCS, vol. 6565, pp. 148–163. Springer, Heidelberg (2011). doi:10.1007/978-3-642-20832-4_10
24. Reiter, R.: A logic for default reasoning. Artif. Intell. **13**, 81–132 (1980)
25. Stenning, K., van Lambalgen, M.: Human Reasoning and Cognitive Science. MIT Press, Boston (2008)
26. van Gelder, A., Ross, K.A., Schlipf, J.S.: The well-founded semantics for general logic programs. J. ACM **38**, 620–650 (1991)

Including Quantification in Defeasible Reasoning for the Description Logic $\mathcal{EL}_\perp$

Maximilian Pensel and Anni-Yasmin Turhan[✉]

Institute for Theoretical Computer Science,
Technische Universität Dresden, Dresden, Germany
{maximilian.pensel,anni-yasmin.turhan}@tu-dresden.de

Abstract. Defeasible Description Logics (DDLs) can state defeasible concept inclusions and often use rational closure according to the KLM postulates for reasoning. If in DDLs with quantification a defeasible subsumption relationship holds between concepts, it can also hold if these concepts appear nested in existential restrictions. Earlier reasoning algorithms did not detect this kind of relationships. We devise a new form of canonical models that extend classical ones for $\mathcal{EL}_\perp$ by elements that satisfy increasing amounts of defeasible knowledge and show that reasoning w.r.t. these models yields the missing rational entailments.

1 Introduction

Description Logics (DLs) *concepts* describe groups of objects by means of other concepts (unary FOL predicates) and roles (binary relations). Such concepts can be related to other concepts in the TBox. Technically, the TBox is a theory constraining the interpretation of concepts. The lightweight DL $\mathcal{EL}$ allows as complex concepts: conjunction and existential restriction which is a form of existential quantification. A prominent DL reasoning problem is to compute subsumption relationships between two concepts. Such a relationship holds, if all instances of one concept must be necessarily instances of the other (w.r.t. the TBox). In $\mathcal{EL}$ subsumption can be computed in polynomial time [1] which made it the choice as the basis for OWL 2 EL, standardised by the W3C. In $\mathcal{EL}$ satisfiability is trivial, since no negation can be expressed. $\mathcal{EL}_\perp$ can express disjointness of concepts and is thus more interesting for non-monotonic reasoning.

Defeasible DLs (DDLs) are non-monotonic extensions of DLs and were intensively investigated [2–6]. Most DDLs allow to state additional relationships between concepts that characterize typical instances and that can get defeated if they cause an inconsistency. The notions of defeasibility and typicality are closely related: the more defeasible information is used for reasoning about a concept, the more typical its instances are regarded. Reasoning problems for DDLs are defined w.r.t. different entailment relations using different forms of closure. We consider rational closure, which is the basis for stronger forms of entailment and

M. Pensel—Supported by DFG in the Research Training Group QuantLA (GRK 1763).

© Springer International Publishing AG 2017
M. Balduccini and T. Janhunen (Eds.): LPNMR 2017, LNAI 10377, pp. 78–84, 2017.
DOI: 10.1007/978-3-319-61660-5_9

based on the KLM postulates [7]. Casini and Straccia lifted these postulates for propositional logic to DLs [4]. In [3] Casini et al. devise a reduction for computing rational closure by materialisation, where the idea is to encode a consistent subset of the defeasible statements as a concept and then use this in classical subsumption queries as additional constraint for the (potential) sub-concept in the query. Their translation of the KLM postulates to DLs yields rational entailments of propositional nature, but does not regard existential restrictions. To be precise, an existential restriction, e.g. $\exists r.B$ requires existence of an element related via role r to an element belonging to concept B—which cannot be expressed in propositional logic. By neglecting quantification in the translation of the postulates, a defeasible implication between concepts need not hold, if these occur nested in existential restrictions. This is at odds with the basic principle of defeasible reasoning: defeasible information *should* be used for reasoning if no more specific knowledge contradicts it. Consequently, the materialisation-based approach [3] misses entailments, since it treats all role successors uniformly as non-typical concept members. Quantified concepts are disregarded in several algorithms for defeasible reasoning in DLs that employ materialisation, such as lexicographic [5] and relevant [3] closure. This even holds for preferential model semantics, by the result of Britz et al. in [2], that entailments of preferential model semantics coincide with those of rational closure in [3].

In this paper we devise a reasoning algorithm for rational entailment for $\mathcal{EL}_\perp$ that derives defeasible knowledge for concepts in existential restrictions by a reduction to classical reasoning. To this end we extend the well-known canonical models for $\mathcal{EL}$ to *typicality* *models* that use domain elements to represent differing levels of typicality of a concept. For a simple form of these typicality models we show that it entails the same consequences as the materialisation-based approach [3]. We devise an extension of these models that capture the maximal typicality of each role successor element individually and thereby facilitates defeasible reasoning also for concepts nested in existential restrictions. These maximal typicality models yield (potentially) more rational entailments. Due to space constraints, we need to refer the reader to the technical report [9] for an introduction to DLs and for the proofs of the claims presented here.

2 Typicality Models for Rational Entailment

In order to achieve rational entailment for quantified concepts, defeasible concept inclusion statements (DCIs) need to hold for concepts in (nested) existential restrictions. We want to decide subsumption between the $\mathcal{EL}_\perp$-concepts C and D w.r.t. the $\mathcal{EL}_\perp$-DKB $\mathcal{K} = (\mathcal{T}, \mathcal{D})$. In order to deal with possible inconsistencies between DBox $\mathcal{D}$ and TBox $\mathcal{T}$, the reasoning procedure needs to consider subsets of the DBox. We use the same sequence of subsets of the DBox as in the materialisation-based approach [3]. Let $\overline{\mathcal{D}} = \bigsqcap_{E \sqsubseteq F \in \mathcal{D}} (\neg E \sqcup F)$ be the materialisation of $\mathcal{D}$. For a DKB $\mathcal{K}$, the set of exceptional DCIs from $\mathcal{D}$ w.r.t. $\mathcal{T}$ is defined as $\mathcal{E}(\mathcal{D}) = \{C \sqsubseteq D \in \mathcal{D} \mid \mathcal{T} \models \overline{\mathcal{D}} \sqcap C \sqsubseteq \perp\}$. The *sequence of DBox subsets* is defined inductively as $\mathcal{D}_0 = \mathcal{D}$ and $\mathcal{D}_i = \mathcal{E}(\mathcal{D}_{i-1})$ for $i \geq 1$.

If $\mathcal{D}$ is finite, there exists a least n with $\mathcal{D}_n = \mathcal{D}_{n+1}$, which we call the *rank of* $\mathcal{D}$ ($rk(\mathcal{D}) = n$). If $\mathcal{D}_n = \emptyset$, then the DKB $\mathcal{K}$ is called *well-separated*, since the set of DCIs can turn consistent by iteratively removing consistent DCIs. A given DKB $\mathcal{K} = (\mathcal{T}, \mathcal{D})$ can be transformed into a well-separated DKB $\mathcal{K}' = (\mathcal{T} \cup \{C \sqsubseteq \bot \mid C \sqsubsetneq D \in \mathcal{D}_n\}, \mathcal{D} \setminus \mathcal{D}_n)$ by deciding a quadratic number of subsumption tests in the size of $\mathcal{D}$. We assume that a given DKB $\mathcal{K} = (\mathcal{T}, \mathcal{D})$ is well-separated and thus the sequence $\mathcal{D}_0, \ldots, \mathcal{D}_n$ for $n = rk(\mathcal{D})$ ends with $\mathcal{D}_n = \emptyset$. The index of a DBox $\mathcal{D}_i$ in this sequence is referred to as its (level of) *typicality*: a lower i indicates higher typicality of $\mathcal{D}_i$.

Canonical models for classical $\mathcal{EL}$ [1], represent every concept F that occurs in an existential restriction in $\mathcal{T}$, i.e., $F \in Qc(\mathcal{K})$ by a single domain element $d_F \in \Delta$. One concept F can be used in several existential restrictions inducing different role-successors with different typicality. There are up to $rk(\mathcal{D}) = n$ different levels of typicality to be reflected in the model.

Definition 1. *Let* $\mathcal{K} = (\mathcal{T}, \mathcal{D})$ *be a DKB with* $rk(\mathcal{D}) = n$. *A complete typicality domain is defined as* $\Delta^{\mathcal{K}} = \bigcup_{i=0}^{n}\{d_F^i \mid F \in Qc(\mathcal{K})\}$. *A domain* Δ *with* $\{d_F^n \mid F \in Qc(\mathcal{K})\} \subseteq \Delta \subseteq \Delta^{\mathcal{K}}$ *is a typicality domain. An interpretation over a typicality domain is a typicality interpretation. A typicality interpretation* $\mathcal{I} = (\Delta^{\mathcal{I}}, \cdot^{\mathcal{I}})$ *is a model of* $\mathcal{K}$ *(written* $\mathcal{I} \models \mathcal{K}$) *iff* $\mathcal{I} \models \mathcal{T}$ *and* $d_F^i \in G^{\mathcal{I}} \implies d_F^i \in H^{\mathcal{I}}$ *for all* $G \sqsubsetneq H \in \mathcal{D}_i$, $0 \le i \le rk(\mathcal{D})$, $F \in Qc(\mathcal{K})$.

We specify when defeasible subsumption relationships hold in typicality interpretations.

Definition 2. *Let* $\mathcal{I}$ *be a typicality interpretation. Then* $\mathcal{I}$ *satisfies a defeasible subsumption* $C \sqsubsetneq D$ *(written* $\mathcal{I} \models C \sqsubsetneq D$) *iff* $d_C^i \in D^{\mathcal{I}}$ *for* $0 \le i \le n$ *s.t.* $d_C^i \in \Delta^{\mathcal{I}}$ *and* $d_C^{i-1} \notin \Delta^{\mathcal{I}}$.

To construct a model for $\mathcal{K}$ by means of a TBox, the auxiliary names from $N_C^{aux} \subseteq N_C \setminus sig(\mathcal{K})$ introduce representatives for $F \in Qc(\mathcal{K})$ on each level of typicality. We use $F_\mathcal{D} \in N_C^{aux}$ to define the *extended TBox* of concept F w.r.t. $\mathcal{D}$: $\mathcal{T}_\mathcal{D}(F) = \mathcal{T} \cup \{F_\mathcal{D} \sqsubseteq F\} \cup \{F_\mathcal{D} \sqcap G \sqsubseteq H \mid G \sqsubsetneq H \in \mathcal{D}\}$. Here $\{F_\mathcal{D} \sqsubseteq F\}$ propagates all constraints on F to $F_\mathcal{D}$. The last set of GCIs is an equivalent $\mathcal{EL}_\bot$-rewriting of $\{F_\mathcal{D} \sqsubseteq \neg G \sqcup H \mid G \sqsubsetneq H \in \mathcal{D}\}$. We have shown in [9] that subsumptions w.r.t. classical TBoxes and subsumptions for auxiliary names w.r.t. TBoxes extended by $D_n = \emptyset$ coincide.

Proposition 3. *Let* $sig(G) \cap N_C^{aux} = \emptyset$. *Then* $F \sqsubseteq_\mathcal{T} G$ *iff* $F_\emptyset \sqsubseteq_{\mathcal{T}_\emptyset(F)} G$.

To use typicality interpretations for reasoning under materialisation-based rational entailment, $\mathcal{D}_i$ needs to be satisfied at the elements on typicality level i, but not (necessarily) for their role successors. It is indeed sufficient to construct a typicality interpretation with minimal typical role successors induced by existential restrictions that satisfy only $\mathcal{T}$ and $\mathcal{D}_n = \emptyset$.

Definition 4. *The minimal typicality model of* $\mathcal{K}$ *is* $\mathfrak{L}_\mathcal{K} = (\Delta^{\mathfrak{L}_\mathcal{K}}, \cdot^{\mathfrak{L}_\mathcal{K}})$, *where the domain is* $\Delta^{\mathfrak{L}_\mathcal{K}} = \{d_F^i \in \Delta^{\mathcal{K}} \mid F_{\mathcal{D}_i} \not\sqsubseteq_{\mathcal{T}_{\mathcal{D}_i}(F)} \bot\}$ *and* $\cdot^{\mathfrak{L}_\mathcal{K}}$ *satisfies for all*

elements $d_F^i \in \Delta^{\mathfrak{L}_\mathcal{K}}$ *with* $0 \leq i \leq n = rk(\mathcal{D})$ *both of the following conditions:*
(i) $d_F^i \in A^{\mathfrak{L}_\mathcal{K}}$ *iff* $F_{\mathcal{D}_i} \sqsubseteq_{\mathcal{T}_{\mathcal{D}_i}(F)} A$*, for* $A \in sig_{N_C}(\mathcal{K})$ *and (ii)* $(d_F^i, d_G^n) \in r^{\mathfrak{L}_\mathcal{K}}$ *iff* $F_{\mathcal{D}_i} \sqsubseteq_{\mathcal{T}_{\mathcal{D}_i}(F)} \exists r.G$*, for* $r \in sig_{N_R}(\mathcal{K})$.

Typicality models need not use the complete typicality domain due to inconsistencies. They are models of DKBs (cf. Definition 1) as we have shown in [9]. Proposition 3 implies that $\mathfrak{L}_\mathcal{K}$, restricted to elements of typicality n, is the canonical model for an $\mathcal{EL}_\perp$-TBox $\mathcal{T}$. Next, we characterise different entailment relations based on different kinds of typicality models which vary in the typicality admitted for required role successors. First, we use minimal typicality models to characterize entailment of propositional nature $\models_p$, where all role successors are uniformly non-typical.

Definition 5. *Let* $\mathcal{K}$ *be a DKB.* $\mathcal{K}$ *propositionally entails a defeasible subsumption relationship* $C \sqsubseteq\!\!\!\sim D$ *(written* $\mathcal{K} \models_p C \sqsubseteq\!\!\!\sim D$*) iff* $\mathfrak{L}_\mathcal{K} \models C \sqsubseteq\!\!\!\sim D$.

Deciding entailments of propositional nature by computing the extended TBox for a concept F and deriving its minimal typicality model, coincides with enriching F with the materialisation of $\mathcal{D}$.

Lemma 6. *Let* $sig(X) \cap N_C^{aux} = \emptyset$ *for* $X \in \{\mathcal{T}, \mathcal{D}, C, D\}$*. Then* $\overline{\mathcal{D}} \sqcap C \sqsubseteq_\mathcal{T} D$ *iff* $C_\mathcal{D} \sqsubseteq_{\mathcal{T}_\mathcal{D}(C)} D$.

Proof (sketch). The proof is by induction on $|\mathcal{D}|$. In the base case $\mathcal{D} = \emptyset$ and Proposition 3 holds. For the induction step, let $\mathcal{D}' = \mathcal{D} \cup \{G \sqsubseteq\!\!\!\sim H\}$ and distinguish: (i) $\overline{\mathcal{D}} \sqcap C \sqsubseteq G$ and (ii) $\overline{\mathcal{D}} \sqcap C \not\sqsubseteq G$. For case (i), we show: reasoning with $C \sqcap H$, yields the same consequences as with C. All elements in $C \sqcap H$ satisfying G, already satisfy H. Hence, $G \sqsubseteq\!\!\!\sim H$ can be removed from $\mathcal{D}'$ and the induction hypothesis holds. In case (ii) $G \sqsubseteq\!\!\!\sim H$ has no effect on reasoning: since $\overline{\mathcal{D}} \sqcap C \not\sqsubseteq G$, some element in C, satisfying all DCIs in D, does not satisfy G. Thus, $G \sqsubseteq\!\!\!\sim H$ does not affect reasoning and allows application of the induction hypothesis directly.

We denote the result of the reasoning algorithm from [3] as *materialisation-based rational entailment* (written $\mathcal{K} \models_m C \sqsubseteq\!\!\!\sim D$). This algorithm and typicality interpretations use the same DBox sequence. Although $\models_p$ and $\models_m$ are defined on different semantics, they derive the same subsumption relationships.

Theorem 7. $\mathcal{K} \models_p C \sqsubseteq\!\!\!\sim D$ *iff* $\mathcal{K} \models_m C \sqsubseteq\!\!\!\sim D$.

The crucial point in the proof of Theorem 7 is the use of Lemma 6. We have established an alternative characterisation for materialisation-based rational entailment by minimal typicality models. It ignores defeasible knowledge relevant to existential restrictions which motivates the extension of typicality models with role successors of the same concept, but of varying typicality.

3 Maximal Typicality Models for Rational Entailment

In materialisation-based rational entailment concepts implying $\exists r.C$ are *all* represented by a single element d_C^i allowing for one degree of (non-)typicality. Now, to obtain models where each role successor is of the highest (consistent) level of typicality, we transform minimal typical models s.t. each role edge is copied, but its endpoint, say d_H^i, is exchanged for a more typical representative d_H^{i-1}.

Definition 8. *Let $\mathcal{I}$ be a typicality interpretation over a DKB $\mathcal{K}$. The set of more typical role edges for a given role r is defined as $TR_{\mathcal{I}}(r) = \{(d_G^i, d_H^j) \in \Delta^{\mathcal{I}} \times \Delta^{\mathcal{I}} \setminus r^{\mathcal{I}} \mid \exists k > 0. (d_G^i, d_H^{j+k}) \in r^{\mathcal{I}}\}$. Let $\mathcal{I}$ and $\mathcal{J}$ be typicality interpretations. $\mathcal{J}$ is a typicality extension of $\mathcal{I}$ iff (i) $\Delta^{\mathcal{J}} = \Delta^{\mathcal{I}}$, (ii) $A^{\mathcal{J}} = A^{\mathcal{I}}$ (for $A \in N_C$), (iii) $r^{\mathcal{J}} = r^{\mathcal{I}} \cup R$ (for $r \in sig_{N_R}(\mathcal{K})$ and $R \subseteq TR_{\mathcal{I}}(r)$), and (vi) $\exists r \in sig_{N_R}(\mathcal{K}). r^{\mathcal{I}} \subset r^{\mathcal{J}}$. The set of all typicality extensions of $\mathcal{I}$ is $typ(\mathcal{I})$.*

Unfortunately, typicality extensions do not preserve the property of being a typicality model, since the increased typicality of the successor can invoke new concept memberships of the role *predecessor* which in turn can necessitate new additions in order to be compliant with all GCIs from $\mathcal{T}$. We formalize the required additions to obtain a model again by *model completions*.

Definition 9. *Let $\mathcal{K}$ be a DKB with $n = rk(\mathcal{D})$ and Δ a typicality domain. An interpretation $\mathcal{I} = (\Delta, \cdot^{\mathcal{I}})$ is a model completion of an interpretation $\mathcal{J} = (\Delta, \cdot^{\mathcal{J}})$ iff (i) $\mathcal{J} \subseteq \mathcal{I}$, (ii) $\mathcal{I} \models \mathcal{K}$, and (iii) $\forall E \in Qc(\mathcal{K}). d_F^i \in (\exists r.E)^{\mathcal{I}} \implies (d_F^i, d_E^n) \in r^{\mathcal{I}}$ (for any $F \in Qc(\mathcal{K})$ and $0 \le i \le n$). The set of all model completions of $\mathcal{J}$ is denoted as $mc(\mathcal{J})$.*

Note, that model completions introduce role successors only on typicality level n which can necessitate typicality extensions again. Thus typicality extensions and model completions are applied alternately until a fixpoint is reached. Clearly neither Definition 8 nor 9 yield a unique extension or completion. Thus, one (full) upgrade step is executed by an operator T working on sets of typicality interpretations. E.g. applying T to the singleton set $\{\mathcal{I}\}$ results in a set of all possible model completions of all possible typicality extensions of $\mathcal{I}$ (one upgrade step). The fixpoint of T is denoted as $typ^{max}()$, it collects the maximal typicality interpretations for a given input (c.f. [9]). There are several ways to use these sets of obtained maximal typicality models for our new entailment. Since in classical DL reasoning entailment considers *all* models, we employ cautious reasoning. We use a single model that is canonical in the sense that it is contained in all maximal typicality models obtained from $\mathfrak{L}_{\mathcal{K}}$. The *rational canonical model* $\mathfrak{R}_{\mathcal{K}}$ is defined as $\mathfrak{R}_{\mathcal{K}} = \bigcap_{\mathcal{I} \in typ^{max}(\{\mathfrak{L}_{\mathcal{K}}\})} \mathcal{I}$. This intersection is well-defined as $typ^{max}(\{\mathfrak{L}_{\mathcal{K}}\})$ is finite and $\mathfrak{L}_{\mathcal{K}} \in mc(\mathfrak{L}_{\mathcal{K}})$ is not empty, see [9].

Lemma 10. *The rational canonical model $\mathfrak{R}_{\mathcal{K}}$ is a model of the DKB $\mathcal{K}$.*

Basis of the proof of Lemma 10 is that the intersection of models satisfying Conditions (i)–(iii) in Definition 9 will satisfy the same conditions. The rational canonical model is used to decide nested rational entailments of the form

$\exists r.C \sqsubseteq_{\mathcal{K}} \exists r.D$, which requires to propagate DCIs to the instances of C and D. For a DKB $\mathcal{K}$, we capture (quantifier aware) *nested rational entailment* as $\mathcal{K} \models_q C \sqsubseteq D$ iff $\mathfrak{R}_{\mathcal{K}} \models C \sqsubseteq D$.

Theorem 11. *Let $\mathcal{K}$ be a DKB and C, D concepts. Then (i) $\mathcal{K} \models_m C \sqsubseteq D \implies \mathcal{K} \models_q C \sqsubseteq D$, and (ii) $\mathcal{K} \models_m C \sqsubseteq D \nLeftarrow \mathcal{K} \models_q C \sqsubseteq D$.*

To show Claim (i) of Theorem 11 observe that $\mathcal{J} \in typ^{max}(\{\mathfrak{L}_{\mathcal{K}}\})$ implies that $\mathfrak{L}_{\mathcal{K}} \subseteq \mathcal{J}$ and thus $\mathfrak{L}_{\mathcal{K}} \subseteq \mathfrak{R}_{\mathcal{K}}$. Hence by Definition 2, $\mathfrak{L}_{\mathcal{K}} \models C \sqsubseteq D$ implies $\mathfrak{R}_{\mathcal{K}} \models C \sqsubseteq D$. Claim (ii) is shown by an example in [9].

Theorem 11 shows that our approach produces results that satisfy the rational reasoning postulates as introduced by KLM [7] and lifted to DLs as presented in [3], as it gives strictly more entailments than the materialisation approach. The additional entailments are compliant with the argument of Lehmann and Magidor [8] that implications inferred from conditional knowledge bases should *at least* satisfy the postulates for rational reasoning.

4   Concluding Remarks

We have proposed a new approach to characterize entailment under rational closure (for deciding subsumption) in the DDL $\mathcal{EL}_\perp$ motivated by the fact that earlier reasoning procedures do not treat existential restrictions adequately. The key idea is to extend canonical models such that for each concept from the DKB, several copies representing different typicality levels of the respective concept are introduced. In minimal typicality models the role successors are "non-typical" in the sense that they satisfy only the GCIs from the TBox. Such models can be computed by a reduction to classical TBox reasoning by extended TBoxes. We showed that the entailments obtained from minimal typical models coincide with those obtained by materialisation. For maximal typicality models, where role successors are of "maximal typicality", DCIs are propagated to each role successor individually, thus allowing for more entailments. Existential restrictions are disregarded in several materialisation-based algorithms for defeasible reasoning in DLs, such as lexicographic [5] and relevant [3] closure. It is future work to extend our approach to these more sophisticated semantics.

References

1. Brandt, S.: Polynomial time reasoning in a description logic with existential restrictions, GCI axioms, and–what else? In: Proceedings of ECAI 2004, pp. 298–302 (2004)
2. Britz, A., Casini, G., Meyer, T., Moodley, K., Varzinczak, I.: Ordered interpretations and entailment for defeasible description logics. CAIR & UKZN, Technical report (2013)
3. Casini, G., Meyer, T., Moodley, K., Nortjé, R.: Relevant closure: a new form of defeasible reasoning for description logics. In: Fermé, E., Leite, J. (eds.) JELIA 2014. LNCS (LNAI), vol. 8761, pp. 92–106. Springer, Cham (2014). doi:10.1007/978-3-319-11558-0_7

4. Casini, G., Straccia, U.: Rational closure for defeasible description logics. In: Janhunen, T., Niemelä, I. (eds.) JELIA 2010. LNCS (LNAI), vol. 6341, pp. 77–90. Springer, Heidelberg (2010). doi:10.1007/978-3-642-15675-5_9
5. Casini, G., Straccia, U.: Lexicographic closure for defeasible description logics. In: Proceedings of the 8th Australasian Ontology Workshop, pp. 28–39 (2012)
6. Giordano, L., Gliozzi, V., Olivetti, N., Pozzato, G.L.: Semantic characterization of rational closure: from propositional logic to description logics. Artif. Intell. **226**, 1–33 (2015)
7. Kraus, S., Lehmann, D.J., Magidor, M.: Nonmonotonic reasoning, preferential models and cumulative logics. Artif. Intell. **44**(1–2), 167–207 (1990)
8. Lehmann, D., Magidor, M.: What does a conditional knowledge base entail? Artif. Intell. **55**, 1–60 (1989)
9. Pensel, M., Turhan, A.-Y.: Including quantification in defeasible reasoning for the description logic $\mathcal{EL}_\perp$. LTCS-Report 17–01, TU Dresden (2017). http://lat.inf.tu-dresden.de/research/reports.html

A Monotonic View on Reflexive Autoepistemic Reasoning

Ezgi Iraz Su$^{(\boxtimes)}$

CMAF-CIO, University of Lisbon, 1749-016 Lisbon, Portugal
eirsu@fc.ul.pt

Abstract. This paper introduces a novel monotonic modal logic, able to characterise reflexive autoepistemic reasoning of the nonmonotonic variant of modal logic **SW5**: we add a second new modal operator into the original language of **SW5**, and show that the resulting formalism called **M**R**AE*** is strong enough to capture the *minimal model* notion underlying some major forms of nonmonotonic logic among which are autoepistemic logic, default logic, and nonmonotonic logic programming. The paper ends with a discussion of a general strategy, naturally embedding several nonmonotonic logics of similar kinds.

Keywords: Nonmonotonic modal logic · **SW5** · Reflexive autoepistemic reasoning · Minimal model

1 Introduction

Autoepistemic reasoning is an important form of nonmonotonic reasoning [10], introduced by Moore in order to allow an agent to reason about his own knowledge [11]. Autoepistemic logic (**AEL**) [3] extends classical logic by a (non-monotonic)n epistemic modal operator $\Box$ (often written 'L' in the literature), in which $\Box\alpha$ is read: "α is believed". Logical consequence in **AEL** then builds on the notion of *stable expansion* [18], which corresponds to an "acceptable" set of possible beliefs of an agent. In 1991, Schwarz introduced *reflexive* autoepistemic logic (R**AEL**), that models knowledge [13,15] rather than belief. Both **AEL** and R**AEL** are strongly related to other important forms of nonmonotonic reasoning such as default logic [8] and nonmonotonic logic programming [7,9]. The minimal model notion [14] underlies all these forms of nonmonotonic reasoning [2,6,16,17].

In his 1991 paper, Schwarz also pointed out that **AEL** and R**AEL** are nothing, but respectively nonmonotonic variants of modal logics **KD45** and **SW5**.

E.I. Su—A similar work was given for nonmonotonic **KD45** in an unpublished paper by Luis Fariñas del Cerro, Andreas Herzig, and Levan Uridia, so special thanks go to them for motivation and some discussions. I am also grateful to the anonymous reviewers for their useful comments. This paper has been supported by the research unit "*Centro de Matemática, Aplicações Fundamentais e Investigação Operacional (CMAF-CIO)*" at the University of Lisbon, Portugal.

© Springer International Publishing AG 2017
M. Balduccini and T. Janhunen (Eds.): LPNMR 2017, LNAI 10377, pp. 85–100, 2017.
DOI: 10.1007/978-3-319-61660-5_10

Schwarz's frames are built as follows: we first determine the set of all possible worlds W. Then, we identify a nonempty subset $C \subseteq W$ (which will characterise '*agent's belief set*'). Finally, we relate each world in W to every world of C via an accessibility relation $\mathcal{T}$, and so define $\mathcal{T} = W \times C$. Thus, C appears to be a (maximal) $\mathcal{T}$-*cluster*[1]. His reflexive frames are then obtained by replacing $\mathcal{T}$ with its reflexive closure. In this paper, we first extend Schwarz's reflexive (**SW5**) frames [15] with a second relation $\mathcal{S}$, interpreting a new modal operator [S]. In our frame structure, $\mathcal{T}$ interprets [T], which is a direct (Gödel) translation of $\square$ in the language of **SW5**. We call the resulting (monotonic) formalism **MRAE**: *modal logic of reflexive autoepistemic reasoning*. We then extend **MRAE** with the *negatable axiom*, resulting in **MRAE***. The negatable axiom ensures that $W \setminus C$ is nonempty. It so turns our frames into exactly *two-floor* structures, among which the second is the maximal $\mathcal{T}$-cluster C. Essentially, this axiom makes it possible to falsify any non-tautological formula of the language because it allows us to have all possible valuations in an **MRAE*** model. Thus, we show that the modal formula $\langle T \rangle [T](\alpha \wedge [S]\neg\alpha)$ characterises maximal α-clusters in **MRAE***. This result paves the way to our final goal in which we capture RAE reasoning in monotonic **MRAE***:

> β is a logical consequence of α in nonmonotonic **SW5**   if and only if
> $$[T]\big(tr(\alpha) \wedge [S]\neg tr(\alpha)\big) \to [T]tr(\beta) \text{ is valid in } \mathbf{MRAE}^*.$$

As a result, the nonmonotonic deduction mechanism turns into a validity proving procedure in a monotonic (modal) logic.

The rest of the paper is organised as follows. Section 2 introduces a new monotonic modal logic called **MRAE**: we first define its bimodal language, and then propose two classes of frames, namely **K** and **F**. They are respectively based on standard Kripke frames, and the component frames in the form of a floor structure. We next axiomatise the validities of our logic, and finally prove that **MRAE** is sound and complete w.r.t. both semantics. In Sect. 3, we extend **MRAE** with the negatable axiom, and call the resulting logic **MRAE***. We introduce two kinds of model structures, **K*** and **F***, and end with the soundness and completeness results. Section 3.2 recalls reflexive autoepistemic logic (or nonmonotonic **SW5**): we define the preference relation over models of an **SW5** formula, and choose its minimal model according to this order. Section 3.3 gives the main result of the paper: an embedding of the consequence relation of **SW5** into **MRAE***. Finally, Sect. 4 makes a brief overview of this paper, and mentions our future goals. They all appear in a line of work that aims to reexamine the logical foundations of nonmonotonic logics.

---

[1] A cluster is simply an **S5** model in which the accessibility relation is an equivalence relation.

2 Modal Logic of Reflexive Autoepistemic Reasoning: MRAE

In this section, we propose a novel monotonic modal logic called *modal logic of reflexive autoepistemic reasoning*, and denote it by **MRAE**. We start with the language.

2.1 Language ($\mathcal{L}_{[\mathrm{T}],[\mathrm{S}]}$)

Throughout the paper we suppose $\mathbb{P}$ an infinite set of propositional variables, and $\mathbb{P}_\varphi$ its restriction to the propositional variables of a formula φ. We also consider *Prop* as the set of all propositional formulas of our language.

The language $\mathcal{L}_{[\mathrm{T}],[\mathrm{S}]}$ is then defined by the following grammar: (for $p \in \mathbb{P}$)

$$\varphi ::= p \mid \neg\varphi \mid \varphi \to \varphi \mid [\mathrm{T}]\varphi \mid [\mathrm{S}]\varphi.$$

$\mathcal{L}_{[\mathrm{T}],[\mathrm{S}]}$ is so a bimodal language with modalities [T] and [S]. When we restrict $\mathcal{L}_{[\mathrm{T}],[\mathrm{S}]}$ to [T] and the Boolean connectives, we call the result $\mathcal{L}_{[\mathrm{T}]}$.

Finally, $\mathcal{L}_{[\mathrm{T}],[\mathrm{S}]}$ has the following standard abbreviations: $\top \stackrel{\mathrm{def}}{=} \varphi \to \varphi$, $\bot \stackrel{\mathrm{def}}{=} \neg(\varphi \to \varphi)$, $\varphi \lor \psi \stackrel{\mathrm{def}}{=} \neg\varphi \to \psi$, $\varphi \land \psi \stackrel{\mathrm{def}}{=} \neg(\varphi \to \neg\psi)$, and $\varphi \leftrightarrow \psi \stackrel{\mathrm{def}}{=} (\varphi \to \psi) \land (\psi \to \varphi)$. Moreover, $\langle\mathrm{T}\rangle\varphi$ and $\langle\mathrm{S}\rangle\varphi$ respectively abbreviate $\neg[\mathrm{T}]\neg\varphi$ and $\neg[\mathrm{S}]\neg\varphi$.

2.2 Kripke Semantics for MRAE

We first describe Kripke frames for **MRAE**, and call this class **K**. A **K**-frame is a Kripke frame $(W, \mathcal{T}, \mathcal{S})$ in which

- W is a non-empty set of possible worlds;
- $\mathcal{T}$ and $\mathcal{S}$ are binary relations on W satisfying the following conditions:

refl($\mathcal{T}$)	for every w, $w\mathcal{T}w$;
trans($\mathcal{T}$)	for every w, u, v, if $w\mathcal{T}u$ and $u\mathcal{T}v$ then $w\mathcal{T}v$;
wEuclid($\mathcal{T}$)	for every w, u, v, if $w\mathcal{T}u$ and $w\mathcal{T}v$ then "$w = u$ or $w = v$ or $u\mathcal{T}v$";
refl$_2$($\mathcal{S}$)	for every w, u, if $w\mathcal{S}u$ then $u\mathcal{S}u$;
wtriv$_2$($\mathcal{S}$)	for every w, u, v, if $w\mathcal{S}u$ and $u\mathcal{S}v$ then $u = v$;
msym($\mathcal{T}, \mathcal{S}$)	for every w, u, if $w\mathcal{T}u$ then "$u\mathcal{T}w$ or $u\mathcal{S}w$";
wmsym($\mathcal{S}, \mathcal{T}$)	for every w, u, if $w\mathcal{S}u$ then "$w = u$ or $u\mathcal{T}w$".

The first three properties are well known to characterise the frames of **SW5** [1], so a **K**-frame is an extension of an **SW5** frame by a second relation $\mathcal{S}$. Note also that trans($\mathcal{T}$) and wEuclid($\mathcal{T}$) give us an important new property ($\mathcal{T}$-constraint) of such frames:

floor$_2$($\mathcal{T}$) : for every w, u, if $w\mathcal{T}u$ then "$u\mathcal{T}w$ or $v\mathcal{T}u$ for every v with $u\mathcal{T}v$".

$$(1)$$

Given a **K**-frame $\mathcal{F} = (W, \mathcal{T}, \mathcal{S})$, a **K**-model is a pair $\mathcal{M} = (\mathcal{F}, V)$ in which $V : W \to 2^{\mathbb{P}}$ is a *valuation* function, assigning to each $w \in W$ a set of propositional variables. Then, a pointed **K**-model is a pair $(\mathcal{M}, w) = ((W, \mathcal{T}, \mathcal{S}, V), w)$ in which $w \in W$.

Truth Conditions. The truth conditions are standard: for $p \in \mathbb{P}$,

$$\mathcal{M}, w \models p \qquad \text{if } p \in V(w);$$
$$\mathcal{M}, w \models \neg\varphi \qquad \text{if } \mathcal{M}, w \not\models \varphi;$$
$$\mathcal{M}, w \models \varphi \to \psi \quad \text{if } \mathcal{M}, w \not\models \varphi \text{ or } \mathcal{M}, w \models \psi;$$
$$\mathcal{M}, w \models [\text{T}]\varphi \qquad \text{if } \mathcal{M}, u \models \varphi \text{ for every } u \text{ such that } w\mathcal{T}u;$$
$$\mathcal{M}, w \models [\text{S}]\varphi \qquad \text{if } \mathcal{M}, u \models \varphi \text{ for every } u \text{ such that } w\mathcal{S}u.$$

We say that φ is *satisfiable* in **MRAE** if φ has a model $\mathcal{M}$ in **MRAE**, i.e., $\mathcal{M}, w \models \varphi$ for some model $\mathcal{M}$ and a world w in $\mathcal{M}$. Then, φ is *valid* in **MRAE** if $\mathcal{M}, w \models \varphi$ for every world w of every model $\mathcal{M}$ (for short, $\models \varphi$). We also say that φ is valid in $\mathcal{M}$ when $\mathcal{M}, w \models \varphi$ for every w in $\mathcal{M}$, and we write $\mathcal{M} \models \varphi$. The following proposition says that while checking satisfiability, it is enough to only consider models with finite valuations.

Proposition 1. *Given an $\mathcal{L}_{[\text{T}],[\text{S}]}$ formula φ and a **K**-model $\mathcal{M} = (W, \mathcal{T}, \mathcal{S}, V)$, let the valuation V^φ be defined as follows: for every $w \in W$,*

$$V_w^\varphi = V_w \cap \mathbb{P}_\varphi.$$

*Then $\mathcal{M}^\varphi = (W, \mathcal{T}, \mathcal{S}, V^\varphi)$ is also a **K**-model and for every $w \in W$,*

$$\mathcal{M}, w \models \varphi \text{ if and only if } \mathcal{M}^\varphi, w \models \varphi.$$

2.3 Cluster-Based Floor Semantics for MRAE

We here define the frames of a floor structure for **MRAE**, and call this class **F**. The underlying idea is due to the property 'floor$_2(\mathcal{T})$' (see $\mathcal{T}$-constraint (1)) of **K**-frames, so **F** is in fact a subclass of **K**. However, **F**-frames with some additional properties play an important role in the completeness proof. We start with the definition of a cluster [1].

Definition 1. *Given a **K**-frame $(W, \mathcal{T}, \mathcal{S})$, let $C \subseteq W$ be a nonempty subset. Then,*

- *C is a $\mathcal{T}$-cluster if $w\mathcal{T}u$ for every $w, u \in C$;*
- *C is maximal if for every $w \in W$, and every $u \in C$, $u\mathcal{T}w$ implies $w \in C$;*
- *C is final if $w\mathcal{T}u$ for every $w \in W$ and every $u \in C$.*

It follows from Definition 1 that the restriction of $\mathcal{T}$ to a $\mathcal{T}$-cluster[2] C (abbreviated $\mathcal{T}|_C$) is a trivial (equivalence) relation. So, $(C, \mathcal{T})$ happens to be an **S5** frame.

---

[2] Unless specified otherwise, any definition of this paper is given w.r.t. the relation $\mathcal{T}$.

Given a **K**-frame $\mathcal{F} = (W, \mathcal{T}, \mathcal{S})$, $\mathcal{T}$ partitions $\mathcal{F}$ into disjoint subframes $\mathcal{F}' = (W', \mathcal{T}, \mathcal{S})$ in which $\mathcal{T}|_{W'} = (W' \times C) \cup \Delta_{W'}{}^3$ for some cluster C such that $C \subseteq W' \subseteq W$. Each $\mathcal{F}'$ contains a unique maximal cluster C, which is final in $\mathcal{F}'$. Thus, for distinct $w, u, v \in W$, if $w\mathcal{T}u$ and $w\mathcal{T}v$, then u and v should belong to the same cluster C, and w is in $\mathcal{F}'$, specified by C. When $W' = C$, then $\mathcal{F}'$ corresponds to the cluster itself, and $\mathcal{T}|_{W'}$ is an equivalence relation. We now define two operators $\mathcal{T}, \mathcal{S} : 2^W \longrightarrow 2^W$ with

$$\mathcal{T}(X) = \{u \in W : w\mathcal{T}u \text{ for some } w \in X \subseteq W\};$$
$$\mathcal{S}(X) = \{u \in W : w\mathcal{S}u \text{ for some } w \in X \subseteq W\}.$$

When $X = \{w\}$, we simply write $\mathcal{T}(w)$, denoting the set of worlds that w can see via $\mathcal{T}$.

Proposition 2. *Given a **K**-frame $\mathcal{F} = (W, \mathcal{T}, \mathcal{S})$ and $w \in W$,*

1. *if $w \in C$ for some cluster C in $\mathcal{F}$, then $\mathcal{T}(w)$ is the maximal cluster containing C;*
2. *if $w \notin C$ for any cluster C in $\mathcal{F}$, then $\mathcal{T}(w) \setminus \{w\} = C$ is a maximal cluster C in $\mathcal{F}$.*

The proof immediately follows from the constraints $\text{refl}(\mathcal{T})$, $\text{trans}(\mathcal{T})$, and $\text{wEuclid}(\mathcal{T})$. It is also easy to check that both $\mathcal{T}(\cdot)$ and $\mathcal{S}(\cdot)$ are homomorphisms under set union:

$$\mathcal{T}(X \cup Y) = \mathcal{T}(X) \cup \mathcal{T}(Y) \text{ and } \mathcal{S}(X \cup Y) = \mathcal{S}(X) \cup \mathcal{S}(Y).$$

We now formally define the above-mentioned partitions ('*components*') of a **K** frame.

Definition 2. *Given a **K**-frame $\mathcal{F} = (W, \mathcal{T}, \mathcal{S})$, let $\mathbb{C} = (C, A)$ be a pair of disjoint subsets of W (i.e., $C, A \subseteq W$ and $C \cap A = \emptyset$). Then, $\mathbb{C}$ is called a component of $\mathcal{F}$ if:*

1. *C is a (nonempty) maximal cluster;*
2. *$\mathcal{T} \cap (A \times C) = A \times C$;*
3. *$\mathcal{T} \cap (A \times A) = \Delta_A$ (where Δ_A is the diagonal of $A \times A$).*

So, a component $\mathbb{C} = (C, A)$ is an 'at most two-layered' structure: if $A \neq \emptyset$, then A is the *first floor*, and C is the *second floor*; but if $A = \emptyset$, then $\mathbb{C}$ is reduced to the cluster C itself. Note that $\mathbb{C}$ is, in fact, a special **K**-frame as it can be transformed into a **K**-frame

$$\mathcal{F}^{\mathbb{C}} = \Big(C \cup A, \; ((C \cup A) \times C) \cup \Delta_A, \; (C \times A) \cup \Delta_A\Big)$$

where Δ_A is the diagonal of $A \times A$. Any component $\mathbb{C}$ of $\mathcal{F}$ is so a subframe of $\mathcal{F}$ (w.r.t. $\mathcal{T}$). Given any two different components $\mathbb{C}_1$ and $\mathbb{C}_2$ of $\mathcal{F}$, $C_1 \cup A_1$ and

---

^{3} $\Delta_{W'}$ is the diagonal of $W' \times W'$, i.e., $\Delta_{W'} = \{(a, a) : a \in W'\}$.

$C_2 \cup A_2$ are disjoint, and $\mathbb{C}_1$ and $\mathbb{C}_2$ are disconnected in the sense that there is no $\mathcal{T}$-access (nor an $\mathcal{S}$-access) from one to the other. As a result, a **K**-frame $\mathcal{F}$ is composed of a (arbitrary) union of components except that $\mathcal{F}$ contains a component of a single cluster structure in which $\mathcal{S} \neq \emptyset$. Finally, an **F**-frame for **MRAE** is simply a component $\mathbb{C} = (C, A)$.

We now define a function $\mu : \mathbf{F} \to \mathbf{K}$, assigning a **K**-frame $\mu(\mathbb{C}) = \mathcal{F}^{\mathbb{C}}$ (as above) to each **F**-frame $\mathbb{C}$. Since two distinct components give rise to two distinct frames of **K**, μ is 1-1, but not onto[4]. As a consequence, **F** is indeed a (proper) subclass of **K**.

Proposition 3. *Given a **K**-frame $\mathcal{F} = (W, \mathcal{T}, \mathcal{S})$, let $C \subseteq W$ be a maximal cluster. Then, take $A = \mathcal{S}(C) \setminus C$, so we have the following:*

1. $(C, A) \in \mathbf{F}$;
2. $\mathcal{S}(C \cup A) \subseteq C \cup A$;
3. $\mathcal{T}(C \cup A) = C \cup A$.

Corollary 1. *Given a pointed **K**-frame $\mathcal{F}_w = (\mathcal{F}, w)$, let $C = \mathcal{T}(w)$ if w is in a cluster structure; otherwise, let $C = \mathcal{T}(w) \setminus \{w\}$. Then, take $A = \mathcal{S}(C) \setminus C$. So, $\mathbb{C}^{\mathcal{F}_w} = (C, A) \in \mathbf{F}$.*

Note that the component generated by $w \in \mathcal{F}$ is exactly the one in which w is placed. Therefore, any point from the same component $\mathbb{C}$, contained in $\mathcal{F}$, forms $\mathbb{C}$ itself. Using Corollary 1, we now define another function ν, assigning to each pointed **K**-frame $\mathcal{F}_w = (\mathcal{F}, w)$ an **F**-frame $\mathbb{C}^{\mathcal{F}_w}$. Clearly, ν is not 1-1, but is onto. Finally, $\{\nu(\mathcal{F}_w) : w \in W\}$ identifies all components $\mathbb{C}$ in $\mathcal{F}$. The next proposition generalises this observation.

Proposition 4. *For $\mathbb{C} = (C, A) \in \mathbf{F}$ and $w \in C \cup A$, we have $\nu(\mu(\mathbb{C}), w) = \mathbb{C}$.*

These transformations between frame structures of **MRAE** enable us to define valuations on the components $\mathbb{C} \in \mathbf{F}$ as well, resulting in an alternative semantics for **MRAE** via **F**-models. The new semantics can be viewed as a reformulation of the Kripke semantics: given a **K**-model $\mathcal{M} = (\mathcal{F}^{\mathbb{C}}, V)$ for some Kripke frame $\mathcal{F}^{\mathbb{C}} \in \mu(\mathbf{F})$ and a valuation V, one can transform $\mathcal{F}^{\mathbb{C}}$ to a component $\nu(\mathcal{F}_w^{\mathbb{C}}) = \nu(\mathcal{F}^{\mathbb{C}}, w) = \mathbb{C} \in \mathbf{F}$ for some $w \in \mathbb{C}$ (see Proposition 4). This discussion allows us to define pairs $(\mathbb{C}, V)$ in which $\mathbb{C} = (C, A) \in \mathbf{F}$, and V is a valuation restricted to $C \cup A$. We call such valuated components '***F**-models*', with which we can transfer **K**-satisfaction to **F**-satisfaction.

Truth Conditions. The truth conditions are the same as before, but we want to further clarify the modal cases: for an **F**-model $\mathcal{M} = (C, A, V)$ and $w \in C \cup A$, we have:

---

[4] Note that there is no **F**-frame being mapped to (i) a **K**-frame containing more than one cluster structure in it, and (ii) a **K**-frame composed of only a single cluster structure in which $\mathcal{S} \neq \emptyset$.

$\mathcal{M}, w \models [\mathrm{T}]\psi$ iff (i) if $w \in C$ then $\mathcal{M}, v \models \psi$ for every $v \in C$ and

(ii) if $w \in A$ then $\mathcal{M}, v \models \psi$ for every $v \in C \cup \{w\}$;

$\mathcal{M}, w \models [\mathrm{S}]\psi$ iff (i) if $w \in A$ then $\mathcal{M}, w \models \psi$ and (ii) if $w \in C$ then

(a) $A \neq \emptyset$ implies $\mathcal{M}, v \models \psi$ for every $v \in A$, yet

(b) $A = \emptyset$ gives us nothing except $\mathcal{M}, v \models \psi$ for every $v \in \mathcal{S}(w)$.

The following proposition shows the relation between two semantics of **MRAE**. This result is an immediate consequence of Proposition 4.

Proposition 5. *Let* $(\mathbb{C}, V)$ *be an* **F**-*model, and* w *be in* $\mathbb{C}$. *Then, we have for* $\varphi \in \mathcal{L}_{[\mathrm{T}], [\mathrm{S}]}$,

$$(\mathbb{C}, V), w \models \varphi \quad \text{if and only if} \quad (\mathcal{F}^{\mathbb{C}}, V), w \models \varphi.$$

2.4 Axiomatisation of MRAE

We here give an axiomatisation of **MRAE**, and then prove its completeness. Recall that K([T]), T([T]), 4([T]), and W5([T]) characterise the normal modal logic **SW5** [1]. The monotonic logic defined by Table 1 is **MRAE**. We observe that W5([T]) is equivalent to the schemas '$(\varphi \wedge \langle \mathrm{T} \rangle \neg \varphi) \rightarrow [\mathrm{T}]\langle \mathrm{T} \rangle \neg \varphi$' and '$\langle \mathrm{T} \rangle \varphi \rightarrow (\varphi \vee [\mathrm{T}]\langle \mathrm{T} \rangle \varphi)$', so we will use them interchangeably in the paper. The schemas $\mathrm{T}_2([\mathrm{S}])$ and $\mathrm{WTriv}_2([\mathrm{S}])$ can be combined into the single axiom $\mathrm{Triv}_2([\mathrm{S}])$, i.e., $[\mathrm{S}]([\mathrm{S}]\varphi \leftrightarrow \varphi)$, referring to the "triviality in the second $\mathcal{S}$-step". Finally, MB([T], [S]) and WMB([S], [T]) are familiar from tense logics.

Table 1. Axiomatisation of **MRAE**

K([T])	the axioms and the inference rules of modal logic **K** for [T]
K([S])	the axioms and the inference rules of modal logic **K** for [S]
T([T])	$[\mathrm{T}]\varphi \rightarrow \varphi$
4([T])	$[\mathrm{T}]\varphi \rightarrow [\mathrm{T}][\mathrm{T}]\varphi$
W5([T])	$\langle \mathrm{T} \rangle [\mathrm{T}]\varphi \rightarrow (\varphi \rightarrow [\mathrm{T}]\varphi)$
$\mathrm{T}_2([\mathrm{S}])$	$[\mathrm{S}]([\mathrm{S}]\varphi \rightarrow \varphi)$
$\mathrm{WTriv}_2([\mathrm{S}])$	$[\mathrm{S}](\varphi \rightarrow [\mathrm{S}]\varphi)$
MB([T], [S])	$\varphi \rightarrow [\mathrm{T}](\langle \mathrm{T} \rangle \varphi \vee \langle \mathrm{S} \rangle \varphi)$
WMB([S], [T])	$\varphi \rightarrow [\mathrm{S}](\varphi \vee \langle \mathrm{T} \rangle \varphi)$

2.5 Soundness and Completeness of MRAE

The axiom schemas in Table 1 precisely characterise **K**: we only show that W5([T]) describes the weak Euclidean property of $\mathcal{T}$, namely $\mathrm{wEuclid}(\mathcal{T})$, but the rest is similar.

– Let $\mathcal{M} = (W, \mathcal{T}, \mathcal{S}, V)$ be a **K**-model, satisfying wEuclid($\mathcal{T}$) ($*$). Let $w \in W$ be such that $\mathcal{M}, w \models \varphi \wedge \langle \mathrm{T} \rangle \neg \varphi$. Then, $\mathcal{M}, w \models \varphi$ and there exists $u \in W$ with $w\mathcal{T}u$ and $\mathcal{M}, u \models \neg \varphi$. Clearly, $u \neq w$. By the hypothesis ($*$), we have $u\mathcal{T}v$ (and also $v\mathcal{T}u$) for every $v \neq w$ such that $w\mathcal{T}v$. Hence, $\mathcal{M}, v \models \langle \mathrm{T} \rangle \neg \varphi$ for every $v \neq w$ with $w\mathcal{T}v$; but $\mathcal{M}, w \models \langle \mathrm{T} \rangle \neg \varphi$ too. As a result, $\mathcal{M}, v \models [\mathrm{T}]\langle \mathrm{T} \rangle \neg \varphi$, and so W5($[\mathrm{T}]$) is valid in $\mathcal{M}$.

– Let $\mathcal{F} = (W, \mathcal{T}, \mathcal{S})$ be a **K**-frame in which wEuclid($\mathcal{T}$) fails. So, there exists $w, u, v \in W$ such that $w \neq u$, $w \neq v$, and $(w, u), (w, v) \in \mathcal{T}$; yet $(u, v) \notin \mathcal{T}$ or $(v, u) \notin \mathcal{T}$. WLOG, take $(u, v) \notin \mathcal{T}$ ($*$). Then, we take a valuation V such that $\mathcal{M}, v \not\models \varphi$, and $\mathcal{M}, x \models \varphi$ for every $x \neq v$. Thus, $\mathcal{M}, w \models \varphi \wedge \langle \mathrm{T} \rangle \neg \varphi$, and also ($*$) implies that $\mathcal{M}, u \models [\mathrm{T}]\varphi$. Thus, $\mathcal{M}, w \models \langle \mathrm{T} \rangle [\mathrm{T}]\varphi$; in other words, $\mathcal{M}, w \not\models [\mathrm{T}]\langle \mathrm{T} \rangle \neg \varphi$. So, $\mathcal{M}, w \not\models (\varphi \wedge \langle \mathrm{T} \rangle \neg \varphi) \rightarrow [\mathrm{T}]\langle \mathrm{T} \rangle \neg \varphi$. As a result, W5($[\mathrm{T}]$) is not valid in $\mathcal{M}$.

Corollary 2. *The modal logic* **MRAE** *is sound w.r.t. the class of* **K**-*frames.*

It suffices to show that our inference rules are validity-preserving.

Theorem 1. *The modal logic* **MRAE** *is complete w.r.t. the class of* **K**-*frames.*

Proof. We use the method of canonical models [1]: we first define the canonical model $\mathcal{M}^c = (W^c, \mathcal{T}^c, \mathcal{S}^c, V^c)$ in which

– W^c is the set of maximal consistent sets of **MRAE**.
– $\mathcal{T}^c$ and $\mathcal{S}^c$ are the accessibility relations on W^c satisfying:

$$\Gamma \mathcal{T}^c \Gamma' \text{ if and only if } \{\psi : [\mathrm{T}]\psi \in \Gamma\} \subseteq \Gamma';$$
$$\Gamma \mathcal{S}^c \Gamma' \text{ if and only if } \{\psi : [\mathrm{S}]\psi \in \Gamma\} \subseteq \Gamma'.$$

– The valuation V^c is defined by $V^c(\Gamma) = \Gamma \cap \mathbb{P}$, for every $\Gamma \in W^c$.

By induction on φ, we prove a truth lemma saying that "$\Gamma \models \varphi$ iff $\varphi \in \Gamma$" for every φ in $\mathcal{L}_{[\mathrm{T}],[\mathrm{S}]}$. Then, it remains to show that $\mathcal{M}^c$ satisfies all constraints of **K** and so, is a legal **K**-model of **MRAE**. We here give the proof only for wtriv$_2$($\mathcal{S}$) and wmsym($\mathcal{S}, \mathcal{T}$).

– The schema WTriv$_2$($[\mathrm{S}]$) guarantees that $\mathcal{M}^c$ satisfies wtriv$_2$($\mathcal{S}$): let $\Gamma_1 \mathcal{S}^c \Gamma_2$ ($*$) and $\Gamma_2 \mathcal{S}^c \Gamma_3$ ($**$). Then, assume for a contradiction that $\Gamma_2 \neq \Gamma_3$. Thus, there exists $\varphi \in \Gamma_2$ with $\neg \varphi \in \Gamma_3$, implying $\langle \mathrm{S} \rangle \neg \varphi \in \Gamma_2$ by the hypothesis ($**$). Since Γ_2 is maximal consistent, $\varphi \wedge \langle \mathrm{S} \rangle \neg \varphi \in \Gamma_2$. Hence, using the hypothesis ($*$), we also get $\langle \mathrm{S} \rangle (\varphi \wedge \langle \mathrm{S} \rangle \neg \varphi) \in \Gamma_1$. However, since Γ_1 is maximal consistent, any instance of WTriv$_2$($[\mathrm{S}]$) is in Γ_1. Thus, $[\mathrm{S}](\varphi \rightarrow [\mathrm{S}]\varphi) \in \Gamma_1$, contradicting the consistency of Γ_1.

– The schema WMB($[\mathrm{S}], [\mathrm{T}]$) ensures that wmsym($\mathcal{S}, \mathcal{T}$) holds in $\mathcal{M}^c$: suppose that $\Gamma \mathcal{S}^c \Gamma'$ ($*$) for $\Gamma, \Gamma' \in W^c$. WLOG, let $\Gamma \neq \Gamma'$. Then, there exists $\psi \in \Gamma'$ with $\neg \psi \in \Gamma$. We need to show that $\Gamma' \mathcal{T}^c \Gamma$. So, let φ be such that $[\mathrm{T}]\varphi \in \Gamma'$. As Γ' is maximal consistent, and $\psi, [\mathrm{T}]\varphi \in \Gamma'$, we have both $\varphi \vee \psi \in \Gamma'$ and $[\mathrm{T}]\varphi \vee [\mathrm{T}]\psi \in \Gamma'$. We know that $[\mathrm{T}]\varphi \vee [\mathrm{T}]\psi \rightarrow [\mathrm{T}](\varphi \vee \psi)$ is a theorem of **MRAE**, so it is also an element of Γ'. Then, by Modus Ponens (MP), we get

$[T](\varphi \vee \psi) \in \Gamma'$, further implying $(\varphi \vee \psi) \wedge [T](\varphi \vee \psi) \in \Gamma'$ since we already have $(\varphi \vee \psi) \in \Gamma'$. Assumption $(*)$ gives us that $\langle S \rangle \big((\varphi \vee \psi) \wedge [T](\varphi \vee \psi) \big) \in \Gamma$. Since Γ is maximal consistent, any instance of WMB$([S], [T])$ is in Γ; in particular, so is $\langle S \rangle \big((\varphi \vee \psi) \wedge [T](\varphi \vee \psi) \big) \to (\varphi \vee \psi)$. Finally, again by MP, we have $\varphi \vee \psi \in \Gamma$. Due to $\neg \psi \in \Gamma$, it follows that $\varphi \in \Gamma$.

Theorem 2. *The modal logic* **MRAE** *is sound and complete w.r.t. the class* **F**.

Proof. Since every component $\mathbb{C} \in \mathbf{F}$ can be converted to a **K**-frame $\mu(\mathbb{C})$, soundness follows from Corollary 2. As for completeness, let $\varphi \in \mathcal{L}_{[T],[S]}$ be a nontheorem of **MRAE**, then $\neg \varphi$ is a consistent formula of our logic. Let $\Gamma_{\neg \varphi}$ be a maximal consistent set in the canonical model $\mathcal{M}^c$ (defined in Theorem 1) such that $\neg \varphi \in \Gamma_{\neg \varphi}$. We know that $\mathcal{M}^c$ is a **K**-model, so we form the component $\mathbb{C}^c = (C^c, A^c)$ with $\Gamma_{\neg \varphi} \in C^c \cup A^c$ as described in Corollary 1. Thus, $\mathbb{C}^c \in \mathbf{F}$. Moreover, by Proposition 4 and Proposition 5, modal satisfaction is preserved between $\mathcal{M}^c$ and $\mathbb{C}^c$. As a result $\mathbb{C}^c, \Gamma_{\neg \varphi} \not\models \varphi$.

3   Modal Logic MRAE*

In this section, we discuss an extension of **MRAE** with a new axiom schema

$$\text{Neg}([S], [T]): \quad [T]\theta \to \langle T \rangle \langle S \rangle \neg \theta$$

where $\theta \in Prop$ is nontautological. We call this schema *negatable axiom*, and represent the resulting formalism by **MRAE***. **MRAE***-models are of 2 kinds, namely **K*** and **F***. They are obtained respectively from **K** and **F** by adding a 'model' constraint

neg$(\mathcal{S}, \mathcal{T})$: for every $P \subseteq \mathbb{P}$, there exists a world w of each model such that $P = V(w)$.

In other words, **MRAE***-models can falsify any propositional nontheorem φ, i.e., for every such φ, there exists a world w of each model such that $w \models \neg \varphi$. As a result, every **F***-model $(\mathbb{C}, V) = (C, A, V)$ has exactly a 'two-floor' form: $A \neq \emptyset$, and A includes any world w at which any propositional nontheorem φ, valid in a cluster structure C, is refuted. Then, a **K***-model is an arbitrary combination of **F***-models. Below we show that Neg$([S], [T])$ precisely corresponds to neg$(\mathcal{S}, \mathcal{T})$.

Proposition 6. *Given a* **K***-model* $\mathcal{M} = (W, \mathcal{T}, \mathcal{S}, V)$,

Neg$([S], [T])$ is valid in $\mathcal{M}$ *if and only if* $\mathcal{M}$ *is a* **K***-model.*

Proof. Let $\mathcal{M} = (W, \mathcal{T}, \mathcal{S}, V)$ be a **K**-model of **MRAE**.
($\Longrightarrow$):   Assume that $\mathcal{M}$ is not a **K***-model. Then, there exists a nontautological propositional formula φ such that $\mathcal{M} \models \varphi$. Clearly, $[T]\varphi$, $[S]\varphi$ and $[T][S]\varphi$ are all valid in $\mathcal{M}$, but then so is $[T]\varphi \wedge [T][S]\varphi$. Thus, Neg$([S], [T])$ is not valid in $\mathcal{M}$.
($\Longleftarrow$):   Let $\mathcal{M}$ be a **K***-model $(\star)$. Let $\varphi \in Prop$ be a nontheorem. Take $\beta =$

$[T]\varphi \rightarrow \langle T\rangle\langle S\rangle\neg\varphi$. We need to show that $\mathcal{M} \models \beta$. Let $w \in W$ be such that $\mathcal{M}, w \models [T]\varphi$. We first consider the **F**-model $(\mathbb{C}, V) = (C, A, V)$ in which (C, A) is the **F**-frame, generated by w as in Corollary 1. By construction, φ is valid in C, and $(\star)$ implies an existence of $u \in A$ such that u refutes φ. By the frame properties of **F**, there exists $v \in C$ satisfying vSu and so $\mathcal{M}, v \models \langle S\rangle\neg\varphi$. No matter to which floor w belongs, $w\mathcal{T}v$. Thus, $\mathcal{M}, w \models \langle T\rangle\langle S\rangle\neg\varphi$.

Corollary 3. *For an* **F**-*model* $(\mathbb{C}, V) = (C, A, V)$,

$Neg([S], [T])$ *is valid in* $(\mathbb{C}, V)$ *if and only if* $(\mathbb{C}, V)$ *is an* **F***-*model.*

Although $Neg([S], [T])$ has an elegant representation, we find it handier to work with

$$Neg'([S], [T]): \quad [T]\varphi \rightarrow \langle T\rangle\langle S\rangle(\neg\varphi \wedge Q)$$

where $\varphi \in Prop$ is a nontheorem, and Q is a conjunction of a finite (possibly empty) set of literals (a propositional letter or its negation) such that the set $\{\neg\varphi, Q\}$ is consistent.

Proposition 7. *Given a* **K***-*model* $\mathcal{M} = (W, \mathcal{T}, \mathcal{S}, V)$ *and* $w \in W$,

$\mathcal{M}, w \models Neg([S], [T])$ *if and only if* $\mathcal{M}, w \models Neg'([S], [T])$.

Proof. The right-to-left direction is straightforward: take $Q = \emptyset$, and the result follows. For the opposite direction, we first assume that $\mathcal{M}, w \models Neg([S], [T])$ $(\star)$. Let $\varphi \in Prop$ be a nontheorem of **MRAE*** such that $\mathcal{M}, w \models [T]\varphi$ $(\star\star)$. Let Q be a conjunction of finite literals such that $\neg\varphi \wedge Q$ is consistent. Then, $\varphi \vee \neg Q \in Prop$ is a nontheorem of **MRAE***. Moreover, from the assumption $(\star\star)$, we also obtain $\mathcal{M}, w \models [T](\varphi \vee \neg Q)$. Lastly, by the hypothesis $(\star)$, we have $\mathcal{M}, w \models \langle T\rangle\langle S\rangle$ $(\neg\varphi \wedge Q)$.

We finally transform a valuated cluster (C, V) into an **F***-model. We first construct a set

$$A = \{x_\varphi : \text{for every } \varphi \in Prop \text{ such that } \neg\varphi \not\vdash \bot, \;\; (C, V) \models \varphi \text{ and } x_\varphi \notin C\}$$

into which we put a point $x_\varphi \notin C$ for every nontheorem $\varphi \in Prop$ that is valid in C. So, $A \cap C = \emptyset$. Then, we extend the equivalence relation on C to $\mathcal{T} = ((C \cup A) \times C) \cup \Delta_A$ on $C \cup A$. The valuation V defined over C is also extended to $V' : C \cup A \longrightarrow \mathbb{P}$ satisfying: "$V'|_C = V$ and $V'(x_\varphi)$ is designed to falsify φ". Obviously, $(C, A, V') \in$ **F***.

3.1   Soundness and Completeness of **MRAE***

We here show that **MRAE*** is sound and complete w.r.t. **F***: soundness follows from Theorem 2 and Corollary 3. To see completeness, we first take a canonical model $\mathcal{M}^c = (W^c, \mathcal{T}^c, \mathcal{S}^c, V^c)$ of **MRAE*** (see Theorem 1 for the details). Then, we define a valuated component $(\mathbb{C}^c, V^c) = (C^c, A^c, V^c)$ for $C^c, A^c \subseteq W^c$ as in

Sect. 2.5. We want to show that $(\mathbb{C}^c, V^c)$ is indeed an $\mathbf{F}^*$-model. So, it is enough to prove that the axiom schema $\mathrm{Neg}([S], [T])$ ensures the property $\mathrm{neg}(\mathcal{S}, \mathcal{T})$. To start with, we know that every $\mathbf{F}$-frame $\mathbb{C}$ corresponds a $\mathbf{K}$-frame $\mu(\mathbb{C}) = \mathcal{F}^{\mathbb{C}}$, and by Proposition 4, $\nu(\mu(\mathbb{C}), w) = \mathbb{C}$ for $w \in C \cup A$. Thus, $(\mu(\mathbb{C}^c), V^c)$ is a $\mathbf{K}$-model. For nontautological $\varphi \in Prop$, let us assume $\varGamma \models \varphi$ (i.e., $\varphi \in \varGamma$) for every $\varGamma \in C^c$. This implies that $(\mathbb{C}^c, V^c), \varGamma \models [T]\varphi$ (i.e., $[T]\varphi \in \varGamma$), for every $\varGamma \in C^c$. Using the fact that $\mu(\mathbb{C}^c)$ is part of the canonical model $\mathcal{M}^c$, we have $\mathcal{T}^c|_{C^c \cup A^c} \supset (C^c \cup A^c) \times C^c$. Thus, $(\mathbb{C}^c, V^c), \varGamma \models \langle T \rangle [T]\varphi$ for every $\varGamma \in C^c \cup A^c$. As any instance of $\mathrm{Neg}([S], [T])$ is valid in $(\mathbb{C}^c, V^c)$, $\langle T \rangle \langle S \rangle \neg \varphi \in \varGamma$ for every $\varGamma \in C^c \cup A^c$. In other words, $(\mathbb{C}^c, V^c) \models \langle T \rangle \langle S \rangle \neg \varphi$. Thus, there exists $\varGamma' \in W^c$ such that $\varGamma \mathcal{T}^c \varGamma'$ and $\varGamma' \models \langle S \rangle \neg \varphi$ (i.e., $\langle S \rangle \neg \varphi \in \varGamma'$). As $\mathcal{T}(C \cup A) = C \cup A$ in $\mathbf{F}$, we get $\varGamma' \in C^c \cup A^c$. Moreover, there also exists $\varGamma'' \in W^c$ such that $\varGamma' \mathcal{S}^c \varGamma''$ and $\varGamma'' \models \neg \varphi$. Remember that $\mathcal{S}(C \cup A) \subseteq C \cup A$ in $\mathbf{F}$. We so have $\varGamma'' \in C^c \cup A^c$; yet from our initial hypothesis, we further obtain that $\varGamma'' \in A^c$. To sum up, $\varGamma''$ is a maximal consistent set in $\mathbb{C}^c$ such that $\varGamma'' \not\models \varphi$, and this ends the proof. Recall that a $\mathbf{K}^*$-model is an arbitrary combination of $\mathbf{F}^*$-models, so we can generalise the results above to $\mathbf{K}^*$.

3.2 Reflexive Autoepistemic Logic or Nonmonotonic SW5

This section recalls the minimal model semantics for nonmonotonic **SW5** [15]. We first define a *preference* relation between **SW5** models, enabling us to check minimisation.

Definition 3. *A model* $\mathcal{N} = (N, R, U)$ *is preferred over a valuated cluster* (C, V) *if*

- $N = C \cup \{r\}$ *in which* $r \notin C$;
- $R = (N \times C) \cup \{(r, r)\}$;
- *The valuations* V *and* U *agree on* C, *i.e.,* $V = U|_C$;
- *There is a propositional formula* φ *such that* $(C, V) \models \varphi$, *but* $\mathcal{N}, r \not\models \varphi$.

We write $\mathcal{N} \succ (C, V)$ to denote that $\mathcal{N}$ is preferred over (C, V). Then, a minimal model of a theory (finite set of formulas) $\varGamma$ in **SW5** is a valuated cluster (C, V) if

- $C \models \varGamma$ *and*
- $\mathcal{N} \not\models \varGamma$ *for every model* $\mathcal{N}$ *such that* $\mathcal{N} \succ (C, V)$.

Finally, a formula φ is a *logical consequence* of a theory $\varGamma$ in **SW5** ($\varGamma \approx_{\mathbf{SW5}} \varphi$) if φ is valid in every minimal model of $\varGamma$. For example, $q \approx_{\mathbf{SW5}} \neg p \vee q$, yet $\neg p \vee q \not\approx_{\mathbf{SW5}} q$. We end up with Schwarz's important result [14], saying that minimal model semantics is an appropriate semantics for nonmonotonic modal logics, in particular, for **SW5**:

Lemma 1 ([14]). $\varGamma \approx_{\mathbf{SW5}} \varphi$ *if and only if* $\varGamma \hspace{1pt}\vdash\!\!\sim_{\mathbf{SW5}} \varphi$.

3.3   Embedding Nonmonotonic SW5 into Monotonic MRAE*

We here embed $RAEL$ into $\mathbf{M}RAE^*$. Along this aim, we translate the language of **SW5** ($\mathcal{L}_{\mathbf{SW5}}$) to $\mathcal{L}_{[T],[S]}$ via a mapping 'tr': we replace $\Box$ by $[T]$. The next proposition proves that this translation is correct, and clarifies how to capture minimal models of **SW5**.

Proposition 8. *Given an $\mathbf{F}^*$-model $(\mathbb{C}, V) = (C, A, V)$, and a formula α in $\mathcal{L}_{\mathbf{SW5}}$,*

1. $(\mathbb{C}, V), w \models tr(\alpha)$, *for every* $w \in C$ *if and only if* $(C, V|_C) \models \alpha$.
2. $(\mathbb{C}, V) \models \langle T \rangle [T](tr(\alpha) \wedge [S]\neg tr(\alpha))$ *if and only if* $(C, V|_C)$ *is a minimal model for* α.

Proof. The proof of the first item is given by induction on α. As to the second item, we first assume that $(\mathbb{C}, V) \models \langle T \rangle [T](tr(\alpha) \wedge [S]\neg tr(\alpha))$ $(\star)$. We want to show that $(C, V|_C)$ is a minimal model for α.

(1) From $(\star)$ we obtain that $(\mathbb{C}, V), u \models tr(\alpha)$ $(\blacktriangle)$, and $(\mathbb{C}, V), u \models [S]\neg tr(\alpha)$ $(\blacktriangledown)$ for every $u \in C$ (consider: for $w \in A$, $(\star)$ implies that there exists $u \in C \cup A$ such that $w \mathcal{T} u$ and $(\mathbb{C}, V), u \models [T](tr(\alpha) \wedge [S]\neg tr(\alpha)$, hence $u \in C$, otherwise it yields a contradiction). Then, using Proposition 8.1 and $(\blacktriangle)$, we get $(C, V|_C) \models \alpha$. So, the first condition holds.

(2) By definition, it remains to show that $\mathcal{N} \not\models \alpha$ for every model $\mathcal{N}$ such that $\mathcal{N} \succ (C, V|_C)$. Let $\mathcal{N} = (N, R, U)$ be a preferred model over the valuated cluster $(C, V|_C)$ satisfying: $N = C \cup \{r\}$ for some $r \notin C$, $R = (N \times C) \cup \{(r, r)\}$ and $U|_C = V|_C$. By Definition 3, we also know that there exists $\psi \in Prop$ such that $\mathcal{N}, c \models \psi$ $(\bullet)$ for every $c \in C$, but $\mathcal{N} \not\models \psi$. Therefore, we conclude that $\mathcal{N}, r \not\models \psi$ (in other words, $\mathcal{N}, r \models \neg \psi$).

(3) By hypothesis, $(\mathbb{C}, V)$ is an $\mathbf{F}^*$-model. Thus, Neg($[S], [T]$) is valid in it; due to Proposition 7 so is Neg'($[S], [T]$). Hence, $(\mathbb{C}, V) \models [T]\varphi \to \langle T \rangle \langle S \rangle (\neg \varphi \wedge Q)$ for a nontautological $\varphi \in Prop$ and a conjunction of a finite set of literals Q such that $\{\neg \varphi, Q\}$ is consistent.

(4) By using $(\bullet)$ of item (2), Proposition 8.1, and the F-frame properties, we have $(\mathbb{C}, V), u \models [T]tr(\psi)$ for every $u \in C$ $(\clubsuit)$. Note that $tr(\psi) \in Prop$, and $tr(\psi)$ is not a tautology (otherwise $\mathcal{N}, r \models \psi$). Take $Q' = \left(\bigwedge_{p \in \left(\mathbb{P}_\alpha \cap U(r) \right)} p \right) \wedge \left(\bigwedge_{q \in \left(\mathbb{P}_\alpha \setminus U(r) \right)} \neg q \right)$. It is clear that $\mathcal{N}, r \models Q'$, but we also know that $\mathcal{N}, r \models \neg \psi$; so they result in $\mathcal{N}, r \models \neg \psi \wedge Q'$. This implies that $\{\neg \psi, Q'\}$ is consistent; but then so is $\{\neg tr(\psi), Q'\}$ (WLOG, we take $tr(p) = p$ for every $p \in \mathbb{P}$, and so $tr(Q') = Q'$). Hence, the instance of the negatable axiom, namely $[T]tr(\psi) \to \langle T \rangle \langle S \rangle (\neg tr(\psi) \wedge Q')$, is valid in $(\mathbb{C}, V)$. Then, from $(\clubsuit)$ by using MP, we obtain that $(\mathbb{C}, V), u \models \langle T \rangle \langle S \rangle (\neg \psi \wedge Q')$ for every $u \in C$. This further means that there exists a point $x_\psi \in A$ such that $(\mathbb{C}, V), x_\psi \models \neg tr(\psi) \wedge Q'$, viz. $(\mathbb{C}, V), x_\psi \models \neg tr(\psi)$ and $(\mathbb{C}, V), x_\psi \models Q'$. So, we deduce that r and x_ψ agree on $\mathbb{P}_\alpha$, i.e., $V(x_\psi) \cap \mathbb{P}_{tr(\alpha)} = U(r) \cap \mathbb{P}_\alpha$.

(5) Finally, $(\blacktriangledown)$ gives us that $(\mathbb{C}, V), x \models \neg tr(\alpha)$ for every $x \in A$; in particular, $(\mathbb{C}, V), x_\psi \models \neg tr(\alpha)$. To summarise the discussion above:

1. The two pointed models $\big((C \cup \{x_\psi\}, V|_{C \cup \{x_\psi\}}), x_\psi\big)$ in **MRAE*** and (N, r) in **SW5** have the same structure: both contain the same maximal valuated cluster $(C, V|_C)$ and one additional point which sees all points in the cluster;
2. $\mathbb{P}_\alpha = \mathbb{P}_{tr(\alpha)}$ and $V(x_\psi) \cap \mathbb{P}_{tr(\alpha)} = U(r) \cap \mathbb{P}_\alpha$;
3. Both α and $tr(\alpha)$ are the exact copies of each other, except that one contains $\square$ wherever the other contains [T] (note that $tr(\alpha)$ contains neither [S] nor $\langle$S$\rangle$).

Then, it follows that $N, r \models \neg\alpha$ (i.e., $N, r \not\models \alpha$). Consequently, $(C, V|_C)$ is a minimal model for α. The other part of the proof follows alike, so we leave it to the reader.

We are now ready to show how we capture logical consequence of **SW5** in **MRAE***.

Theorem 3. *For formulas α and β in $\mathcal{L}_{\text{SW5}}$,*

$$\alpha \mathrel{\approx\!\!\!\!\!\sim}_{\text{SW5}} \beta \quad \text{if and only if} \quad \models_{\text{MRAE}^*} [\text{T}]\big(tr(\alpha) \wedge [\text{S}]\neg tr(\alpha)\big) \to [\text{T}]tr(\beta).$$

Proof. We first take $\zeta = [\text{T}]\big(tr(\alpha) \wedge [\text{S}]\neg tr(\alpha)\big) \to [\text{T}]tr(\beta)$.

($\Longrightarrow$): Suppose that $\alpha \mathrel{\approx\!\!\!\!\!\sim}_{\text{SW5}} \beta$. We want to prove that ζ is valid in **MRAE***. Let $(\mathbb{C}, V) = (C, A, V)$ be an **F***-model in which C is a maximal cluster. Then, $(C, V|_C)$ is a valuated cluster over C. Thus, it suffices to show that $(\mathbb{C}, V) \models \zeta$. To begin with, for every $w \in A$, $(\mathbb{C}, V), w \models \zeta$ trivially holds: as $\mathcal{T}$ and $\mathcal{S}$ are both reflexive in A, $(\mathbb{C}, V), w \not\models [\text{T}](tr(\alpha) \wedge [\text{S}]\neg tr(\alpha))$ for any $w \in A$ (otherwise, it gives a contradiction). Let $x \in C$ be such that $(\mathbb{C}, V), x \models [\text{T}](tr(\alpha) \wedge [\text{S}]\neg tr(\alpha))$. Then, by the **F**-frame properties, $(\mathbb{C}, V), y \models \langle\text{T}\rangle[\text{T}](tr(\alpha) \wedge [\text{S}]\neg tr(\alpha))$ immediately follows for every $y \in C \cup A$. Hence, using Proposition 8.2, we conclude that $(C, V|_C)$ is a minimal model for α. Moreover, since $\alpha \mathrel{\approx\!\!\!\!\!\sim}_{\text{SW5}} \beta$ by hypothesis, β is valid in $(C, V|_C)$ (i.e., $(C, V|_C) \models \beta$). Again, by Proposition 8.1, we have $(\mathbb{C}, V), y \models tr(\beta)$ for every $y \in C$. Clearly, $\mathbb{C}, y \models [\text{T}]tr(\beta)$ for every $y \in C$ (see Sect. 2.3, truth conditions); in particular, $\mathbb{C}, x \models [\text{T}]tr(\beta)$.

($\Longleftarrow$): Suppose that ζ is valid in **MRAE*** ($\star$). We need to prove that $\alpha \mathrel{\approx\!\!\!\!\!\sim}_{\text{SW5}} \beta$. Let (C, V) be a minimal model of α. Then, it suffices to show that $(C, V) \models \beta$. To see this, we first construct an **F***-model $\mathbb{C} = (C, A, \overline{V})$: let C be exactly the same cluster as in (C, V), so C is our maximal α-cluster. Take $A = \{r : (C \cup \{r\}, R, U) \succ (C, V)\}$. Choose $\overline{V}|_C = V$, and $\overline{V}(r) = U(r)$ for every r in any preferred model $(C \cup \{r\}, R, U)$ over (C, V). Thus, ($\star$) implies that $(\mathbb{C}, V) \models \zeta$ ($\blacktriangle$). By the minimal model definition, $(C, V) \models \alpha$. Then, Proposition 8.1 further gives us that $(\mathbb{C}, V), x \models tr(\alpha)$ for every $x \in C$. Thanks to our construction, $(\mathbb{C}, V), y \not\models tr(\alpha)$, for any $y \in A$. Then, we have $(\mathbb{C}, V), x \models [\text{S}]\neg tr(\alpha)$ for every $x \in C$ (see Sect. 2.3, truth conditions). As a result, $(\mathbb{C}, V), x \models tr(\alpha) \wedge [\text{S}]\neg tr(\alpha)$ for every $x \in C$. By using truth conditions again, we get $(\mathbb{C}, V), x \models [\text{T}](tr(\alpha) \wedge [\text{S}]\neg tr(\alpha))$ for all $x \in C$. Hence, from ($\blacktriangle$) by MP, it follows that $(\mathbb{C}, V), x \models [\text{T}]tr(\beta)$ for every $x \in C$. Clearly, $tr(\beta)$ is also valid in C. Finally, Proposition 8.1 gives us the desired result, viz. $(C, V) \models \beta$ in **SW5**.

Corollary 4. *For $\alpha \in \mathcal{L}_{\mathbf{SW5}}$, α has a minimal model   if and only if*

$$[\mathrm{T}](tr(\alpha) \wedge [\mathrm{S}]\neg tr(\alpha)) \text{ is satisfiable in } \mathbf{MRAE^*}.$$

The proof immediately follows from Theorem 3 by taking $\beta = \bot$. Nonmonotonic inference of **SW5** reflects inference of *RAEL*, and it is complete w.r.t. minimal model semantics. We end our discussion with the following characterisation theorem.

Corollary 5. *For $\alpha, \beta \in \mathcal{L}_{\mathbf{SW5}}$, $\alpha \hspace{0.1em}\vdash_{RAEL} \beta$ if and only if*

$$\models_{\mathbf{MRAE^*}} [\mathrm{T}](tr(\alpha) \wedge [\mathrm{S}]\neg tr(\alpha)) \to [\mathrm{T}]tr(\beta).$$

Proof. First of all, by Theorem 3, $\models_{\mathbf{MRAE^*}} [\mathrm{T}](tr(\alpha) \wedge [\mathrm{S}]\neg tr(\alpha)) \to [\mathrm{T}]tr(\beta)$ if and only if $\alpha \hspace{0.1em}\approx_{\mathbf{SW5}} \beta$. Then, by Lemma 1, $\alpha \hspace{0.1em}\approx_{\mathbf{SW5}} \beta$ if and only if $\alpha \hspace{0.1em}\vdash_{RAEL} \beta$. Finally, by a well known fact, $\alpha \hspace{0.1em}\vdash_{RAEL} \beta$ if and only if $\alpha \hspace{0.1em}\vdash_{\mathbf{SW5}} \beta$. So, the result follows.

4　Conclusion and Further Research

In this paper, we design a novel monotonic modal logic, namely **MRAE***
that captures nonmonotonic **SW5**. We demonstrate this embedding by mapping reflexive autoepistemic logic (**SW5**) into **MRAE***. This way, we see that **MRAE*** is able to characterise the existence of a minimal model as well as nonmonotonic consequence in **SW5**.

Our work provides an alternative to Levesque's monotonic bimodal logic of only knowing [4,5], by which he captures **AEL**. His language has two modal operators, namely B and N: B is similar to [T]. N is characterised by the complement of the relation, interpreting B. Levesque's frame constraints on the accessibility relation differ from ours, and he identifies nonmonotonic consequence with the implication

$$\bigl(\mathsf{B}\ tr(\alpha) \wedge \mathsf{N}\ \neg tr(\alpha)\bigr) \to \mathsf{B}\ tr(\beta).$$

Note that our implication $[\mathrm{T}]\bigl(tr(\alpha) \wedge [\mathrm{S}]\neg tr(\alpha)\bigr) \to [\mathrm{T}]tr(\beta)$, capturing nonmonotonic consequence of **SW5**, and the formula $\langle\mathrm{T}\rangle[\mathrm{T}]\bigl(tr(\alpha) \wedge [\mathrm{S}]\neg tr(\alpha)\bigr)$ characterising minimal models of **SW5** perfectly work for the nonmonotonic variant of **KD45** as well.

As a future work we will attempt to create a general methodology to capture minimal model reasoning, underlying several nonmonotonic formalisms. This paper stands a strong initiative by its possible straightforward implementations to different kinds of nonmonotonic formalisms of similar type. This is, in particular, the case for nonmonotonic variant of **S4F** [12,16], which also have a cluster-based two-floor semantics similar to that of **SW5**. Such research will then enable us to compare various forms of nonmonotonic formalisms in a single monotonic modal logic. Levesque attacked the same problem with an emphasis on the only knowing notion. However, his reasoning does not have a unifying attitude, nor a general mechanism.

References

1. Blackburn, P., de Rijke, M., Venema, Y.: Modal logic. Cambridge Tracts in Theoretical Computer Science, vol. 53. Cambridge University Press (2002)
2. Halpern, J.-Y., Moses, Y.: Towards a theory of knowledge and ignorance: preliminary report. In: Apt, K.-R. (ed.) Logics and Models of Concurrent Systems, vol. 13, pp. 459–476. Springer (1985)
3. Konolige, K.: Autoepistemic logic. In: Gabbay, D.-M., Hogger, C.-J., Robinson, J.-A. Handbook of Logic in Artificial Intelligence and Logic Programming, vol. 3, pp. 217–295. Oxford University Press, Inc. (1994)
4. Lakemeyer, G., Levesque, H.-J.: Only knowing meets nonmonotonic modal logic. In: Brewka, G., Eiter, T., McIlraith, S.-A. (eds.) Proceedings of the 13th International Conference on the Principles of Knowledge Representation and Reasoning (KR 2012), pp. 350–357. AAAI Press (2012)
5. Levesque, H.-J.: All i know: a study in autoepistemic logic. Artif. Intell. J. **42**, 263–309 (1990)
6. Lifschitz, V.: Minimal belief and negation as failure. Artif. Intell. J. **70**(1–2), 53–72 (1994). Oxford University Press
7. Lifschitz, V., Schwarz, G.: Extended logic programs as autoepistemic theories. In: Proceedings of the 2nd Intenational Workshop (LPNMR 1993), pp. 101–114. MIT Press (1993)
8. Marek, V.-W., Truszczyński, M.: Relating autoepistemic and default logics. In: Brachman, R.-J., Levesque, H.-J., Reiter, R. (eds.) Proceedings of the 1st International Conference on the Principles of Knowledge Representation and Reasoning (KR 1989), pp. 276–288. Morgan Kaufmann, Toronto (1989)
9. Marek, V.-W., Truszczyński, M.: Reflexive autoepistemic logic and logic programming. In: Proceedings of the 2nd International Workshop on Logic Programming and Non-Monotonic Reasoning, pp. 115–131. MIT Press (1993)
10. Marek, V.-W., Truszczyński, M.: Nonmonotonic Logic: Context-Dependent Reasoning. Springer, Heidelberg (1993)
11. Moore, R.-C.: Semantical considerations on nonmonotonic logic. Artif. Intell. J. **25**(1), 75–94 (1985). Elsevier
12. Pearce, D., Uridia, L.: An approach to minimal belief via objective belief. In: Walsh, T. (ed.) Proceedings of the International Conference on Artificial Intelligence (IJCAI 2011), vol. 22(1), pp. 1045–1050. AAAI Press (2011)
13. Schwarz, G.-F.: Autoepistemic logic of knowledge. In: Nerode, A., Marek, V.-W., Subrahmanian, V.-S. (eds.) Proceedings of the 1st International Workshop on Logic Programming and Nonmonotonic Reasoning (LPNMR 1991), pp. 260–274. MIT Press (1991)
14. Schwarz, G.-F.: Minimal model semantics for nonmonotonic modal logics. In: Proceedings of the 7th Annual IEEE Symposium on Logic in Computer Science (LICS 1992), pp. 34–43. IEEE Computer Society Press (1992)
15. Schwarz, G.-F.: Reflexive autoepistemic logic. Fundamenta Informaticae **17**(1–2), 157–173 (1992). IOS Press
16. Schwarz, G.-F., Truszczynski, M.: Modal logic S4F and the minimal knowledge paradigm. In: Moses, Y. (ed.) Proceedings of the 4th Conference on Theoretical Aspects of Reasoning about Knowledge (TARK 1992), pp. 184–198. Morgan Kaufmann (1992)

17. Shoham, Y.: A semantical approach to nonmonotonic logics. In: Proceedings of the 10th International Joint Conference on Artificial Intelligence (IJCAI 1987), pp. 1413–1419. Morgan Kaufmann Publishers Inc. (1987)
18. Stalnaker, R.: What is a nonmonotonic consequence relation? Fundamenta Informaticae **21**(1–2), 7–21 (1994). IOS Press

Minimal Inference Problem Over Finite Domains: The Landscape of Complexity

Michał Wrona[(⊠)]

Theoretical Computer Science Department,
Faculty of Mathematics and Computer Science,
Jagiellonian University, Kraków, Poland
wrona@tcs.uj.edu.pl

Abstract. The complexity of the general inference problem for propositional circumscription in Boolean logic (or equivalently over the two-element domain) has been recently classified. This paper generalizes the problem to arbitrary finite domains. The problem we study here is parameterized by a set of relations (a constraint language), from which we are allowed to build a knowledge base, and a linear order on the domain, which is used to compare models.

We use the algebraic approach provided originally in order to understand the complexity of the constraint satisfaction problem to give first non-trivial dichotomies and tractability results for the minimal inference problem over finite domains.

1 Introduction

The need for logics that could capture human way of thinking triggered the development of an area of Artificial Intelligence called nonmonotonic reasoning. A number of formalisms emerged. One of the most important and best studied is circumscription introduced by McCarthy [17]. The circumscription of a formula is the set of its minimal models that are supposed to represent possible situations that are consistent with common sense.

It is often the case [6,9,11,18] that models are compared according to the preorder ($\leq_{(P,Z)}$) induced by a partition of variables V into three subsets P, Z, Q (possibly empty) where P — variables that are subject to minimizing, Q — variables that maintain the fixed value, and Z — variables whose value can vary. Now, for two assignments $\alpha, \beta : V \to D$ we will have ($\alpha \leq_{(P,Z)} \beta$) if $\alpha[Q] = \beta[Q]$ (α is equal to β on variables in Q) and $\alpha[P] \leq \beta[P]$ (α is less than or equal to β on variables in P) where $\leq$ is the coordinatewise extension of the natural order on $\{0,1\}$ with $0 < 1$.

As every logical formalism does, circumscription gives rise to two main computational problems: the model-checking problem and the inference problem. In this paper we concentrate on the inference problem for propositional circumscription, called also *the minimal inference problem*. In the most general formulation

The research was partially supported by NCN grant number 2014/14/A/ST6/00138.

© Springer International Publishing AG 2017
M. Balduccini and T. Janhunen (Eds.): LPNMR 2017, LNAI 10377, pp. 101–113, 2017.
DOI: 10.1007/978-3-319-61660-5_11

an instance of this problem in the Boolean case consists of two CNF formulas: a *knowledge base* φ and a *query* ψ over the same set of variables V partitioned into P, Z, Q. In the minimal inference problem we ask if every $(\leq_{(P,Z)})$-min model (minimal model wrt. $(\leq_{(P,Z)})$) of φ is a model of ψ. Since this task can be performed for every clause of ψ independently and we are interested in the complexity of the problem up to polynomial time reduction, we can assume that ψ is a disjunction of propositional literals (a clause). This problem is in general Π_2^P-complete [10]. Thus, one considers the general minimal inference problem GMININF(Γ) parameterized by a set of relations Γ over $\{0, 1\}$.

- *Instance of GMININF(Γ)*: a conjunction of atomic formulas φ of the form: $R_1(x_1^1, \ldots, x_{k_1}^1) \wedge \cdots \wedge R_l(x_1^l, \ldots, x_{k_l}^l)$, where every R_i with $i \in [l]$ is a relation symbol in a signature of Γ, over variables V partitioned into P, Z, Q, and a propositional clause ψ over V.
- *Question:* is every $(\leq_{(P,Z)})$-min model of φ a model of ψ?

The complete complexity classification of GMININF(Γ) with respect to Γ has been obtained after a series of papers, e.g. [6,8,10,15] in [9]. Under usual complexity theoretical assumptions, in this case that: P $\subsetneq$ coNP $\subsetneq$ Π_2^P, it is shown that GMININF(Γ) is either Π_2^P-complete, or coNP-complete, or in P. This raises a question about a similar classification in many-valued logics, or, more generally, over arbitrary finite domains. Especially that circumscription over larger finite domains has been studied in the literature, e.g. in [7,20].

The same course of events took place in the case of the problem CSP(Γ) where a question is whether a given conjunction of atomic formulas over the signature of Γ is satisfiable. In [22], Schaefer established the dichotomy between NP-complete and in P in the case where Γ is over the two-element domain. Then researchers turned to a so-called Feder-Vardi conjecture that states that a similar dichotomy holds for arbitrary finite domains. The understanding and many advanced partial results[1], see [16] for a recent survey, were possible thanks to the development of the so-called algebraic approach [5,12]. This approach has been also already applied to the model checking and the inference problem in propositional circumscription over arbitrary finite domains in [19]. Here we use algebra for the inference problem in a bit different formulation.

Certainly every relation $R \subseteq D^n$ over any finite domain D can be defined by a conjunction of disjunctions of disequalities (a CNF of disequalities) of the form $(x \neq d)$ where x is a variable and $d \in D$ simply by the formula: $\bigwedge_{(d_1,\ldots,d_n)\notin R}(x_1 \neq d_1 \vee \cdots \vee x_n \neq d_n)$. This implies that in the most general version of the minimal inference problem the input may consist of a CNF of disequalities that states for a knowledge base, a CNF of disequalities that states for a query and a *linear order* $\mathcal{O} = (D; \leq^{\mathcal{O}})$. The preorder $\leq_{(P,Z)}^{\mathcal{O}}$ is defined as in

---

[1] Recently, three different groups of researchers announced a proof of the dichotomy.

the two-element case with a difference that we use $\leq^{\mathcal{O}}$ on coordinates instead of $0 < 1$. Since CNFs of disequalities coincide with clauses if $|D| = 2$, we have that the minimal inference problem in this formulation is Π_2^P-hard. It is straightforward to show that the problem is in fact Π_2^P-complete. Thus, it is natural to ask about the complexity of the parametrized version GMININF$(\Gamma, \mathcal{O})$ defined below. As in the Boolean case we can assume that a query consists of a single disjunction of disequalities.

- *Instance of GMININF$(\Gamma, \mathcal{O})$:* a conjunction of atomic formulas φ of the form: $R_1(x_1^1, \ldots, x_{k_1}^1) \wedge \cdots \wedge R_l(x_1^l, \ldots, x_{k_l}^l)$, where every R_i with $i \in [l]$ is a relation symbol in a signature of Γ, over variables V partitioned into P, Z, Q, and a disjunction of disequalities ψ of the form $(x_1 \neq d_1 \vee \cdots \vee x_k \neq d_k)$ over V.
- *Question:* is every $(\leq_{(P,Z)}^{\mathcal{O}})$-min model of φ a model of ψ?

For an example, consider Γ over $D = \{1, 2, 3\}$ containing the relation $R_{\neq} := \{(d_1, d_2) \in D^2 \mid d_1 \neq d_2\}$ and the order $1 <^{\mathcal{O}} 2 <^{\mathcal{O}} 3$. A formula $R_{\neq}(x_1, x_2) \wedge R_{\neq}(x_2, x_3) \wedge R_{\neq}(x_3, x_4) \wedge R_{\neq}(x_2, x_3)$ and a disjunction $(x_1 \neq 2 \vee x_4 \neq 1)$ form an instance of GMININF$(\Gamma, \mathcal{O})$. This problem is Π_2^P-complete. Throughout the paper we give parametrizations of GMININF$(\Gamma, \mathcal{O})$ of lower complexity.

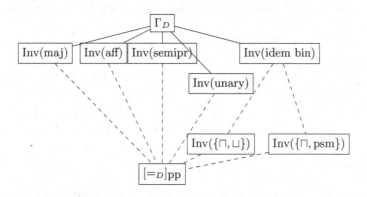

Fig. 1. An illustrative presentation of the lattice of relational clones over a domain D.

Contribution. Our attack on the complexity classification of GMININF$(\Gamma, \mathcal{O})$ is based on the algebraic approach. All the notions we use in this paper are defined carefully in Sect. 2. As explained there in order to complete the classification task, it is enough to establish the complexity only for relational clones that are constraint languages closed under primitive positive definitions. Equivalently, a relational clone is the set, denoted by Inv(F), of all relations invariant under (preserved by) all operations in some set F. The relational clones over D are organized in the lattice ordered by inclusion. The larger is the relational clone, the complexity of the problem is harder. An illustrative presentation of the lattice over some D containing only kinds of relational clones we look at in this paper is presented in Fig. 1. For a pair of relational clones linked by a line the one placed higher contains the one which is below. The solid lines indicate that there are no relational clones inbetween, the dashed ones that there might be some. The top element of the lattice is the set of all relations Γ_D over D. The only operations that preserve all relations in Γ_D are projections. The bottom of the lattice is the set of relations primitively positively definable by means of equality. Such are preserved by all operations over D.

The problem GMININF($\Gamma_D, \mathcal{O}$) is Π_2^P-complete for all $|D| \geq 2$. To find sub-problems of lower complexity we climb down the lattice of relational clones. The natural choice for the languages to be studied first are *maximal constraint languages* that in the lattice lay directly below Γ_D. According to Rosenberg's Five Types Theorem [21], it is enough to consider languages of the form $\mathrm{Inv}(\{f\})$ where f is a unary operation, a semiprojection, a binary idempotent operation, a majority operation or an affine operation, see Fig. 1. Their complexity was analyzed for many problems parametrized by constraint languages: for CSPs the result can be found in [1,4] We show that GMININF($\Gamma, \mathcal{O}$) is coNP-hard for all such languages. Furthermore for maximal conservative (i.e., containing all unary relations) languages the problem is either Π_2^P-complete or coNP-complete. See [3] for the CSP classification over conservative languages.

In order to find tractable (polynomial-time decidable) classes we climb down the lattice of relational clones even further. In particular we give dichotomies between coNP-complete and P for:

– GMININF($\mathrm{Inv}(\{\sqcap, \sqcup\}), \mathcal{O}$) where $\sqcap, \sqcup$ are the join and the meet of some lattice, and
– GMININF($\mathrm{Inv}(\{\sqcap, f\}), \mathcal{O}$) where $\sqcap$ is the meet of some semilattice and f a newly defined *pms*-operation.

This gives us new large tractability classes and new coNP-hardness results. Futhermore, our algorithms are based on polymorphisms which is crucial for further generalizations, consult again [16] to see the importance of polymorphisms in providing algorithms for the CSP.

Related Work. Classifications such as a trichotomy for GMININF(Γ) for Γ over the two-element domain as one presented in [9] are much easier to be obtained than analogical results over arbitrary finite domains. The reason is that in the two-element case, the lattice of relational clones (so-called Post's lattice) has countably many and very well described elements. Thus, in order to obtain a classification it is enough to consider the problem for each of them. The situation is very different already over the three-element domain where there are uncountably many relational clones and the lattice is not comprehensible.

The results on the minimal inference problem over arbitrary finite domains included in [19] concern a version of GMININF($\Gamma, \mathcal{O}$) where the query ψ is a single atomic formula $R(x_1, \ldots, x_k)$ where R corresponds to some k-ary relation over the domain of Γ, and the order $\mathcal{O}$ is a part of the input. The authors of [19] provide mainly preliminary results that may be seen as tools for classifying the problem. Some of these results such as Theorem 1 we reprove here for our version of the problem and use heavily in our paper. The only complexity result in [19], however, is a dichotomy between Π_2^P-complete and in coNP for conservative languages Γ over the three-element domain[2] and even this specific result follows in a rather straightforward way from the dichotomy for CSP(Γ) over the three-element domain [2].

In this paper, for the first time, we provide complexity results characteristic for the minimal inference problem over larger domains. We expect the classification for the problem GMININF to be completely different than and not easily obtainable from the one for CSP. This is already true over two elements. For the first time, we provide here polynomial algorithms. The tractability results for GMININF(Inv($\{\sqcap, \sqcup\}$), $\mathcal{O}$) and GMININF(Inv($\{\sqcap, f\}$), $\mathcal{O}$) substantially generalize these for GMININF(Inv($\wedge, \vee$)) and GMININF(Inv($\{\wedge, (x \wedge (y \vee \neg z))\}$)) from the two-element world. Our dichotomy: between Π_2^P-complete and in coNP for conservative maximal languages generalize a dichotomy between these two classes for the two-element domain from [9] and for the three-element conservative case (up to a small difference in the definition of the problems) from [19].

2 Preliminaries

We write $t = (t[1], \ldots, t[n])$ for a tuple of elements and $[n]$ to denote the set $\{1, \ldots, n\}$. The reverse of the order $\mathcal{O}_1 = (D; \leq^{\mathcal{O}})$ is $\mathcal{O}_2 = (D; (\leq^{\mathcal{O}})^{-1})$ where $(\leq^{\mathcal{O}})^{-1}$ is the relation $\{(a, b) \in D \mid (b, a) \in \leq^{\mathcal{O}}\}$.

Constraint Languages. In this paper a *(constraint) language* over (always finite) domain D, denoted by capital Greek letters such as Γ, is a set of relations over D. A signature of Γ denoted usually by τ is a set of relation symbols associated to relations in Γ. For the sake of simplicity we usually use the same symbols to denote both a relation symbol and the corresponding relation. We also assume that the domain of a relation or a language under consideration is the set D.

A *primitive positive formula (pp-formula)* over a signature τ is a first-order formula built exclusively from conjunction, existential quantifiers and atomic formulas over τ and equalities, that is atomic formulas of the form $(x = y)$. We say that a relation R has a pp-definition over a set of relations Γ if there exists a pp-formula over the signature of Γ that holds exactly on those tuples that are contained in R. We say that a set of relations Δ has a pp-definition over Γ if every relation in Δ has a pp-definition over Γ. A set of relations with a pp-definition over Γ is denoted by $[\Gamma]_{pp}$ and called a *relational clone*.

---

[2] The paper claims that the dichotomy is for all languages over the three-element domain. However, this is not true since the proof of Theorem 3.6 is flawed for domains with more than two elements [14].

We define a Γ-formula to be a conjunction of atomic formulas over a signature of Γ. Observe that a Γ-formula is a special form of a pp-formula where quantifiers and equality are not in use. Let $\psi := R(x_1, \ldots, x_k)$ be an atomic Γ-formula with $x_1, \ldots, x_k$ not necessarily different and $W \subseteq \{x_1, \ldots, x_k\}$. We write $\psi|_W$ for an atomic formula $R|_W(y_1, \ldots, y_l)$ where $R|_W$ is a projection of R to coordinates corresponding to variables in W and $y_1, \ldots, y_l$ is a subsequence of $x_1, \ldots, x_k$ containing only variables in W. For a Γ-formula $\varphi = \psi_1 \wedge \ldots \wedge \psi_n$ over V and $W \subseteq V$ we write $\varphi|_W$ to indicate $\psi_1|_W \wedge \ldots \wedge \psi_n|_W$.

Operations, Polymorphisms. Let Γ be a language over domain D. An operation $f : D^n \to D$ is a *polymorphism* of an m-ary relation R if for all m-tuples $t_1, \ldots, t_n \in R$, it holds that the tuple $(f(t_1[1], \ldots, t_n[1]), \ldots, f(t_1[m], \ldots, t_n[m]))$ is also in R. An operation f is a polymorphism of a language Γ if it is a polymorphism of every relation in Γ. If $f : D^n \to D$ is a polymorphism of Γ, R, we say that f *preserves* Γ, R. The set of relations preserved by a set of operations F is denoted by $\mathrm{Inv}(F)$. The following Galois correspondence links sets of polymorphisms and relational clones, see e.g. [12].

Lemma 1. *Let Γ be a constraint language. Then $\Gamma_1 \subseteq [\Gamma_2]_{pp}$ if and only if $Pol(\Gamma_2) \subseteq Pol(\Gamma_1)$.*

Here we list some kinds of operations that are of use for this paper. We say that an operation $f : D^n \to D$ is *idempotent* if for all $x \in D$ we have $f(x, \ldots, x) = x$. An operation $f : D \to D$ that is bijective is said to be a *permutation*. An operation $f : D^n \to D$ is a *projection* if there exists $i \in [n]$ such that $f(x_1, \ldots, x_i, \ldots, x_n) = x_i$ for all $x_1, \ldots, x_n \in D$. We say that a ternary operation $f : D^3 \to D$ is a *majority operation* if for all $x, y \in D$ we have $f(x, x, y) = f(x, y, x) = f(y, x, x) = x$, and that f is *affine* if for all $x_1, x_2, x_3 \in D$ we have $f(x_1, x_2, x_3) = x_1 - x_2 + x_3$, where $+$ and $-$ are the operations of an Abelian (commutative) group $(D; +, -)$. An operation $f : D^n \to D$ with $D \geq 3$ is said to be a *semiprojection* if there exists $i \in [n]$ such that for all $x_1, \ldots, x_n \in D$ we have $f(x_1, \ldots, x_n) = x_i$ whenever $|\{x_1, \ldots, x_n\}| < n$ and f is not a projection. A *semilattice operation* s on a set D is an idempotent operation satisfying universally the identities $s(x, y) = s(y, x)$ and $s(s(x, y), z) = s(x, s(y, z))$. The first of these identities implies that s is commutative and the other that s is associative. We define a ternary operation $f : D^3 \to D$ to be a *pms*-operation compliant with a semilattice operation s if for all $x, y \in D$ it satisfies: $f(x, y, y) = x$, $f(y, x, y) = y$ and $f(y, y, x) = s(x, y)$.

The General Minimal Inference Problem. We now give a careful definition of the GMININF problem and provide basic results that help classifying the complexity of the problem. Some of them for a variant of the problem we study here are already available in the literature [8,9,19].

Let V be a set of variables and D a finite set. We use small Greek letters: α, β, γ to denote assignments of the type $V \to D$. We say that $\alpha : V_1 \to D$ is a restriction of $\beta : V_2 \to D$ to (variables in) V_1 if $V_1 \subseteq V_2$ and for all $v \in V_1$, we have $\alpha(v) = \beta(v)$. In this case we also say that β is an extension of α to (variables in) V_2. For $\alpha : V \to D$ and $V_1 \subseteq V$, we write $\alpha[V_1]$ to indicate the restriction of α to V_1.

Let $\mathcal{O} = (D; \leq^{\mathcal{O}})$ be a linear order. We extend the order to assignments in the natural way. For $\alpha, \beta : V \to D$ we have $\alpha \leq^{\mathcal{O}} \beta$ if for all $v \in V$ it holds $\alpha(v) \leq^{\mathcal{O}} \beta(v)$. We write $\alpha <^{\mathcal{O}} \beta$ if $\alpha \leq^{\mathcal{O}} \beta$ and for at least one $v \in V$ it holds $\alpha(v) <^{\mathcal{O}} \beta(v)$. For the purposes of this paper we need also a special preorder on assignments denoted by $(\leq^{\mathcal{O}}_{(P,Z)})$ and defined as follows. Let $\alpha, \beta : V \to D$ and P, Z, Q be a partition of V. We have $(\alpha \leq^{\mathcal{O}}_{(P,Z)} \beta)$ if $\alpha[Q] = \beta[Q]$ and $\alpha[P] \leq^{\mathcal{O}} \beta[P]$ and $(\alpha <^{\mathcal{O}}_{(P,Z)} \beta)$ if $\alpha[Q] = \beta[Q]$ and $\alpha[P] <^{\mathcal{O}} \beta[P]$. Let Γ be a constraint language over D and φ be a Γ-formula over variables V. An assignment $\alpha : V \to D$ is a model of φ if for every conjunct $R(x_1, \ldots, x_k)$ of φ we have $(\alpha(x_1), \ldots, \alpha(x_k)) \in R$. We say that a model α of φ is a $(\leq^{\mathcal{O}}_{(P,Z)})$-minimal $((\leq^{\mathcal{O}}_{(P,Z)})$-min) model of φ if there is no model β of φ such that $\beta <^{\mathcal{O}}_{(P,Z)} \alpha$.

We now rephrase the definition of GMININF$(\Gamma, \mathcal{O})$ from the introduction using new notions introduced in this section. The definitions are equivalent.

Definition 1. *[GMININF$(\Gamma, \mathcal{O})$]*

- *INSTANCE: A Γ-formula φ over variables V partitioned into three sets P, Z, Q and a disjunction of disequalities ψ over V.*
- *QUESTION: Is every $(\leq^{\mathcal{O}}_{(P,Z)})$-min model of φ a model of ψ?*

For finite Γ we measure the complexity of GMININF$(\Gamma, \mathcal{O})$ as a function of the length of a Γ-formula. In this paper we consider Γ and $\mathcal{O}$ that make the problem Π_2^P-complete, coNP-complete or in P. For an infinite set of relations Γ we use the usual convention. We say that GMININF$(\Gamma, \mathcal{O})$ is Π_2^P-hard (coNP-hard) if there is a finite $\Gamma' \subseteq \Gamma$ such that GMININF$(\Gamma', \mathcal{O})$ is Π_2^P-hard (coNP-hard) and that GMININF$(\Gamma, \mathcal{O})$ is in Π_2^P, coNP or P if for every finite $\Gamma' \subseteq \Gamma$ the problem GMININF$(\Gamma', \mathcal{O})$ is in Π_2^P, coNP or P, respectively.

The computational complexity of GMININF$(\Gamma, \mathcal{O})$ is fully captured by the relational clone $[\Gamma]_{pp}$, or equivalently the set of polymorphisms of Γ.

Theorem 1. *Let Γ_1, Γ_2 be constraint languages such that $\Gamma_1 \subseteq [\Gamma_2]_{pp}$ (or equivalently $Pol(\Gamma_2) \subseteq Pol(\Gamma_1)$), then there is a polynomial-time reduction from GMININF$(\Gamma_1, \mathcal{O})$ to GMININF$(\Gamma_2, \mathcal{O})$.*

This is usually easier to look at GMINEXT than GMININF.

Definition 2. *[GMINEXT$(\Gamma, \mathcal{O})$]*

- *INSTANCE: A Γ-formula φ over variables V partitioned into three sets P, Z, Q and a partial assignment $\alpha : V_1 \to D$ with $V_1 \subseteq V$.*
- *QUESTION: Is there an extension $\beta : V \to D$ of α such that β is a $(\leq^{\mathcal{O}}_{(P,Z)})$-min model of φ?*

The following proposition reveals the connection between the problems. We have that the complement of GMININF($\Gamma, \mathcal{O}$) and GMINEXT($\Gamma, \mathcal{O}$) are polynomially equivalent. The reduction from the complement of GMININF($\Gamma, \mathcal{O}$) to GMINEXT($\Gamma, \mathcal{O}$) comes to replacing a disjunction $\psi := (x_1 \neq d_1 \vee \cdots \vee x_k \neq d_k)$ with $\alpha : \{x_1, \ldots, x_k\} \mapsto D$ such that $\alpha(x_i) = d_i$ for $i \in [k]$. Now, if there is a $(\leq^{\mathcal{O}}_{(P,Z)})$-min model β of φ (the same for both instances) that is not a model of ψ it satisfies $\beta(x_i) = d_i$ for $i \in [k]$, and hence β extends α. On the other hand, if there is a $(\leq^{\mathcal{O}}_{(P,Z)})$-min model β of φ extending α, then it certainly does not satisfy ψ. For the reduction from GMINEXT($\Gamma, \mathcal{O}$) to the complement of GMININF($\Gamma, \mathcal{O}$), we replace α with $\psi := (x_1 \neq \alpha(x_1) \vee \cdots \vee x_k \neq \alpha(x_k))$, where $\{x_1, \ldots, x_k\}$ is the domain of α. The proof is analogous.

Proposition 1. *Let $\mathcal{O} = (D; \leq^{\mathcal{O}})$ be a linear order and Γ a constraint language over D. The problem GMINEXT($\Gamma, \mathcal{O}$) is Σ^P_2-hard, NP-hard, in NP, in P if and only if GMININF($\Gamma, \mathcal{O}$), is Π^P_2-hard, coNP-hard, in coNP, in P, respectively.*

3   Maximal Constraint Languages

In this section we give a lower bound for GMININF($\Gamma, \mathcal{O}$) over maximal constraint languages Γ and a dichotomy for conservative maximal languages.

Definition 3. *Let Γ_D be the set of all relations over domain D. A constraint language $\Gamma \subseteq \Gamma_D$ is maximal if $[\Gamma]_{pp} \subsetneq \Gamma_D$ and for every $R \notin \Gamma$, we have that $[\Gamma \cup R]_{pp} = \Gamma_D$. A constraint language Γ over D is conservative if Γ contains all subsets of D as unary relations.*

To build the classification we use some methods [4] and some results [6,9] known from the literature. We start with Rosenberg's theorem.

Theorem 2 (Rosenberg Theorem). *Every maximal constraint language has the form $Inv(\{f\})$ where the operation f is of one of the following types:*

1. *a unary operation which is either a permutation or else acts identically on its range;*
2. *a binary operation which is not a projection;*
3. *a majority operation;*
4. *an affine operation;*
5. *a semiprojection.*

We need to know the complexity of GMININF($Inv(f), \mathcal{O}$) for every type of operations from Theorem 2. In the case where $f : D^3 \to D$ is an affine operation we focus on two particular relations R_+ and R_{++} defined as follows. Let $\bot, \square \in D$ be the two least elements in $\mathcal{O}$, i.e., $\bot <^{\mathcal{O}} \square <^{\mathcal{O}} x$ for all $x \in D \setminus \{\bot, \square\}$. We will have:

- $R_+ = \{(x, y) \mid x + y = \bot + \square\}$,
- $R_{++} = \{(x, y, z) \mid x + y + z = \bot + \bot + \square\}$,

where $+$ comes from an Abelian group $(D; +, -)$. Since R_+ and R_{++} are defined by a single equation, it is straightforward to show that they are both in $\mathrm{Inv}(f)$.

Lemma 2. *Let $(D; +, -)$ be an Abelian group and $\mathcal{O} = (D; \leq^{\mathcal{O}})$ a linear order. Then GMININF$(\Gamma, \mathcal{O})$ with $\Gamma = \{R_+, R_{++}\}$ is coNP-hard.*

We first prove that the problem for maximal constraint languages is coNP-hard.

Theorem 3. *Let $\mathcal{O} = (D; \leq^{\mathcal{O}})$ be a linear order and Γ a maximal constraint language over Γ, then GMININF$(\Gamma, \mathcal{O})$ is coNP-hard.*

We are able to prove the full dichotomy only under an additional assumption that languages are conservative.

Theorem 4. *Let $\mathcal{O} = (D; \leq^{\mathcal{O}})$ be a linear order and $\mathrm{Inv}(\{f\})$, for some operation f, a conservative maximal constraint language over D. Then we have exactly one of the following:*

1. *f is a unary operation, or there is a two-element $D' \subseteq D$ such that $f|_{D'}$ is a projection and then GMININF$(\mathrm{Inv}(\{f\}), \mathcal{O})$ is Π_2^P-complete.*
2. *f is a commutative binary operation, a majority operation, or an affine operation and then GMININF$(\mathrm{Inv}(\{f\}), \mathcal{O})$ is coNP-complete.*

4   The Minimal Inference Problem and Semilattice Operations

Let $\mathcal{O} = (D; \leq^{\mathcal{O}})$ be a linear order, Γ a constraint language over D and φ a Γ-formula. We say that a model α of φ is the least (the greatest) model of φ wrt. $\leq^{\mathcal{O}}$ if for every model β of φ, it holds $\alpha \leq^{\mathcal{O}} \beta$ ($\beta \leq^{\mathcal{O}} \alpha$). The least (the greatest) model does not have to exist. However, if Γ is preserved by some well-behaved semilattice operation, then we have the following.

Observation 1. *Let $\mathcal{O} = (D; \leq^{\mathcal{O}})$ be a linear order and Γ a constraint language over D preserved by the meet $\sqcap$ (the join $\sqcup$) of some meet-semilattice (join-semilattice) $\mathcal{L} = (D; \leq^{\mathcal{L}})$ such that $\leq^{\mathcal{L}} \subseteq \leq^{\mathcal{O}}$. Let φ be a Γ-formula. Then there exists a model α of φ such that α is the least (the greatest) model of φ wrt the order $\leq^{\mathcal{L}}$ and at the same time α is the least (the greatest) model of φ wrt the order $\leq^{\mathcal{O}}$.*

In the case described by the previous observation we can quickly compute the least (the greatest) wrt both $\leq^{\mathcal{L}}$ and $\leq^{\mathcal{O}}$ model of a Γ-formula extending a given assignment α. The procedure is very well known [12] and may be performed by enforcing *generalized arc consistency*. We refer to this procedure by leastext$(\mathcal{O}, \varphi, \alpha, V_2)$ (greatext$(\mathcal{O}, \varphi, \alpha, V_2)$).

Proposition 2. *Let $\mathcal{O} = (D; \leq^{\mathcal{O}})$ be a linear order and Γ a constraint language preserved by the meet $\sqcap$ (the join $\sqcup$) of some meet-semilattice (join-semilattice) $\mathcal{L} = (D; \leq^{\mathcal{L}})$ such that $\leq^{\mathcal{L}} \subseteq \leq^{\mathcal{O}}$. Then the procedure leastext$(O, \varphi, \alpha, V_2)$ (the procedure greatext $(O, \varphi, \alpha, V_2)$) for a Γ-formula φ and a partial assignment $\alpha : V_1 \to D$ with $V_1 \subseteq V_2 \subseteq V$ returns:*

- *false* if there is no model of $\varphi|_{V_2}$ extending α;
- the least (the greatest), wrt $\leq^{\mathcal{O}}$, model of $\varphi|_{V_2}$ extending α.

The procedures work in polynomial time.

In the case we consider here $\alpha[P]$ of a $(\leq^{\mathcal{O}}_{(P,Z)})$-min model α is uniquely determined by $\alpha[Q]$.

Observation 2. *Let $\mathcal{O} = (D; \leq)$ be a linear order and Γ a constraint language preserved by the meet $\sqcap : D^2 \to D$ of some meet-semilattice $\mathcal{L} = (D; \leq^{\mathcal{L}})$ such that $\leq^{\mathcal{L}} \subseteq \leq^{\mathcal{O}}$. Let φ be a Γ-formula over variables V partitioned into P, Z, Q. Then α is a $(\leq^{\mathcal{O}}_{(P,Z)})$-min model of φ if and only if $\alpha[P] = \beta[P]$ where $\beta = leastext(\mathcal{O}, \varphi, \alpha[Q], V)$.*

Theorem 5.2 in [13] states that a relation R is preserved by the meet $\sqcap$ of a linear order $\mathcal{L} = (D; \leq^{\mathcal{L}})$ iff it can be defined by a conjunction of clauses of the form $(x_1 \geq^{\mathcal{L}} a_1 \vee \cdots \vee x_k \geq^{\mathcal{L}} a_k) \to (x_i >^{\mathcal{L}} b_i)$ where $x_1, \ldots, x_k$ are variables; $a_1, \ldots, a_k, b_i$ with $i \leq k$ are in D. We note that this result can be extended to all *semilattices*.

5   Lattice

In this section we consider relations in $\mathrm{Inv}(\{\sqcap, \sqcup\})$ where $\sqcap, \sqcup$ are the meet and the join of some lattice $\mathcal{L} = (D; \leq^{\mathcal{L}})$. For an example of such a relation consider $D = \{\bot, a_1, a_2, a_3, b_2, \top\}$ and the order $\leq^{\mathcal{L}}$ such that $\bot \leq^{\mathcal{L}} a_1 \leq^{\mathcal{L}} a_2 \leq^{\mathcal{L}} b_2 \leq^{\mathcal{L}} \top$ and $a_1 \leq^{\mathcal{L}} a_3 \leq^{\mathcal{L}} \top$. Thus, in particular a_2, b_2 are not comparable with a_3. It is now straightforward to show that $R := ((x_1 \geq^{\mathcal{L}} a_1 \wedge x_2 \geq^{\mathcal{L}} a_2) \to x_2 \geq^{\mathcal{L}} b_2) \wedge (x_1 \geq a_1 \to x_2 \geq a_3)$ is preserved by $\sqcup$ and $\sqcap$.

In Fig. 2, we present an algorithm for $\mathrm{GMINEXT}(\mathrm{Inv}(\{\sqcap, \sqcup\}), \mathcal{O})$ for the case where $\sqcap, \sqcup$ are the meet and the join of some lattice $\mathcal{L} = (D; \leq^{\mathcal{L}})$ that can be extended to $\mathcal{O}$. The algorithm works in polynomial time. By Proposition 1, it gives us a polynomial algorithm for the problem GMININF.

Lemma 3. *Let $\mathcal{O} = (D; \leq^{\mathcal{O}})$ be a linear order and Γ a constraint language preserved by $\sqcap, \sqcup$ that are the meet and the join of some lattice $\mathcal{L} = (D; \leq^{\mathcal{L}})$ such that $\leq^{\mathcal{L}} \subseteq \leq^{\mathcal{O}}$. For a given Γ-formula φ over variables V partitioned into P, Z, Q and a partial assignment $\alpha : V_1 \to D$ with $V_1 \subseteq V$ we have that α can be extended to a $(\leq^{\mathcal{O}}_{(P,Z)})$-min model of φ iff the algorithm Lattice returns **true**. The algorithm Lattice works in polynomial time.*

We now turn to the hardness result. We use the notation $Z^{c,d}_{a,b}$, where $a \neq c$ and $b \neq d$, for the relation $\{(a, b), (a, d), (c, d)\}$.

Lemma 4. *Let $\mathcal{O} = (D; \leq^{\mathcal{O}})$ and $a, b, c, d, e, f \in D$ such that $a \neq b$, $c <^{\mathcal{O}} d$ and $e <^{\mathcal{O}} f$. Then $\mathrm{GMININF}(\Gamma, \mathcal{O})$ where $\Gamma = \{Z^{a,d}_{b,c}, Z^{b,f}_{a,e}\}$ is coNP-hard.*

Finally we give the dichotomy.

Parameters: a linear order $\mathcal{O} = (D; \leq^{\mathcal{O}})$, a constraint language Γ over D such that Γ is preserved by both the meet $\sqcap$ and the join $\sqcup$ of the lattice $\mathcal{L} = (D; \leq^{\mathcal{L}})$ with $\leq^{\mathcal{L}} \subseteq \leq^{\mathcal{O}}$.

Data: A Γ-formula φ over variables V partitioned into P, Z, Q and a partial assignment $\alpha : V_1 \to D$ for some $V_1 \subseteq V$.

Result: If there exists a $(\leq^{\mathcal{O}}_{(P,Z)})$-min model of φ extending α, then `true`, else `false`.

1. $Q_2 := Q \setminus V_1$
2. $\alpha_G := \text{greatext}(\mathcal{O}, \varphi, \alpha, V_1 \cup Q_2)$
3. If $(\alpha_G == \text{false})$ return `false`
4. $\beta_G := \text{leastext}(\mathcal{O}, \varphi, \alpha_G, V)$
5. If $(\beta_G == \text{false})$ return `false`
6. $\beta_{CG} := \text{leastext}(\mathcal{O}, \varphi, \beta_G[Q], V)$
7. If $(\beta_G[P] == \beta_{CG}[P])$ return `true`
8. Return `false`

Fig. 2. Algorithm lattice

Theorem 5. *Let $\mathcal{O} = (D; \leq^{\mathcal{O}})$ be a linear order and $\sqcup, \sqcap$ the meet and the join of some lattice $\mathcal{L} = (D, \leq^{\mathcal{L}})$. Then we have one of the following.*

- *If $\leq^{\mathcal{L}} \subseteq \leq^{\mathcal{O}}$ or $(\leq^{\mathcal{L}})^{-1} \subseteq \leq^{\mathcal{O}}$, then GMININF($\text{Inv}(\{\sqcap, \sqcup\}), \mathcal{O}$) is in P.*
- *If neither $\leq^{\mathcal{L}} \subseteq \leq^{\mathcal{O}}$ nor $(\leq^{\mathcal{L}})^{-1} \subseteq \leq^{\mathcal{O}}$, then GMININF($\text{Inv}(\{\sqcap, \sqcup\}), \mathcal{O}$) is coNP-hard.*

6 Semilattice and a *pms*-operation

In this section we consider GMININF($\text{Inv}(\{\sqcap, f\}), \mathcal{O}$) in the case where $\sqcap$ is the meet of some meet-semilattice $\mathcal{L} = (D, \leq^{\mathcal{L}})$ and f is a *pms*-operation compliant with $\sqcap$. When it comes to examples of relations preserved by both operations, it is straightforward to prove that all relations definable by conjunctions of clauses of the form $(x \geq^{\mathcal{L}} d)$ and $(\neg x_1 \geq^{\mathcal{L}} d_1 \vee \cdots \vee \neg x_k \geq^{\mathcal{L}} d_k)$ are in $\text{Inv}(\{\sqcap, f\})$ where f is a pms-operation that for three pairwise different $d_1, d_2, d_3 \in D$ returns $(d_1 \sqcap d_2 \sqcap d_3)$.

We now turn to the complexity of the problem. Again we work rather with GMINEXT($\text{Inv}, (\{\sqcap, f\}), \mathcal{O}$) than GMININF($\text{Inv}(\{\sqcap, f\}), \mathcal{O}$). The procedure in Fig. 3 solves the problem in P on the condition that $\leq^{\mathcal{L}} \subseteq \leq^{\mathcal{O}}$.

Lemma 5. *Let $\mathcal{O} = (D; \leq^{\mathcal{O}})$ be a linear order and Γ a constraint language preserved by both the meet $\sqcap$ of some meet-semilattice $\mathcal{L} = (D; \leq^{\mathcal{L}})$ such that $\leq^{\mathcal{L}} \subseteq \leq^{\mathcal{O}}$ and a pms-operation f compliant with $\sqcap$. For a given Γ-formula φ over variables V partitioned into P, Z, Q and a partial assignment $\alpha : V_1 \to D$ with $V_1 \subseteq V$ we have that α can be extended to a $(\leq^{\mathcal{O}}_{(P,Z)})$-min model of φ iff the algorithm MeetPMS returns `true`. The algorithm MeetPMS works in polynomial time.*

Parameters: a linear order $\mathcal{O} = (D; \leq^{\mathcal{O}})$, a constraint language Γ over D preserved by both the meet $\sqcap$ of some meet-semilattice $\mathcal{L} = (D; \leq^{\mathcal{L}})$ such that $\leq^{\mathcal{L}} \subseteq \leq^{\mathcal{O}}$ and a *pms*-operation compliant with $\sqcap$.

Data: A Γ-formula φ over variables V partitioned into P, Z, Q and a partial assignment $\alpha : V_1 \to D$ for some $V_1 \subseteq V$.

Result: If there exists a $(\leq^{\mathcal{O}}_{(P,Z)})$-min model of φ extending α, then `true`, else `false`.

1. $\beta_L := \text{leastext}(\mathcal{O}, \varphi, \alpha, V)$
2. If $(\beta_L == \text{false})$ return `false`
3. $\beta_{CL} := \text{leastext}(\mathcal{O}, \varphi, \beta_L[Q], V)$
4. If $(\beta_L[P] == \beta_{CL}[P])$ return `true`
5. Return `false`

Fig. 3. Algorithm MeetPMS

We close this section by complementing Lemma 5 with a hardness result.

Theorem 6. *Let $\mathcal{O} = (D; \leq^{\mathcal{O}})$ be a linear order, $\sqcap$ the meet of some meet-semilattice $\mathcal{L} = (D; \leq^{\mathcal{L}})$ and f a pms-operation compliant with $\sqcap$. Then we have one of the following:*

- *If $\leq^{\mathcal{L}} \subseteq \leq^{\mathcal{O}}$, then GMININF($Inv(\{\sqcap, f\}), \mathcal{O}$) is in P.*
- *If $\leq^{\mathcal{L}} \nsubseteq \leq^{\mathcal{O}}$, then GMININF($Inv(\{\sqcap, f\}), \mathcal{O}$) is coNP-hard.*

7 Summary and Future Work

In this article we have systematically studied the complexity of the minimal inference problem over arbitrary finite domains. We considered a version of the problem parameterized by a constraint language and a linear order. We provided a dichotomy for maximal conservative languages: between Π_2^P-complete and coNP-complete and two tractability results complemented by coNP-hardness results. This gives two dichotomies: between coNP-complete and in P. Furthermore, one of the tractability results is based on a newly discovered operation: a pms-operation. Identifying tractable classes and appropriate polymorphisms is crucial when one works in algebraic approach.

We believe that our research will soon result in more advanced classifications, e.g., for all conservative languages or all languages over the three-element domain. Both classifications were provided for the CSP, see [2,3].

References

1. Bulatov, A.A.: A graph of a relational structure and constraint satisfaction problems. In: Proceedings of the Symposium on Logic in Computer Science (LICS), Turku, Finland (2004)

2. Bulatov, A.A.: A dichotomy theorem for constraint satisfaction problems on a 3-element set. J. ACM **53**(1), 66–120 (2006)
3. Bulatov, A.A.: Complexity of conservative constraint satisfaction problems. ACM Trans. Comput. Log. **12**(4), 24 (2011)
4. Bulatov, A.A., Krokhin, A.A., Jeavons, P.: The complexity of maximal constraint languages. In: Proceedings of the Symposium on Theory of Computing (STOC), pp. 667–674 (2001)
5. Bulatov, A.A., Krokhin, A.A., Jeavons, P.G.: Classifying the complexity of constraints using finite algebras. SIAM J. Comput. **34**, 720–742 (2005)
6. Cadoli, M., Lenzerini, M.: The complexity of propositional closed world reasoning and circumscription. J. Comput. Syst. Sci. **48**(2), 255–310 (1994)
7. Doherty, P., Kachniarz, J., Szałas, A.: Using contextually closed queries for local closed-world reasoning in rough knowledge databases. In: Skowron, A., Polkowski, L., Pal, S.K. (eds.) Rough-Neural Computing: Techniques for Computing with Words. Cognitive Technologies, pp. 219–250. Springer, Heidelberg (2004)
8. Durand, A., Hermann, M.: The inference problem for propositional circumscription of affine formulas is coNP-complete. In: STACS, pp. 451–462 (2003)
9. Durand, A., Hermann, M., Nordh, G.: Trichotomies in the complexity of minimal inference. Theor. Comput. Syst. **50**(3), 446–491 (2012)
10. Eiter, T., Gottlob, G.: Propositional circumscription and extended closed-world reasoning are IIp2-complete. Theor. Comput. Sci. **114**(2), 231–245 (1993)
11. Gelfond, M., Przymusinska, H., Przymusinski, T.C.: On the relationship between circumscription and negation as failure. Artif. Intell. **38**(1), 75–94 (1989)
12. Jeavons, P., Cohen, D., Gyssens, M.: Closure properties of constraints. J. ACM **44**(4), 527–548 (1997)
13. Jeavons, P., Cooper, M.C.: Tractable constraints on ordered domains. Artif. Intell. **79**(2), 327–339 (1995)
14. Jonsson, P.: The complexity of mincsp and csp, personal communication
15. Kirousis, L.M., Kolaitis, P.G.: A dichotomy in the complexity of propositional circumscription. Theor. Comput. Syst. **37**(6), 695–715 (2004)
16. Krokhin, A.A., Zivny, S. (eds.): The Constraint Satisfaction Problem: Complexity and Approximability, Dagstuhl Follow-Ups, vol. 7. Schloss Dagstuhl - Leibniz-Zentrum fuer Informatik (2017)
17. McCarthy, J.: Circumscription - a form of non-monotonic reasoning. Artif. Intell. **13**(1–2), 27–39 (1980)
18. McCarthy, J.: Applications of circumscription to formalizing common-sense knowledge. Artif. Intell. **28**(1), 89–116 (1986)
19. Nordh, G., Jonsson, P.: An algebraic approach to the complexity of propositional circumscription. In: Proceedings of the Symposium on Logic in Computer Science (LICS), pp. 367–376 (2004)
20. Przymusinski, T.C.: Three-valued nonmonotonic formalisms and semantics of logic programs. Artif. Intell. **49**(1–3), 309–343 (1991)
21. Rosenberg, I.G.: Minimal Clones I: The Five Types. Lectures in Universal Algebra (Proc. Conf. Szeged, 1983). Colloq. Math. Soc. J. Bolyai, vol. 43, pp. 405–427 (1986)
22. Schaefer, T.J.: The complexity of satisfiability problems. In: Proceedings of the Symposium on Theory of Computing (STOC), pp. 216–226 (1978)

Answer Set Programming

Gelfond-Zhang Aggregates as Propositional Formulas

Pedro Cabalar[1], Jorge Fandinno[1], Torsten Schaub[2(✉)],
and Sebastian Schellhorn[2]

[1] University of Corunna, A Coruña, Spain
{cabalar,jorge.fandino}@udc.es
[2] University of Potsdam, Potsdam, Germany
{torsten,seschell}@cs.uni-potsdam.de

Abstract. We show that any ASP aggregate interpreted under Gelfond and Zhang's (GZ) semantics can be replaced (under strong equivalence) by a propositional formula. Restricted to the original GZ syntax, the resulting formula is reducible to a disjunction of conjunctions of literals but the formulation is still applicable even when the syntax is extended to allow for arbitrary formulas (including nested aggregates) in the condition. Once GZ-aggregates are represented as formulas, we establish a formal comparison (in terms of the logic of Here-and-There) to Ferraris' (F) aggregates, which are defined by a different formula translation involving nested implications. In particular, we prove that if we replace an F-aggregate by a GZ-aggregate in a rule head, we do not lose answer sets (although more can be gained). This extends the previously known result that the opposite happens in rule bodies, i.e., replacing a GZ-aggregate by an F-aggregate in the body may yield more answer sets. Finally, we characterise a class of aggregates for which GZ- and F-semantics coincide.

1 Introduction

Answer Set Programming (ASP [3]) has become an established problem-solving paradigm and a prime candidate for practical Knowledge Representation and Reasoning (KRR). The reasons for this success are manifold. The most obvious one is the availability of effective solvers [8,12] and a growing list of covered application domains [6]. A probably equally important reason is its declarative semantics, having been generalized from the original *stable models* [13] of normal logic programs up to arbitrary first-order [11,18] and infinitary [15] formulas. Several logical characterizations of ASP have been obtained, among which *Equilibrium Logic* [17] and its monotonic basis, the intermediate logic of *Here-and-There* (HT), are arguably the most prominent ones. These generalisations have allowed us to understand ASP as a general logical framework for Non-Monotonic Reasoning. Finally, a third relevant cause of ASP's success lies in its flexible specification

Partially supported by grants GPC 2016/035 (Xunta de Galicia, Spain), TIN 2013-42149-P (MINECO, Spain), and SCHA 550/9 (DFG, Germany).

M. Balduccini and T. Janhunen (Eds.): LPNMR 2017, LNAI 10377, pp. 117–131, 2017.
DOI: 10.1007/978-3-319-61660-5_12

language [5], offering constructs especially useful for practical KRR. Some of its distinctive constructs are *aggregates*, allowing for operations on sets of elements such as counting the number of instances for which a formula holds, or adding all the integer values for some predicate argument. Unfortunately, there is no clear agreement on the expected behavior of aggregates in ASP, and several alternative semantics have been defined [7,10,14,19–21], among which perhaps Ferraris' [10] and Faber et al's [7] are the two more consolidated ones due to their respective implementations in the ASP solvers clingo [12] and DLV [8]. Although these two approaches may differ when the aggregates are in the scope of default negation, they coincide for the rest of cases (like all the examples in this paper), even when aggregates are involved in recursive definitions. Ferraris' (F-)aggregates additionally show a remarkable feature: they can be expressed as propositional formulas in the logic of HT, something that greatly simplifies their formal treatment. To illustrate this, let us explore the simple rule:

$$p(a) \leftarrow \text{count}\{X : p(X)\} \geq n. \tag{1}$$

where $p(a)$ recursively depends on the number of atoms of the form $p(X)$. Suppose first that $n = 1$. Since the domain only contains a, $\text{count}\{X : p(X)\} \geq 1$ is true iff $p(a)$ holds. This is captured in Ferraris' translation of (1) for $n = 1$ that amounts to $p(a) \leftarrow p(a)$, a tautology whose only stable model is $\emptyset$. Suppose now that $n = 0$. Then, the aggregate is considered as tautological and the HT-translation of (1) corresponds to $p(a) \leftarrow \top$ whose unique stable model is $\{p(a)\}$. Finally, as one more elaboration, assume $n = 1$ and suppose we add the fact $p(b)$. Then, (1) becomes the formula $p(a) \leftarrow p(a) \vee p(b)$ that, together with fact $p(b)$, is HT-equivalent to:

$$p(b). \qquad p(a) \leftarrow p(a). \qquad p(a) \leftarrow p(b). \tag{2}$$

This results in the unique stable model $\{p(a), p(b)\}$.

Recently, Gelfond and Zhang [14] (GZ) proposed a more restrictive interpretation of recursive aggregates that imposes the so-called *Vicious Circle Principle*, namely, *"no object or property can be introduced by the definition referring to the totality of objects satisfying this property."* According to this principle, if we have a program whose only definition for $p(a)$ is (1), we may leave $p(a)$ false, but we cannot be forced to derive its truth, since it depends on a set of atoms $\{X : p(X)\}$ that includes $p(a)$ itself. In this way, if $n = 1$, the GZ-stable model for (1) is also $\emptyset$, as there is no need to assume $p(a)$. However, if we have $n = 0$, we cannot leave $p(a)$ false any more (the rule would have a true body and a false head) and, at the same time, $p(a)$ cannot become true because it depends on a vicious circle. Something similar happens for $n = 1$ when adding fact $p(b)$. As shown in [2], GZ-programs are stronger than F-programs in the sense that, when they represent the same problem, any GZ-stable model is also an F-stable model, but the opposite may not hold (as we saw in the examples above). Without entering a discussion of which semantics is more intuitive or suitable for practical purposes, one objective disadvantage of GZ-aggregates is that they lacked a logical representation so far; they were exclusively defined in terms of a reduct, something that made their formal analysis more limited and the comparison to F-aggregates more cumbersome.

In this paper, we show that, in fact, it is also possible to understand a GZ-aggregate as a propositional formula, classically equivalent to the F-aggregate translation, but with a *different meaning* in HT. For instance, the GZ-translation for (1) with $n = 1$ coincides with the F-encoding $p(a) \leftarrow p(a)$, but if we change to $n = 0$ we get the formula $p(a) \leftarrow p(a) \vee \neg p(a)$ whose antecedent is valid in classical logic, but not in HT. In fact, the whole formula is HT-equivalent to the program:

$$p(a) \leftarrow p(a). \qquad p(a) \leftarrow \neg p(a).$$

This makes it now obvious that there is no stable model. Similarly, when we add fact $p(b)$ and $n = 1$, the GZ-translation eventually leads to the propositional program:

$$p(b). \quad p(a) \leftarrow p(a) \wedge \neg p(b). \quad p(a) \leftarrow \neg p(a) \wedge p(b). \quad p(a) \leftarrow p(a) \wedge p(b). \quad (3)$$

Again, it is classically equivalent to the F-translation (2), but quite different in logic programming, where the third rule and fact $p(b)$ enforce the non-existence of stable models.

The rest of the paper is organized as follows. In the next section, we review some basic definitions that will be needed through the paper. In Sect. 3, we present a generalisation of Ferraris' reduct that covers GZ-aggregates and show that the latter can be replaced, under strong equivalence, by a propositional formula. In Sect. 4, we show that, in general, GZ-aggregates are stronger than F-aggregates in HT and, as a result, characterise the effect of replacing some occurrence of a GZ-aggregate by a corresponding F-aggregate. We also identify a family of aggregates in which both semantics coincide. Finally, Sect. 5 concludes the paper.

2 Background

We begin by introducing some basic definitions used in the rest of the paper. Let $\mathcal{L}$ be some syntactic language and assume we have a definition of *stable model* for any theory $\Gamma \subseteq \mathcal{L}$ in that syntax. Moreover, let $SM(\Gamma)$ denote the stable models of Γ. Two theories Γ, Γ' are *strongly equivalent*, written $\Gamma \equiv_s \Gamma'$, iff $SM(\Gamma \cup \Delta) = SM(\Gamma' \cup \Delta)$ for any theory Δ. We will provide a stronger definition of $\equiv_s$ for expressions in $\mathcal{L}$. Let $\Gamma(\varphi)$ denote some theory with a *distinguished occurrence* of a subformula φ and let $\Gamma(\psi)$ be the result of replacing that occurrence φ by ψ in $\Gamma(\varphi)$. Two expressions $\varphi, \psi \in \mathcal{L}$ are said to be *strongly equivalent*, also written $\varphi \equiv_s \psi$, when $\Gamma(\varphi) \equiv_s \Gamma(\psi)$ for an arbitrary[1] $\Gamma(\varphi) \subseteq \mathcal{L}$. We also recall

---

[1] Note that, for arbitrary languages and semantics, this definition is stronger than usual, as it refers to *any subformula replacement* and not just conjunctions of formulas, as usual. For instance, $\varphi \equiv_s \psi$ also implies $\{\varphi \otimes \alpha\} \equiv_s \{\psi \otimes \alpha\}$ for any binary operator $\otimes$ in our language. When $\mathcal{L}$ is a logical language and $\equiv_s$ amounts to equivalence in HT (or any logic with substitution of equivalents) the distinction becomes irrelevant.

next some basic definitions and results related to the logic of *Here-and-There* (HT). Let *At* be a set of ground atoms called the *propositional signature*. A *propositional formula* φ is defined using the grammar:

$$\varphi ::= \bot \mid a \mid \varphi \wedge \varphi \mid \varphi \vee \varphi \mid \varphi \rightarrow \varphi \qquad \text{for any } a \in At.$$

We use Greek letters φ and ψ and their variants to stand for propositional formulas. We define the derived operators $\neg\varphi \stackrel{\text{def}}{=} (\varphi \rightarrow \bot)$, $\varphi \leftrightarrow \psi \stackrel{\text{def}}{=} (\varphi \rightarrow \psi) \wedge (\psi \rightarrow \varphi)$ and $\top \stackrel{\text{def}}{=} \neg\bot$. A propositional formula in which every occurrence of an implication is of the form $\neg\varphi = \varphi \rightarrow \bot$ is called a *nested expression*.

A *classical interpretation* T is a set of atoms $T \subseteq At$. We write $\models_{cl}$ to stand both for classical satisfaction and entailment, and $\equiv_{cl}$ represents classical equivalence. An HT-*interpretation* is a pair $\langle H, T \rangle$ of sets of atoms $H \subseteq T \subseteq At$. An interpretation $\langle H, T \rangle$ *satisfies* a formula φ, written $\langle H, T \rangle \models \varphi$, if any of the following recursive conditions holds:

- $\langle H, T \rangle \not\models \bot$
- $\langle H, T \rangle \models p$ iff $p \in H$
- $\langle H, T \rangle \models \varphi \wedge \psi$ iff $\langle H, T \rangle \models \varphi$ and $\langle H, T \rangle \models \psi$
- $\langle H, T \rangle \models \varphi \vee \psi$ iff $\langle H, T \rangle \models \varphi$ or $\langle H, T \rangle \models \psi$
- $\langle H, T \rangle \models \varphi \rightarrow \psi$ iff both (i) $T \models_{cl} \varphi \rightarrow \psi$ and (ii) $\langle H, T \rangle \not\models \varphi$ or $\langle H, T \rangle \models \psi$

It is not difficult to see that $\langle T, T \rangle \models \phi$ iff $T \models_{cl} \phi$. A *(propositional) theory* is a set of propositional formulas. An interpretation $\langle H, T \rangle$ is a *model* of a theory Γ when $\langle H, T \rangle \models \varphi$ for all $\varphi \in \Gamma$. A theory Γ *entails* a formula φ, written $\Gamma \models \varphi$, when all models of Γ satisfy φ. Two theories Γ, Γ' are (HT)-*equivalent*, written $\Gamma \equiv \Gamma'$, if they share the same set of models.

Definition 1 (equilibrium/stable model). *A total interpretation* $\langle T, T \rangle$ *is an equilibrium model of a formula* φ *iff* $\langle T, T \rangle \models \varphi$ *and there is no* $H \subset T$ *such that* $\langle H, T \rangle \models \varphi$. *If so, we say that* T *is a* stable model *of* φ. $\qquad\square$

Proposition 1. *The following are general properties of HT:*

1. *if* $\langle H, T \rangle \models \varphi$ *then* $\langle T, T \rangle \models \varphi$ *(i.e.,* $T \models_{cl} \varphi$*)*
2. $\langle H, T \rangle \models \neg\varphi$ *iff* $T \models_{cl} \neg\varphi$

$\qquad\square$

Definition 2 (Ferraris' reduct). *The reduct of a formula* φ *with respect to an interpretation* T, *written* φ^T, *is defined as:*

$$\varphi^T \stackrel{\text{def}}{=} \begin{cases} \bot & \text{if } T \not\models_{cl} \varphi \\ a & \text{if } \varphi \text{ is some atom } a \in T \\ \varphi_1^T \otimes \varphi_2^T & \text{if } T \models_{cl} \varphi \text{ and } \varphi = (\varphi_1 \otimes \varphi_2) \text{ for some } \otimes \in \{\wedge, \vee, \rightarrow\} \end{cases}$$

That is, φ^T is the result of replacing by $\bot$ each maximal subformula ψ of φ s.t. $T \not\models_{cl} \psi$.

Proposition 2 (Lemma 1 in [9]). *For any pair of interpretations* $H \subseteq T$ *and any* φ: $H \models_{cl} \varphi^T$ *iff* $\langle H, T \rangle \models \varphi$. $\qquad\square$

As is well-known, strong equivalence for propositional formulas corresponds to HT-equivalence [16], that is, $\varphi \equiv_s \psi$ iff $\varphi \equiv \psi$ in that language. The following result follows from Corollary 3 in [1].

Proposition 3. *If $\varphi \models \psi$ and $\varphi \equiv_{cl} \psi$, then $SM(\varphi) \supseteq SM(\psi)$.*   □

In other words, if φ is stronger than ψ in HT (and so, in classical logic too), but they further happen to be classically equivalent, then φ is weaker with respect to stable models. As an example, note that $(p \lor q) \models (\neg p \to q)$. As they are classically equivalent, $SM(p \lor q) \supseteq SM(\neg p \to q)$ which is not such a strong result. However, since HT-entailment is monotonic with respect to conjunction, it follows that $(p \lor q) \land \gamma \models (\neg p \to q) \land \gamma$ also holds for any γ, and thus, if we replace a disjunctive rule $(p \lor q)$ by $(\neg p \to q)$ in any program we may lose some stable models, but the remaining are still applicable to $(p \lor q)$. We can generalize this behavior not only on conjunctions, but also to cover the replacement of any subformula φ. We say that an occurrence φ of a formula is *positive* in a theory $\Gamma(\varphi)$ if the number of implications in $\Gamma(\varphi)$ containing occurrence φ in the antecedent is even. It is called *negative* otherwise.

Proposition 4. *Let φ and ψ be two formulas satisfying $\varphi \models \psi$ and $\varphi \equiv_{cl} \psi$. Then:*

1. *$SM(\Gamma(\varphi)) \supseteq SM(\Gamma(\psi))$ for any theory $\Gamma(\varphi)$ where occurrence φ is positive;*
2. *$SM(\Gamma(\varphi)) \subseteq SM(\Gamma(\psi))$ for any theory $\Gamma(\varphi)$ where occurrence φ is negative.*   □

Back to the example, note that $(p \lor q)$ occurs positively in $(p \lor q) \land \gamma$ and so, $(\neg p \to q) \land \gamma$ has a subset of stable models. On the other hand, it occurs negatively in $(p \lor q) \to \gamma$, and so, $(\neg p \to q) \to \gamma$ has a superset of stable models.

3   Aggregates as Formulas

To deal with aggregates, we consider a simplified first order signature $\Sigma = \langle \mathcal{C}, \mathcal{A}, \mathcal{P} \rangle$ formed by three pairwise disjoint sets respectively called *constants*, *aggregate symbols* and *predicate symbols*. Since we focus on translations to propositional expressions, we do not consider functions or arithmetic operations[2] (other than aggregates). As a result, a *term* can only be either a constant $c \in \mathcal{C}$ or a variable X. We use boldface symbols to represent tuples of terms, such as $\boldsymbol{t}$, and write $|\boldsymbol{t}|$ to stand for the tuple's arity. As usual, a *predicate atom* (or *atom* for short) is an expression of the form $p(\boldsymbol{t})$ where $\boldsymbol{t}$ is a tuple of terms; moreover, $p(\boldsymbol{t})$ is said to be *ground* iff all its terms are constants $\boldsymbol{t} \subseteq \mathcal{C}^{|\boldsymbol{t}|}$. We write $At(\mathcal{C}, \mathcal{P})$ to stand for the set of ground atoms for predicates $\mathcal{P}$ and constants $\mathcal{C}$. A *literal* is either an atom a (positive literal) or its default negation *not* a (negative literal). An *(aggregate) formula* φ is recursively defined by the following grammar:

$$\varphi ::= \perp \mid p(\boldsymbol{t}) \mid f\{\boldsymbol{X} : \varphi\} \prec n \mid f\{\boldsymbol{c} : \varphi, \ldots, \boldsymbol{c} : \varphi\} \prec n \mid \varphi \land \varphi \mid \varphi \lor \varphi \mid \varphi \to \varphi$$

---

[2] An extension to a full first-order language is under development.

where $p(t)$ is an atom, $f \in \mathcal{A}$ is an aggregate symbol, X is a non-empty tuple of variables, c is a non-empty tuple of constants, $\prec \in \{=, \neq, \leq, \geq, <, >\}$ is an *arithmetic relation*, and $n \in \mathbb{Z}$ is an integer constant. A *(aggregate) theory* is a set of aggregate formulas. As we can see, we have two types of *aggregates*: $f\{X : \varphi\} \prec n$ called *GZ-aggregates* (or *set atoms*); and $f\{c_1 : \varphi_1, \ldots, c_m : \varphi_m\} \prec n$, with c_i of same arity, called *F-aggregates*. This syntactic distinction respects the original syntax[3] and also turns out to be convenient for comparison purposes, since we can assign a different semantics to each type of aggregate without ambiguity. An important observation is that, in our general language, it is possible to nest GZ and F-aggregates in a completely arbitrary way, since φ and $\varphi_1, \ldots, \varphi_m$ inside brackets are aggregate formulas in their turn. Achieving this generalisation is not surprising, once aggregates can be seen as propositional formulas. A *GZ-formula* (resp. *F-formula*) is one in which all its aggregates are GZ-aggregates (resp. F-aggregates). We sometimes informally talk about *rules* (resp. *programs*) instead of formulas (resp. theories) when the syntax coincides with the usual in logic programming.

The technical treatment of F-aggregates is directly extracted from [10], so the focus in this section is put on GZ-aggregates, where our contribution lies. One of their distinctive features is the use of variables X. In fact, a formula φ inside $A = f\{X : \varphi\} \prec n$ (called the *condition* of A) normally contains occurrences of X, so we usually write it as $cond(X)$. Moreover, the occurrences of variables X in A are said to be *bound* to A. A variable occurrence X in a formula φ is *free* if it is not bound to any GZ-aggregate in φ. A formula or theory is *closed* iff it contains no free variables. We define the grounding of a formula $\varphi(X)$ with free variables X as expected: $\mathrm{Gr}(\varphi(X)) \stackrel{\mathrm{def}}{=} \{\varphi(c) \mid c \in \mathcal{C}^{|X|}\}$. In the rest, we assume that all theories are closed. This is not a limitation, since a non-closed formula $\varphi(X)$ in some theory can be understood as an abbreviation of its grounding $\mathrm{Gr}(\varphi(X))$, as usual. Given a set of formulas S, we write $\bigwedge S$ and $\bigvee S$ to stand for their conjunction and disjunction, respectively; we let $\bigvee \emptyset = \bot$ and $\bigwedge \emptyset = \top$.

To define the semantics, we assume that for all aggregate symbols $f \in \mathcal{A}$ and arities $m \geq 1$, there exists a predefined associated partial function $\hat{f}_m : 2^{\mathcal{C}^m} \rightarrow \mathbb{Z}$ that, for each set S of m-tuples of constants, either returns a number $\hat{f}_m(S)$ or is undefined. This predefined value is the expected one for the usual aggregate functions $\mathtt{sum}, \mathtt{count}, \mathtt{max}$, etc. For example, for aggregate symbol $\mathtt{sum}$ and arity $m = 1$ the function returns the aggregate addition when the set consists of (1-tuples of) integer numbers. For instance, $\widehat{\mathtt{sum}}_1(\{7, 2, -4\}) = 5$ and $\widehat{\mathtt{sum}}_1(\emptyset) = 0$ but $\widehat{\mathtt{sum}}_1(\{3, a, 3, b\})$ is undefined. For integer aggregate functions of arity $m > 1$, we assume that the aggregate is applied on the leftmost elements in the tuples when all of them are integer, so that, for instance, $\widehat{\mathtt{sum}}_2(\{\langle 7, a\rangle, \langle 2, b\rangle, \langle 2, a\rangle\}) = 11$. We omit the arity when clear from the context.

A *classical interpretation* T is a set of ground atoms $T \subseteq At(\mathcal{C}, \mathcal{P})$.

Definition 3. *A classical interpretation T satisfies a formula φ, denoted by $T \models_{cl} \varphi$, iff*

---

[3] Ferraris actually uses $\varphi_i = w$ rather than $c : \varphi$, but this is not a substantial difference, assuming w is the first element in tuple c.

1. $T \not\models_{cl} \bot$
2. $T \models_{cl} p(\boldsymbol{c})$ iff $p(\boldsymbol{c}) \in T$ for any ground atom $p(\boldsymbol{c})$
3. $T \models_{cl} f\{\boldsymbol{X} : cond(\boldsymbol{X})\} \prec n$ iff $\hat{f}_{|\boldsymbol{X}|}(\{\ \boldsymbol{c} \in C^{|\boldsymbol{X}|}\ |\ T \models_{cl} cond(\boldsymbol{c})\ \})$ has some value $k \in \mathbb{Z}$ and $k \prec n$ holds for the usual meaning of arithmetic relation $\prec$
4. $T \models_{cl} f\{\boldsymbol{c}_1 : \varphi_1, \ldots, \boldsymbol{c}_m : \varphi_m\} \prec n$ iff $\hat{f}_{|\boldsymbol{c}_1|}(\{\ \boldsymbol{c}_i\ |\ T \models_{cl} \varphi_i\ \})$ has some value $k \in \mathbb{Z}$ and, again, $k \prec n$ holds for its usual meaning
5. $T \models_{cl} \varphi \wedge \psi$ iff $T \models_{cl} \varphi$ and $T \models_{cl} \psi$
6. $T \models_{cl} \varphi \vee \psi$ iff $T \models_{cl} \varphi$ or $T \models_{cl} \psi$
7. $T \models_{cl} \varphi \rightarrow \psi$ iff $T \not\models_{cl} \varphi$ or $T \models_{cl} \psi$.

We say that T is a model *of a theory Γ iff $T \models_{cl} \varphi$ for all $\varphi \in \Gamma$.*   □

Given interpretation T, we divide any theory Γ into the two disjoint subsets:

$$\Gamma_T^+ \stackrel{\text{def}}{=} \{\varphi \in \Gamma \mid T \models_{cl} \varphi\} \qquad \Gamma_T^- \stackrel{\text{def}}{=} \Gamma \setminus \Gamma_T^+$$

that is, the formulas in Γ satisfied by T and not satisfied by T, respectively. When set Γ is parametrized, say $\Gamma(z)$, we write $\Gamma_T^+(z)$ and $\Gamma_T^-(z)$ instead of $\Gamma(z)_T^+$ and $\Gamma(z)_T^-$. For instance, $\mathtt{Gr}_T^+(\varphi)$ collects the formulas from $\mathtt{Gr}(\varphi)$ satisfied by T.

Definition 4 (reduct). *The* reduct *of a GZ-aggregate $A = f\{\boldsymbol{X} : cond(\boldsymbol{X})\} \prec n$ with respect to a classical interpretation T is the formula:*

$$A^T \stackrel{\text{def}}{=} \begin{cases} \bot & \text{if } T \not\models_{cl} A \\ \big(\bigwedge \mathtt{Gr}_T^+(cond(\boldsymbol{X})) \big)^T & \text{otherwise} \end{cases}$$

The reduct of an F-aggregate $B = f\{\boldsymbol{c}_1 : \varphi_1, \ldots, \boldsymbol{c}_m : \varphi_m\} \prec n$ is the formula:

$$B^T \stackrel{\text{def}}{=} \begin{cases} \bot & \text{if } T \not\models_{cl} B \\ f\{\boldsymbol{c}_1 : \varphi_1^T, \ldots, \boldsymbol{c}_m : \varphi_m^T\} \prec n & \text{otherwise} \end{cases}$$

The reduct of any other formula is just as in Definition 2. The reduct of a theory is the set of reducts of its formulas.   □

Definition 5 (stable model). *A classical interpretation T is a* stable model *of a theory Γ iff T is a $\subseteq$-minimal model of Γ^T.*   □

Note that, when restricted to F-formulas, Definition 3 exactly matches the reduct definition for aggregate theories by Ferraris [10]. On the other hand, when restricted to GZ-formulas, it generalises the reduct definition by Gelfond-Zhang [14] allowing arbitrary formulas in $cond(\boldsymbol{X})$, including nested aggregates. For this reason, in our setting, the reduct is inductively applied to $(\bigwedge \mathtt{Gr}_T^+(cond(\boldsymbol{X})))^T$. In the original case [14], $cond(\boldsymbol{X})$ was a conjunction of atoms, but it is straightforward to see that, then, $(\bigwedge \mathtt{Gr}_T^+(cond(\boldsymbol{X})))^T = \bigwedge \mathtt{Gr}_T^+(cond(\boldsymbol{X}))$. To sum up, the above definitions of stable model and reduct correspond to the original ones for Ferraris [10] and Gelfond-Zhang [14] when restricted to their respective syntactic fragments.

Proposition 5. *Stable models are classical models.* □

Example 1. Let P_1 be the program consisting of fact $p(b)$ and rule (1) with $n = 1$, and let A_1 be the GZ-aggregate in that rule. We show that P_1 has no stable models. Given ground atoms $p(a)$ and $p(b)$, the only model of the program is $T = \{p(a), p(b)\}$, since $p(b)$ is fixed as a fact, and so, A_1 must be true, so (1) entails $p(a)$ too. Since $T \models_{cl} A_1$, the reduct becomes $A_1^T = p(a) \land p(b)$, and so, P_1^T contains the rules $p(b)$ and $p(a) \leftarrow p(a) \land p(b)$, being their least model $\{p(b)\}$, so T is not stable. As another example of the aggregate reduct, if we took $T = \emptyset$, then $T \not\models_{cl} A_1$ and $A_1^T = \bot$. □

Example 2. As an example of nested GZ-aggregates, consider:

$$A_2 = \text{count}\{\ X : \text{sum}\{Y : owns(X, Y)\} \geq 10\ \} \geq 2$$

and imagine that $owns(X, Y)$ means that X owns some item Y whose cost is also Y. Accordingly, A_2 checks whether there are 2 or more persons X that own items for a total cost of at least 10. Suppose we have the interpretation:

$$T = \{owns(a, 6), owns(a, 8), owns(b, 2), owns(b, 3), owns(c, 12)\}$$

Then A_2 holds in T since both a and c have total values greater than 10: 14 for a and 12 for c. Therefore, A_2^T corresponds to $(\bigwedge \text{Gr}_T^+(\text{sum}\{Y : owns(X, Y)\} \geq 10))^T$. After grounding free variable X, we obtain:

$$(\text{sum}\{Y : owns(a, Y)\} \geq 10 \land \text{sum}\{Y : owns(c, Y)\} \geq 10)^T$$

Note that b does not occur, since its total sum is lower than 10 in T. If we apply again the reduct to the conjuncts above, we eventually obtain the conjunction: $owns(a, 6) \land owns(a, 8) \land owns(c, 12)$.

□

Proposition 6. *Given formulas φ and ψ, the following conditions hold:*
1. *$H \models_{cl} \varphi^T$ implies $T \models_{cl} \varphi$, and*
2. *$T \models_{cl} \varphi^T$ iff $T \models_{cl} \varphi$*

for any pair of interpretations $H \subseteq T$. Furthermore, the following condition also holds
1. *if $H \models_{cl} \varphi^T$ iff $H \models_{cl} \psi^T$ for all interpretations $H \subseteq T$, then φ and ψ are strongly equivalent, $\varphi \equiv_s \psi$.* □

Proposition 6 generalises results from [10] to our extended language combining GZ and F-aggregates. In particular, item 1 provides a sufficient condition for strong equivalence that, in the case of propositional formulas, amounts to HT-equivalence.

We now move to consider propositional translations of aggregates. As said in the introduction, any F-aggregate can be understood as a propositional formula. Take any F-aggregate $B = f\{c_1 : \varphi_1, \ldots, c_m : \varphi_m\} \prec n$ where all formulas in

$\{\varphi_1, \ldots, \varphi_m\}$ are propositional – moreover, let us call *conds* (the *conditions*) to this set of formulas. By $\Phi[B]$, we denote the propositional formula:

$$\Phi[B] \quad \overset{\text{def}}{=} \quad \bigwedge_{T:T \not\models_{cl} B} \left(\left(\bigwedge conds_T^+ \right) \to \left(\bigvee conds_T^- \right) \right) \qquad (4)$$

The following result directly follows from Proposition 12 in [10].

Proposition 7. *For any F-aggregate B with propositional conditions, $B \equiv_s \Phi[B]$.*   □

Given an F-formula φ, we can define its (strongly equivalent) propositional translation $\Phi[\varphi]$ as the result of the exhaustive replacement of non-nested aggregates B by $\Phi[B]$ until all aggregates are eventually removed.

Our main contribution is to provide an analogous propositional encoding for GZ-aggregates. Given a GZ-aggregate $A = f\{\boldsymbol{X} : cond(\boldsymbol{X})\} \prec n$ with a propositional condition $cond(\boldsymbol{X})$, we define the propositional formula $\Phi[A]$ as

$$\Phi[A] \quad \overset{\text{def}}{=} \quad \bigvee_{T:T \models_{cl} A} \left(\bigwedge \mathsf{Gr}_T^+(cond(\boldsymbol{X})) \wedge \neg \bigvee \mathsf{Gr}_T^-(cond(\boldsymbol{X})) \right) \qquad (5)$$

Proposition 8. *For any GZ-aggregate A with a propositional condition, $A \equiv_s \Phi[A]$.*   □

This result allows us to define, for any *arbitrary* aggregate formula φ, its (strongly equivalent) propositional translation $\Phi[\varphi]$, again by exhaustive replacement of non-nested aggregates (now of any kind) C by their propositional formulas $\Phi[C]$. For any theory Γ, its (strongly equivalent) propositional translation is defined as $\Phi[\Gamma] \overset{\text{def}}{=} \{ \Phi[\varphi] \mid \varphi \in \Gamma \}$.

Example 3. Take again the aggregate A_1 in the body of rule (1) with $n = 1$ and assume we have constants a, b. The classical models of A_1 are $\{p(a)\}$, $\{p(b)\}$ and $\{p(a),\ p(b)\}$, since some atom $p(X)$ must hold. As a result:

$$\Phi[A_1] \quad = \quad p(a) \wedge \neg p(b) \ \vee \ p(b) \wedge \neg p(a) \ \vee \ p(a) \wedge p(b)$$

and $\Phi[(1)]$ amounts to the last three rules in (3).   □

Example 4. Take GZ-aggregate $A_3 = \mathtt{count}\{X : p(X)\} = 1$ and assume we have constants $\mathcal{C} = \{a, b, c\}$. The classical models of A_3 are $\{p(a)\}$, $\{p(b)\}$ and $\{p(c)\}$. Accordingly:

$$\Phi[A_3] \quad = \quad \begin{array}{l} p(a) \wedge \neg(p(b) \vee p(c)) \\ \vee\ p(b) \wedge \neg(p(a) \vee p(c)) \\ \vee\ p(c) \wedge \neg(p(a) \vee p(b)) \end{array} \quad \equiv \quad \begin{array}{l} p(a) \wedge \neg p(b) \wedge \neg p(c) \\ \vee\ p(b) \wedge \neg p(a) \wedge \neg p(c) \\ \vee\ p(c) \wedge \neg p(a) \wedge \neg p(b) \end{array}$$

Theorem 1 (Main Result). *Any aggregate theory Γ is strongly equivalent to its propositional translation $\Phi[\Gamma]$, that is, $\Gamma \equiv_s \Phi[\Gamma]$.*   □

4 Relation to Ferraris Aggregates

In this section, we study the relation between GZ and F-aggregates. One first observation is that GZ-aggregates are first-order structures with quantified variables, while F-aggregates allow sets of propositional expressions. Encoding a GZ-aggregate as an F-aggregate is easy: we can just ground the variables. The other direction, however, is not always possible, since the set of conditions in the F-aggregate may not have a regular representation in terms of variable substitutions. Given a GZ-aggregate $A = f\{\boldsymbol{X} : cond(\boldsymbol{X})\} \prec n$ we define its corresponding F-aggregate $\mathrm{F}[A]$:

$$\mathrm{F}[A] \overset{\text{def}}{=} f\{\ \boldsymbol{c} : cond(\boldsymbol{c}) \mid cond(\boldsymbol{c}) \in \mathrm{Gr}(cond(\boldsymbol{X}))\ \} \prec n \tag{6}$$

This correspondence is analogous to the process of *instantiation* used in [7] to ground aggregates with variables. It is easy to check that A and $\mathrm{F}[A]$ are classically equivalent. This can be checked using satisfaction from Definition 3 or classical logic for their propositional representations $\Phi[A] \equiv_{cl} \Phi[\ \mathrm{F}[A]]$. Moreover, it can be observed that these two logical representations are somehow *dual*. Indeed, $\Phi[\ \mathrm{F}[A]]$ eventually amounts to:

$$\Phi[\ \mathrm{F}[A]] \quad \equiv \quad \bigwedge_{T:T\nvDash_{cl}A}\left(\left(\bigwedge\mathrm{Gr}_T^{+}(cond(\boldsymbol{X}))\right) \to \left(\bigvee\mathrm{Gr}_T^{-}(cond(\boldsymbol{X}))\right)\right)$$

$$\tag{7}$$

which is a conjunction of formulas like $\alpha \to \beta$ for *countermodels* of A, whereas $\Phi[A]$, formula (5), is a disjunction of formulas like $\alpha \wedge \neg\beta$ for *models* of A. Another interesting consequence of the classical equivalence of A and $\mathrm{F}[A]$ is that, due to Propositions 1 and 2, we can safely replace one by another when negated. In other words:

Proposition 9. *GZ-aggregate A and its corresponding F-aggregate $\mathrm{F}[A]$ are strongly equivalent when occurring in the scope of negation.* □

However, as we saw in the introduction examples, replacing a GZ-aggregate A by its F-aggregate version $\mathrm{F}[A]$ may change the program semantics. Still, the stable models obtained after such replacement are not arbitrary. As we said, [2] proved that if the GZ-aggregate A occurs in a positive rule body, then the replacement by the F-aggregate $\mathrm{F}[A]$ preserves the stable models, but may yield more. Next, we generalise this result to aggregate theories without nested GZ-aggregates. To this aim, we make use of the following proposition asserting that, indeed, $\Phi[A]$ is stronger than $\Phi[\ \mathrm{F}[A]]$ in HT.

Proposition 10. *For any GZ-aggregate A, $\Phi[A] \models \Phi[\ \mathrm{F}[A]]$ in HT.* □

Theorem 2. *For any occurrence A of a GZ-aggregate without nested aggregates:*
1. *$SM(\Gamma(A)) \supseteq SM(\Gamma(\ \mathrm{F}[A]))$ for any theory $\Gamma(A)$ where occurrence A is positive;*

2. $SM(\Gamma(A)) \subseteq SM(\Gamma(\ F[A]))$ *for any theory* $\Gamma(A)$ *where occurrence* A *is negative.*   □

Proof. From $\Phi[A] \equiv_{cl} \Phi[\ F[A]]$ and Proposition 10 we directly apply Proposition 4.   □

In particular, this means that if we replace a (non-nested) GZ-aggregate A by its F-version $F[A]$ in the positive head of some rule, we still get stable models of the original program, but perhaps not all of them. Theorem 2 is not directly applicable to theories with nested aggregates because applying operator $\Phi[\cdot]$ produces a new formula in which nested aggregates may occur both positively and negatively.

It is well known that GZ and F-semantics do not agree even in the case of monotonic aggregates as illustrated by the example in the introduction. Nevertheless, we identify next a more restricted family of aggregates for which both semantics coincide.

Proposition 11. *Any GZ-aggregate* A *of the following types satisfies* $A \equiv_s F[A]$:
1. $A = (\mathtt{count}\{\boldsymbol{X} : cond(\boldsymbol{X})\} = n)$
2. $A = (\mathtt{sum}\{\boldsymbol{X} : cond(\boldsymbol{X})\} = n)$ *when:* $\mathcal{C} \cap \mathbb{Z}_{\leq 0} = \emptyset$, *or* $\mathcal{C} \cap \mathbb{Z}_{\geq 0} = \emptyset$.   □

Note that the result *ii)* of Proposition 11 does not apply when 0 is in the domain. For instance, the program consisting of the rule $p(0) \leftarrow \mathtt{sum}\{\boldsymbol{X} : p(\boldsymbol{X})\} = 0$ has a unique stable model $\{p(0)\}$ under Ferraris' semantic but no stable model under GZ's one.

5 Conclusions

We have provided a (strong equivalence preserving) translation from logic programs with GZ-aggregates to propositional theories in *Equilibrium Logic*. Once we understand aggregates as propositional formulas, it is straightforward to extend the syntax to arbitrary nesting of aggregates (both GZ and F-aggregates) plus propositional connectives, something we called *aggregate theories*. We have provided two alternative semantics for these theories: one based on a direct, combined extension of GZ and F-reducts, and the other on a translation to propositional formulas. The propositional formula translation has helped us to characterise the effect (with respect to the obtained stable models) of replacing a GZ-aggregate by its corresponding F-aggregate. Moreover, we have been able to prove that both aggregates have the same behaviour in the scope of negation. Finally, we identified a class of aggregates in which the GZ and F-semantics coincide. It is worth to mention that a close look at the proof of this result also suggests an extension to a broader class, something that will be studied in the future.

We expect that the current propositional formula translations will open new possibilities to explore formal properties and potential implementations of both GZ and F-aggregates, possibly extending the idea of [2] to our general aggregate theories. Finally, an extension of the current approach to a full first-order language with partial, evaluable functions (such as [4]), is currently under development.

Proofs of Propositions 8 and 11

Proof of Proposition 8. Note that, from Proposition 6, if $H \models_{cl} A^T$ iff $H \models_{cl} \Phi[A]^T$ holds then $A \equiv_s \Phi[A]$ holds. Hence, it is enough to show $H \models_{cl} A^T$ iff $H \models_{cl} \Phi[A]^T$.

Let us show that $H \models_{cl} A^T$ implies $H \models_{cl} \Phi[A]^T$ for any pair of classical interpretations such that $H \subseteq T$. Note that $T \not\models_{cl} A$ implies that $A^T = \bot$ and, thus, $H \not\models_{cl} A^T$ and the statement holds vacuous.

Then, we may assume without loss of generality that $T \models_{cl} A$ and, thus, to show that $H \models_{cl} \Phi[A]^T$, it is enough to show that the following two conditions hold:

1. $H \models_{cl} \quad cond(\mathbf{c})^T$ for every $\mathbf{c} \in \mathcal{C}^{\mathbf{X}}$ s.t. $T \models_{cl} cond(\mathbf{c})$, and
2. $H \models_{cl} (\neg cond(\mathbf{c}))^T$ for every $\mathbf{c} \in \mathcal{C}^{\mathbf{X}}$ s.t. $T \not\models_{cl} cond(\mathbf{c})$

Furthermore, by definition, $T \models_{cl} A$ implies

$$A^T = \bigwedge \{ cond(\mathbf{c})^T \mid T \models_{cl} cond(\mathbf{c}) \text{ with } \mathbf{c} \in \mathcal{C}^{\mathbf{X}} \} \tag{8}$$

and, thus, $H \models_{cl} A^T$ implies that 1 holds. Moreover, $T \not\models_{cl} cond(\mathbf{c})$ implies $cond(\mathbf{c})^T = \bot$ which, in its turn, implies $H \models_{cl} (\neg cond(\mathbf{c}))^T$ and, thus, 2 follows. The other way around. Assume that $H \models_{cl} \Phi[A]^T$. Then, there is $I \models_{cl} A$ satisfying the following two conditions:

1. $H \models_{cl} \quad cond(\mathbf{c})^T$ for every $\mathbf{c} \in \mathcal{C}^{\mathbf{X}}$ such that $I \models_{cl} cond(\mathbf{c})$, and
2. $H \models_{cl} (\neg cond(\mathbf{c}))^T$ for every $\mathbf{c} \in \mathcal{C}^{\mathbf{X}}$ such that $I \not\models_{cl} cond(\mathbf{c})$

From Proposition 6 and the fact that $H \subseteq T$, it follows that $H \models_{cl} cond(\mathbf{c})^T$ implies that $T \models_{cl} cond(\mathbf{c})$ and, thus, 1 implies

1. $T \models_{cl} cond(\mathbf{c})$ for every $\mathbf{c} \in \mathcal{C}^{\mathbf{X}}$ s.t. $I \models_{cl} cond(\mathbf{c})$

Similarly, $H \models_{cl} (\neg cond(\mathbf{c}))^T$ implies $T \models_{cl} (\neg cond(\mathbf{c}))$ which, in its turn, implies that $T \not\models_{cl} cond(\mathbf{c})$. Thus, 2 implies

1. $T \not\models_{cl} cond(\mathbf{c})$ for every $\mathbf{c} \in \mathcal{C}^{\mathbf{X}}$ such that $I \not\models_{cl} cond(\mathbf{c})$

From 1 and 1, it follows that

1. $T \models_{cl} cond(\mathbf{c})$ iff $I \models_{cl} cond(\mathbf{c})$ holds for every $\mathbf{c} \in \mathcal{C}^{\mathbf{X}}$

In its turn, this implies that $T \models_{cl} A$ iff $I \models_{cl} A$ and, since $I \models_{cl} A$, it follows that $T \models_{cl} A$ and, thus, we have that (8) holds (note that $T \not\models_{cl} A$ would imply that $A^T = \bot$). Then, to show that $H \models_{cl} A^T$ is enough to show that $H \models_{cl} cond(\mathbf{c})^T$ for every $\mathbf{c} \in \mathcal{C}^{\mathbf{X}}$ such that $T \models cond(\mathbf{c})$ which follows from 1 and 1. $\square$

Proof of Proposition 11

We will need the following notation. Let $\prec$ denote any relation symbol. We say that $A = (f\{\mathbf{X} : cond(\mathbf{X})\} \prec n)$ is *monotone* (resp. *antimonotone*) iff $\hat{f}(W_1) \prec n$ implies $\hat{f}(W_2) \prec n$ for all sets of tuples $W_1 \subseteq W_2 \subseteq \mathcal{C}^{|\mathbf{X}|}$ (resp. $W_2 \subseteq W_1 \subseteq \mathcal{C}^{|\mathbf{X}|}$). It is *regular* iff for any pair W_1, W_2 of sets of tuples of constants s.t. $W_1 \subset W_2$ satisfies that either $\hat{f}(W_1) \not\prec n$ or $\hat{f}(W_2) \not\prec n$. Furthermore, by $A^{\geq}$ we denote the GZ-aggregate $f^{\geq}\{\mathbf{X} : cond(\mathbf{X})\} \geq n$ "testing" the greater or equal relation, that is, its function $\hat{f}^{\geq}$ is defined so that $\hat{f}^{\geq}(W') \geq n$ iff $\hat{f}(W) \prec n$ and $W' \supseteq W$. Analogolusly, $A^{\leq}$ stands for $f^{\leq}\{X : cond(X)\} \leq n$ whose function $\hat{f}^{\leq}$ is defined so that $\hat{f}^{\leq}(W') \leq n$ iff $\hat{f}(W) \prec n$ and $W' \subseteq W$. Then, by $\Phi^D[A]$, we denote the following formula:

$$\bigwedge_{T \notin [\![A^{\leq}]\!]} \left(\neg \bigwedge \mathtt{Gr}_T^+(cond(X)) \right) \wedge \bigwedge_{T \notin [\![A^{\geq}]\!]} \left(\bigvee \mathtt{Gr}_T^-(cond(X)) \right) \tag{9}$$

Proposition 12. *Any regular set atom A satisfies:*

1. *set atoms $A^{\geq}$ and $A^{\leq}$ are respectively monotone and antimonotone,*
2. $T \models_{cl} A$ *iff* $T \models_{cl} A^{\geq} \wedge A^{\leq}$,
3. $\Phi[\; F[A]] \equiv_s \Phi[\; F[A^{\geq}]] \wedge \Phi[\; F[A^{\leq}]] \equiv_s \Phi^D[\; F[A]].$ $\qquad\square$

For $A = f\{X : cond(X)\} \prec n$, we define $W_{\langle H,T \rangle}(A) \stackrel{\text{def}}{=} \{\, c \mid \langle H,T \rangle \models cond(c) \,\}$. We also use $W_T(A)$ to stand for $W_{\langle T,T \rangle}(A)$.

Proposition 13. *Any set atom $A = f\{X : cond(X)\} \prec c$ with propositional $cond(X)$ satisfies that (i) $\langle H,T \rangle \models \Phi[\; F[A]]$ iff (ii) $H \models_{cl} F[A]^T$ iff (iii) $\hat{f}(W_{\langle H,T \rangle}(A)) \prec n$ and $\hat{f}(W_T(A)) \prec n$.* $\qquad\square$

Lemma 1. *Any regular GZ-aggregate A satisfies: $\Phi[\; F[A]] \models \Phi[A]$.* $\qquad\square$

Proof. Suppose, for the sake of contradiction, that there is some HT-interpretation such that $\langle H,T \rangle \models \Phi[\; F[A]]$, but $\langle H,T \rangle \not\models \Phi[A]$. Suppose also that $T \models_{cl} A$. Then, from $\langle H,T \rangle \not\models \Phi[A]$, it follows that one of the following conditions must hold

1. $\langle H,T \rangle \not\models cond(c)$ for some $c \in C^{|X|}$ s.t. $T \models_{cl} cond(c)$,
2. $\langle H,T \rangle \not\models \neg cond(c)$ for some $c \in C^{|X|}$ s.t. $T \not\models_{cl} cond(c)$

On the one hand, $\langle H,T \rangle \not\models \neg cond(c)$ implies $T \not\models_{cl} \neg cond(c)$ which, in its turn, implies $T \models_{cl} cond(c)$. Thus, 2 is a contradiction. On the other hand, from Proposition 1, it follows that $W_{\langle H,T \rangle} \subseteq W_T$ holds for any HT-interpretation $\langle H,T \rangle$. Then, 1 implies that $W_{\langle H,T \rangle} \subset W_T$ which, since A is regular, implies that either $\hat{f}(W_{\langle H,T \rangle}) \not\prec n$ or $\hat{f}(W_T) \not\prec n$ hold. If the former, then $\langle H,T \rangle \not\models \Phi[\; F[A]]$ (Proposition 13) which is a contradiction with the assumption $\langle H,T \rangle \models \Phi[\; F[A]]$. If the latter, $T \not\models_{cl} A \equiv_{cl} \Phi[A]$ which is a contradiction with the facts $T \models_{cl} \Phi[\; F[A]]$ (because $\langle H,T \rangle \models \Phi[\; F[A]]$) and $\Phi[A] \equiv_{cl} \Phi[\; F[A]]$. Hence, it must be that $T \not\models A$ and, thus, Proposition 12, implies:

1. either $T \not\models A^{\geq}$ or $T \not\models A^{\leq}$, and
2. $\langle H,T \rangle \models \Phi[\; F[A]] \equiv_s \Phi^D[\; F[A]]$

In its turn, these conditions imply that one of the following two contradictions must hold:

1. $\langle H,T \rangle \models \neg cond(c)$ for some $c \in C^{|X|}$ s.t. $T \models cond(c)$ (if $T \not\models A^{\leq}$)
2. $\langle H,T \rangle \models cond(c)$ for some $c \in C^{|X|}$ s.t. $T \not\models cond(c)$ (if $T \not\models A^{\geq}$),

Consequently, $\langle H,T \rangle \models \Phi[\; F[A]]$ implies $\langle H,T \rangle \models \Phi[A]$.

Proof of Proposition 11. From Proposition 10 and Lemma 1, it respectively follows $\Phi[A] \models \Phi[\ F[A]]$ and $\Phi[\ F[A]] \models \Phi[A]$. Thus, $\Phi[A] \equiv_s \Phi[\ F[A]]$. It only remains to note that $\widehat{\mathrm{count}}(W_1) < \widehat{\mathrm{count}}(W_2)$ for all $W_1 \subset W_2$ and, thus, $\mathrm{count}\{\boldsymbol{X} : cond(\boldsymbol{X})\} = n$ is regular. Similarly, if there is no $c \in \mathcal{C}$ such that $c \leq 0$ (resp. $c \geq 0$), then $\widehat{\mathrm{sum}}(W_1) < \widehat{\mathrm{sum}}(W_2)$ (resp. $\widehat{\mathrm{sum}}(W_1) > \widehat{\mathrm{sum}}(W_2)$)

References

1. Aguado, F., Cabalar, P., Pearce, D., Pérez, G., Vidal, C.: A denotational semantics for equilibrium logic. Theory Pract. Log. Program. **15**(4–5), 620–634 (2015)
2. Alviano, M., Faber, W.: Stable model semantics of abstract dialectical frameworks revisited: a logic programming perspective. In: Proceedings of the Twenty-Fourth International Joint Conference on Artificial Intelligence, pp. 2684–2690. AAAI Press (2015)
3. Baral, C.: Knowledge Representation, Reasoning and Declarative Problem Solving. Cambridge University Press, New York (2003)
4. Cabalar, P.: Functional answer set programming. Theory Pract. Log. Program. **11**(2–3), 203–233 (2011)
5. Calimeri, F., Faber, W., Gebser, M., Ianni, G., Kaminski, R., Krennwallner, T., Leone, N., Ricca, F., Schaub, T.: ASP-Core-2: input language format (2012). https://www.mat.unical.it/aspcomp2013/ASPStandardization/
6. Erdem, E., Gelfond, M., Leone, N.: Applications of ASP. AI Mag. **37**(3), 53–68 (2016)
7. Faber, W., Pfeifer, G., Leone, N.: Semantics and complexity of recursive aggregates in answer set programming. Artif. Intell. **175**(1), 278–298 (2011)
8. Faber, W., Pfeifer, G., Leone, N., Dell'Armi, T., Ielpa, G.: Design and implementation of aggregate functions in the DLV system. Theory Pract. Log. Program. **8**(5–6), 545–580 (2008)
9. Ferraris, P.: Answer sets for propositional theories. In: Baral, C., Greco, G., Leone, N., Terracina, G. (eds.) LPNMR 2005. LNCS (LNAI), vol. 3662, pp. 119–131. Springer, Heidelberg (2005). doi:10.1007/11546207_10
10. Ferraris, P.: Logic programs with propositional connectives and aggregates. ACM Trans. Comput. Log. **12**(4), 25 (2011)
11. Ferraris, P., Lee, J., Lifschitz, V.: A new perspective on stable models. In: Proceedings of the Twentieth International Joint Conference on Artificial Intelligence (IJCAI 2007), pp. 372–379. AAAI/MIT Press (2007)
12. Gebser, M., Kaufmann, B., Schaub, T.: Conflict-driven answer set solving: from theory to practice. Artif. Intell. **187–188**, 52–89 (2012)
13. Gelfond, M., Lifschitz, V.: The stable model semantics for logic programming. In: Proceedings of the Fifth International Conference and Symposium of Logic Programming (ICLP 1988), pp. 1070–1080. MIT Press (1988)
14. Gelfond, M., Zhang, Y.: Vicious circle principle and logic programs with aggregates. Theory Pract. Log. Program. **14**(4–5), 587–601 (2014)
15. Harrison, A., Lifschitz, V., Truszczynski, M.: On equivalence of infinitary formulas under the stable model semantics. Theory Pract. Log. Program. **15**(1), 18–34 (2015)
16. Lifschitz, V., Pearce, D., Valverde, A.: Strongly equivalent logic programs. ACM Trans. Comput. Log. **2**(4), 526–541 (2001)

17. Pearce, D.: A new logical characterisation of stable models and answer sets. In: Dix, J., Pereira, L.M., Przymusinski, T.C. (eds.) NMELP 1996. LNCS, vol. 1216, pp. 57–70. Springer, Heidelberg (1997). doi:10.1007/BFb0023801

18. Pearce, D.: Equilibrium logic. Ann. Math. Artif. Intell. **47**(1–2), 3–41 (2006)

19. Pelov, N., Denecker, M., Bruynooghe, M.: Well-founded and stable semantics of logic programs with aggregates. Theory Pract. Log. Program. **7**(3), 301–353 (2007)

20. Simons, P., Niemel, I., Soininen, T.: Extending and implementing the stable model semantics. Artif. Intell. **138**(1–2), 181–234 (2002)

21. Son, T., Pontelli, E.: A constructive semantic characterization of aggregates in answer set programming. Theory Pract. Log. Program. **7**(3), 355–375 (2007)

Answer Set Solving with Bounded Treewidth Revisited

Johannes K. Fichte[✉][iD], Markus Hecher[iD], Michael Morak,
and Stefan Woltran

Institute of Information Systems,, TU Wien Wien, Austria
{fichte,hecher,morak,woltran}@dbai.tuwien.ac.at

Abstract. Parameterized algorithms are a way to solve hard problems more efficiently, given that a specific parameter of the input is small. In this paper, we apply this idea to the field of answer set programming (ASP). To this end, we propose two kinds of graph representations of programs to exploit their treewidth as a parameter. Treewidth roughly measures to which extent the internal structure of a program resembles a tree. Our main contribution is the design of parameterized dynamic programming algorithms, which run in linear time if the treewidth and weights of the given program are bounded. Compared to previous work, our algorithms handle the full syntax of ASP. Finally, we report on an empirical evaluation that shows good runtime behaviour for benchmark instances of low treewidth, especially for counting answer sets.

Keywords: Parameterized algorithms · Tree decompositions

1 Introduction

Parameterized algorithms [5] have attracted considerable interest in recent years and allow to tackle hard problems by directly exploiting a small parameter of the input problem. One particular goal in this field is to find guarantees that the runtime is exponential exclusively in the parameter, and polynomial in the input size (so-called fixed-parameter tractable algorithms). A parameter that has been researched extensively is treewidth [2]. Generally speaking, treewidth measures the closeness of a graph to a tree, based on the observation that problems on trees are often easier than on arbitrary graphs. A parameterized algorithm exploiting small treewidth takes a tree decomposition, which is an arrangement of a graph into a tree, and evaluates the problem in parts, via dynamic programming (DP) on the tree decomposition.

ASP [3] is a logic-based declarative modelling language and problem solving framework where solutions, so called answer sets, of a given logic program

---

For additional details and proofs, we refer to an extended self-archived version [8]. A preliminary version of the paper was presented on the workshop TAASP'16. Research was supported by the Austrian Science Fund (FWF), Grant Y698. The first and second author are also affiliated with the University of Potsdam, Germany.

M. Balduccini and T. Janhunen (Eds.): LPNMR 2017, LNAI 10377, pp. 132–145, 2017.
DOI: 10.1007/978-3-319-61660-5_13

directly represent the solutions of the modelled problem. Jakl et al. [9] give a DP algorithm for disjunctive rules only, whose runtime is linear in the input size of the program and double exponential in the treewidth of a particular graph representation of the program structure. However, modern ASP systems allow for an extended syntax that includes, among others, weight rules and choice rules. Pichler et al. [10] investigated the complexity of programs with weight rules. They also presented DP algorithms for programs with cardinality rules (i.e., restricted version of weight rules), but without disjunction.

In this paper, we propose DP algorithms for finding answer sets that are able to directly treat all kinds of ASP rules. While such rules can be transformed into disjunctive rules, we avoid the resulting polynomial overhead with our algorithms. In particular, we present two approaches based on two different types of graphs representing the program structure. Firstly, we consider the primal graph, which allows for an intuitive algorithm that also treats the extended ASP rules. While for a given disjunctive program the treewidth of the primal graph may be larger than treewidth of the graph representation used by Jakl et al. [9], our algorithm uses simpler data structures and lays the foundations to understand how we can handle also extended rules. Our second graph representation is the incidence graph, a generalization of the representation used by Jakl et al. Algorithms for this graph representation are more sophisticated, since weight and choice rules can no longer be completely evaluated in the same computation step. Our algorithms yield upper bounds that are linear in the program size, double-exponential in the treewidth, and single-exponential in the maximum weights. We extend our two algorithms to count optimal answer sets. For this particular task, experiments show that we are able to outperform existing systems from multiple domains, given input instances of low treewidth, both randomly generated and obtained from real-world graphs of traffic networks. Our system is publicly available on github[1].

2   Formal Background

Answer Set programming (ASP).   *ASP* is a declarative modeling and problem solving framework; for a full introduction, see, e.g., [3]. State-of-the-art ASP grounders support the full ASP-Core-2 language [4] and output smodels input format [13], which we will use for our algorithms. Let ℓ, m, n be non-negative integers such that $\ell \leq m \leq n$, a_1, ..., a_n distinct propositional atoms, w, w_1, ..., w_n non-negative integers, and $l \in \{a_1, \neg a_1\}$. A *choice rule* is an expression of the form, $\{a_1; \ldots; a_\ell\} \leftarrow a_{\ell+1}, \ldots, a_m, \neg a_{m+1}, \ldots, \neg a_n$, a *disjunctive rule* is of the form $a_1 \vee \cdots \vee a_\ell \leftarrow a_{\ell+1}, \ldots, a_m, \neg a_{m+1}, \ldots, \neg a_n$ and a *weight rule* is of the form $a_\ell \leftarrow w \leqslant \{a_{\ell+1} = w_{\ell+1}, \ldots, a_m = w_m, \neg a_{m+1} = w_{m+1}, \ldots, \neg a_n = w_n\}$. Finally, an *optimization rule* is an expression of the form $\leftsquigarrow l[w]$. A *rule* is either a disjunctive, a choice, a weight, or an optimization rule. For a choice, disjunctive, or weight rule r, let $H_r := \{a_1, \ldots, a_\ell\}$, $B_r^+ := \{a_{\ell+1}, \ldots, a_m\}$, and $B_r^- := \{a_{m+1}, \ldots, a_n\}$. For a weight rule r, let $wght(r, a)$ map atom a to its

---
[1] See https://github.com/daajoe/dynasp/tree/v2.0.0.

corresponding weight w_i in rule r if $a = a_i$ for $\ell + 1 \leq i \leq n$ and to 0 otherwise, let $\mathrm{wght}(r, A) := \sum_{a \in A} \mathrm{wght}(r, a)$ for a set A of atoms, and let $\mathrm{bnd}(r) := w$ be its *bound*. For an optimization rule r, let $\mathrm{cst}(r) := w$ and if $l = a_1$, let $B_r^+ := \{a_1\}$ and $B_r^- := \emptyset$; or if $l = \neg a_1$, let $B_r^- := \{a_1\}$ and $B_r^+ := \emptyset$. For a rule r, let $\mathrm{at}(r) := H_r \cup B_r^+ \cup B_r^-$ denote its *atoms* and $B_r := B_r^+ \cup \{\neg b \mid b \in B_r^-\}$ its *body*. A *program* Π is a set of rules. Let $\mathrm{at}(\Pi) := \{\mathrm{at}(r) \mid r \in \Pi\}$ and let $\mathrm{CH}(\Pi), \mathrm{DISJ}(\Pi), \mathrm{OPT}(\Pi)$ and $\mathrm{WGT}(\Pi)$ denote the set of all choice, disjunctive, optimization and weight rules in Π, respectively. A set $M \subseteq \mathrm{at}(\Pi)$ *satisfies* a rule r if (i) $(H_r \cup B_r^-) \cap M \neq \emptyset$ or $B_r^+ \not\subseteq M$ for $r \in \mathrm{DISJ}(\Pi)$, (ii) $H_r \cap M \neq \emptyset$ or $\sum_{a_i \in M \cap B_r^+} w_i + \sum_{a_i \in B_r^- \setminus M} w_i < \mathrm{bnd}(r)$ for $r \in \mathrm{WGT}(\Pi)$, or (iii) $r \in \mathrm{CH}(\Pi) \cup \mathrm{OPT}(\Pi)$. M is a model of Π, denoted by $M \vDash \Pi$, if M satisfies every rule $r \in \Pi$. Further, let $\mathrm{Mod}(\mathcal{C}, \Pi) := \{C \mid C \in \mathcal{C}, C \vDash \Pi\}$ for $\mathcal{C} \subseteq 2^{\mathrm{at}(\Pi)}$. The *reduct* r^M (i) of a choice rule r is the set $\{a \leftarrow B_r^+ \mid a \in H_r \cap M, B_r^- \cap M = \emptyset\}$ of rules, (ii) of a disjunctive rule r is the singleton $\{H_r \leftarrow B_r^+ \mid B_r^- \cap M = \emptyset\}$, and (iii) of a weight rule r is the singleton $\{H_r \leftarrow w' \leqslant [a = \mathrm{wght}(r, a) \mid a \in B_r^+]\}$ where $w' = \mathrm{bnd}(r) - \sum_{a \in B_r^- \setminus M} \mathrm{wght}(r, a)$. $\Pi^M := \{r' \mid r' \in r^M, r \in \Pi\}$ is called *GL reduct* of Π with respect to M. A set $M \subseteq \mathrm{at}(\Pi)$ is an *answer set* of program Π if (i) $M \vDash \Pi$ and (ii) there is no $M' \subsetneq M$ such that $M' \vDash \Pi^M$, that is, M is *subset minimal with respect to* Π^M. We call $\mathrm{cst}(\Pi, M, A) := \sum_{r \in \mathrm{OPT}(\Pi),\ A \cap [(B_r^+ \cap M) \cup (B_r^- \setminus M)] \neq \emptyset} \mathrm{cst}(r)$ the *cost* of model M for Π with respect to the set $A \subseteq \mathrm{at}(\Pi)$. An answer set M of Π is *optimal* if its cost is minimal over all answer sets.

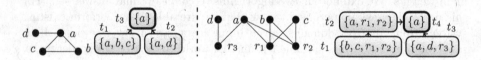

Fig. 1. Graph G_1 with a TD of G_1 (left) and graph G_2 with a TD of G_2 (right).

Example 1. Let $\Pi := \{\overbrace{\{a; b\} \leftarrow c}^{r_1};\ \overbrace{c \leftarrow 1 \leqslant \{b = 1, \neg a = 1\}}^{r_2};\ \overbrace{d \vee a \leftarrow}^{r_3}\}$. Then, the sets $\{a\}$, $\{c, d\}$ and $\{b, c, d\}$ are answer sets of Π. ∎

Given a program Π, we consider the problems of computing an answer set (called AS) and outputting the number of optimal answer sets (called #AspO).

Tree Decompositions. Let $G = (V, E)$ be a graph, $T = (N, F, n)$ a rooted tree, and $\chi : N \to 2^V$ a function that maps each node $t \in N$ to a set of vertices. We call the sets $\chi(\cdot)$ *bags* and N the set of nodes. Then, the pair $\mathcal{T} = (T, \chi)$ is a *tree decomposition (TD)* of G if the following conditions hold: (i) all vertices occur in some bag, that is, for every vertex $v \in V$ there is a node $t \in N$ with $v \in \chi(t)$; (ii) all edges occur in some bag, that is, for every edge $e \in E$ there is a node $t \in N$ with $e \subseteq \chi(t)$; and (iii) the *connectedness condition*: for any three nodes $t_1, t_2, t_3 \in N$, if t_2 lies on the unique path from t_1 to t_3, then $\chi(t_1) \cap \chi(t_3) \subseteq \chi(t_2)$. We call $\max\{|\chi(t)| - 1 \mid t \in N\}$ the *width* of the TD.

The *treewidth* $tw(G)$ of a graph G is the minimum width over all possible TDs of G. For some arbitrary but fixed integer k and a graph of treewidth at most k, we can compute a TD of width $\leq k$ in time $2^{\mathcal{O}(k^3)} \cdot |V|$ [2]. Given a TD (T, χ) with $T = (N, \cdot, \cdot)$, for a node $t \in N$ we say that type(t) is *leaf* if t has no children; *join* if t has children t' and t'' with $t' \neq t''$ and $\chi(t) = \chi(t') = \chi(t'')$; *int* ("introduce") if t has a single child t', $\chi(t') \subseteq \chi(t)$ and $|\chi(t)| = |\chi(t')| + 1$; *rem* ("removal") if t has a single child t', $\chi(t) \subseteq \chi(t')$ and $|\chi(t')| = |\chi(t)| + 1$. If every node $t \in N$ has at most two children, type(t) $\in \{leaf, join, int, rem\}$, and bags of leaf nodes and the root are empty, then the TD is called *nice*. For every TD, we can compute a nice TD in linear time without increasing the width [2]. In our algorithms, we will traverse a TD bottom up, therefore, let post-order(T, t) be the sequence of nodes in post-order of the induced subtree $T' = (N', \cdot, t)$ of T rooted at t.

Algorithm 1. Algorithm $\mathcal{DP}_\mathcal{A}(\mathcal{T})$ for Dynamic Programming on TD $\mathcal{T}$ for ASP.

In: Table algorithm $\mathcal{A}$, nice TD $\mathcal{T} = (T, \chi)$ with $T = (N, \cdot, n)$ of $G(\Pi)$ according to $\mathcal{A}$.
Out: Table: maps each TD node $t \in T$ to some computed table τ_t.
1 **for** iterate t in post-order(T,n) **do**
2   Child-Tabs := {Tables[t'] | t' is a child of t in T}
3   Tables[t] := $\mathcal{A}(t, \chi(t), \Pi_t, \text{at}_{\leq t}, \text{Child-Tabs})$

Example 2. Figure 1 (left) shows a graph G_1 together with a TD of G_1 that is of width 2. Note that G_1 has treewidth 2, since it contains a clique on the vertices $\{a, b, c\}$. Further, the TD $\mathcal{T}$ in Fig. 2 is a nice TD of G_1.   ∎

Graph Representations of Programs. In order to use TDs for ASP solving, we need dedicated graph representations of ASP programs. The *primal graph* $P(\Pi)$ of program Π has the atoms of Π as vertices and an edge $a\,b$ if there exists a rule $r \in \Pi$ and $a, b \in \text{at}(r)$. The *incidence graph* $I(\Pi)$ of Π is the bipartite graph that has the atoms and rules of Π as vertices and an edge $a\,r$ if $a \in \text{at}(r)$ for some rule $r \in \Pi$. These definitions adapt similar concepts from SAT [11].

Example 3. Recall program Π of Example 1. We observe that graph G_1 (G_2) in the left (right) part of Fig. 1 is the primal (incidence) graph of Π.   ∎.

Sub-Programs. Let $\mathcal{T} = (T, \chi)$ be a nice TD of graph representation $H \in \{I(\Pi), P(\Pi)\}$ of a program Π. Further, let $T = (N, \cdot, n)$ and $t \in N$. The *bag-rules* are defined as $\Pi_t := \{r \mid r \in \Pi, \text{at}(r) \subseteq \chi(t)\}$ if H is the primal graph and as $\Pi_t := \Pi \cap \chi(t)$ if H is the incidence graph. Further, the set $\text{at}_{\leq t} := \{a \mid a \in \text{at}(\Pi) \cap \chi(t'), t' \in \text{post-order}(T, t)\}$ is called *atoms below t*, the *program below t* is defined as $\Pi_{\leq t} := \{r \mid r \in \Pi_{t'}, t' \in \text{post-order}(T, t)\}$, and the *program strictly below t* is $\Pi_{<t} := \Pi_{\leq t} \setminus \Pi_t$. It holds that $\Pi_{\leq n} = \Pi_{<n} = \Pi$ and $\text{at}_{\leq n} = \text{at}(\Pi)$.

Example 4. Intuitively, TDs of Fig. 1 enable us to evaluate Π by analyzing sub-programs ($\{r_1, r_2\}$ and $\{r_3\}$) and combining results agreeing on a. Indeed, for the given TD of Fig. 1 (left), $\Pi_{\leq t_1} = \{r_1, r_2\}$, $\Pi_{\leq t_2} = \{r_3\}$ and $\Pi = \Pi_{\leq t_3} = \Pi_{<t_3} = \Pi_{t_1} \cup \Pi_{t_2}$. For the TD of Fig. 1 (right), we have $\Pi_{\leq t_1} = \{r_1, r_2\}$ and $\text{at}_{\leq t_1} = \{b, c\}$, as well as $\Pi_{\leq t_3} = \{r_3\}$ and $\text{at}_{\leq t_3} = \{a, d\}$. Moreover, for TD $\mathcal{T}$ of Fig. 2, $\Pi_{\leq t_1} = \Pi_{\leq t_2} = \Pi_{\leq t_3} = \Pi_{<t_4} = \emptyset$, $\text{at}_{\leq t_3} = \{a, b\}$ and $\Pi_{\leq t_4} = \{r_1, r_2\}$.   ∎

3   ASP via Dynamic Programming on TDs

In the next two sections, we propose two dynamic programming (DP) algorithms, DP_{PRIM} and DP_{INC}, for ASP without optimization rules based on two different graph representations, namely the primal and the incidence graph. Both algorithms make use of the fact that answer sets of a given program Π are (i) models of Π and (ii) subset minimal with respect to Π^M. Intuitively, our algorithms compute, for each TD node t, (i) sets of atoms—(local) *witnesses*—representing parts of potential models of Π, and (ii) for each local witness M subsets of M—(local) *counterwitnesses*—representing subsets of potential models of Π^M which (locally) contradict that M can be extended to an answer set of Π. We give the basis of our algorithms in Algorithm 1 (DP_A), which sketches the general DP scheme for ASP solving on TDs. Roughly, the algorithm splits the search space based on a given nice TD and evaluates the input program Π in parts. The results are stored in so-called tables, that is, sets of all possible tuples of witnesses and counterwitnesses for a given TD node. To this end, we define the *table algorithms* PRIM and INC, which compute tables for a node t of the TD using the primal graph $P(\Pi)$ and incidence graph $I(\Pi)$, respectively. To be more concrete, given a table algorithm $A \in \{PRIM, INC\}$, algorithm DP_A visits every node $t \in T$ in post-order; then, based on Π_t, computes a table τ_t for node t from the tables of the children of t, and stores τ_t in Tables[t].

Algorithm 2. Table algorithm $\mathrm{PRIM}(t, \chi_t, \Pi_t, \cdot, \text{Child-Tabs})$.

In: Bag χ_t, bag-rules Π_t and child tables Child-Tabs of node t. **Out:** Table τ_t.

1   **if** type(t) = *leaf* **then** $\tau_t := \{\langle \emptyset, \emptyset \rangle\}$ /* Abbreviations see Footnote a. */

2   **else if** type(t) = *int*, $a \in \chi_t$ is introduced and $\tau' \in$ *Child-Tabs* **then**

3      $\tau_t := \{\langle M_a^+, \text{Mod}(\{M\} \cup [\mathcal{C} \sqcup \{a\}] \cup \mathcal{C}, \Pi_t^{M_a^+}) \rangle$ $| \ \langle M, \mathcal{C} \rangle \in \tau', M_a^+ \vDash \Pi_t\} \bigcup$

4      $\{\langle M, \text{Mod}(\mathcal{C}, \Pi_t^M) \rangle$ $| \ \langle M, \mathcal{C} \rangle \in \tau', M \vDash \Pi_t\}$

5   **else if** type(t) = *rem*, $a \notin \chi_t$ is removed and $\tau' \in$ *Child-Tabs* **then**

6     $\tau_t := \{\langle M_a^-, \{C_a^- \mid C \in \mathcal{C}\}\rangle$ $| \ \langle M, \mathcal{C} \rangle \in \tau'\}$

7   **else if** type(t) = *join* and $\tau', \tau'' \in$ *Child-Tabs* with $\tau' \neq \tau''$ **then**

8     $\tau_t := \{\langle M, (\mathcal{C}' \cap \mathcal{C}'') \cup (\mathcal{C}' \cap \{M\}) \cup (\{M\} \cap \mathcal{C}'') \rangle$ $| \ \langle M, \mathcal{C}' \rangle \in \tau', \langle M, \mathcal{C}'' \rangle \in \tau''\}$

   ^a $\mathcal{S} \sqcup \{e\} := \{S \cup \{e\} \mid S \in \mathcal{S}\}$, $S_e^+ := S \cup \{e\}$, and $S_e^- := S \setminus \{e\}$

3.1   Using Decompositions of Primal Graphs

In this section, we present our algorithm PRIM in two parts: (i) finding models of Π and (ii) finding models which are subset minimal with respect to Π^M. For sake of clarity, we first present only the first tuple positions (red parts) of Algorithm 2 (PRIM) to solve (i). We call the resulting table algorithm MOD.

Example 5. Consider program Π from Example 1 and in Fig. 2 (left) TD $\mathcal{T} = (\cdot, \chi)$ of $P(\Pi)$ and the tables $\tau_1, \ldots, \tau_{12}$, which illustrate computation results obtained during post-order traversal of $\mathcal{T}$ by DP_{MOD}. Table $\tau_1 = \{\langle \emptyset \rangle\}$ as type(t_1) = *leaf*. Since type(t_2) = *int*, we construct table τ_2 from τ_1 by taking $M_{1.i}$ and $M_{1.i} \cup \{a\}$ for each $M_{1.i} \in \tau_1$ (corresponding to a guess on a).

Then, t_3 introduces b and t_4 introduces c. $\Pi_{t_1} = \Pi_{t_2} = \Pi_{t_3} = \emptyset$, but since $\chi(t_4) \subseteq \mathrm{at}(r_1) \cup \mathrm{at}(r_2)$ we have $\Pi_{t_4} = \{r_1, r_2\}$ for t_4. In consequence, for each $M_{4.i}$ of table τ_4, we have $M_{4.i} \models \{r_1, r_2\}$ since MOD enforces satisfiability of Π_t in node t. We derive tables τ_7 to τ_9 similarly. Since $\mathrm{type}(t_5) = rem$, we remove atom b from all elements in τ_4 to construct τ_5. Note that we have already seen all rules where b occurs and hence b can no longer affect witnesses during the remaining traversal. We similarly construct $\tau_{t_6} = \tau_{10} = \{\langle\emptyset\rangle, \langle a\rangle\}$. Since $\mathrm{type}(t_{11}) = join$, we construct table τ_{11} by taking the intersection $\tau_6 \cap \tau_{10}$. Intuitively, this combines witnesses agreeing on a. Node t_{12} is again of type rem. By definition (primal graph and TDs) for every $r \in \Pi$, atoms $\mathrm{at}(r)$ occur together in at least one common bag. Hence, $\Pi = \Pi_{\leq t_{12}}$ and since $\tau_{12} = \{\langle\emptyset\rangle\}$, we can construct a model of Π from the tables. For example, we obtain the model $\{a, d\} = M_{11.2} \cup M_{4.2} \cup M_{9.3}$. ∎

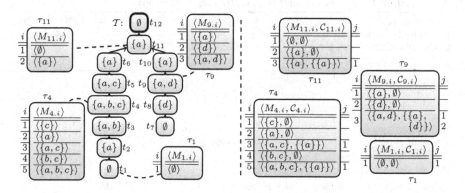

Fig. 2. Selected DP tables of MOD (left) and PRIM (right) for nice TD $\mathcal{T}$.

PRIM is given in Algorithm 2. Tuples in τ_t are of the form $\langle M, \mathcal{C}\rangle$. Witness $M \subseteq \chi(t)$ represents a model of Π_t witnessing the existence of $M' \supseteq M$ with $M' \models \Pi_{\leq t}$. The family $\mathcal{C} \subseteq 2^M$ contains sets of models $C \subseteq M$ of the GL reduct $(\Pi_t)^M$. $\mathcal{C}$ witnesses the existence of a set C' with counterwitness $C \subseteq C' \subsetneq M'$ and $C' \models (\Pi_{\leq t})^{M'}$. There is an answer set of Π if table t_n for root n contains $\langle\emptyset, \emptyset\rangle$. Since in Example 5 we already explained the first tuple position and thus the witness part, we only briefly describe the parts for counterwitnesses. In the introduce case, we want to store only counterwitnesses for not being minimal with respect to the GL reduct of the bag-rules. Therefore, in Line 3 we construct for M_a^+ counterwitnesses from either some witness M ($M \subsetneq M_a^+$), or of any $C \in \mathcal{C}$, or of any $C \in \mathcal{C}$ extended by a (every $C \in \mathcal{C}$ was already a counterwitness before). Line 4 ensures that only counterwitnesses that are models of the GL reduct Π_t^M are stored (via $\mathrm{Mod}(\cdot, \cdot)$). Line 6 restricts counterwitnesses to its bag content, and Line 8 enforces that child tuples agree on counterwitnesses.

Example 6. Consider Example 5, its TD $\mathcal{T} = (\cdot, \chi)$, Fig. 2 (right), and the tables $\tau_1, \ldots, \tau_{12}$ obtained by $\mathcal{DP}_{\mathsf{PRIM}}$. Since we have $\mathrm{at}(r_1) \cup \mathrm{at}(r_2) \subseteq \chi(t_4)$, we require $C_{4.i.j} \models \{r_1, r_2\}^{M_{4.i}}$ for each counterwitness $C_{4.i.j} \in \mathcal{C}_{4.i}$ in tuples of τ_4. For $M_{4.5} = \{a, b, c\}$ observe that the only counterwitness of $\{r_1, r_2\}^{M_{4.5}} = \{a \leftarrow c, b \leftarrow c, c \leftarrow 1 \leq \{b = 1\}\}$ is $C_{4.5.1} = \{a\}$. Note that witness $M_{11.2}$ of table τ_{11} is the result of joining $M_{4.2}$ with $M_{9.1}$ and witness $M_{11.3}$ (counterwitness $C_{11.3.1}$) is the result of joining $M_{4.3}$ with $M_{9.3}$ ($C_{4.3.1}$ with $C_{9.3.1}$), and $M_{4.5}$ with $M_{9.3}$ ($C_{4.5.1}$ with $C_{9.3.2}$). $C_{11.3.1}$ witnesses that neither $M_{4.3} \cup M_{9.3}$ nor $M_{4.5} \cup M_{9.3}$ forms an answer set of Π. Since τ_{12} contains $\langle \emptyset, \emptyset \rangle$ there is no counterwitness for $M_{11.2}$, we can construct an answer set of Π from the tables, e.g., $\{a\}$ can be constructed from $M_{4.2} \cup M_{9.1}$. ∎

Theorem 1. *Given a program Π, the algorithm $\mathcal{DP}_{\mathsf{PRIM}}$ is correct and runs in time $\mathcal{O}(2^{2^{k+2}} \cdot \|P(\Pi)\|)$ where k is the treewidth of the primal graph $P(\Pi)$.*

Algorithm 3. Table algorithm $\mathsf{INC}(t, \chi_t, \Pi_t, \mathrm{at}_{\leq t}, \text{Child-Tabs})$.

In: Bag χ_t, bag-rules Π_t, atoms-below $\mathrm{at}_{\leq t}$, child tables Child-Tabs of t. **Out:** Tab. τ_t.

1 **if** type$(t) = leaf$ **then** $\tau_t := \{\langle \emptyset, \emptyset, \ \emptyset \rangle\}$ /* Abbreviations see Footnote 2. */
2 **else if** type$(t) = int$, $a \in \chi_t \setminus \Pi_t$ is introduced and $\tau' \in$ Child-Tabs **then**
3      $\tau_t := \{\langle M_a^+, \sigma] \ \mathrm{SatRules}(\dot{\Pi}_t^{(t,\sigma)}, M_a^+), \ \{\langle M, \sigma] \ \mathrm{SatRules}(\dot{\Pi}_t^{(t,\sigma,M_a^+)}, M)\rangle\} \cup$
4          $\{\langle C_a^+, \rho] \ \mathrm{SatRules}(\dot{\Pi}_t^{(t,\rho,M_a^+)}, C_a^+)\rangle \mid \langle C, \rho \rangle \in \mathcal{C}\} \cup$
5          $\{\langle C, \rho] \ \mathrm{SatRules}(\dot{\Pi}_t^{(t,\rho,M_a^+)}, C)\rangle \mid \langle C, \rho \rangle \in \mathcal{C}\})$ $\mid \langle M, \sigma, \mathcal{C} \rangle \in \tau'\} \cup$
6          $\{\langle M, \sigma] \ \mathrm{SatRules}(\dot{\Pi}_t^{(t,\sigma)}, M),$
7          $\{\langle C, \rho] \ \mathrm{SatRules}(\dot{\Pi}_t^{(t,\rho,M)}, C)\rangle \mid \langle C, \rho \rangle \in \mathcal{C}\})$ $\mid \langle M, \sigma, \mathcal{C} \rangle \in \tau'\}$
8 **else if** type$(t) = int$, $r \in \chi_t \cap \Pi_t$ is introduced and $\tau' \in$ Child-Tabs **then**
9      $\tau_t := \{\langle M, \sigma_r^+] \ \mathrm{SatRules}(\{\dot{r}\}^{(t,\sigma_r^+)}, M),$
10          $\{\langle C, \rho_r^+] \ \mathrm{SatRules}(\{\dot{r}\}^{(t,\rho_r^+,M)}, C)\rangle \mid \langle C, \rho \rangle \in \mathcal{C}\})$ $\mid \langle M, \sigma, \mathcal{C} \rangle \in \tau'\}$
11 **else if** type$(t) = rem$, $a \notin \chi_t$ is removed atom and $\tau' \in$ Child-Tabs **then**
12      $\tau_t := \{\langle M_a^-, \sigma] \ \mathrm{UpdtWgt}(\Pi_t, M, a),$
13          $\{\langle C_a^-, \rho] \ \mathrm{UpdtWgt\&Ch}(\Pi_t, M, C, a)\rangle \mid \langle C, \rho \rangle \in \mathcal{C}\})$ $\mid \langle M, \sigma, \mathcal{C} \rangle \in \tau'\}$
14 **else if** type$(t) = rem$, $r \notin \chi_t$ is removed rule and $\tau' \in$ Child-Tabs **then**
15      $\tau_t := \{\langle M, \sigma_{\{r\}}^-, \{\langle C, \rho_{\{r\}}^-] \mid \langle C, \rho \rangle \in \mathcal{C}, \rho(r) = \infty\}\rangle \mid \langle M, \sigma, \mathcal{C} \rangle \in \tau', \sigma(r) = \infty\}$
16 **else if** type$(t) = join$ and $\tau', \tau'' \in$ Child-Tabs with $\tau' \neq \tau''$ **then**
17      $\tau_t := \{\langle M, \sigma'] \ \sigma'', \{\langle C, \rho'] \ \rho''\rangle \mid \langle C, \rho' \rangle \in \mathcal{C}', \langle C, \rho'' \rangle \in \mathcal{C}''\} \cup$
18          $\{\langle M, \rho] \ \sigma''\rangle \mid \langle M, \rho \rangle \in \mathcal{C}'\} \cup$
19          $\{\langle M, \sigma'] \ \rho\rangle \mid \langle M, \rho \rangle \in \mathcal{C}''\})$ $\mid \langle M, \sigma', \mathcal{C} \rangle \in \tau', \langle M, \sigma'', \mathcal{C}'' \rangle \in \tau''\}$

3.2 Using Decompositions of Incidence Graphs

Our next algorithm ($\mathcal{DP}_{\mathsf{INC}}$) takes the incidence graph as graph representation of the input program. The treewidth of the incidence graph is smaller than the treewidth of the primal graph plus one, cf., [7,11]. More importantly, the primal graph contains a clique on $\mathrm{at}(r)$ for each rule r. The incidence graph, compared to the primal graph, contains rules as vertices and its relationship to the atoms in terms of edges. By definition, we have no guarantee that all atoms of a rule occur together in the same bag of TDs of the incidence graph. For that reason, we *cannot* locally check the satisfiability of a rule when traversing the TD without additional stored information (so-called *rule-states* that intuitively represent how much of a rule is already (dis-)satisfied). We only know that for each rule r there is a path $p = t_{int}, t_1, \ldots, t_m, t_{rem}$ where t_{int} introduces r and t_{rem} removes r and when considering t_{rem} in the table algorithm we have seen all atoms that occur in rule r. Thus, on removal of r in t_{rem} we ensure that r is satisfied while taking rule-states for choice and weight rules into account. Consequently, our tuples will contain a witness, its rule-state, and counterwitnesses and their rule-states.

A tuple in τ_t for Algorithm 3 (INC) is a triple $\langle M, \sigma, \mathcal{C} \rangle$. The set $M \subseteq \mathrm{at}(\Pi) \cap \chi(t)$ represents again a witness. A *rule-state* σ is a mapping $\sigma : \Pi_t \to \mathbb{N}_0 \cup \{\infty\}$. A rule state for M represents whether rules of $\chi(t)$ are either (i) satisfied by a superset of M or (ii) undecided for M. Formally, the set $SR(\Pi_t, \sigma)$ of satisfied bag-rules Π_t consists of each rule $r \in \Pi_t$ such that $\sigma(r) = \infty$. Hence, M witnesses a model $M' \supseteq M$ where $M' \vDash \Pi_{<t} \cup SR(\Pi_t, \sigma)$. $\mathcal{C}$ concerns counterwitnesses.

We compute a new rule-state σ from a rule-state, "updated" bounds for weight rules (UpdtWgt), and satisfied rules (SatRules, defined below). We define $\mathrm{UpdtWgt}(\Pi_t, M, a) := \sigma'$ depending on an atom a with $\sigma'(r) := \mathrm{wght}(r, \{a\} \cap [(B_r^- \setminus M) \cup (B_r^+ \cap M)])$, if $r \in \mathrm{WGT}(\Pi_t)$. We use binary operator $\uplus^2$ to combine rule-states, which ensures that rules satisfied in at least one operand remain satisfied. Next, we explain the meaning of rule-states.

Example 7. Consider program Π from Example 1 and TD $\mathcal{T}' = (\cdot, \chi)$ of $I(\Pi)$ and the tables $\tau_1, \ldots, \tau_{18}$ in Fig. 3 (left). We are only interested in the first two tuple positions (red and green parts) and implicitly assume that "i" refers to Line i in the respective table. Consider $M_{4.1} = \{c\}$ in table τ_4. Since $H_{r_2} = \{c\}$, witness $M_{4.1} = \{c\}$ satisfies rule r_2. As a result, $\sigma_{4.1}(r_2) = \infty$ remembering satisfied rule r_2 for $M_{4.1}$. Since $c \notin M_{4.2}$ and $B_{r_1}^+ = \{c\}$, $M_{4.2}$ satisfies rule r_1, resulting in $\sigma_{4.2}(r_1) = \infty$. Rule-state $\sigma_{4.1}(r_1)$ represents that r_1 is undecided for $M_{4.2}$. For weight rule r_2, rule-states remember the sum of body weights involving removed atoms. Consider $M_{6.2} = M_{6.3} = \emptyset$ of table τ_6. We have $\sigma_{6.2}(r_2) \neq \sigma_{6.3}(r_2)$, because $M_{6.2}$ was obtained from some $M_{5.i}$ of table τ_5 with $b \notin M_{5.i}$ and b occurs in $B_{r_2}^+$ with weight 1, resulting in $\sigma_{6.3}(r_2) = 1$; whereas $M_{6.3}$ extends some $M_{5.j}$ with $b \notin M_{5.j}$. ∎

In order to decide in node t whether a witness satisfies rule $r \in \Pi_t$, we check satisfiability of program $\dot{\mathcal{R}}(r)$ constructed by $\dot{\mathcal{R}}$, which maps rules to state-programs. Formally, for $M \subseteq \chi(t) \setminus \Pi_t$, $\mathrm{SatRules}(\dot{\mathcal{R}}, M) := \sigma$ where $\sigma(r) := \infty$ if $(r, \mathcal{R}) \in \dot{\mathcal{R}}$ and $M \vDash \mathcal{R}$.

Definition 1. Let Π be a program, $\mathcal{T} = (\cdot, \chi)$ be a TD of $I(\Pi)$, t be a node of $\mathcal{T}$, $\mathcal{P} \subseteq \Pi_t$, and $\sigma : \Pi_t \to \mathbb{N}_0 \cup \{\infty\}$ be a rule-state. The state-program $\mathcal{P}^{(t, \sigma)}$ is obtained from $\mathcal{P} \cup \{\leftarrow B_r \mid r \in \mathrm{CH}(\mathcal{P}), H_r \subsetneq \mathrm{at}_{\leq t}\}$ [3] by

---

[2] $\sigma \uplus \rho := \{(x, \underset{(x, c_1) \in \sigma}{\Sigma} c_1 + \underset{(x, c_2) \in \rho}{\Sigma} c_2) \mid (x, \cdot) \in \sigma \cup \rho\}$; $\sigma_r^+ := \sigma \cup \{(r, 0)\}$; $\sigma_S^- := \{(x, y) \in \sigma \mid x \notin S\}$.

[3] We require to add $\{\leftarrow B_r \mid r \in \mathrm{CH}(\mathcal{P}), H_r \subsetneq \mathrm{at}_{\leq t}\}$ in order to decide satisfiability for corner cases of choice rules involving counterwitnesses of Line 3 in Algorithm 3.

1. *removing rules r with $\sigma(r) = \infty$ ("already satisfied rules");*
2. *removing from every rule all literals $a, \neg a$ with $a \notin \chi(t)$; and*
3. *setting new bound $\max\{0, \text{bnd}(r) - \sigma(r) - \text{wght}(r, at(r) \setminus at_{\leq t})\}$ for weight rule r.*

We define $\dot{\mathcal{P}}^{(t,\sigma)} : \mathcal{P} \to 2^{\mathcal{P}^{(t,\sigma)}}$ by $\dot{\mathcal{P}}^{(t,\sigma)}(r) := \{r\}^{(t,\sigma)}$ for $r \in \mathcal{P}$.

Example 8. Observe $\Pi_{t_1}^{(t_1, \emptyset)} = \{\{b\} \leftarrow c, \leftarrow c, c \leftarrow 0 \leq \{b = 1\}\}$ and $\Pi_{t_2}^{(t_2, \emptyset)} = \{\{a\} \leftarrow, \leftarrow 1 \leq \{\neg a = 1\}\}$ for Π_{t_1}, Π_{t_2} of Fig. 1 (right).   ∎

The following example provides an idea how we compute models of a given program using the incidence graph. The resulting algorithm IMOD is the same as INC, except that only the first two tuple positions (red and green parts) are considered.

Example 9. Again, we consider Π of Example 1 and in Fig. 3 (left) $\mathcal{T}'$ as well as tables $\tau_1, \ldots, \tau_{18}$. Table $\tau_1 = \{\langle \emptyset, \emptyset \rangle\}$ as type$(t_1) = \textit{leaf}$. Since type$(t_2) = \textit{int}$ and t_2 introduces atom c, we construct τ_2 from τ_1 by taking $M_{2.1} := M_{1.1} \cup \{c\}$ and $M_{2.2} := M_{1.1}$ as well as rule-state $\emptyset$. Because type$(t_3) = \textit{int}$ and t_3 introduces rule r_1, we consider state program $L_3 := \{r_1\}^{(t_3, \{(r_1, 0)\})} = \{\leftarrow c\}$ for SatRules$(\dot{L}_3, M_{2.1}) = \{(r_1, 0)\}$ as well as SatRules$(\dot{L}_3, M_{2.2}) = \{(r_1, \infty)\}$ (according to Line 9 of Algorithm 3). Because type$(t_4) = \textit{int}$ and t_4 introduces rule r_2, we consider $M_{3.1} := M_{2.1}$ and $M_{3.2} := M_{2.2}$ and state program $L_4 := \{r_2\}^{(t_4, \{(r_2, 0)\})} = \{c \leftarrow 0 \leqslant \{\}\} = \{c \leftarrow\}$ for SatRules$(\dot{L}_4, M_{3.1}) = \{(r_2, \infty)\}$ as well as SatRules$(\dot{L}_4, M_{3.2}) = \{(r_2, 0)\}$ (see Line 9). Node t_5 introduces b (table not shown) and node t_6 removes b. Table τ_6 was discussed in Example 7. When we remove b in t_6 we have decided the "influence" of b on the satisfiability of r_1 and r_2 and thus all rules where b occurs. Tables τ_7 and τ_8 can be derived similarly. Then, t_9 removes rule r_2 and we ensure that every witness $M_{9.1}$ can be extended to a model of r_2, i.e., witness candidates for τ_9 are $M_{8.i}$ with $\sigma_{8.i}(r_2) = \infty$. The remaining tables are derived similarly. For example, table τ_{17} for join node t_{17} is derived analogously to table τ_{17} for algorithm PRIM in Fig. 2, but, in addition, also combines the rule-states as specified in Algorithm 3.   ∎

Since we already explained how to obtain models, we only briefly describe how we handle the counterwitness part. Family $\mathcal{C}$ consists of tuples (C, ρ) where $C \subseteq \text{at}(\Pi) \cap \chi(t)$ is a *counterwitness* in t to M. Similar to the rule-state σ the rule-state ρ for C under M represents whether rules of the GL reduct Π_t^M are either (i) satisfied by a superset of C or (ii) undecided for C. Thus, C witnesses the existence of $C' \subsetneq M'$ satisfying $C' \vDash (\Pi_{<t} \cup SR(\Pi_t, \rho))^{M'}$ since M witnesses a model $M' \supseteq M$ where $M' \vDash \Pi_{<t} \cup SR(\Pi_t, \rho)$. In consequence, there exists an answer set of Π if the root table contains $\langle \emptyset, \emptyset, \emptyset \rangle$. In order to locally decide rule satisfiability for counterwitnesses, we require state-programs under witnesses.

Definition 2. *Let Π be a program, $\mathcal{T} = (\cdot, \chi)$ be a TD of $I(\Pi)$, t be a node of $\mathcal{T}$, $\mathcal{P} \subseteq \Pi_t$, $\rho : \Pi_t \to \mathbb{N}_0 \cup \{\infty\}$ be a rule-state and $M \subseteq \text{at}(\Pi)$. We define state-program $\mathcal{P}^{(t,\rho,M)}$ by $[S^{(t,\rho)}]^M$ where $S := \mathcal{P} \cup \{\leftarrow B_r \mid r \in \text{CH}(\mathcal{P}), \rho(r) > 0\}$, and $\dot{\mathcal{P}}^{(t,\rho,M)} : \mathcal{P} \to 2^{\mathcal{P}^{(t,\rho,M)}}$ by $\dot{\mathcal{P}}^{(t,\rho,M)}(r) := \{r\}^{(t,\rho,M)}$ for $r \in \mathcal{P}$.*

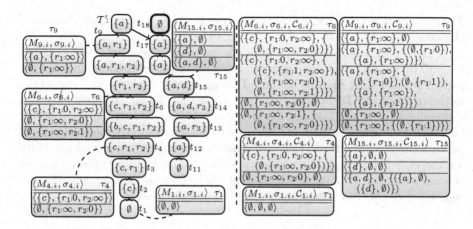

Fig. 3. Selected DP tables of IMOD (left) and INC (right) for nice TD $\mathcal{T}'$.

We compute a new rule-state ρ for a counterwitness from an earlier rule-state, satisfied rules (SatRules), and both (a) "updated" bounds for weight rules or (b) "updated" value representing whether the head can still be satisfied ($\rho(r) \leq$ 0) for choice rules r (UpdtWgt&Ch). Formally, UpdtWgt&Ch(Π_t, M, C, a) := σ' depending on an atom a with (a) $\sigma'(r) := $ wght($r, \{a\} \cap [(B_r^- \setminus M) \cup (B_r^+ \cap C)]$), if $r \in$ WGT(Π_t); and (b) $|\{a\} \cap H_r \cap (M \setminus C)|$, if $r \in$ CH(Π_t).

Algorithm 4. Algorithm #OINC($t, \chi_t, \Pi_t, \text{at}_{\leq t}$, Child-Tabs).

In: Bag χ_t, bag-rules Π_t, atoms-below $\text{at}_{\leq t}$, child tables Child-Tabs of t. **Out:** Tab. τ_t.
/* For $\langle M, \sigma, \mathcal{C}, c, n \rangle$, we only state affected parts (cost c and count n);
   "..." indicates computation as before. $\{...\}$ denotes a multiset.   */

1 **if** type(t) = *leaf* **then** $\tau_t := \{\langle \emptyset, \ldots, 0, 1 \rangle\}$
2 **else if** type(t) = *int*, $a \in \chi_t \setminus \Pi_t$ *is introduced and* $\tau' \in$ *Child-Tabs* **then**
3 $\quad \tau_t := \{\langle M, \ldots, \text{cst}(\Pi, \emptyset, \{a\}) + c, n\rangle \qquad\qquad | \ \langle M, \sigma, \mathcal{C}, c, n \rangle \in \tau'\} \bigcup$
4 $\quad\quad \{\langle M_a^+, \ldots, \text{cst}(\Pi, \{a\}, \{a\}) + c, n\rangle \qquad | \ \langle M, \sigma, \mathcal{C}, c, n \rangle \in \tau'\}$
5 **else if** type(t) = *int or rem*, *removed or introduced* $r \in \Pi_t$, $\tau' \in$ *Child-Tabs* **then**
6 $\quad \tau_t := \{\langle M, \ldots, c, n\rangle \qquad\qquad\qquad\quad | \ \langle M, \sigma, \mathcal{C}, c, n \rangle \in \tau', \ldots\}$
7 **else if** type(t) = *rem*, $a \notin \chi_t$ *is removed atom and* $\tau' \in$ *Child-Tabs* **then**
8 $\quad \tau_t := \text{cnt}(\text{kmin}(\{\langle M_a^-, \ldots, c, n\rangle \qquad | \ \langle M, \sigma, \mathcal{C}, c, n \rangle \in \tau'\}))$
9 **else if** type(t) = *join and* $\tau', \tau'' \in$ *Child-Tabs with* $\tau' \neq \tau''$ **then**
10 $\quad \tau_t := \text{cnt}(\text{kmin}(\{\langle M, \ldots, c' + c'' - \text{cst}(\Pi, M, \chi_t), n' \cdot n''\rangle$
11 $\quad\quad | \ \langle M, \sigma', \mathcal{C}', c', n' \rangle \in \tau', \langle M, \sigma'', \mathcal{C}'', c'', n'' \rangle \in \tau''\}))$

Theorem 2. *The algorithm* $\mathcal{DP}_{\text{INC}}$ *is correct.*

Proof. (Idea) A tuple at a node t guarantees that there exists a model for the sub-program induced by the subtree rooted at t, which works for all node types. While traversing the tree decomposition, every answer set is indeed considered.

Theorem 3. *Given a program* Π, *algorithm* $\mathcal{DP}_{\text{INC}}$ *runs in time* $\mathcal{O}(2^{2^{k+2} \cdot \ell^{k+1}} \cdot \|I(\Pi)\|)$, *where* $k := tw(I(\Pi))$, *and* $\ell := max\{3, \text{bnd}(r) \mid r \in$ WGT(Π)$\}$.

The runtime bounds stated in Theorem 3 appear to be worse than in Theorem 1. However, $tw(I(\Pi)) \leq tw(P(\Pi)) + 1$ and $tw(P(\Pi)) \geq \max\{|at(r)| \mid r \in \Pi\}$ for a given program Π. Further, there are programs where $tw(I(\Pi)) = 1$, but $tw(P(\Pi)) = k$, e.g., a program consisting of a single rule r with $|at(r)| = k$. Consequently, worst-case runtime bounds of $\mathcal{DP}_{\mathsf{PRIM}}$ are at least double-exponential in the rule size and $\mathcal{DP}_{\mathsf{PRIM}}$ will perform worse than $\mathcal{DP}_{\mathsf{INC}}$ on input programs containing large rules. However, due to the rule-states, data structures of $\mathcal{DP}_{\mathsf{INC}}$ are much more complex than of $\mathcal{DP}_{\mathsf{PRIM}}$. In consequence, we expect $\mathcal{DP}_{\mathsf{PRIM}}$ to perform better in practice if rules are small and incidence and primal treewidth are therefore almost equal. In summary, we have a trade-off between (i) a more general parameter decreasing the theoretical worst-case runtime and (ii) less complex data structures decreasing the practical overhead to solve AS.

3.3   Extensions for Optimization and Counting

In order to find an answer set of a program with optimization statements or the number of optimal answer sets (#AsPO), we extend our algorithms PRIM and INC. Therefore, we augment tuples stored in tables with an integers c and n describing the cost and the number of witnessed sets. Due to space restrictions, we only present adaptions for INC. We state which parts of INC we adapt to compute the number of optimal answer sets in Algorithm 4 (#OINC). To slightly simplify the presentation of optimization rules, we assume without loss of generality that whenever an atom a is introduced in bag $\chi(t)$ for some node t of the TD, the optimization rule r, where a occurs, belongs to the bag $\chi(t)$. First, we explain how to handle costs making use of function $\mathrm{cst}(\Pi, M, A)$ as defined in Sect. 2. In a leaf (Line 1) we set the (current) cost to 0. If we introduce an atom a (Line 2–4) the cost depends on whether a is set to true or false in M and we add the cost of the "child" tuple. Removal of rules (Line 5–6) is trivial, as we only store the same values. If we remove an atom (Line 7–8), we compute the minimum costs only for tuples $\langle M_a^-, \sigma, \mathcal{C}, c, n \rangle$ where c is minimal among M_a^-, σ, $\mathcal{C}$, that is, for a multiset $\mathcal{S}$ we let $\mathrm{kmin}(\mathcal{S}) := \{\langle M_a^-, \sigma, \mathcal{C}, c, n \rangle \mid c = \min\{c' : \langle M_a^-, \sigma, \mathcal{C}, c', \cdot \rangle \in \mathcal{S}\}, \langle M_a^-, \sigma, \mathcal{C}, c, n \rangle \in \mathcal{S}\}$. We require a multiset notation for counting (see below). If we join two nodes (Line 9–11), we compute the minimum value in the table of one child plus the minimum value of the table of the other child minus the value of the cost for the current bag, which is exactly the value we added twice. Next, we explain how to handle the number of witnessed sets that are minimal with respect to the cost. In a leaf (Line 1), we set the counter to 1. If we introduce/remove a rule or introduce an atom (Line 2–6), we can simply take the number n from the child. If we remove an atom (Line 7–8) we first obtain a multiset from computing kmin, which can contain several tuples for $M_a^-, \sigma, \mathcal{C}, c$ as we obtained M_a^- either from $M \setminus \{a\}$ if $a \in M$ or M if $a \notin M$ giving rise multiple solutions, that is, $\mathrm{cnt}(\mathcal{S}) := \{\langle M, \sigma, \mathcal{C}, c, \sum_{\langle M, \sigma, \mathcal{C}, c, n' \rangle \in \mathcal{S}} n' \rangle \mid \langle M, \sigma, \mathcal{C}, c, n \rangle \in \mathcal{S}\}$. If we join nodes (Line 7–9), we multiply the number n' from the tuple of one child with the number n'' from the tuple of the other child, restrict results with respect to minimum costs, and sum up the resulting numbers.

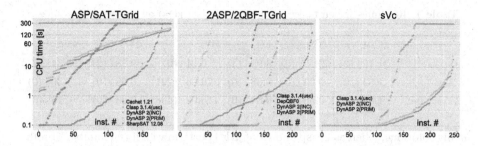

Fig. 4. Results of randomly generated and selected real-world instances.

Corollary 1. *Given a program Π, algorithm #OINC runs in time $\mathcal{O}(log(m) \cdot 2^{2^{k+2} \cdot \ell^{k+1}} \|I(\Pi)\|^2)$, where $k := tw(I(\Pi))$, $\ell := max\{3, bnd(r) : r \in \mathrm{WGT}(\Pi)\}$, and $m := \Sigma_{r \in \mathrm{OPT}(\Pi)} \mathrm{wght}(r)$.*

4   Experimental Evaluation

We implemented the algorithms $\mathcal{DP}_{\mathsf{PRIM}}$ and $\mathcal{DP}_{\mathsf{INC}}$ into a prototypical solver DynASP2($\cdot$) and performed experiments to evaluate its runtime behavior. Clearly, we *cannot* hope to solve programs with graph representations of high treewidth. However, programs involving real-world graphs such as graph problems on transit graphs admit TDs of small width. We used both random and structured instances for our benchmarks[4], see also [8]. The random instances (SAT-TGRID, 2QBF-TGRID, ASP-TGRID, 2ASP-TGRID) were designed to have a high number of variables and solutions and treewidth at most three. The structured instances model various graph problems (2COL, 3COL, DS, ST CVC, sVC) on real world mass transit graphs. For a graph, program 2COL counts all 2-colorings, 3COL counts all 3-colorings, DS counts all minimal dominating sets, ST counts all Steiner trees, CVC counts all cardinality-minimal vertex covers, and sVC counts all subset-minimal vertex covers. In order to draw conclusions about the efficiency of DynASP2, we mainly inspected the cpu running time and number of timeouts using the average over three runs per instance (three fixed seeds allow certain variance [1] for heuristic TD computation). We limited available memory (RAM) to 4GB (to run SharpSAT on large instances), and cpu time to 300 s, and then compared DynASP2 with the dedicated #SAT solvers SharpSAT [14] and Cachet [12], the QBF solver DepQBF0, and the ASP solver Clasp. Fig. 4 illustrates runtime results as a cactus plot. Table 1 reports on the average running times, numbers of solved instances and timeouts on the structured instance sets.

Summary. Our empirical benchmark results confirm that DynASP2 exhibits competitive runtime behavior if the input instance has small treewidth. Compared to modern ASP and QBF solvers, DynASP2 has an advantage in case of

---

[4] https://github.com/daajoe/dynasp_experiments.

Table 1. Runtimes (given in sec.; #timeouts in brackets) on real-world instances.

	2Col		3Col		Ds		St		cVc		sVc	
Clasp(usc)	31.72	(21)	0.10	(0)	8.99	(3)	0.21	(0)	29.88	(21)	98.34	(71)
DynASP2(PRIM)	1.54	(0)	0.53	(0)	0.68	(0)	79.36	(221)	0.99	(0)	1.30	(0)
DynASP2(INC)	1.43	(0)	0.58	(0)	0.54	(0)	115.02	(498)	0.68	(0)	0.78	(0)

many solutions, whereas Clasp and DepQBF0 perform well if the number of solutions is relatively small. However, DynASP2 is still reasonably fast on structured instances with few solutions as it yields the result mostly within less than 10 s. We observed that INC seems to be the better algorithm in our setting, indicating that the smaller width obtained by decomposing the incidence graph generally outweighs the benefits of simpler solving algorithms for the primal graph. However, if INC and PRIM run with graphs of similar width, PRIM benefits from its simplicity. A comparison to #SAT solvers suggests that, on random instances, they have a lower overhead (which is not surprising, since our algorithms are built for ASP), but after about 150 s. Our algorithms solve more instances.

5 Conclusion

We presented novel DP algorithms for ASP, extending previous work [9] in order to cover the full ASP syntax. Our algorithms are based on two graph representations of programs and run in linear time with respect to the treewidth of these graphs and weights used in the program. Experiments indicate that our approach seems to be suitable for practical use, at least for certain classes of instances with low treewidth, and hence could fit into a portfolio solver. A further use of our techniques could be for extensions of ASP such as HEX [6].

References

1. Abseher, M., Dusberger, F., Musliu, N., Woltran, S.: Improving the efficiency of dynamic programming on tree decompositions via machine learning. In: IJCAI 2015 (2015)
2. Bodlaender, H., Koster, A.M.C.A.: Combinatorial optimization on graphs of bounded treewidth. Comput. J. **51**(3), 255–269 (2008)
3. Brewka, G., Eiter, T., Truszczyński, M.: Answer set programming at a glance. Commun. ACM **54**(12), 92–103 (2011)
4. Calimeri, F., Faber, W., Gebser, M., Ianni, G., Kaminski, R., Krennwallner, T., Leone, N., Ricca, F., Schaub, T.: ASP-core-2 input language format (2013)
5. Cygan, M., Fomin, F.V., Kowalik, L., Lokshtanov, D., Marx, D., Pilipczuk, M., Saurabh, S.: Parameterized Algorithms. Springer, Heidelberg (2015)
6. Eiter, T., Fink, M., Ianni, G., Krennwallner, T., Schüller, P.: Pushing efficient evaluation of HEX programs by modular decomposition. In: Delgrande, J.P., Faber, W. (eds.) LPNMR 2011. LNCS (LNAI), vol. 6645, pp. 93–106. Springer, Heidelberg (2011). doi:10.1007/978-3-642-20895-9_10

7. Fichte, J.K., Szeider, S.: Backdoors to tractable answer-set programming. Artif. Intell. **220**, 64–103 (2015)
8. Fichte, J.K., Hecher, M., Morak, M., Woltran, S.: Answer set solving with bounded treewidth revisited. CoRR, arXiv:1702.02890 (2017)
9. Jakl, M., Pichler, R., Woltran, S.: Answer-set programming with bounded treewidth. In: IJCAI 2009, vol. 2 (2009)
10. Pichler, R., Rümmele, S., Szeider, S., Woltran, S.: Tractable answer-set programming with weight constraints: bounded treewidth is not enough. Theory Pract. Log. Program. **14**(2), 141–164 (2014)
11. Samer, M., Szeider, S.: Algorithms for propositional model counting. J. Discrete Algorithms **8**(1), 50–64 (2010)
12. Sang, T., Bacchus, F., Beame, P., Kautz, H.A., Pitassi, T.: Combining component caching and clause learning for effective model counting. In: SAT 2004 (2004)
13. Syrjänen,   T.:   Lparse   1.0   user's   manual   (2002). tcs.hut.fi/Software/smodels/lparse.ps
14. Thurley, M.: sharpSAT - counting models with advanced component caching and implicit BCP. In: SAT 2006 (2006)

Vicious Circle Principle and Formation of Sets in ASP Based Languages

Michael Gelfond and Yuanlin Zhang[✉]

Texas Tech University, Lubbock, TX, USA
{michael.gelfond,y.zhang}@ttu.edu

Abstract. The paper continues the investigation of Poincare and Russel's Vicious Circle Principle (VCP) in the context of the design of logic programming languages with sets. We expand previously introduced language $\mathcal{A}log$ with aggregates by allowing infinite sets and several additional set related constructs useful for knowledge representation and teaching. In addition, we propose an alternative formalization of the original VCP and incorporate it into the semantics of new language, $\mathcal{S}log^+$, which allows more liberal construction of sets and their use in programming rules. We show that, for programs without disjunction and infinite sets, the formal semantics of aggregates in $\mathcal{S}log^+$ coincides with that of several other known languages. Their intuitive and formal semantics, however, are based on quite different ideas and seem to be more involved than that of $\mathcal{S}log^+$.

1 Introduction

This paper is the continuation of work started in [12] with introduction of $\mathcal{A}log$ – a version of Answer Set Prolog (ASP) with aggregates. The semantics of $\mathcal{A}log$ combines the Rationality Principle of ASP [9] with the adaptation of the Vicious Circle Principle (VCP) introduced by Poincare and Russel [21,22] in their attempt to resolve paradoxes of set theory. In $\mathcal{A}log$, the latter is used to deal with formation of sets and their legitimate use in program rules. To understand the difficulty addressed by $\mathcal{A}log$ consider the following programs:

Example 1. P_0 consisting of a rule:
    p(1) :- card{X: p(X)} != 1.
P_1 consisting of rules:
    p(1) :- p(0). p(0) :- p(1). p(1) :- card{X: p(X)} != 1.
P_2 consisting of rules:
    p(1) :- card{X: p(X)} >= 0.
Even for these seemingly simple programs, there are different opinions about their meaning. To the best of our knowledge all ASP based semantics, including that of [4,12,24]) view P_0 as a bad specification. It is inconsistent, i.e., has no answer sets. Opinions differ, however, about the meaning of the other two programs. [4] views P_1 as a reasonable specification having one answer set – $\{p(0), p(1)\}$. According to [12,24] P_1 is inconsistent. According to most semantics P_2 has one answer set, $\{p(1)\}$. $\mathcal{A}log$, however, views it as inconsistent.

© Springer International Publishing AG 2017
M. Balduccini and T. Janhunen (Eds.): LPNMR 2017, LNAI 10377, pp. 146–159, 2017.
DOI: 10.1007/978-3-319-61660-5_14

As in the naive set theory, the difficulty in interpretations seems to be caused by self-reference. In both P_1 and P_2, the definition of $p(1)$ references the set described in terms of p. It is, of course, not entirely clear how this type of differences can be resolved. Sometimes, further analysis can find convincing arguments in favor of one of the proposals. Sometimes, the analysis discovers that different approaches really model different language or world phenomena and are, hence, all useful in different contexts. We believe that the difficulty can be greatly alleviated if the designers of the language provide its users with as clear intuitive meaning of the new constructs as possible. Accordingly, the *set name* construct $\{X : p(X)\}$ of $\mathcal{A}log$ denotes *the set of all objects believed by the rational agent associated with the program to satisfy property p.* (This reading is in line with the epistemic view of ASP connectives shared by the authors.) The difficulties with self-reference in $\mathcal{A}log$ are resolved by putting the following intuitive restriction on the formation of sets[1]:

An expression $\{X : p(X)\}$ denotes a set S only if for every t rational belief in $p(t)$ can be established without a reference to S, or equivalently, *the reasoner's belief in $p(t)$ can not depend on existence of a set denoted by $\{X : p(X)\}$.*

We view this restriction as a possible interpretation of VCP and refer to it as *Strong VCP*. Let us illustrate the intuition behind $\mathcal{A}log$ set constructs.

Example 2. Let us consider programs from Example 1. P_0 clearly has no answer set since $\emptyset$ does not satisfy its rule and there is no justification for believing in $p(1)$. P_1 is also inconsistent. To see that notice that the first two rules of the program limit our possibilities to $A_1 = \emptyset$ and $A_2 = \{p(0), p(1)\}$. In the first case $\{X : p(X)\}$ denotes $\emptyset$. But this contradicts the last rule of the program. A_1 cannot be an answer set of P_1. In A_2, $\{X : p(X)\}$ denotes $S = \{0, 1\}$. But this violates our form of VCP since the reasoner's beliefs in both, $p(0)$ and $p(1)$, cannot be established without reference to S. A_2 is not an answer set either. Now consider program P_2. There are two candidate answer sets[2]: $A_1 = \emptyset$ and $A_2 = \{p(1)\}$. In A_1, $S = \emptyset$ which contradicts the rule. In A_2, $S = \{1\}$ but this would contradict the $\mathcal{A}log$'s VCP. The program is inconsistent[3].

We hope that the examples are sufficient to show how the informal semantics of $\mathcal{A}log$ can give a programmer some guidelines in avoiding formation of sets problematic from the standpoint of VCP. In what follows, we

---

[1] It is again similar to set theory where the difficulty is normally avoided by restricting comprehension axioms guaranteeing existence of sets denoted by expressions of the form $\{X : p(X)\}$. In ASP such restrictions are encoded in the definition of answer sets.

[2] By a candidate answer set we mean a consistent set of ground regular literals satisfying the rules of the program.

[3] There is a common argument for the semantics in which $\{p(1)\}$ would be the answer set of P_2: "Since $card\{X : p(X)\} \geq 0$ is always true it can be dropped from the rule without changing the rule's meaning". But the argument assumes existence of the set denoted by $\{X : p(X)\}$ which is not always the case in $\mathcal{A}log$.

- Expand $\mathcal{A}log$ by allowing infinite sets and several additional set related constructs useful for knowledge representation and teaching.
- Propose an alternative formalization of the original VCP and incorporate it into the semantics of new language, $\mathcal{S}log^+$, which allows more liberal construction of sets and their use in programming rules. (The name of the new language is explained by its close relationship with language $\mathcal{S}log$ [24] – see Theorem 2).
- Show that, for programs without disjunction and infinite sets, the formal semantics of aggregates in $\mathcal{S}log^+$ coincides with that of several other known languages. Their intuitive and formal semantics, however, are based on quite different ideas and seem to be more involved than that of $\mathcal{S}log^+$.
- Prove some basic properties of programs in (extended) $\mathcal{A}log$ and $\mathcal{S}log^+$.

2 Syntax and Semantics of $\mathcal{A}log$

In what follows we retain the name $\mathcal{A}log$ for the new language and refer to the earlier version as "original $\mathcal{A}log$".

2.1 Syntax

Let Σ be a (possibly sorted) signature with a finite collection of predicate and function symbols and (possibly infinite) collection of object constants, and let $\mathcal{A}$ be a finite collection of symbols used to denote functions from sets of terms of Σ into integers. Terms and literals over signature Σ are defined as usual and referred to as *regular*. Regular terms are called *ground* if they contain no variables and no occurrences of symbols for arithmetic functions. Similarly for literals. We refer to an expression $\{\bar{X} : cond\}$ where $cond$ is a finite collection of regular literals and $\bar{X}$ is the list of variables occurring in $cond$, as a *set name*. It is read as *the set of all objects of the program believed to satisfy cond*. Variables from $\bar{X}$ are often referred to as *set variables*. An occurrence of a set variable in a set name is called *bound* within the set name. Since treatment of variables in extended $\mathcal{A}log$ is the same as in the original language we limit our attention to programs in which every occurrence of a variable is bound. Rules containing non-bound occurrences of variables are considered as shorthands for their ground instantiations (for details see [12]).

A *set atom* of $\mathcal{A}log$ is an expression of the form $f_1(S_1) \odot f_2(S_2)$ or $f(S) \odot k$ where f, f_1, f_2 are functions from $\mathcal{A}$, S, S_1, S_2 are set names, k is a number, and $\odot$ is an arithmetic relation $>, \geq, <, \leq, =$ or $!=$, or of the form $S_1 \otimes S_2$ where $\otimes$ is $\subset, \subseteq$, or $=$. We often write $f(\{\bar{X} : p(\bar{X})\})$ as $f\{\bar{X} : p(\bar{X})\}$ and $\{\bar{X} : p(\bar{X})\} \otimes S$ and $S \otimes \{\bar{X} : p(\bar{X})\}$ as $p \otimes S$ and $S \otimes p$ respectively. Regular and set atoms are referred to as *atoms*. A *rule* of $\mathcal{A}log$ is an expression of the form

$$head \leftarrow body \tag{1}$$

where *head* is a disjunction of regular literals or a set atom of the form $p \subseteq S$, $S \subseteq q$, or $p = S$, and *body* is a collection of regular literals (possibly preceded by *not*) and set atoms. A rule with set atom in the head is called *set introduction rule*. Note that both head and body of a rule can be infinite. All parts of *Alog* rules, including *head*, can be empty. A *program* of *Alog* is a collection of *Alog*'s rules.

2.2 Semantics

To define the semantics of *Alog* programs we first notice that the *standard definition of answer set from* [10] *is applicable to programs with infinite rules*. Hence we already have the definition of answer set for *Alog* programs not containing occurrences of set atoms. We also need the satisfiability relation for set atoms. Let A be a set of ground regular literals. If $f(\{\bar{t} : cond(\bar{t}) \subseteq A\})$ is defined then $f(\{\bar{X} : cond\}) \geq k$ is satisfied by A (is *true* in A) iff $f(\{\bar{t} : cond(\bar{t}) \subseteq A\}) \geq k$. Otherwise, $f(\{\bar{X} : cond\}) \geq k$ is falsified (is *false* in A). If $f(\{\bar{t} : cond(\bar{t}) \subseteq A\})$ is not defined then $f(\{\bar{X} : cond\}) \geq k$ is *undefined* in A. (For instance, atom $card\{X : p(X)\} \geq 0$ is undefined in A if A contains an infinite collection of atoms formed by p.) Similarly for other set atoms. Finally a rule is *satisfied* by S if its head is *true* in S or its body is *false* or *undefined* in S.

Answer Sets for Programs Without Set Introduction Rules. To simplify the presentation we first give the definition of answer sets for programs whose rules contain no set atoms in their heads. First we need the following definition:

Definition 1 (Set Reduct of *Alog*). *Let Π be a ground program of *Alog*. The set reduct of Π with respect to a set of ground regular literals A is obtained from Π by*

1. *removing rules containing set atoms which are* false *or* undefined *in A.*
2. *replacing every remaining set atom SA by the union of $cond(\bar{t})$ such that $\{\bar{X} : cond(\bar{X})\}$ occurs in SA and $cond(\bar{t}) \subseteq A$.*

The first clause of the definition removes rules that are useless because of the truth values of their aggregates in A. The next clause reflects the principle of avoiding vicious circles. Clearly, set reducts do not contain set atoms.

Definition 2 (Answer Set). *A set A of ground regular literals over the signature of a ground *Alog* program Π is* an answer set *of Π if A is an answer set of the set reduct of Π with respect to A.*

It is easy to see that for programs of the original *Alog* our definition coincides with the old one. Next several examples demonstrate the behavior of our semantics for programs not covered by the original syntax.

Infinite Universe

Example 3 (Aggregates on infinite sets). Consider a program E_1 consisting of the following rules:

```
even(0). even(I+2) :- even(I). q :- min{X : even(X)} = 0.
```
The program has one answer set, $S_{E_1} = \{q, even(0), even(2), \dots\}$. Indeed, the reduct of E_1 with respect to S_{E_1} is the infinite collection of rules

```
even(0).
even(2) :- even(0).
...
q :- even(0),even(2),even(4)...
```

The last rule has the infinite body constructed in the last step of Definition 1. Clearly, S_{E_1} is a subset minimal collection of ground literals satisfying the rules of the reduct (i.e. its answer set). Hence S_{E_1} is an answer set of E_1.

Example 4 (Programs with undefined aggregates). Now consider a program E_2 consisting of the rules:
```
even(0). even(I+2) :- even(I). q :- card{X : even(X)} > 0.
```
This program has one answer set, $S_{E_2} = \{even(0), even(2), \dots\}$. Since the aggregate *card*, ranging over natural numbers, is not defined on the set $\{t : even(t) \in S_{E_2}\}$. This means that the body of the last rule is undefined. According to clause one of Definition 1 this rule is removed. The reduct of E_2 with respect to S_{E_2} is
```
even(0). even(2) :- even(0). even(4) :- even(2). ......
```
Hence S_{E_2} is the answer set of E_2.[4] It is easy to check that, since every set A satisfying the rules of E_2 must contain all even numbers, S_{E_1} is the only answer set.

Programs with Set Atoms in the Bodies of Rules

Example 5 (Set atoms in the rule body). Consider a knowledge base containing two complete lists of atoms:

```
taken(mike,cs1).  taken(mike,cs2).  taken(john,cs2).
required(cs1).    required(cs2).
```

Set atoms allow for a natural definition of the new relation, *ready_to_graduate*(S), which holds if student S has taken all the required classes from the second list:

```
ready_to_graduate(S) :- {C: required(C)} ⊆ {C:taken(S,C)}.
```

The intuitive meaning of the rule is reasonably clear. The program consisting of this rule and the closed world assumption:

```
-ready_to_graduate(S) :- not ready_to_graduate(S)
```

---

[4] Of course this is true only because of our (somewhat arbitrary) decision to limit aggregates of $\mathcal{A}log$ to those ranging over natural numbers. We could, of course, allow aggregates mapping sets into ordinals. In this case the body of the last rule of E_2 will be defined and the only answer set of E_2 will be S_{E_1}.

implies that Mike is ready to graduate while John is not. If the list of classes taken by a student is incomplete the closed world assumption should be removed but the first rule still can be useful to determine people who are definitely ready to graduate. Even though the story can be represented in ASP without the set atoms, such representations are substantially less intuitive and less elaboration tolerant. Here is a simplified example of alternative representation suggested to the authors by a third party:

```
ready_to_graduate :- not -ready_to_graduate.
-ready_to_graduate :- not taken(c).
```

(Here student is eliminated from the parameters and we are limited to only one required class, c.) Even though in this case the answers are correct, unprincipled use of default negation leads to some potential difficulties. Suppose, for instance, that a student may graduate if given a special permission. This can be naturally added as a rule

```
ready_to_graduate :- permitted.
```

If the program is expanded by `permitted` it becomes inconsistent. This, of course, is unintended and contradicts our intuition. No such problem exists for the original representation.

The next example shows how the semantics deals with vicious circles.

Example 6 (Set atoms in the rule body). Consider a program P_4

```
p(a) :- p ⊆ {X : q(X)}.     q(a).
```

in which definition of $p(a)$ depends on the existence of the set denoted by $\{X : p(X)\}$. In accordance with the vicious circle principle no answer set of this program can contain $p(a)$. There are only two candidates for answer sets of P_4: $S_1 = \{q(a)\}$ and $S_2 = \{q(a), p(a)\}$. The set atom reduct of P_4 wrt S_1 is

```
p(a) :- q(a).     q(a).
```

while set atom reduct of P_4 with respect to S_2 is

```
p(a) :- p(a),q(a).     q(a).
```

Clearly, neither S_1 nor S_2 is an answer set of P_4. As expected, the program is inconsistent.

Programs with Set Introduction Rules. A set introduction rule with head $p \subseteq S$ (where p is a predicate symbol and S is a set name) defines set p as an arbitrary subset of S; rule with head $p = S$ simply gives S a different name; $S \subseteq p$ defines p as an arbitrary superset of S.

Example 7 (Set introduction rule). According to this intuitive reading the program P_9:

```
q(a).     p ⊆ {X:q(X)}.
```

has answer sets $A_1 = \{q(a)\}$ where the set p is empty and $A_2 = \{q(a), p(a)\}$ where $p = \{a\}$.

The formal definition of answer sets of programs with set introduction rules is given via a notion of *set introduction reduct*. (The definition is similar to that presented in [7]).

Definition 3 (Set Introduction Reduct). *The set introduction reduct of a ground $\mathcal{A}log$ program Π with respect to a set of ground regular literals A is obtained from Π by (1) replacing every set introduction rule of Π whose head is not true in A by "$\leftarrow$ body", and (2) replacing every set introduction rule of Π whose head $p \subseteq \{\bar{X} : q(\bar{X})\}$ (or $p = \{\bar{X} : q(\bar{X})\}$ or $\{\bar{X} : q(\bar{X})\} \subseteq p$) is true in A by "$p(\bar{t}) \leftarrow$ body" for each $p(\bar{t}) \in A$.*

Set A is an answer set *of Π if it is an answer set of the set introduction reduct of Π with respect to A.*

Example 8 (Set introduction rule). Consider a program P_9 from Example 7. The reduct of this program with respect to $A_1 = \{q(a)\}$ is $\{q(a).\}$ and hence A_1 is an answer set of P_9. The reduct of P_9 with respect to $A_2 = \{q(a), p(a)\}$ is $\{q(a). \ p(a).\}$ and hence A_2 is also an answer set of P_9. There are no other answer sets.

The use of a set introduction rule $p \subseteq S \leftarrow$ *body* is very similar to that of choice rule $\{p(\bar{X}) : q(\bar{X})\} \leftarrow$ *body* of [19] implemented in Clingo and other similar systems. In fact, if p from the set introduction rule does not occur in the head of any other rule of the program, the two rules have the same meaning. However if this condition does not hold the meaning is different. An $\mathcal{A}log$ program consisting of rules $p \subseteq \{X : q_1(X)\}$ and $p \subseteq \{X : q_2(X)\}$ defines an arbitrary set p from the intersection of q_1 and q_2. With choice rules it is not the case. We prefer the set introduction rule because of its more intuitive reading (after all everyone is familiar with the statement "p is an arbitrary subset of q") and relative simplicity of the definition of its formal semantics as compared with that of the choice rule. Our last example shows how subset introduction rule with equality can be used to represent synonyms:

Example 9 (Synonyms). Suppose we have a set of cars represented by atoms formed by a predicate symbol *car*, e.g., $\{car(a). \ car(b).\}$ The following rule

```
carro = {X:car(X)} :- spanish.
```

allows to introduce a new name of this set for Spanish speaking people. Clearly, *car* and *carro* are synonyms. Hence, program $P_9 \cup \{spanish.\}$ has one answer set: $\{spanish, car(a), car(b), carro(a), carro(b)\}$.

3   Alternative Formalization of VCP – Language $\mathcal{S}log^+$

In this section we introduce an alternative interpretation of VCP (referred to as *weak VCP*) and incorporate it in the semantics of a new logic programming language with set, called $\mathcal{S}log^+$. The syntax of $\mathcal{S}log^+$ coincides with that of $\mathcal{A}log$. Its informal semantics is based on weak VCP. By $C(T)$ we denote a set atom containing an occurrence of set term T. The *instantiation* of $C(\{X : p(X)\})$ in a set A of regular literals obtained from $C(\{X : p(X)\})$ by replacing $\{X : p(X)\}$ by $\{t : p(t) \in A\}$. The weak VCP is: *belief in $p(t)$ (i.e. inclusion of $p(t)$ in an answer set A) must be established without reference to the instantiation of a set atom C in A unless the truth of this instantiation can be demonstrated without reference to $p(t)$.*

Example 10. To better understand the weak VCP, let us consider program

   p(0) :- C.   :- not p(0).

First we assume C be $card\{X : p(X)\} > 0$. There is only one candidate answer set $A = \{p(0)\}$ for this program. Belief in $p(0)$ (i.e. its membership in answer set A) can only be established by checking if instantiation $card\{t : p(t) \in A\} > 0$ of C in A holds. This is prohibited by weak VCP unless the truth of this instantiation can be demonstrated without reference to $p(0)$. But this cannot be so demonstrated because $card\{t : p(t) \in A\} > 0$ holds only when $p(0)$ is in A. Hence, A is not an answer set. Now let C be $card\{X : p(X)\} \geq 0$. This time the truth of instantiation $card\{t : p(t) \in A\} \geq 0$ of C can be demonstrated without reference to $p(0)$ – the instantiation would be true even if A were empty. Hence $p(0)$ must be believed and thus the program has one answer set, $\{p(0)\}$.

To make weak VCP based semantics precise we need the following notation and definitions: By $\bar{W}^n, \bar{V}^n$ we denote n-ary vectors of sets of ground regular literals and by W_i, V_i their i-th coordinates. $\bar{W}^n \leq \bar{V}^n$ if for every i, $W_i \subseteq V_i$. $\bar{W}^n < \bar{V}^n$ if $\bar{W}^n \leq \bar{V}^n$ and $\bar{W}^n \neq \bar{V}^n$. A set atom $C(\{X : p_1(X)\}, \ldots, \{X : p_n(X)\})$ is *satisfied* by $\bar{W}^n$ if $C(\{t : p_1(t) \in W_1\}, \ldots, \{t : p_n(t) \in W_n\})$ is true.

Definition 4 (Minimal Support). *Let A be a set of ground regular literals of Π, and C be a set atom with n parameters. $\bar{W}^n$ is a minimal support for C in A if*

- *for every $1 \leq i \leq n$, $W_i \subseteq A$.*
- *every $\bar{V}^n$ such that for every $1 \leq i \leq n$, $W_i \subseteq V_i \subseteq A$ satisfies C.*
- *no $\bar{U}^n$, such that $\bar{U}^n < \bar{W}^n$, satisfies the first two conditions.*

Intuitively, the weak VCP says that set atom C can be safely used to support the reasoner's beliefs iff the existence of a minimal support of C can be established without reference to those beliefs. Precise definition of answer sets of $Slog^+$ is obtained by replacing Definition 1 of set reduct of $Alog$ by Definition 5 below and combining it with Definition 3.

Definition 5 (Set-reduct of $Slog^+$). *A set reduct of $Slog^+$ program Π with respect to a set A of ground regular literals is obtained from Π by (1) removing rules containing set atoms which are false or undefined in A, and (2) Replacing every remaining set atom C in the body of the rule by the union of coordinates of one of its minimal supports. A is an answer set of a $Slog^+$ program Π if A is an answer set of a weak set reduct of Π with respect to A.*

Example 11. Consider now an $Slog^+$ program P_3

```
p(3) :- card{X : p(X)} >= 2.   p(2) :- card{X : p(X)} >= 2.    p(1).
```

It has two candidate answer sets: $A_1 = \{p(1)\}$ and $A_2 = \{p(1), p(2), p(3)\}$. In A_1 the corresponding condition is not satisfied and, hence, the weak set reduct of the program with respect to A_1 is $p(1)$. Consequently, A_1 is an answer set of P_3. In A_2 the condition has three minimal supports: $M_1 = \{p(1), p(2)\}$, $M_2 = \{p(1), p(3)\}$, and $M_3 = \{p(2), p(3)\}$. Hence, the program has nine weak set reducts of P_3 with respect to A_2. Each reduct is of the form

```
    p(3) :- Mi.   p(2) :- Mj.     p(1).
```

where M_i and M_j are minimal supports of the condition. Clearly, the first two rules of such a reduct are useless and hence A_2 is not an answer set of this reduct. Consequently A_2 is not an answer set of P_3.

The following two results help to better understand the semantics of $Slog^+$.

Theorem 1. *If a set A is an Alog answer set of Π then A is an $Slog^+$ answer set of Π.*

As an $Slog^+$ program, P_2 has an answer set of $\{p(1)\}$, but it has no answer set as an $Alog$ program. The following result shows that there are many such programs and justifies our name for the new language.

Theorem 2. *Let Π be a program which, syntactically, belongs to both $Slog$ and $Slog^+$. A set A is an $Slog$ answer set of Π iff it is an $Slog^+$ answer set of Π.*

As shown in [24] $Slog$ has sufficient expressive power to formalize complex forms of recursion, including that used in the Company Control Problem [4]. Theorem 2 guarantees that the same representations will work in $Slog^+$. Of course, in many respects $Slog^+$ substantially increases the expressive power of $Slog$. Most importantly it expands the $Slog$ semantics to programs with epistemic disjunction – something which does not seem to be easy to do using the original definition of $Slog$ answer sets. Of course, new set constructs and rules with infinite number of literals are available in $Slog^+$ but not in $Slog$. On another hand, $Slog$ allows multisets – a feature we were not trying to include in our language. The usefulness of multisets and the analysis of its cost in terms of growing complexity of the language due to its introduction is still under investigation.

Unfortunately, the additional power of $Slog^+$ as compared with $Alog$ comes at a price. Part of it is a comparative complexity of the definition of $Slog^+$ set reduct. But, more importantly, the formalization of the weak VCP does not eliminate all the known paradoxes of reasoning with sets. Consider, for instance the following example:

Example 12. Recall program P_2:

```
p(1) :- card{X:p(X)} >= 0.
```

from Example 1 and assume, for simplicity, that parameters of p are restricted to $\{0, 1\}$. Viewed as a program of $\mathcal{A}log$, P_2 is inconsistent. In $\mathcal{S}log^+$ (and hence in $\mathcal{S}log$ and $\mathcal{F}log$ (the language defined in [4])) it has an answer set $\{p(1)\}$. The latter languages therefore admit existence of set $\{X : p(X)\}$. Now let us look at program P_5:

```
p(1) :- card{X : p(X)} = Y, Y >=0.
```

and its grounding P_6:

```
 p(1) :- card{X:p(X)} = 1, 1>=0.     p(1) :- card{X:p(X)} = 0, 0>=0.
```

They seem to express the same thought as P_2, and it is natural to expect all these programs to be equivalent. It is indeed true in $\mathcal{A}log$ – none of the programs is consistent. According to the semantics of $\mathcal{S}log^+$ (and $\mathcal{S}log$ and $\mathcal{F}log$), however, P_5 and P_6 are inconsistent. To see that notice that there are two candidate answer sets for P_6: $A_1 = \emptyset$ and $A_2 = \{p(1)\}$. The minimal support of $card\{X : p(X)\} = 0$ in A_1 is $\emptyset$ and hence the only weak set reduct of P_6 with respect to A_1 is $\{\text{p(1) :- 0>=0}\}$. A_1 is not an answer set of P_6. The minimal support of $card\{X : p(X)\} = 1$ in A_2 is $\{p(1)\}$. The only weak set reduct is $\{\text{p(1) :- p(1),1>=0}\}$. A_2 is not an answer set of P_6 either. It could be that this paradoxical behavior will be in the future explained from some basic principles but currently authors are not aware of such an explanation.

4 Properties of VCP Based Extensions of ASP

In this section we give some basic properties of $\mathcal{A}log$ and $\mathcal{S}log^+$ programs. Propositions 1 and 2 ensure that, as in regular ASP, answer sets of $\mathcal{A}log$ program are formed using the program rules together with the rationality principle. Proposition 3 is the $\mathcal{A}log/\mathcal{S}log^+$ version of the Splitting Set Theorem – basic technical tool used in theoretical investigations of ASP and its extensions [11,15,27].

Proposition 1 (Rule Satisfaction and Supportedness). *Let A be an $\mathcal{A}log$ or $\mathcal{S}log^+$ answer set of a ground program Π. Then*

- *A satisfies every rule r of Π.*
- *If $p(\bar{t}) \in A$ then there is a rule r from Π such that the body of r is satisfied by A and*
  - *$p(\bar{t})$ is the only atom in the head of r which is true in A or*
  - *the head of r is of the form $p \odot \{\bar{X} : q(\bar{X})\}$ and $q(\bar{t}) \in A$. (It is often said that rule r supports atom p.)*

By the intuitive and formal meaning of set introduction rules, the anti-chain property no longer holds. However, the anti-chain property still holds for programs without set atoms in the heads of their rules.

Proposition 2 (Anti-chain Property). *If Π is a program without set atoms in the heads of its rules then there are no $\mathcal{A}log$ answer sets A_1, A_2 of Π such that $A_1 \subset A_2$. Similarly for its $\mathcal{S}log^+$ answer sets.*

Before formulating the next result we need some terminology. We say that a ground literal l *occurs in a set atom* C if there is a set name $\{X : cond(X)\}$ occurring in C and l is a ground instance of some literal in *cond*. If B is a set of ground literals possibly preceded by default negation *not* then l occurs in B if $l \in B$, or *not* $l \in B$, or l occurs in some set atom from B. Let Π be a program with signature Σ. A set S of ground regular literals of Σ is called a *splitting set* of Π if, for every rule r of Π, if l occurs in the head of r then every literal occurring in the body of r belongs to S. The set of rules of Π constructed from literals of S is called *the bottom* of Π relative to S; the remaining rules are referred to as *the top* of Π relative to S. Note that the definition implies that no literal occurring in the bottom of Π relative to S can occur in the heads of rules from the top of Π relative to S.

Proposition 3 (Splitting Set Theorem). *Let Π be a ground program, S be its splitting set, and Π_1 and Π_2 be the bottom and the top of Π relative to S respectively. Then a set A is an answer set of Π iff $A \cap S$ is an answer set of Π_1 and A is an answer set of $(A \cap S) \cup \Pi_2$.*

Note that this formulation differs from the original one in two respects. First, rules of the program can be infinite. Second, the definition of occurrence of a regular literal in a rule changes to accommodate the presence of set atoms.

5   Related Work

There are multiple approaches to introducing aggregates in logic programming languages under the answer sets semantics [4–7, 12–14, 16–20, 23, 24]. In addition to this work our paper was significantly influenced by the original work on VCP in set theory and principles of language design advocated by Dijkstra, Hoare, Wirth and others. Harrison et al.'s work [13] explaining the semantics of some constructs of gringo in terms of infinitary formulas of Truszczynski [26] led to their inclusion in $\mathcal{A}log$ and $\mathcal{S}log^+$. The notion of set reduct of $\mathcal{A}log$ was influenced by the reduct introduced for defining the semantics of Epistemic Specification in [8]. Recent work by Alviano and Faber [1] helped us to realize the close relationship between $\mathcal{A}log$ and $\mathcal{S}log$ and Argumentation theory [25] which certainly deserves further investigation, as well as provided us with additional knowledge about $\mathcal{A}log$. More information about $\mathcal{S}log$ and $\mathcal{S}log^+$ can be found in Sect. 3. Shen et al. [23] and Liu et al. [16] propose equivalent semantics for disjunctive constraint programs (i.e., programs with rules whose bodies are built from constraint atoms and whose heads are epistemic disjunctions of such atoms). This generalizes the standard ASP semantics for disjunctive programs. We conjecture that when we adapt our definition of $\mathcal{S}log^+$ semantics to disjunctive constraint programs, it will coincide with that of [16, 23]. However, our definition seems to be simpler and is based on clear, VCP related intuition. Finally, it is interesting to see if results similar to that in [2] can be obtained for Alog.

6   Conclusion

The paper belongs to the series of works aimed at the development of an answer set based knowledge representation language. Even though we want to have a language suitable for serious applications our main emphasis is on teaching. This puts additional premium on clarity and simplicity of the language design. In particular we believe that the constructs of the language should have a simple syntax and a clear intuitive semantics based on understandable informal principles. In our earlier paper [12] we concentrated on a language $\mathcal{A}log$ expanding standard Answer Set Prolog by aggregates with a simpler syntax than the existing work and a semantics based on a particularly simple and restrictive formalization of VCP. In this paper we:

- Expanded syntax and semantics of the original $\mathcal{A}log$ by allowing
  - rules with an infinite number of literals – a feature of theoretical interest also useful for defining aggregates on infinite sets;
  - subset relation between sets in the bodies of rules concisely expressing a specific form of universal quantification;
  - set introduction – a feature with functionality somewhat similar to that of the choice rule of clingo but with different intuitive semantics.

  Our additional set constructs are aimed at showing that our original languages can be expanded in a natural and technically simple ways. Other constructs such as set operations and rules with variables ranging over sets (in the style of [3]), etc. are not discussed. Partly this is due to space limitations – we do not want to introduce any new constructs without convincing examples of their use. The future will show if such extensions are justified.
- Introduced a new KR language, $\mathcal{S}log^+$, with the same syntax as $\mathcal{A}log$ but different semantics for the set related constructs. The new language is less restrictive and allows formation of substantially larger collection of sets. Its semantics is based on the alternative, weaker formalization of VCP.
- Proved that (with the exception of multisets) $\mathcal{S}log^+$ is an extension of a well known aggregate language $\mathcal{S}log$. The semantics of the new language is based on the intuitive idea quite different from that of $\mathcal{S}log$ and the definition of its semantics is simpler. We point out some paradoxes of $\mathcal{S}log^+$ (and $\mathcal{F}log$) which prevent us from advocating them as standard ASP language with aggregates.
- Proved a number of basic properties of programs of $\mathcal{A}log$ and $\mathcal{S}log^+$.

References

1. Alviano, M., Faber, W.: Stable model semantics of abstract dialectical frameworks revisited: a logic programming perspective. In: Proceedings of the 21st International Joint Conference on Artificial Intelligence, IJCAI Organization, Buenos Aires, Argentina, pp. 2684–2690 (2015)
2. Alviano, M., Faber, W., Gebser, M.: Rewriting recursive aggregates in answer set programming: back to monotonicity. Theory Pract. Logic Program. **15**(4–5), 559–573 (2015)

3. Dovier, A., Pontelli, E., Rossi, G.: Intensional sets in *CLP*. In: Palamidessi, C. (ed.) ICLP 2003. LNCS, vol. 2916, pp. 284–299. Springer, Heidelberg (2003). doi:10.1007/978-3-540-24599-5_20

4. Faber, W., Pfeifer, G., Leone, N.: Semantics and complexity of recursive aggregates in answer set programming. Artif. Intell. **175**(1), 278–298 (2011)

5. Ferraris, P., Lifschitz, V.: Weight constraints as nested expressions. TPLP **5**(1–2), 45–74 (2005)

6. Gebser, M., Harrison, A., Kaminski, R., Lifschitz, V., Schaub, T.: Abstract gringo. Theory Pract. Logic Program. **15**(4–5), 449–463 (2015)

7. Gelfond, M.: Representing knowledge in A-Prolog. In: Kakas, A.C., Sadri, F. (eds.) Computational Logic: Logic Programming and Beyond. LNCS (LNAI), vol. 2408, pp. 413–451. Springer, Heidelberg (2002). doi:10.1007/3-540-45632-5_16

8. Gelfond, M.: New semantics for epistemic specifications. In: Delgrande, J.P., Faber, W. (eds.) LPNMR 2011. LNCS, vol. 6645, pp. 260–265. Springer, Heidelberg (2011). doi:10.1007/978-3-642-20895-9_29

9. Gelfond, M., Kahl, Y.: Knowledge Representation, Reasoning, and the Design of Intelligent Agents. Cambridge University Press, Cambridge (2014)

10. Gelfond, M., Lifschitz, V.: Classical negation in logic programs and disjunctive databases. New Gener. Comput. **9**(3/4), 365–386 (1991)

11. Gelfond, M., Przymusinska, H.: On consistency and completeness of autoepistemic theories. Fundam. Inf. **16**(1), 59–92 (1992)

12. Gelfond, M., Zhang, Y.: Vicious circle principle and logic programs with aggregates. TPLP **14**(4–5), 587–601 (2014). http://dx.doi.org/10.1017/S1471068414000222

13. Harrison, A.J., Lifschitz, V., Yang, F.: The semantics of gringo and infinitary propositional formulas. In: KR (2014)

14. Lee, J., Lifschitz, V., Palla, R.: A reductive semantics for counting and choice in answer set programming. In: Proceedings of the Twenty-Third AAAI Conference on Artificial Intelligence, AAAI 2008, Chicago, Illinois, USA, 13–17 July 2008, pp. 472–479 (2008). http://www.aaai.org/Library/AAAI/2008/aaai08-075.php

15. Lifschitz, V., Turner, H.: Splitting a logic program. In: Proceedings of the 11th International Conference on Logic Programming (ICLP 1994), pp. 23–38 (1994)

16. Liu, G., Goebel, R., Janhunen, T., Niemelä, I., You, J.-H.: Strong equivalence of logic programs with abstract constraint atoms. In: Delgrande, J.P., Faber, W. (eds.) LPNMR 2011. LNCS, vol. 6645, pp. 161–173. Springer, Heidelberg (2011). doi:10.1007/978-3-642-20895-9_15

17. Liu, L., Pontelli, E., Son, T.C., Truszczynski, M.: Logic programs with abstract constraint atoms: the role of computations. Artif. Intell. **174**(3–4), 295–315 (2010)

18. Marek, V.W., Remmel, J.B.: Set constraints in logic programming. In: Lifschitz, V., Niemelä, I. (eds.) LPNMR 2004. LNCS, vol. 2923, pp. 167–179. Springer, Heidelberg (2003). doi:10.1007/978-3-540-24609-1_16

19. Niemela, I., Simons, P., Soininen, T.: Extending and implementing the stable model semantics. Artif. Intell. **138**(1–2), 181–234 (2002)

20. Pelov, N., Denecker, M., Bruynooghe, M.: Well-fouded and stable semantics of logic programs with aggregates. Theory Pract. Logic Program. **7**, 355–375 (2007)

21. Poincare, H.: Les mathematiques et la logique. Rev. de metaphysique et de morale **14**, 294–317 (1906)

22. Russell, B.: Mathematical logic as based on the theory of types. Am. J. Math. **30**(3), 222–262 (1908)

23. Shen, Y., You, J., Yuan, L.: Characterizations of stable model semantics for logic programs with arbitrary constraint atoms. TPLP **9**(4), 529–564 (2009)

24. Son, T.C., Pontelli, E.: A constructive semantic characterization of aggregates in answer set programming. TPLP **7**(3), 355–375 (2007)
25. Strass, H.: Approximating operators and semantics for abstract dialectical frameworks. Artif. Intell. **205**, 39–70 (2013)
26. Truszczynski, M.: Connecting first-order ASP and the logic FO(ID) through reducts. In: Erdem, E., Lee, J., Lierler, Y., Pearce, D. (eds.) Correct Reasoning. LNCS, vol. 7265, pp. 543–559. Springer, Heidelberg (2012). doi:10.1007/978-3-642-30743-0_37
27. Turner, H.: Splitting a default theory. In: Proceedings of AAAI 1996, pp. 645–651 (1996)

Answer Set Programs with Queries over Subprograms

Christoph Redl[(⊠)]

Institut für Informationssysteme, Technische Universität Wien,
Favoritenstraße 9-11, A-1040 Vienna, Austria
redl@kr.tuwien.ac.at

Abstract. Answer-Set Programming (ASP) is a declarative programming paradigm. In this paper we discuss two related restrictions and present a novel modeling technique to overcome them: (1) Meta-reasoning about the collection of answer sets of a program is in general only possible by external postprocessing, but not within the program. This prohibits the direct continuation of reasoning based on the answer to the query over a (sub)program's answer sets. (2) The saturation programming technique exploits the minimality criterion for answer sets of a disjunctive ASP program to solve co-NP-hard problems, which typically involve checking if a property holds *for all* objects in a certain domain. However, the technique is advanced and not easily applicable by average ASP users; moreover, the use of default-negation within saturation encodings is limited.

In this paper, we present an approach which allows for brave and cautious query answering over normal subprograms within a disjunctive program in order to address restriction (1). The query answer is represented by a dedicated atom within each answer set of the overall program, which paves the way also for a more intuitive alternative to saturation encodings and allows also using default-negation within such encodings, which addresses restriction (2).

Keywords: Answer Set Programming · Nonmonotonic reasoning · FLP semantics · Meta programming · Saturation

1 Introduction

Answer-Set Programming (ASP) is a declarative programming paradigm based on nonmonotonic programs and a multi-model semantics [13]. The problem at hand is encoded as an ASP program whose models, called *answer sets*, correspond one-to-one to the solutions of the problem. In this paper we discuss two reasoning resp. modeling restrictions, which turn out to be related.

The first restriction concerns *meta-reasoning* about the answer sets of a *(sub)program* within another *(meta-)program*, such as aggregation of results.

This research has been supported by the Austrian Science Fund (FWF) project P27730.

M. Balduccini and T. Janhunen (Eds.): LPNMR 2017, LNAI 10377, pp. 160–175, 2017.
DOI: 10.1007/978-3-319-61660-5_15

This is usually done during postprocessing, i.e., the answer sets are inspected after the reasoner terminates. Some simple reasoning tasks, such as brave or cautious query answering, are directly supported by some systems. However, even then the answer to a brave or cautious query is not represented *within* the program but appears only as output on the command-line, which prohibits the direct continuation of reasoning based on the query answer. An existing approach, which allows for meta-reasoning *within* a program over the answer sets of another program, are *manifold programs*. They compile the calling and the called program into a single one [8]. The answer sets of the called program are then represented within each answer set of the calling program. However, this approach uses weak constraints, which are not supported by all systems. Moreover, the encoding requires a separate copy of the subprogram for each atom occurring in it, which appears to be impractical. Another approach are *nested* HEX-*programs*. Here, dedicated atoms access answer sets of a subprograms and their literals explicitly as accessible objects [4]. However, this approach is based on HEX-programs [6] – an extension of ASP – and not applicable if an ordinary ASP solver is used. Moreover, the meta- and the subprogram are evaluated by two isolated reasoner instances, which may harm efficient evaluation.

The second restriction concerns the *saturation technique* (cf. e.g. [3]), which is a modeling technique that allows for solving co-NP-hard problems within disjunctive ASP. To this end, minimality of answer sets is exploited to check if a property holds *for all* objects in a certain domain. However, the technique is advanced and not easily applicable by average ASP users. Moreover, the use of default-negation for checking properties within saturation encodings is restricted as it may harm the support of atoms. Then, default-negation needs to be rewritten, but it is not always obvious how this can be done. This calls for an alternative to saturation, which hides this rewriting from the user.

In this paper, we first present an **encoding which allows for deciding inconsistency of a normal logic program within a disjunctive program**. Inconsistency resp. consistency of the subprogram is represented by a dedicated atom within each answer set of the overall program. This encoding is then exploited to realize **query answering over normal subprograms within disjunctive ASP**; in contrast to related approaches (e.g. [1], see Sect. 6), ours makes such queries more explicit, which is easier to understand for users. While the encoding itself is based on the saturation technique, once it is defined, it can be flexibly used for query answering without deep knowledge about the saturation technique. This results in a new modeling technique as alternative to saturation, which supports unrestricted use of default-negation.

We proceed as follows:

- In Sect. 2 we recapitulate answer set programming and the saturation technique.
- In Sect. 3 we discuss restrictions of saturation and point out that using default-negation within saturation encodings would be convenient but is not easily possible.

- In Sect. 4 we show how inconsistency of a normal logic program can be decided within another (disjunctive) program. To this end, we present a saturation encoding which simulates the computation of answer sets of the subprogram and represents the existence of an answer set by a single atom of the meta-program.
- In Sect. 5 we discuss query answering based on this encoding. To this end, we first realize brave and cautious query answering over a subprogram in Sect. 5.1. This feature is then further exploited in Sect. 5.2 for realizing a new modeling technique as an alternative to saturation, but which supports default-negation. The encoding can be used as a black box at this point such that the user does not need to have deep knowledge about the underlying ideas. Instead, checking if a property holds *for all* objects in a domain can be naturally expressed as a cautious query.
- In Sect. 6 we discuss related work.
- In Sect. 7 we conclude and give an outlook on future work.

Proofs are outsourced to http://www.kr.tuwien.ac.at/research/projects/inthex/qa-ext.pdf.

2   Preliminaries

We first recapitulate answer set programming and the saturation technique.

Answer Set Programming. Our alphabet consists of possibly infinite sets of constant symbols $\mathscr{C}$ (including all integers), variables $\mathscr{V}$, function symbols $\mathscr{F}$, and predicate symbols $\mathscr{P}$. We assume that $\mathscr{V}$ is disjoint from all other sets, while symbols may be shared between the other sets. We let the set of terms $\mathscr{T}$ be the least set such that $\mathscr{C} \subseteq \mathscr{T}$, $\mathscr{V} \subseteq \mathscr{T}$, and $f \in \mathscr{F}$, $T_1, \ldots, T_\ell \in \mathscr{T}$ implies $f(T_1, \ldots, T_\ell) \in \mathscr{T}$. An (ordinary) atom is of form $p(t_1, \ldots, t_\ell)$ with predicate symbol $p \in \mathscr{P}$ and terms $t_1, \ldots, t_\ell \in \mathscr{T}$, abbreviated as $p(\mathbf{t})$; we write $t \in \mathbf{t}$ if $t = t_i$ for some $1 \leq i \leq \ell$. A term resp. atom is called *ground* if it does not contain variables.

An *interpretation* over the (finite) set $\mathscr{A}$ of ground atoms is a set $I \subseteq \mathscr{A}$, where $a \in I$ expresses that a is true and $a \notin I$ that a is false. A builtin atom is of form $t_1 \circ t_2$ with terms $t_1, t_2 \in \mathscr{T}$ and comparison operator $\circ \in \{=, \neq, <, \leq, \geq, >\}$. For a ground builtin atom $t_1 \circ t_2$ and an interpretation I we have that $I \models t_1 = t_2$ if t_1 is equal to t_2 and $I \not\models t_1 = t_2$ otherwise; conversely for $I \models t_1 \neq t_2$. Operators $<, \leq, \geq$ and $>$ have the standard semantics and are defined only if t_1 and t_2 are integers. Similarly, arithmetic atoms are of form $t_1 \circ t_2 \circ' t_3$ with integer terms $t_1, t_2, r_3 \in \mathscr{T}$, comparison operator $\circ \in \{=, \neq, <, \leq, \geq, >\}$ and arithmetic operator $\circ' \in \{+, -, *, /\}$, which have the standard semantics.

We now recall disjunctive logic programs under the answer set semantics [13].

Definition 1. *An* answer set program P *consists of rules*

$$a_1 \vee \cdots \vee a_k \leftarrow b_1, \ldots, b_m, \text{not } b_{m+1}, \ldots, \text{not } b_n \,, \tag{1}$$

where each a_i is an atom, and each b_j is an atom or a builtin atom. A program is called normal *if $k \leq 1$ for all rules, and* disjunctive *otherwise.*

A rule resp. program is ground if it contains only ground atoms. Interpretations I are over the atoms $A(P)$ occurring in the ground program P at hand. A ground rule r of form (1) is satisfied under I, denoted $I \models r$, if $a_i \in I$ for some $1 \leq i \leq k$, or $b_i \notin I$ for some $1 \leq i \leq m$, or $b_i \in I$ for some $m+1 \leq i \leq n$. A ground program P is satisfied under I, denoted $I \models P$, if each $r \in P$ is satisfied under I. For such a rule r we let $H(r) = \{a_1, \ldots, a_k\}$ be its *head*, $B^+(r) = \{b_1, \ldots, b_m\}$ be its *positive body* and $B^-(r) = \{b_{m+1}, \ldots, b_n\}$ be its *negative body*.

The answer sets of a ground program P are defined as follows. The *(GL-)reduct* [13] of P wrt. interpretation I is the set $P^I = \{H(r) \leftarrow B^+(r) \mid r \in \Pi, I \not\models b$ for all $b \in B^-(r)\}$.

Definition 2. *An interpretation I is an answer set of a ground program P, if I is a $\subseteq$-minimal model of P^I.*

Note that for a normal program P and an interpretation I, the reduct P^I is a positive program. This allows for an alternative characterization of answer sets of normal logic programs based on fixpoint iteration. For a positive normal program P, we let $T_P(S) = \{a \in H(r) \mid r \in P, B^+(r) \subseteq S\}$ be the monotonic *immediate consequence operator*, which derives the consequences of a set S of atoms when applying the positive rules in P. Then the least fixpoint of T_P over the empty set, denoted $lfp(T_P)$, is the *unique* least model of P. Hence, an interpretation I is an answer set of a normal logic program P if $I = lfp(T_{P^I})$.

The answer sets of a non-ground program P are given by those of its *grounding* $grnd(P)$, which results from P if all variables are replaced by terms in all possible ways.

Saturation Technique. The saturation technique dates back to the Σ_2^P-hardness proof of disjunctive ASP [2], but was later exploited as a modeling technique, cf. e.g. [3]. It is applied for solving co-NP-hard problems, which typically involve checking a condition *for all* objects in a domain. Importantly, such a check *cannot* be encoded in a *normal* logic program such that the program has an answer set iff the condition holds for all guesses (unless $NP = coNP$). Instead, one can only write a normal program which has *no* answer set if the property holds for all guesses. This limitation inhibits that reasoning continues *within* the program after checking the property. Instead, non-existence of answer sets needs to be determined in the postprocessing.

A concrete example is checking if a given graph is *not* 3-colorable. Consider

$$P_{3col} = F \cup \{c_1(X) \leftarrow node(X), \text{not } c_2(X), \text{not } c_3(X) \mid \{c_1, c_2, c_3\} = \{r, g, b\}\}$$
$$\cup \{\leftarrow c(X), c(Y), edge(X, Y) \mid c \in \{r, g, b\}\},$$

where the graph is supposed to be defined by facts F over predicates *node* and *edge*. Its answer sets correspond one-to-one to valid 3-colorings. Thus, the program does *not* have an answer set if and only if there is no valid 3-coloring. However, it is not possible to define a normal program with an answer set that represents that there is *no* such coloring.

This is only possible with disjunctive programs and the saturation technique. To this end, the search space is defined in a program component P_{guess} using disjunctions. Another program component P_{check} checks if the current guess satisfies the property (e.g., being *not* a valid 3-coloring) and derives a dedicated saturation atom *sat* in this case. A third program component P_{sat} derives all atoms from P_{guess} whenever *sat* is true, i.e., it *saturates the model*. This has the following effect: if all guesses fulfill the property, all atoms in P_{guess} are derived for all guesses and the so-called *saturation model* $I_{sat} = A(P_{guess} \cup P_{check})$ is an answer set of $P_{guess} \cup P_{check} \cup P_{sat}$. On the other hand, if there is at least one guess which does not fulfill it, then *sat* – and possibly further atoms – are not derived. Then, by minimality of answer sets, I_{sat} is not an answer set.

Example 1. The program $P_{non3col} = F \cup P_{guess} \cup P_{check} \cup P_{sat}$ where

$$P_{guess} = \{r(X) \vee g(X) \vee b(X) \leftarrow node(X)\}$$
$$P_{check} = \{sat \leftarrow c(X), c(Y), edge(X, Y) \mid c \in \{r, g, b\}\}$$
$$P_{sat} = \{c(X) \leftarrow node(X), sat \mid c \in \{r, g, b\}\}$$

has the answer set $I_{sat} = A(P_{non3col})$ iff the graph specified by facts F is not 3-colorable. Otherwise its answer sets are proper subsets of I_{sat} which represent valid 3-colorings.   □

3   Restrictions of the Saturation Technique

For complexity reasons, any problem in co-NP can be polynomially reduced to brave reasoning over disjunctive ASP (the latter is Σ_2^P-complete [7]), but the reduction is not always obvious. In particular, the saturation technique is difficult to apply if the property to check cannot be easily expressed without default-negation. This is because saturation works only if I_{sat} is an answer set of $P_{guess} \cup P_{check} \cup P_{sat}$ whenever no proper subset is one. While this is guaranteed if no default-negation occurs in I_{sat}, it might be unstable otherwise.

Example 2. A *vertex cover* of a graph $\langle V, E \rangle$ is a subset $S \subseteq V$ of its nodes s.t. each edge in E is incident with at least one node in S. Deciding if a graph has *no* vertex cover S with size $|S| \leq k$ for some integer k is co-NP-complete. Consider P_{vc} consisting of facts F over *node* and *edge* and the following parts:

$$P_{guess} = \{in(X) \vee out(X) \leftarrow node(X)\}$$
$$P_{check} = \{sat \leftarrow edge(X, Y), not\ in(X), not\ in(Y);$$
$$sat \leftarrow in(X_1), \ldots, in(X_{k+1}), X_1 \neq X_2, \ldots, X_k \neq X_{k+1}\}$$
$$P_{sat} = \{in(X) \leftarrow node(X), sat;\ out(X) \leftarrow node(X), sat\}$$

Program P_{guess} guesses a candidate vertex cover S, P_{check} derives *sat* whenever for some edge $(u, v) \in E$ neither u nor v is in S (thus S is invalid), and P_{sat} saturates in this case.   □

Observe that for inconsistent instances F (e.g. $\langle \{a, b, c, d\}, \{(a, b), (b, c),$ $(c, d)\} \rangle$ with $k = 1$), this encoding does not work as desired because model $I_{sat} = A(P_{vc})$ is unstable. More specifically, the instances of the first rule of P_{check} are eliminated from $P_{vc}^{I_{sat}}$ due to default-negation. But then, the least model of the reduct does not contain sat or any atom $in(\cdot)$. Then, $I_< := I_{sat} \setminus (\{sat\} \cup \{in(x) \mid x \in V\})$ is a smaller model of the reduct and I_{sat} is not an answer set of $P_{vc} \cup F$.

In this example, the problem may be fixed by replacing literals not $in(X)$ and not $in(Y)$ by $out(X)$ and $out(Y)$, respectively. That is, instead of checking if a node is not in the vertex cover, one explicitly checks if it is out. However, the situation is more cumbersome if default-negation does not directly concern the guessed atoms but derived ones.

Example 3. A *Hamiltonian cycle* in a graph $\langle V, E \rangle$ is a cycle that visits each node in V exactly once. Deciding if a given graph has a Hamiltonian cycle is a well-known NP-complete problem; deciding if a graph does not have such a cycle is therefore co-NP-complete. A natural attempt to solve the problem using saturation is as follows:

$$P_{guess} = \{in(X, Y) \lor out(X, Y) \leftarrow arc(X, Y)\} \tag{2}$$

$$P_{check} = \{sat \leftarrow in(Y_1, X), in(Y_2, X), Y1 \neq Y2;$$

$$\qquad sat \leftarrow in(X, Y_1), in(X, Y_2), Y1 \neq Y2 \tag{3}$$

$$\qquad sat \leftarrow node(X), not\, hasIn(X); \quad sat \leftarrow node(X), not\, hasOut(X) \tag{4}$$

$$\qquad hasIn(X) \leftarrow node(X), in(Y, X); \quad hasOut(X) \leftarrow node(X), in(X, Y)\} \tag{5}$$

$$P_{sat} = \{in(X, Y) \leftarrow arc(X, Y), sat; \quad out(X, Y) \leftarrow arc(X, Y), sat\} \tag{6}$$

Program P_{guess} guesses a candidate Hamiltonian cycle as a set of arcs. Program P_{check} derives sat whenever some node in V does not have exactly one incoming and exactly one outgoing arc, and P_{sat} saturates in this case. The check is split into two checks for at most (rules (3)) and at least (rules (4)) one incoming/outgoing arc. While the check if a node has at most one incoming/outgoing arcs is possible using the positive rules (3), the check if a node has at least one incoming/outgoing edge is more involved. In contrast to the check in Example 2, one cannot reasonably perform it based on the atoms from P_{guess} alone. Instead, auxiliary predicates $hasIn$ and $hasOut$ are defined by rules (5). Unlike $in(\cdot, \cdot)$, the negation of $hasIn(\cdot)$ and $hasOut(\cdot)$ is not explicitly represented, thus default-negation is used in rules (4) of P_{check}. However, this harms stability of I_{sat}: the graph $\langle \{a, b, c\}, \{(a, b), (b, a), (b, c), (c, b)\} \rangle$, which does not have a Hamiltonian cycle, causes $P_{guess} \cup P_{check} \cup P_{sat}$ to be inconsistent. This is due to default-negation in P_{check}, which eliminates rules (4) from the reduct wrt. I_{sat}, which in turn has a smaller model. □

Note that in the previous example, for a fixed node X the literal not $hasOut(X)$ is used to determine if all atoms $in(X, Y)$ are false (or equivalently: if all atoms $out(X, Y)$ are true). Here, default-negation can be eliminated on the ground level by replacing rule $sat \leftarrow node(X), not\, hasOut(X)$ by

$sat \leftarrow node(x), out(x, y_1), \ldots, out(x, y_n)$ for all nodes $x \in V$ and all nodes y_i for $1 \leq i \leq n$ such that $(x, y_i) \in E$.[1] But this is not always possible:

Example 4. Deciding if a ground normal ASP program P is inconsistent is co-NP-complete. An attempt to apply the saturation technique is as follows:

$$P' = \{true(a) \lor false(a) | a \in A(P)\} \tag{7}$$

$$\cup \{inReduct(r) \leftarrow \{false(b) \mid b \in B^-(r)\} \mid r \in P\} \tag{8}$$

$$\cup \{leastModel(a) \leftarrow inReduct(r),$$
$$\{leastModel(b) \mid b \in B^+(r)\} \mid r \in P, a \in H(r)\} \tag{9}$$

$$\cup \{noAS \leftarrow false(a), leastModel(a) \mid a \in A(P)\} \tag{10}$$

$$\cup \{noAS \leftarrow true(a), \text{not } leastModel(a) \mid a \in A(P)\} \tag{11}$$

$$\cup \{true(a) \leftarrow noAS; \; false(a) \leftarrow noAS \mid a \in A(P)\} \tag{12}$$

$$\cup \{inReduct(r) \leftarrow noAS\} \tag{13}$$

The idea is to guess all possible interpretations I over the atoms $A(P)$ in P (rules (7)). Next, rules (8) identify the rules $r \in P$ which are in P^I (modulo $B^-(r)$); these are all rules $r \in P$ whose atoms $B^-(r)$ are all false. Rules (9) compute the least model of the reduct by simulating fixpoint iteration under operator T_P. Rules (10) and (11) compare the least model of the reduct to I: if this comparison fails, then I is not an answer set and rules (12) and (13) saturate. However, the comparison of the least model of the reduct to the original guess in rule (11) uses default-negation. In contrast to Example 3, it is not straightforward how to eliminate the negation, even on the ground level. □

We conclude that some problems involve checks which can easily be expressed using negation, but such a check within a saturation encoding may harm stability of the saturation model. In the next section, we present a valid encoding for checking inconsistency of normal programs, as discussed in Example 4, within disjunctive ASP.

4   Deciding Inconsistency of Normal Programs in Disjunctive ASP

We reduce the check for inconsistency of a normal logic program P to brave reasoning over a disjunctive meta-program. The major part M of the meta-program is static and consists of proper rules which are independent of P. The concrete program P is then specified by facts M^P which are added to the static part. The overall program $M \cup M^P$ is constructed such that it is consistent for all P and its answer sets either represent the answer sets of P, or a dedicated answer sets represents that P is inconsistent.

---

[1] On the non-ground level, this might be simulated using *conditional literals* as supported by some reasoners, cf. [9] and below.

4.1   A Meta-Program for Propositional Programs

In this subsection we restrict the discussion to ground programs P. Moreover, we assume that all predicates in P are of arity 0. This is w.l.o.g. because any atom $p(t_1, \ldots, t_\ell)$ can be replaced by an atom consisting only of a new predicate p' without any parameters. In the meta-program defined in the following, we let all atoms be new atoms which do not occur in P. We further use each rule $r \in P$ also as a new atom in the meta-program. For simplicity, we further disallow constraints $\leftarrow b_1, \ldots, b_m, \text{not } b_{m+1}, \ldots, \text{not } b_n$ (rules with empty head) in P. This is also w.l.o.g. because such a constraint can be seen as an abbreviation for $x \leftarrow b_1, \ldots, b_m, \text{not } b_{m+1}, \ldots, \text{not } b_n, \text{not } x$, where x is a new ground atom which does not appear elsewhere in the program.

The static part consists of component $M_{extract}$ for the extraction of various information from the program encoding M^P, which we call M_{gr}^P in this section to stress that P must be ground, and a saturation encoding $M_{guess} \cup M_{check} \cup M_{sat}$ for the actual inconsistency check. We first show the complete encoding and then discuss its components.

Definition 3. *We define the* meta-program $M = M_{extract} \cup M_{guess} \cup M_{check} \cup M_{sat}$, *where:*

$$M_{extract} = \{atom(X) \leftarrow head(R, X); \ atom(X) \leftarrow bodyP(R, X);$$
$$atom(X) \leftarrow bodyN(R, X)\} \tag{14}$$
$$\cup \ \{rule(R) \leftarrow head(R, X); \ rule(R) \leftarrow bodyP(R, X);$$
$$rule(R) \leftarrow bodyN(R, X)\} \tag{15}$$
$$M_{guess} = \{true(X) \vee false(X) \leftarrow atom(X)\} \tag{16}$$
$$M_{check} = \{inReduct(R) \leftarrow rule(R), (false(X) : bodyN(R, X))\} \tag{17}$$
$$\cup \ \{outReduct(R) \leftarrow rule(R), bodyN(R, X), true(X)\} \tag{18}$$
$$\cup \ \{iter(X, I) \vee niter(X, I) \leftarrow true(X), int(I);$$
$$niter(X, I) \leftarrow false(X), int(I)\} \tag{19}$$
$$\cup \ \{notApp(R) \leftarrow outReduct(R)\} \tag{20}$$
$$\cup \ \{notApp(R) \leftarrow inReduct(R), bodyP(R, X), false(X)\} \tag{21}$$
$$\cup \ \{notApp(R) \leftarrow head(R, X_1), bodyP(R, X_2),$$
$$iter(X_1, I_1), iter(X_2, I_2), I_2 \geq I_1\} \tag{22}$$
$$\cup \ \{noAS \leftarrow true(X), (notApp(R) : head(R, X))\} \tag{23}$$
$$\cup \ \{noAS \leftarrow inReduct(R), head(R, X), false(X),$$
$$(true(Y) : bodyP(R, Y))\} \tag{24}$$
$$\cup \ \{noAS \leftarrow true(X), (niter(X, I) : int(I))\} \tag{25}$$
$$\cup \ \{noAS \leftarrow iter(X, I_1), iter(X, I_2), I_1 \neq I_2\} \tag{26}$$
$$\cup \ \{iter_<(X, I) \leftarrow false(X), int(I); \ iter_<(X, I_2) \leftarrow true(X),$$
$$iter(X, I_1), int(I_2), I_2 > I_1\} \tag{27}$$

$$\cup \{notApp(R) \leftarrow head(R, X_1), iter(X_1, I), I > 0,$$
$$(iter_<(X_2, I) : bodyP(R, X_2))\} \tag{28}$$
$$M_{sat} = \{true(X) \leftarrow atom(X), noAS; \; false(X) \leftarrow atom(X), noAS\} \tag{29}$$
$$\cup \{iter(X, I) \leftarrow atom(X), int(I), noAS;$$
$$niter(X, I) \leftarrow atom(X), int(I), noAS\} \tag{30}$$
$$\cup \{inReduct(R) \leftarrow rule(R), noAS; \; outReduct(R) \leftarrow rule(R), noAS\} \tag{31}$$

Before we come to an explanation of M, we discuss the specification of the program-dependent part M_{gr}^P, which is expected to encode the rules of P as facts which are added to M. To this end, we first set the domain of integers to $|A(P)|$ (which is a sufficiently high value as explained below). Then, each rule of P is represented by atoms $head(r, a)$, $bodyP(r, a)$, and $bodyN(r, a)$, where r is a rule from P (used as new atom representing the respective rule). Here, $head(r, a)$, $bodyP(r, a)$ and $bodyN(r, a)$ denote that a is an atom that occurs in the head, positive and negative body of rule r, respectively. Rules 14 and 15 then extract for the sets of all rules and atoms in P. Formally:

Definition 4. *For a ground normal logic program P we let:*

$$M_{gr}^P = \{int(c) \mid 0 \leq c < |A(P)|\} \cup \{head(r, h) \mid r \in P, h \in H(r)\}$$
$$\cup \{bodyP(r, b) \mid r \in P, h \in B^+(r)\} \cup \{bodyN(r, b) \mid r \in P, h \in B^-(r)\}$$

The structure of the static programs M_{guess}, M_{check} and M_{sat} follows then the basic architecture of saturation encodings presented in Sect. 3. The idea is as follows. Program M_{guess} guesses an answer set candidate I of program P, M_{check} simulates the computation of the reduct P^I and checks if its least model coincides with I, and M_{sat} saturates the model whenever this is *not* the case. If all guesses fail to be answer sets, then every guess leads to saturation and the saturation model is an answer set. On the other hand, if at least one guess represents a valid answer set of P, then the saturation model is not an answer set due to subset-minimality. Hence, $M \cup M^P$ has exactly one (saturated) answer set if P is inconsistent, and it has answer sets which are not saturated if P is consistent, but none of them contains $noAS$.

We turn to the checking part M_{check}. Rules (17) and (18) compute for the current candidate I the rules in P^I: a rule r is in the reduct iff all atoms from $B^-(r)$ are false in I. Here, $(false(X) : bodyN(R, X))$ is a *conditional literal* which evaluates to true iff $false(X)$ holds for all X such that $bodyN(R, X)$ is true, i.e., all atoms in the negative body are false. Rules (19) simulate the computation of the least model $lfp(T_{P^I})$ of P^I using fixpoint iteration. To this end, each atom $a \in I$ is assigned a guessed integer to represent an ordering of derivations during fixpoint iteration under T_P. We need at most $|A(P)|$ iterations because the least model of P contains only atoms from $A(P)$ and the fixpoint iteration stops if no new atoms are derivable. However, since not all instances need the maximum of $|A(P)|$ iterations, there can be gaps in this sequence. For instances which need fewer than $|A(P)|$ iterations. Rules (20)–(26) check if the current interpretation is *not* an answer set of P which can be justified by the guessed derivation sequence, and derive $noAS$ in this case. Importantly, $noAS$ both if (i) I is not an answer set, and if (ii) I *is* an answer set, but one that cannot be reproduced using the guessed derivation sequence. Rules (27) and (28) ensure that true atoms are derived in the earliest possible iteration, which eliminates redundant solutions.

As a preparation for both checks (i) and (ii), rules (20)–(22) determine the rules $r \in P$ which are *not* applicable in the fixpoint iteration (wrt. the current derivation sequence) to justify their head atom $H(r)$ being true. A rule is not applicable if it is not in the reduct (rules (20)), if at least one positive body atom is false (rules (21)), or if it has a positive body atom which is derived in the same or a later iteration (rules (22)) because then the rule cannot fire (yet) in the iteration the head atom was guessed to be derived.

We can then perform the actual checks (i) and (ii). (i) For checking if I is an answer set, rules (23) check if all atoms in I are derived by some rule in P^I (i.e., $I \subseteq lfp(T_{P^I})$). Conversely, rules (24) check if all rules derived by some rule in P^I are also in I (i.e., $I \supseteq lfp(T_{P^I})$). Overall, the rules (23)–(24) check if $I = lfp(T_{P^I})$. This check compares I and $lfp(T_{P^I})$ only under the assumption that the guessed derivation sequence is valid.

(ii) This validity remains to be checked. To this end, rules (25) ensure that an iteration number is specified for all atoms which are true in I; in order to avoid default-negation we explicitly check if all atoms $niter(a, i)$ for $0 \le i \le |A(P)| - 1$ are true. Rules (26) guarantee that this number is unique for each atom. If one of these conditions does not apply, then the *currently* guessed derivation order does not justify that I is accepted as an answer set, hence it is dismissed by deriving $noAS$, even if the same interpretation might be a valid answer set justified by another (valid) derivation sequence. This is by intend because all real answer sets I are justified by some valid derivation sequence.

One can show that atom $noAS$ correctly represents inconsistency of P.

Proposition 1. *For any ground normal logic program P, we have that*

(1) if P is inconsistent, $M \cup M_{gr}^P$ has exactly one answer set which contains $noAS$; and

(2) if P is consistent, $M \cup M_{gr}^P$ has at least one answer set and none of the answer sets of M^P contains $noAS$.

4.2   A Meta-Program for Non-ground Programs

We extend the encoding of a ground normal logic programs as facts as by Definition 4 to non-ground programs. The program-specific part is called M_{ng}^P to stress that P can now be non-ground. In the following, for a rule r let $\mathbf{V}_r$ be the vector of unique variables occurring in r in the order of appearance.

The main idea of the following encoding is to interpret atoms with an arity >0 as function terms. That is, for an atom $p(t_1, \dots, t_\ell)$ we see p as function symbol rather than predicate (recall that Sect. 2 allows that $\mathscr{P} \cap \mathscr{F} \neq \emptyset$). Then, atoms, interpreted as function terms, can occur as parameters of other atoms.

Definition 5. *For a (ground or non-ground) normal logic program P we let:*

$$M_{ng}^{P} = \{int(c) \mid 0 \leq c < |A(P)|\} \cup \{head(r(\mathbf{V}_r), h)$$
$$\leftarrow \{head(R, d) \mid d \in B^+(r)\} \mid r \in P, h \in H(r)\}\}$$
$$\cup \{bodyP(r(\mathbf{V}_r), b) \leftarrow \{head(R, d) \mid d \in B^+(r)\} \mid r \in P, b \in B^+(r)\}$$
$$\cup \{bodyN(r(\mathbf{V}_r), b) \leftarrow \{head(R, d) \mid d \in B^+(r)\} \mid r \in P, b \in B^-(r)\}$$

For each possibly non-ground rule $r \in P$, we construct a unique identifier $r(\mathbf{V}_r)$ for each ground instance of r. It consists of r as unique function symbol and all variables in r as parameters. As for the ground case, the head, the positive and the negative body are extracted from r. However, since variables may occur in any atom of r, we have to add a body to the rules of the representation to ensure safety. To this end, we add a *domain atom head(R, d)* for all positive body atoms $d \in B^+(r)$ to the body of the rule in the meta-program in order to instantiate it with all derivable ground instances. Informally, we create an instance of r for all variable substitutions such that all body atoms of the instance are potentially derivable in the meta-program.

Example 5. Let $P = \{f \colon d(a); \ r_1 \colon q(X) \leftarrow d(X), \text{not } p(X); \ r_2 \colon p(X) \leftarrow d(X),$ $\text{not } q(X)\}$. We have:

$$M_{ng}^{P} = \{head(f, d(a)) \leftarrow; \ head(r_1(X), q(X)) \leftarrow head(R, d(X))\}$$
$$\cup \{bodyP(r_1(X), d(X)) \leftarrow head(R, d(X)); \ bodyN(r_1(X), p(X)) \leftarrow head(R, d(X))\}$$
$$\cup \{head(r_2(X), p(X)) \leftarrow head(R, d(X)); \ bodyP(r_2(X), d(X)) \leftarrow head(R, d(X))\}$$
$$\cup \{bodyN(r_2(X), q(X)) \leftarrow head(R, d(X))\}$$

We explain the encoding with the example of r_1. Since r_1 is non-ground, it may represent multiple ground instances, which are determined by the substitutions of X. We use $r_1(X)$ as identifier and define that, for any substitution of X, atom $q(X)$ appears in the head, $d(X)$ in the positive body and $p(X)$ in the negative body. This is denoted by $head(r_1(X), q(X))$, $bodyP(r_1(X), d(X)) \leftarrow head(R, d(X))$ and $bodyN(r_1(X), d(X)) \leftarrow head(R, p(X))$, respectively. The domain of X is defined by all atoms $d(X)$ which are potentially derivable, i.e., by atoms $head(R, d(X))$. □

One can show that the encoding is still sound and complete for non-ground programs:

Proposition 2. *For any normal logic program P, we have that*

(1) if P is inconsistent, $M \cup M_{ng}^{P}$ has exactly one answer set which contains noAS; and

(2) if P is consistent, $M \cup M_{ng}^{P}$ has at least one answer set and none of the answer sets of M^P contains noAS.

5   Query Answering over Subprograms

In this section we first introduce a technique for query answering over subprograms based on the inconsistency check from the previous section. We then introduce a language extension with dedicated *query atoms* which allow for easy expression of such queries within a program. Finally we demonstrate this language extension with an example.

5.1   Encoding Query Answering

In the following, a query q is a set of ground literals (atoms or default-negated atoms) interpreted as conjunction; for simplicity we restrict the further discussion to ground queries. For an atom or default-negated atom l, let $\bar{l}$ be its negation, i.e., $\bar{l} = a$ if $l = \text{not } a$ and $\bar{l} = \text{not } a$ if $l = a$. We say that an interpretation I satisfies a query q, denoted $I \models q$, if $a \in I$ for all atoms $a \in q$ and not $a \notin I$ for all default-negated atoms not $a \in q$. A logic program P *bravely entails* a query q, denoted $P \models_b q$, if $I \models q$ for some answer set I of P; it *cautiously entails* a query q, denoted $P \models_c q$, if $I \models q$ for all answer sets I of P.

We can reduce query answering to (in)consistency checking as follows:

Proposition 3. *For a normal logic program P and a query q we have that (1) $P \models_b q$ iff $P \cup \{\leftarrow \bar{l} \mid l \in q\}$ is consistent; and (2) $P \models_c q$ iff $P \cup \{\leftarrow q\}$ is inconsistent.*

We now can exploit our encoding for (in)consistency checking for query answering.

Proposition 4. *For a normal logic program P and query q we have that*

(1) $M \cup M_{ng}^{P \cup \{\leftarrow \bar{l}\mid l \in q\}}$ *is consistent and each answer set contains noAS iff*
    $P \not\models_b q$; *and*
(2) $M \cup M_{ng}^{P \cup \{\leftarrow q\}}$ *is consistent and each answer set contains noAS iff $P \models_c q$.*

Based on the previous proposition, we introduce a new language feature which allows for expressing queries more conveniently.

Definition 6. *A query atom is of form $S \vdash_t q$, where $t \in \{b, c\}$ determines the type of the query, S is a normal logic (sub)program, and q is a query over S.*

We allow query atoms to occur in bodies of ASP programs in place of ordinary atoms (in implementations, S may be specified by its filename). The intuition is that for a program P containing a query atom, we have that $S \vdash_b q$ resp. $S \vdash_c q$ is true (wrt. all interpretations I of P) if $S \models_b q$ resp. $S \models_c q$.

Formally we define the semantics of such a program P using the following translation to an ordinary ASP program, based on Proposition 4. We let the answer sets of a program P with query atoms be given by the answer sets of the program $[P]$ defined as follows:

$$[P] = P|_{S \vdash_t q \to noAS_{S \vdash_t q}} \cup \bigcup_{S \vdash_b q \text{ in } P} \left(M \cup M_{ng}^{S \cup \{\leftarrow \bar{l} | l \in q\}}\right)\Big|_{a \to a_{S \vdash_b q}}$$

$$\cup \bigcup_{S \vdash_c q \text{ in } P} \left(M \cup M_{ng}^{S \cup \{\leftarrow q\}}\right)\Big|_{a \to a_{S \vdash_c q}}$$

Here, we let $P|_{S \vdash_t q \to noAS_{S \vdash_t q}}$ denote program P after replacing *every* query atom of kind $S \vdash_t q$ (for some t, S and q) by the new ordinary atom $noAS_{S \vdash_t q}$. Moreover, in the unions, expression $|_{a \to a_{S \vdash_b q}}$ denotes that each atom a is replaced by $a_{S \vdash_b q}$ (likewise for $S \vdash_c q$). This ensures that for every query atom $S \vdash_b q$ resp. $S \vdash_c q$ in P, a separate copy of M and $M_{ng}^{S \cup \{\leftarrow \bar{l} | l \in q\}}$ resp. $M_{ng}^{S \cup \{\leftarrow q\}}\big)\big|_{a \to a_{S \vdash_c q}}$ is generated whose vocabularies are disjoint. In particular, each such copy uses a separate atom $noAS_{S \vdash_b q}$ resp. $noAS_{S \vdash_c q}$ which represents by Proposition 4 the answer to query q. Thus, after replacing each $S \vdash_t q$ in the original program P by the respective atom $noAS_{S \vdash_c q}$, the program behaves as desired. One can show that $[P]$ resembles the aforementioned intuition:

Proposition 5. *For a logic program P with query atoms we have that the answer sets of P and $[P]$, projected to the atoms in P, coincide.*

Note that, while the definition of the above construction of $[P]$ may not be trivial, this does not harm usability from user's perspective. This is because the above rewriting needs to be implemented only once, while the user can simply use query atoms.

Example 6. Consider the check for Hamiltonian cycles in Example 3. As observed, the presented attempt does not work due to default-negation. For $P = \{noHamiltonian \leftarrow P_{guess} \cup P_{check} \vdash_c sat\}$ (and thus for $[P]$) we have that there is an answer set which contains *noHamiltonian* if and only if the graph at hand does not contain a Hamiltonian cycle. On the other hand, if there are Hamiltonian cycles, then the program has at least one answer set but none of the answer sets contains atom *noHamiltonian*.

Note that the subprogram S, over which query answering is performed, can access atoms from program P. Thus, it is possible to perform computations both before and after query answering. For instance, in the previous example the graph which is checked for Hamiltonian cycles may be the result of preceding computations in P.

5.2   Checking Conditions with Default-Negation

Query answering over subprograms can be exploited as a modeling technique to check a criterion for all objects in a domain. As observed in Sect. 3, saturation may fail in such cases. Moreover, saturation is an advanced technique which might be not intuitive for less experienced ASP users (it was previously called 'hardly usable by ASP laymen' [10]). Thus, even for problems whose conditions can be expressed by positive rules, an encoding based on query answering might be easier to understand. To this end, one starts with a program P_{guess} which spans a search space of all objects to check. As with saturation, P_{check} checks if the current guess satisfies the criteria and derives a dedicated atom ok in this case. However, instead of saturating the interpretation whenever ok is true, one now checks if ok is cautiously entailed by $P_{guess} \cup P_{check}$. To this end, one constructs the program $[\{allOk \leftarrow P_{guess} \cup P_{check} \vdash_c ok\}]$. This program is always consistent, has a unique answer set containing $allOk$ whenever the property holds for all guesses in the search space, and has other answer sets none of which contains $allOk$ otherwise.

6   Discussion and Related Work

Related to our approach are *nested* HEX-*programs*, which allow for accessing answer sets of subprograms using dedicated *external atoms* [4]. However, HEX is beyond plain ASP and requires a more sophisticated solver. Similar extensions of ordinary ASP exist [15], but unlike our approach, they did not come with a compilation approach into a single program. Instead, *manifold programs* compile both the meta and the called program into a single one, similarly to our approach [8]. But this work depends on weak constraints, which are not supported by all systems. Moreover, the encoding requires a separate copy of the subprogram for each atom. The idea of representing a subprogram by atoms in the meta-program is similar to approaches for ASP debugging (cf. [12,14]). But the actual computation (as realized by program M) is different: while debugging approaches explain why a particular interpretation is not an answer set (and print the explanation to the user), we aim at detecting the inconsistency and continuing reasoning afterwards. Also the *stable-unstable semantics* supports an explicit interface to (possibly even nested) oracles [1]. However, there are no query atoms but the relation between the guessing and checking programs is realized via an extension of the semantics.

Our encoding is related to a technique towards automated integration of guess and check programs [5], but based on a different characterization of answer sets. Also, their approach can only handle ground programs. Moreover and most importantly, they focus on integrating programs, but does not discuss inconsistency checking or query answering over subprograms. We go a step further and introduce a language extension towards query answering over general subprograms, which is more convenient for average users.

7   Conclusion and Outlook

Saturation is an advanced modeling technique in ASP, which allows for exploiting disjunctions for solving co-NP-hard problems that involve checking a property all objects in a given domain. The use of default-negation in saturation encodings turns out to be problematic and a rewriting is not always straightforward. On the other hand, complexity results imply that any co-NP-hard problem can be reduced to brave reasoning over disjunctive ASP. In this paper, based on an encoding for consistency checking for normal programs, we realized query answering over subprograms.

Future work includes the application of the extension to non-ground queries. Currently, a separate copy of the subprogram is created for every query atom. However, it might be possible, at least in some cases, to answer multiple queries simultaneously. Another possible starting point for future work is the application of our encoding for more efficient evaluation of nested HEX-programs. Currently, nested HEX-programs are evaluated by separate instances of the reasoner for the calling and the called program. While this approach is strictly more expressive (and thus the evaluation also more expensive) due to the possibility to nest programs up to an arbitrary depth, it is possible in some cases to apply the technique from the paper as an evaluation technique (e.g. if the called program is normal and does not contain further nested calls).

Another issue is that if the subprogram is satisfiable, then the meta-program has *multiple* answer sets, each of which representing an answer set of the subprogram. If only consistency resp. inconsistency of the subprogram is relevant for the further reasoning in the meta-program, this leads to the repetition of solutions. In an implementation, this problem can be tackled using projected solution enumeration [11].

References

1. Bogaerts, B., Janhunen, T., Tasharrofi, S.: Stable-unstable semantics: beyond NP with normal logic programs. TPLP **16**(5–6), 570–586 (2016). doi:10.1017/S1471068416000387
2. Eiter, T., Gottlob, G.: On the computational cost of disjunctive logic programming: propositional case. Ann. Math. Artif. Intell. **15**(3–4), 289–323 (1995)
3. Eiter, T., Ianni, G., Krennwallner, T.: Answer set programming: a primer. In: Tessaris, S., Franconi, E., Eiter, T., Gutierrez, C., Handschuh, S., Rousset, M.-C., Schmidt, R.A. (eds.) Reasoning Web 2009. LNCS, vol. 5689, pp. 40–110. Springer, Heidelberg (2009). doi:10.1007/978-3-642-03754-2_2
4. Eiter, T., Krennwallner, T., Redl, C.: HEX-programs with nested program calls. In: Tompits, H., Abreu, S., Oetsch, J., Pührer, J., Seipel, D., Umeda, M., Wolf, A. (eds.) INAP/WLP -2011. LNCS (LNAI), vol. 7773, pp. 269–278. Springer, Heidelberg (2013). doi:10.1007/978-3-642-41524-1_15
5. Eiter, T., Polleres, A.: Towards automated integration of guess and check programs in answer set programming: a meta-interpreter and applications. TPLP **6**(1–2), 23–60 (2006)
6. Eiter, T., Redl, C., Schüller, P.: Problem solving using the HEX family. In: Computational Models of Rationality, pp. 150–174. College Publications (2016)
7. Faber, W., Leone, N., Pfeifer, G.: Semantics and complexity of recursive aggregates in answer set programming. Artif. Intell. **175**(1), 278–298 (2011)

8. Faber, W., Woltran, S.: Manifold answer-set programs and their applications. In: Balduccini, M., Son, T.C. (eds.) Logic Programming, Knowledge Representation, and Nonmonotonic Reasoning. LNCS (LNAI), vol. 6565, pp. 44–63. Springer, Heidelberg (2011). doi:10.1007/978-3-642-20832-4_4

9. Gebser, M., Kaminski, R., Kaufmann, B., Schaub, T.: Answer Set Solving in Practice. Synthesis Lectures on AI and Machine Learning. Morgan and Claypool Publishers (2012)

10. Gebser, M., Kaminski, R., Schaub, T.: Complex optimization in answer set programming. CoRR abs/1107.5742 (2011)

11. Gebser, M., Kaufmann, B., Schaub, T.: Solution enumeration for projected boolean search problems. In: Hoeve, W.-J., Hooker, J.N. (eds.) CPAIOR 2009. LNCS, vol. 5547, pp. 71–86. Springer, Heidelberg (2009). doi:10.1007/978-3-642-01929-6_7

12. Gebser, M., Pührer, J., Schaub, T., Tompits, H.: A meta-programming technique for debugging answer-set programs. In: AAAI, pp. 448–453. AAAI Press (2008)

13. Gelfond, M., Lifschitz, V.: Classical negation in logic programs and disjunctive databases. New Gener. Comput. 9(3–4), 365–386 (1991)

14. Oetsch, J., Pührer, J., Tompits, H.: Catching the ouroboros: on debugging nonground answer-set programs. TPLP 10(4–6), 513–529 (2010)

15. Tari, L., Baral, C., Anwar, S.: A language for modular answer set programming: application to ACC tournament scheduling. In: Vos, M.D., Provetti, A. (eds.) Answer Set Programming, Advances in Theory and Implementation, Proceedings of the 3rd International ASP 2005 Workshop, Bath, UK, 27–29 September 2005, vol. 142. CEUR Workshop Proceedings (2005)

Explaining Inconsistency in Answer Set Programs and Extensions

Christoph Redl[✉]

Institut Für Informationssysteme, Technische Universität Wien,
Favoritenstraße 9-11, 1040 Vienna, Austria
redl@kr.tuwien.ac.at

Abstract. Answer Set Programming (ASP) is a well-known problem solving approach based on nonmonotonic logic programs. HEX-programs extend ASP with *external atoms* for accessing arbitrary external information. In this paper we study inconsistent ASP- and HEX-programs, i.e., programs which do not possess answer sets, and introduce a novel notion of *inconsistency reasons* for characterizing their inconsistency depending on the input facts. This problem is mainly motivated by upcoming applications for optimizations of the evaluation algorithms for HEX-programs. Further applications can be found in ASP debugging. We then analyze the complexity of reasoning problems related to the computation of such inconsistency reasons. Finally, we present a meta-programming encoding in disjunctive ASP which computes inconsistency reasons for given normal logic programs, and a basic procedural algorithm for computing inconsistency reasons for general HEX-programs.

1 Introduction

Answer-Set Programming (ASP) is a declarative programming paradigm based on nonmonotonic programs and a multi-model semantics [11]. The problem at hand is encoded as an ASP-program, which consists of rules, such that its models, called *answer sets*, correspond to the solutions to the original problem. HEX-programs are an extension of ASP with external sources such as description logic ontologies and Web resources. So-called external atoms pass information from the logic program (given by predicate extensions and constants), to an external source, which in turn returns values to the program. Notably, *value invention* allows for domain expansion, i.e., external sources might return values which do not appear in the program. For instance, the external atom $\&synonym[car](X)$ might be used to find the synonyms X of *car*, e.g. *automobile*. Also recursive data exchange between the program and external sources is supported, which leads to high expressiveness of the formalism.

Inconsistent programs are programs which do not possess any answer sets. A natural question about such a program is *why* it is inconsistent. In this paper, we

---

This research has been supported by the Austrian Science Fund (FWF) project P27730.

M. Balduccini and T. Janhunen (Eds.): LPNMR 2017, LNAI 10377, pp. 176–190, 2017.
DOI: 10.1007/978-3-319-61660-5_16

address this question by characterizing inconsistency in terms of sets of atoms occurring in the *input facts (EDB; extensional database)* of the program. Such a characterization is motivated by programs whose *proper rules (IDB; intensional database)* are fixed, but which are evaluated with many different sets of facts. This typically occurs when the proper rules encode the general problem, while the current instance is specified as facts. It is then interesting to identify sets of instances which lead to inconsistency.

A concrete application of inconsistency reasons can be found in the evaluation algorithm for HEX-programs [3]. Here, in order to handle value invention, the overall program is partitioned into multiple *program components* which are arranged in an acyclic *evaluation graph*. Roughly, the program is evaluated by recursively computing answer sets of a program component and adding them as facts to the successor components until the final answer sets of the leaf units are computed; they correspond to the answer sets of the overall program. Then, an explanation for the inconsistency of a lower program component in terms of its input facts can be exploited for skipping evaluations which are known not to yield any answer sets. Other applications can be found in ASP debugging by guiding the user to find reasons why a particular instance or class of instances is inconsistent.

Previous related work in context of answer set program debugging focused on explaining why a *particular interpretation* fails to be an answer set, while we aim at explaining why the overall program is inconsistent. Moreover, previous debugging approaches also focus on explanations which support the *human user* to find errors in the program. Such reasons can be, for instance, in terms of minimal sets of constraints which need to be removed in order to regain consistency, in terms of odd loops (i.e., cycles of mutually depending atoms which involve negation and are of odd length), or in terms of unsupported atoms, see e.g. [1] and [10] for some approaches. To this end, one usually assumes that a single fixed program is given whose undesired behavior needs to be explained. In contrast, we consider a program whose input facts are subject to change and identify classes of instances which lead to inconsistency, which can help users to get better understanding of the structure of the problem.

Inconsistency management has also been studied for more specialized formalisms such as for multi-context systems (MCSs) [5] and for DL-programs [6]. However, the notions of inconsistencies are not directly comparable to ours as they refer to specific elements of the respective formalism (such as bridge rules of MCSs in the former case and the Abox of ontologies in the latter), which do not exist in general logic programs.

In this paper we focus on the study of inconsistency reasons, including the development of a suitable definition, characterizing them, their complexity properties, and techniques for computing them. The results are intended to serve as foundation for the aforementioned applications in future work.

In more detail, our contributions and the organization of this paper is as follows:

- In Sect. 2 we present the necessary preliminaries on ASP and HEX-programs.
- In Sect. 3 we define the novel notion of *inconsistency reasons (IRs)* for HEX-programs wrt. a set of atoms. We then provide a characterization of IRs depending on the models and unfounded sets of the program at hand.
- In Sect. 4 we analyze the complexity of reasoning tasks related to the computation of inconsistency reasons.
- In Sect. 5 we show how inconsistency reasons can be computed for normal logic programs resp. general HEX-programs using a meta-programming technique resp. a procedural algorithm.
- In Sect. 6 we discuss envisaged applications, related work and differences to our approach in more detail.
- In Sect. 7 we conclude and give an outlook on future work.

2   Preliminaries

In this section we recapitulate the syntax and semantics of HEX-programs, which generalize (disjunctive) logic programs under the answer set semantics [11]; for more details and background we refer to [3]. For simplicity, we restrict the formal discussion to ground (variable-free) programs.

A ground atom a is of the form $p(c_1, \ldots, c_\ell)$ with predicate symbol p and constant symbols $c_1, \ldots, c_\ell$ from a finite set $\mathcal{C}$, abbreviated as $p(\mathbf{c})$; we write $c \in \mathbf{c}$ if $c = c_i$ for some $1 \leq i \leq \ell$. An *assignment* A over the (finite) set $\mathcal{A}$ of atoms is a set $A \subseteq \mathcal{A}$; here $a \in A$ expresses that a is true, also denoted $A \models a$, and $a \notin A$ that a is false, also denoted $A \not\models a$. An assignment satisfies a set S of atoms, denoted $A \models S$, if $A \models a$ for some $a \in S$; it does not satisfy S, denoted $A \not\models S$, otherwise.

HEX-**Program Syntax.** HEX-programs extend ordinary ASP-programs by *external atoms*, which enable a bidirectional interaction between a HEX-program and external sources of computation. A *ground external atom* is of the form $\&g[\mathbf{p}](\mathbf{c})$, where $\mathbf{p} = p_1, \ldots, p_k$ is a list of input parameters (predicate names or object constants), called *input list*, and $\mathbf{c} = c_1, \ldots, c_l$ are constant output terms.[1] For such an external atom e with a predicate parameter p, all atoms of form $p(c_1, \ldots, c_\ell)$, which appear in the program, are called *input atoms* of the external atom e.

Definition 1. *A ground HEX-program P consists of rules*

$$a_1 \vee \cdots \vee a_l \leftarrow b_1, \ldots, b_m, \text{not } b_{m+1}, \ldots, \text{not } b_n,$$

where each a_i is a ground atom and each b_j is either an ordinary ground atom or a ground external atom.

---

[1] The distinction is mainly relevant for nonground programs, which we disregard in this paper.

For a rule r, its *head* is $H(r) = \{a_1, \ldots, a_l\}$, its *body* is $B(r) = \{b_1, \ldots, b_m, \text{not } b_{m+1}, \ldots, \text{not } b_n\}$, its *positive body* is $B^+(r) = \{b_1, \ldots, b_m\}$ and its *negative body* is $B^-(r) = \{b_{m+1}, \ldots, b_n\}$. For a HEX-program P we let $X(P) = \bigcup_{r \in P} X(r)$ for $X \in \{H, B, B^+, B^-\}$.

In the following, we call a program *ordinary* if it does not contain external atoms, i.e., if it is a standard ASP-program. Moreover, a rule as by Definition 1 is called *normal* if $k = 1$ and a program is called *normal* if all its rules are normal. A rule $\leftarrow b_1, \ldots, b_m, \text{not } b_{m+1}, \ldots, \text{not } b_n$ (i.e., with $k = 0$) is called a *constraint* and can be seen as normal rule $f \leftarrow b_1, \ldots, b_m, \text{not } b_{m+1}, \ldots, \text{not } b_n, \text{not } f$ where f is a new atom which does not appear elsewhere in the program.

HEX-**Program Semantics.** In the following, assignments are over the set $\mathcal{A}(P)$ of ordinary atoms that occur in the program P at hand. The semantics of a ground external atom $\&g[\mathbf{p}](\mathbf{c})$ wrt. an assignment A is given by the value of a $1+k+l$-ary *two-valued (Boolean) oracle function* $f_{\&g}$ that is defined for all possible values of A, $\mathbf{p}$ and $\mathbf{c}$. Thus, $\&g[\mathbf{p}](\mathbf{c})$ is true relative to A if $f_{\&g}(A, \mathbf{p}, \mathbf{c}) = \mathbf{T}$ and false otherwise. Satisfaction of ordinary rules and ASP-programs [11] is then extended to HEX-rules and -programs as follows. An assignment A satisfies an atom a, denoted $A \models a$, if $a \in A$, and it does not satisfy it, denoted $A \not\models a$, otherwise. It satisfies a default-negated atom not a, denoted $A \models \text{not } a$, if $A \not\models a$, and it does not satisfy it, denoted $A \not\models \text{not } a$, otherwise. A rule r is satisfied under assignment A, denoted $A \models r$, if $A \models a$ for some $a \in H(r)$ or $A \not\models a$ for some $a \in B(r)$. A HEX-program P is satisfied under assignment A, denoted $A \models P$, if $A \models r$ for all $r \in P$.

The answer sets of a HEX-program P are defined as follows. Let the *FLP-reduct* of P wrt. an assignment A be the set $fP^A = \{r \in P \mid A \models b \text{ for all } b \in B(r)\}$ of all rules whose body is satisfied by A. Then

Definition 2. *An assignment A is an answer set of a HEX-program P, if A is a subset-minimal model of fP^A.*[2]

Example 1. Consider the HEX-program $P = \{p \leftarrow \&id[p]()\}$, where $\&id[p]()$ is true iff p is true. Then P has the answer set $A_1 = \emptyset$; indeed it is a subset-minimal model of $fP^{A_1} = \emptyset$.

For a given HEX-program P we let $\mathcal{AS}(P)$ denote the set of all answer sets of P.

3   Explaining Inconsistency of HEX-**Programs**

In this section we consider programs P that are extended by a set of (input) atoms $I \subseteq \mathcal{D}$ from a given domain $\mathcal{D}$, which are added as facts. More precisely, for a given set $I \subseteq \mathcal{D}$, we consider $P \cup facts(I)$, where $facts(I) = \{a \leftarrow \mid a \in I\}$ is the representation of I transformed to facts. This is in spirit of the typical usage of (ASP- and HEX-)programs, where proper rules (IDB; intensional database)

---

[2] For ordinary Π, these are Gelfond & Lifschitz's answer sets.

encode the problem at hand and are fixed, while facts (EDB; extensional database) may be subject to change as they are used to specify a concrete instance. Our goal is to express the reasons for inconsistency of HEX-program $P \cup facts(I)$ in terms of I. That is, we want to find a sufficient criterion wrt. I which guarantees that $P \cup facts(I)$ does not possess any answer sets.

Formalizing Inconsistency Reasons. Inspired by inconsistency explanations for multi-context systems [5], we propose to make such a reason dependent on atoms from $\mathcal{D}$ which *must occur* resp. *must not occur* in I such that $P \cup facts(I)$ is inconsistent, while the remaining atoms from $\mathcal{D}$ might either occur or not occur in I without influencing (in)consistency. We formalize this idea as follows:

Definition 3 (Inconsistency Reason (IR)). *Let P be a HEX-program and $\mathcal{D}$ be a domain of atoms. An* inconsistency reason (IR) *of P wrt. $\mathcal{D}$ is a pair $R = (R^+, R^-)$ of sets of atoms $R^+ \subseteq \mathcal{D}$ and $R^- \subseteq \mathcal{D}$ with $R^+ \cap R^- = \emptyset$ s.t. $P \cup facts(I)$ is inconsistent for all $I \subseteq \mathcal{D}$ with $R^+ \subseteq I$ and $R^- \cap I = \emptyset$.*

Here, R^+ resp. R^- define the sets of atoms which must be present resp. absent in I such that inconsistency of $P \cup facts(I)$ is inconsistent, while atoms from $\mathcal{D}$ which are neither in R^+ nor in R^- might be arbitrarily added or not without affecting inconsistency.

Example 2. An IR of the program $P = \{\leftarrow a, \text{not } c; \ d \leftarrow b.\}$ wrt. $\mathcal{D} = \{a, b, c\}$ is $R = (\{a\}, \{c\})$ because $P \cup facts(I)$ is inconsistent for all $I \subseteq \mathcal{D}$ whenever $a \in I$ and $c \notin I$, while b can be in I or not without affecting (in)consistency.

In general there are multiple IRs, some of which might not be minimal. For instance, the program $\{\leftarrow a; \leftarrow b\}$ has inconsistency reasons $R_1 = (\{a\}, \emptyset)$, $R_2 = (\{b\}, \emptyset))$ and $R_3 = (\{a, b\}, \emptyset)$ wrt. $\mathcal{D} = \{a, b\}$. On the other hand, a program P might not have any IR at all if $P \cup facts(I)$ is consistent for all $I \subseteq \mathcal{D}$. This is the case, for instance, for the empty HEX-program $P = \emptyset$. However, one can show that there is always at least one IR if $P \cup facts(I)$ is inconsistent for some $I \subseteq \mathcal{D}$.

Proposition 1. *For all HEX-programs P and domains $\mathcal{D}$ such that $P \cup facts(I)$ is inconsistent for some set $I \subseteq \mathcal{D}$ of atoms, then there is an IR of P wrt. $\mathcal{D}$.*

Proof. Take some I such that $P \cup facts(I)$ is inconsistent and consider $R = (R^+, R^-)$ with $R^+ = I$ and $R^- = \mathcal{D} \setminus I$. Then the only I' with $R^+ \subseteq I'$ and $R^- \cap I' = \emptyset$ is I itself. But $P \cup facts(I)$ is inconsistent by assumption. $\square$

Characterizing Inconsistency Reasons. Next, we present an alternative characterization of IRs based on the models of a HEX-program P and unfounded sets of P wrt. these models. To this end we use the following definition and result from [9]:

Definition 4 (Unfounded Set). *Given a HEX-program P and an assignment A, let U be any set of atoms appearing in P. Then, U is an* unfounded set *for P wrt. A if, for each rule $r \in P$ with $H(r) \cap U \neq \emptyset$, at least one of the following conditions holds:*

(i) some literal of $B(r)$ is false wrt. A; or
(ii) some literal of $B(r)$ is false wrt. $A \setminus U$; or
(iii) some atom of $H(r) \setminus U$ is true wrt. A.

A model M of a HEX-program P is called *unfounded-free* if it does not intersect with an unfounded set of P wrt. M. One can then show [9]:

Proposition 2. *A model M of a HEX-program P is an answer set of P iff it is unfounded-free.*

In the following, for sets of atoms $\mathcal{H}$ and $\mathcal{B}$, let $\mathcal{P}^e_{\langle \mathcal{H}, \mathcal{B} \rangle}$ denote the set of all HEX-programs whose head atoms come only from $\mathcal{H}$ and whose body atoms and external atom input atoms come only from $\mathcal{B}$. We then use the following result from [12].

Proposition 3. *Let P be a HEX-program. Then $P \cup R$ is inconsistent for all $R \in \mathcal{P}^e_{\langle \mathcal{H}, \mathcal{B} \rangle}$ iff for each classical model A of P there is a nonempty unfounded set U of P wrt. A such that $U \cap A \neq \emptyset$ and $U \cap \mathcal{H} = \emptyset$.*

Intuitively, the result states that a program is inconsistent, iff each (classical) model is dismissed as answer set because it is not unfounded-free. Our concept of an inconsistency reason amounts to a special case of this result. More precisely, (R^+, R^-) for sets of atoms $R^+ \subseteq \mathcal{D}$ and $R^- \subseteq \mathcal{D}$ is an IR iff the program is inconsistent for all additions of facts that contain all atoms from R^+ but none from R^-; this can be expressed by applying the previous result for $\mathcal{B} = \emptyset$ and an appropriate selection of $\mathcal{H}$.

Lemma 1. *Let P be a HEX-program and $\mathcal{D}$ be a domain of atoms. A pair (R^+, R^-) of sets of atoms $R^+ \subseteq \mathcal{D}$ and $R^- \subseteq \mathcal{D}$ with $R^+ \cap R^- = \emptyset$ is an IR of a HEX-program P wrt. $\mathcal{D}$ iff $P \cup facts(R^+) \cup R$ is inconsistent for all $R \in \mathcal{P}^e_{\langle \mathcal{D} \setminus R^-, \emptyset \rangle}$.*

Proof. The claim follows basically from the observation that the sets of programs allowed to be added on both sides are the same. In more detail:

($\Rightarrow$) Let (R^+, R^-) be an IR of P with $R^+ \subseteq \mathcal{D}$ and $R^- \subseteq \mathcal{D}$ with $R^+ \cap R^- = \emptyset$. Let $R \in \mathcal{P}^e_{\langle \mathcal{D} \setminus R^-, \emptyset \rangle}$. We have to show that $P \cup facts(R^+) \cup R$ is inconsistent. Take $I = R^+ \cup \{a \mid a \leftarrow\, \in R\}$; then $R^+ \subseteq I$ and $R^- \cap I = \emptyset$ and thus by our precondition that (R^+, R^-) is an IR we have that $P \cup facts(I)$, which is equivalent to $P \cup facts(R^+) \cup R$, is inconsistent.

($\Leftarrow$) Suppose $P \cup facts(R^+) \cup R$ is inconsistent for all $R \in \mathcal{P}^e_{\langle \mathcal{D} \setminus R^-, \emptyset \rangle}$. Let $I \subseteq \mathcal{D}$ such that $R^+ \subseteq I$ and $R^- \cap I = \emptyset$. We have to show that $P \cup facts(I)$ is inconsistent. Take $R = \{a \leftarrow\, \mid a \in I \setminus R^+\}$ and observe that $R \in \mathcal{P}^e_{\langle \mathcal{D} \setminus R^-, \emptyset \rangle}$. By our precondition we have that $P \cup facts(R^+) \cup R$ is inconsistent. The observation that the latter is equivalent to $P \cup facts(I)$ proves the claim. □

Now Proposition 3 and Lemma 1 in combination allow for characterizing IRs in terms of models and unfounded sets.

Proposition 4. *Let P be a ground HEX-program and $\mathcal{D}$ be a domain. Then a pair of sets of atoms (R^+, R^-) with $R^+ \subseteq \mathcal{D}$, $R^- \subseteq \mathcal{D}$ and $R^+ \cap R^- = \emptyset$ is an IR of P iff for all classical models M of P either (i) $R^+ \not\subseteq M$ or (ii) there is a nonempty unfounded set U of P wrt. M such that $U \cap M \neq \emptyset$ and $U \cap (\mathcal{D} \setminus R^-) = \emptyset$.*

Proof. ($\Rightarrow$) Let (R^+, R^-) be an IR of P wrt. $\mathcal{D}$. Consider a classical model M of P. We show that one of (i) or (ii) is satisfied.

If M is not a classical model of $P \cup facts(R^+)$, then, since M is a model of P, we have that $R^+ \not\subseteq M$ and thus Condition (i) is satisfied.

In case that M is a classical model of $P \cup facts(R^+)$, first observe that, since (R^+, R^-) is an IR, by Lemma 1 $P \cup facts(R^+) \cup R$ is inconsistent for all $R \in \mathcal{P}^e_{\langle \mathcal{D} \setminus R^-, \emptyset \rangle}$. By Proposition 3, this is further the case iff for each classical model M' of $P \cup facts(R^+)$ there is a nonempty unfounded set U such that $U \cap M' \neq \emptyset$ and $U \cap (\mathcal{D} \setminus R^-) = \emptyset$. Since M is a model of $P \cup facts(R^+)$ it follows that an unfounded set as required by Condition (ii) exists.

($\Leftarrow$) Let (R^+, R^-) be a pair of sets of atoms such that for all classical models M of P either (i) or (ii) holds. We have to show that it is an IR of P, i.e., $P \cup facts(I)$ is inconsistent for all $I \subseteq \mathcal{D}$ with $R^+ \subseteq I$ and $R^- \cap I = \emptyset$.

Assignments which are no classical models of $P \cup facts(I)$ cannot be answer sets, thus it suffices to show for all classical models of $P \cup facts(I)$ that they are no answer sets. Consider an arbitrary but fixed $I \subseteq \mathcal{D}$ with $R^+ \subseteq I$ and $R^- \cap I = \emptyset$ and let M be an arbitrary classical model of $P \cup facts(I)$. Then M is also a classical model of P and thus, by our precondition, either (i) or (ii) holds.

If (i) holds, then there is an $a \in R^+$ such that $a \notin M$. But since $R^+ \subseteq I$ we have $a \leftarrow \in facts(I)$ and thus M cannot be a classical model and therefore no answer set of $P \cup facts(I)$. If (ii) holds, then there is a nonempty unfounded set U of P wrt. M such that $U \cap M \neq \emptyset$ and $U \cap (\mathcal{D} \setminus R^-) = \emptyset$, i.e., U does not contain elements from $\mathcal{D} \setminus R^-$. Then U is also an unfounded set of $P \cup facts(R^+)$ wrt. M. Then by Proposition 3 we have that $P \cup facts(R^+) \cup R$ is inconsistent for all $R \in \mathcal{P}^e_{\langle \mathcal{D} \setminus R^-, \emptyset \rangle}$. The latter is, by Lemma 1, the case iff (R^+, R^-) be an IR of P wrt. $\mathcal{D}$. $\square$

Intuitively, each classical model M must be excluded from being an answer set of $P \cup facts(I)$ for any $I \subseteq \mathcal{D}$ with $R^+ \subseteq I$ and $R^- \cap I = \emptyset$. This can either be done by ensuring that M does not satisfy R^+ or that some atom $a \in M$ is unfounded if I is added because $a \notin \mathcal{D} \setminus R^-$.

While the previous result characterizes inconsistency of a program precisely, it is computationally expensive to apply because one does not only need to check a condition for all models of the program at hand, but also for all unfounded sets of all models. We therefore present a second, simpler condition, which refers only to the program's models but not the unfounded sets wrt. these models.

Proposition 5. *Let P be a ground HEX-program and $\mathcal{D}$ be a domain such that $H(P) \cap \mathcal{D} = \emptyset$. For a pair of sets of atoms (R^+, R^-) with $R^+ \subseteq \mathcal{D}$ and $R^- \subseteq \mathcal{D}$, if for all classical models M of P we either have (i) $R^+ \not\subseteq M$ or (ii) $R^- \cap M \neq \emptyset$ then (R^+, R^-) is an IR of P.*

Proof. Let (R^+, R^-) be a pair of sets of atoms such that for all classical models M of P we either have $R^+ \not\subseteq M$ or $R^- \cap M \neq \emptyset$. We have to show that it is an IR of P, i.e., $P \cup \mathit{facts}(I)$ is inconsistent for all $I \subseteq \mathcal{D}$ with $R^+ \subseteq I$ and $R^- \cap I = \emptyset$.

Assignments which are no classical models of $P \cup \mathit{facts}(I)$ cannot be answer sets, thus it suffices to show for all classical models that they are no answer sets. Let M be an arbitrary classical model of $P \cup \mathit{facts}(I)$. Then M is also a classical model of P and thus, by our precondition, either (i) or (ii) holds.

If (i) holds, then there is an $a \in R^+$ such that $a \notin M$. But since $R^+ \subseteq I$ we have $a \leftarrow \in \mathit{facts}(I)$ and thus M cannot be a classical model and therefore no answer set of $P \cup \mathit{facts}(I)$. If (ii) holds, then there is an $a \in R^- \cap M$. Since $R^- \cap I = \emptyset$ we have that $a \leftarrow \notin \mathit{facts}(I)$. Moreover, we have that $a \notin H(P)$ by our precondition that $H(P) \cap \mathcal{D} = \emptyset$. But then $\{a\}$ is an unfounded set of $P \cup \mathit{facts}(I)$ wrt. M such that $\{a\} \cap M \neq \emptyset$ and thus, by Proposition 2, M is not an answer set of P. □

However, in contrast to Proposition 4, note that Proposition 5 does *not* hold in reverse direction. That is, the proposition does not characterize inconsistency exactly but provides a practical means for identifying inconsistency in some cases. This is demonstrated by the following example.

Example 3. Let $\mathcal{D} = \{x\}$ and $P = \{\leftarrow \text{not } a\}$. Then $R = (\emptyset, \emptyset)$ is an IR of P because P is inconsistent and no addition of any atoms as facts is allowed. However, the classical model $M = \{a\}$ contains R^+ but does not intersect with R^-, hence the precondition of Proposition 5 is not satisfied.

4   Computational Complexity

Next, we discuss the computational complexity of reasoning problems in context of computing inconsistency reasons. For our results we assume that oracle functions can be evaluated in polynomial time wrt. the size of their input.

We start with the problem of checking if a candidate IR of a program is a true IR.

Proposition 6. *Given a* HEX*-program P, a domain $\mathcal{D}$, and a pair $R = (R^+, R^-)$ of sets of atoms with $R^+ \subseteq \mathcal{D}$, $R^- \subseteq \mathcal{D}$ and $R^+ \cap R^- = \emptyset$. Deciding if R is an IR of P wrt. $\mathcal{D}$ is (i) Π_2^P-complete in general and (ii) coNP-complete if P is an ordinary disjunction-free program.*

Proof. **Hardness:** Checking if P does not have any answer set is a well-known problem that is Π_2^P-complete for (i) and *coNP*-complete for (ii).

We reduce the problem to checking if a given $R = (R^+, R^-)$ is an IR of P wrt. a domain $\mathcal{D}$, which shows that the latter cannot be easier. Consider $R = (\emptyset, \emptyset)$ and domain $\mathcal{D} = \emptyset$. Then the only $I \subseteq \mathcal{D}$ with $\emptyset \subseteq I$ and $\mathcal{D} \cap I = \emptyset$ is $I = \emptyset$. Then R is an IR iff $P \cup \mathit{facts}(\emptyset) = P$ is inconsistent, which allows for reducing the inconsistency check to the check of R for being an IR of P wrt. $\mathcal{D}$.

Membership: Consider $P' = P \cup \{x \leftarrow | \ x \in R^+\} \cup \{x \leftarrow \text{not } nx; \ nx \leftarrow \text{not } x \ |$ $x \in \mathcal{D} \setminus R^-\}$, where nx is a new atom for all $x \in \mathcal{D}$. Then P' has an answer set iff for some $I \subseteq \mathcal{D}$ with $R^+ \subseteq I$ and $R^- \cap I = \emptyset$ we have that $P \cup facts(I)$ has an answer set. Conversely, P' does not have an answer set, iff for all $I \subseteq \mathcal{D}$ with $R^+ \subseteq I$ and $R^- \cap I = \emptyset$ we have that $P \cup facts(I)$ has no answer set, which is by Definition 3 exactly the case iff R is an IR of P wrt. $\mathcal{D}$. The observation that checking if P' has no answer set – which is equivalent to deciding if R is an IR – is in Π_2^P resp. $coNP$ for (i) resp. (ii), because the property of being ordinary and disjunction-free carries over from P to P', concludes the proof.   □

The complexity of deciding if a program has some IR is then as follows:

Proposition 7. *Given a* HEX-*program P and a domain $\mathcal{D}$. Deciding if there is an IR of P wrt. $\mathcal{D}$ is (i) Σ_3^P-complete in general and (ii) Σ_2^P-complete if P is an ordinary disjunction-free program.*

Proof (Sketch). The hardness proofs are done by reducing satisfiability checking of the QBF formulas $\forall \mathbf{X} \exists \mathbf{Y} \forall \mathbf{Z} \phi(\mathbf{X}, \mathbf{Y}, \mathbf{Z})$ for case (i) and $\forall \mathbf{X} \exists \mathbf{Y} \phi(\mathbf{X}, \mathbf{Y})$ for case (ii) to checking (non-)existence of an IR of P; to this end, a saturation encoding is used. The membership proofs are based on the previous results. Due to limited space, we provide the details of this lengthy proof in an extended version at http://www.kr.tuwien.ac.at/research/projects/inthex/inconsistencyanalysis-ext.pdf.

We summarize the complexity results in Table 1.

Table 1. Summary of complexity results

Reasoning problem	Program class	
	Normal ASP-program	General HEX-program
Checking a candidate IR	$coNP$-c	Π_2^P-c
Checking existence of an IR	Π_2^P-c	Π_3^P-c

For our envisaged application IRs do not need to be minimal, which is why we leave the formal analysis for future work. However, clearly there is a minimal IR iff there is an IR, while the complexity of checking a candidate minimal IR is less obvious.

5   Computing Inconsistency Reasons

In this section we discuss various methods for computing IRs. For ordinary normal programs we present a *meta-programming* approach which encodes the computation of IRs as a disjunctive program. For general HEX-programs this is not easily possible due to complexity reasons and we present a procedural method instead.

Inconsistency reasons for normal ASP-programs. If the program P at hand is normal and does not contain external atoms, then one can construct a positive disjunctive meta-program M^P which checks consistency of P. There are multiple encodings for M^P (see e.g. [8]) based on the simulation of an answer set solver for normal programs. However, the details of M^P are not relevant for the further understanding and it suffices to assume that for a given normal ASP-program P, program M^P has the following properties:

- M^P is always consistent;
- if P is inconsistent, then M^P has a single answer set $A_{sat} = A(M^P)$ containing all atoms in M^P including a dedicated atom *incons* which does not appear in P; and
- if P is consistent, then the answer sets of M^P correspond one-to-one to those of P and none of them contains *incons*.

Then, the atom *incons* in the answer set(s) of M^P represents inconsistency of the original program P. One can then extend M^P in order to compute the inconsistency reasons of P as follows. We construct

$$\tau(\mathcal{D}, P) = M^P \tag{1}$$

$$\cup \; \{a^+ \vee a^- \vee a^x \mid a \in \mathcal{D}\} \cup \tag{2}$$

$$\cup \; \{a \leftarrow a^+; \; \leftarrow a, a^-; \; a \vee \bar{a} \leftarrow a^x; \; \mid a \in \mathcal{D}\} \tag{3}$$

$$\cup \; \{\leftarrow \text{not } incons\}, \tag{4}$$

where a^+, a^-, a^x and $\bar{a}$ are new atoms for all atoms $a \in \mathcal{D}$.

Informally, the idea is that rules (2) guess all possible IRs $R = (R^+, R^-)$, where a^+ represents $a \in R^+$, a^- represents $a \in R^-$ and a^x represents $a \notin R^+ \cup R^-$. Rules (3) guess all possible sets of input facts wrt. the currently guessed IR, where $\bar{a}$ represents that a is not a fact. We know that M^P derives *incons* iff P together with the facts from rules (3) is inconsistent. This allows the constraint (4) to eliminate all candidate IRs, for which not all possible sets of input facts lead to inconsistency.

One can show that the encoding is sound and complete wrt. the computation of IRs:

Proposition 8. *Let P be an ordinary normal program and $\mathcal{D}$ be a domain. Then (R^+, R^-) is an IR of P wrt. $\mathcal{D}$ iff $\tau(\mathcal{D}, P)$ has an answer set $A \supseteq \{a^\sigma \mid \sigma \in \{+, -\}, a \in R^\sigma\}$.*

Proof. The rules (2) guess all possible explanations (R^+, R^-), where a^+ represents $a \in R^+$, a^- represents $a \in R^-$ and a^x represents $a \notin R^+ \cup R^-$. The rules (3) then guess all possible sets of input facts $I \subseteq \mathcal{D}$ with $R^+ \subseteq I$ and $R^- \cap I = \emptyset$ wrt. the previous guess for (R^+, R^-): if $a \in R^+$ then a must be true, if $a \in R^-$ then a cannot be true, and if $a \notin R^+ \cup R^-$ then a can either be true or not.

Now by the properties of M^P we know that $A(M^P)$ is an answer set of M^P iff P is inconsistent wrt. the current facts computed by rules (3). By minimality of answer sets, $A(\tau(\mathcal{D}, P))$ is an answer set of $\tau(\mathcal{D}, P)$.

Rules (4) ensure that also the atoms a^+, a^- and a^x are true for all $a \in \mathcal{D}$. Since M^P is positive, this does not harm the property of being an answer set wrt. M^P.   □

We provide a tool which allows for computing inconsistency reasons of programs, which is available from https://github.com/hexhex/inconsistencyanalysis.

Inconsistency reasons for general HEX-programs. Suppose the IRs of a program P wrt. a domain $\mathcal{D}$ can be computed by a meta-program; then this meta-program is consistent iff an IR of P wrt. $\mathcal{D}$ exists. Therefore, consistency checking over the meta-program must necessarily have a complexity not lower than the one of deciding existence of an IR of P. However, we have shown in Proposition 7 that deciding if a general HEX-program has an IR is Σ_3^P-complete. On the other hand, consistency checking over a general HEX-program is only Σ_2^P-complete. Thus, unless $\Sigma_3^P = \Sigma_2^P$, computing the IRs of a general HEX-program cannot be polynomially reduced to a meta-HEX-program (using a non-polynomial reduction is possible but uncommon, which is why we do not follow this possibility). We present two possible remedies.

Giving up completeness. For HEX-programs P without disjunctions we can specify an encoding for computing its IRs, which is sound but not complete. To this end, we make use of a rewriting of P to an ordinary ASP-program $\hat{P}$, which was previously used for evaluating HEX-programs. In a nutshell, each external atom $\&g[\mathbf{p}](\mathbf{c})$ in P is replaced by an ordinary *replacement atom* $e_{\&g[\mathbf{p}]}(\mathbf{c})$ and rules $e_{\&g[\mathbf{p}]}(\mathbf{c}) \leftarrow \text{not } ne_{\&g[\mathbf{p}]}(\mathbf{c})$ and $ne_{\&g[\mathbf{p}]}(\mathbf{c}) \leftarrow \text{not } e_{\&g[\mathbf{p}]}(\mathbf{c})$ are added to guess the truth value of the former external atom. However, the answer sets of the resulting *guessing program* $\hat{P}$ do not necessarily correspond to answer sets of the original program P. Instead, for each answer set it must be checked if the guesses are correct. An answer set $\hat{A}$ of the guessing program $\hat{P}$ is called a *compatible set* of P, if $f_{\&g}(\hat{A}, \mathbf{p}, \mathbf{c}) = \mathbf{T}$ iff $\mathbf{T}e_{\&g[\mathbf{p}]}(\mathbf{c}) \in \hat{A}$ for all external atoms $\&g[\mathbf{p}](\mathbf{c})$ in P. Each answer set of P is the projection of some compatible set of P. For details about the construction of $\hat{P}$ we refer to [4].

One can exploit the rewriting $\hat{P}$ to compute (some) IRs of P. To this end, one constructs $\tau(\mathcal{D}, \hat{P})$ and computes its answer sets. This yields explanations for the inconsistency of the guessing program $\hat{P}$ rather than the actual HEX-program P. The HEX-program P is inconsistent whenever the guessing program $\hat{P}$ is, and every inconsistency reason of $\hat{P}$ is also one of P. Hence, the approach is sound. However, since $\hat{P}$ might be consistent even if P is inconsistent and $\tau(\mathcal{D}, \hat{P})$ might have no answer set and actual explanations are not found, it is not complete. Formally:

Proposition 9. *Let P be a HEX-program and $\mathcal{D}$ be a domain. Then each IR of $\hat{P}$ wrt. $\mathcal{D}$ is also an IR of P wrt. $\mathcal{D}$, i.e., the use of $\hat{P}$ is sound wrt. the computation of IRs.*

Proof. As each answer set of P is the projection of a compatible set of $\hat{P}$ to the atoms in P, inconsistency of $\hat{P} \cup \mathit{facts}(I)$ implies inconsistency of $\hat{P} \cup \mathit{facts}(I)$ for all $I \subseteq \mathcal{D}$.   □

Algorithm 1. IRComputation

Input: A general HEX-program P and a domain $\mathcal{D}$
Output: All IRs of P wrt. $\mathcal{D}$

for all classical models M of P **do**
  for all $R^+ \subseteq \mathcal{D}$ such that $R^+ \not\subseteq M$ **do**
    for all $R^- \subseteq \mathcal{D}$ such that $R^+ \cap R^- \neq \emptyset$ **do**
      Output·(R^+, R^-)

  for all $R^- \subseteq \mathcal{D}$ such that there is a UFS U of P wrt. M with $U \cap M \neq \emptyset$ and $U \cap (\mathcal{D} \setminus R^-) = \emptyset$ **do**
    for all $R^+ \subseteq \mathcal{D}$ such that $R^+ \cap R^- \neq \emptyset$ **do**
      Output (R^+, R^-)

The following example shows that using $\hat{P}$ for computing IRs is not complete.

Example 4. Consider the inconsistent HEX-program $P = \{p \leftarrow q, \&neg[p]()\}$ and domain $\mathcal{D} = \{q\}$. An IR is $(\{q\}, \emptyset)$. However, the guessing program $\hat{P} \cup \{q \leftarrow\} = \{e_{\&neg[p]} \vee ne_{\&neg[p]}; \ p \leftarrow q, e_{\&neg[p]}; \ q \leftarrow\}$ is consistent and has the answer set $\hat{A} = \{e_{\&neg[p]}, p, q\}$; therefore $(\{q\}, \emptyset)$ is not an IR (actually there is none).

In the previous example, the reason why no inconsistency reason is found is that $\hat{A}$ is an answer set of $\hat{P}$ but the value of $e_{\&neg[p]}$ guessed for the external replacement atom differs from the actual value of $\&neg[p]()$ under $\hat{A}$, i.e., $\hat{A}$ is not compatible.

Using a procedural algorithm. A sound and complete alternative is using a procedural algorithm which implements Proposition 4; a naive one is shown in Algorithm 1.

Proposition 10. *For a general* HEX-*program* P *and a domain* $\mathcal{D}$, *IRComputation*$(P, \mathcal{D})$ *outputs all IRs of* P *wrt.* $\mathcal{D}$.

Proof. The algorithm enumerates all pairs of sets (R^+, R^-) as by Proposition 4. $\square$

However, in general efficient computation of IRs is a challenging problem by itself, which calls for algorithmic optimizations. Since this paper focuses on the notion of IRs and properties, such optimizations are beyond its scope and are left for future work.

6 Discussion and Related Work

The notion of inconsistency reasons is motivated by applications which evaluate programs with fixed proper rules under different sets of facts. Then, inconsistency reasons can be exploited to skip evaluations which are known to yield no answer sets.

A concrete application can be found in optimizations of the evaluation algorithm for HEX-programs [3]. Here, in order to handle programs with expanding domains (value invention), the overall program is partitioned into multiple *program components* which are arranged in an acyclic *evaluation graph*, which is evaluated top-down. To this end, each answer set of a program component is recursively added as facts to the successor component until the final answer sets of the leaf units are computed, which correspond to the overall answer sets. We envisage to exploit inconsistency reasons by propagating them as constraints to predecessor components in order to eliminate interpretations, which would fail to be answer sets of the overall program anyway, already earlier. While we have constructed synthetic examples where the technique leads to an exponential speedup, an empirical analysis using real-world applications is left for future work.

Previous work on inconsistency analysis was mainly in the context of debugging of answer set programs and faulty systems in general, based on symbolic diagnosis [13]. For instance, the approach in [14] computes inconsistency explanations in terms of either minimal sets of constraints which need to be removed in order to regain consistency, or of odd loops (in the latter case the program is called *incoherent*). The realization is based on another (meta-)ASP-program. Also the more general approaches in [1] and [10] rewrite the program to debug into a meta-program using dedicated control atoms. The goal is to support the human user to find reasons for undesired behavior of the program. Possible queries are, for instance, why a certain interpretation is not an answer set or why a certain atom is not true in an answer set. Explanations are then in terms of unsatisfied rules, only cyclically supported atoms, or unsupported atoms. Furthermore, previous work introduced decision criteria on the inconsistency of programs [12] which we exploited in Sect. 3, but did not introduce a notion of IRs.

Because these approaches are based on meta-programming they are, in this respect, similar to ours. However, an important difference is that we do not compute explanations why a particular interpretation fails to be answer set of a fixed program, but we rather analyze inconsistency of a program wrt. a set of possible extensions by facts. Another difference concerns the goal of the approaches: while the aforementioned ones aim at answer set debugging and therefore at human-readable explanations of the inconsistency, ours is *not* intended to be used as a debugging technique for humans to find errors in a program. In view of our envisaged application, it rather aims at machine-processable explanations wrt. input facts (i.e., it identifies unsatisfiable instances).

In contrast to meta-programming approaches, also procedural algorithms have been proposed which explain why a set of atoms is contained in a given answer set resp. not contained in any answer set [2]. However, they do not introduce a formal definition of an explanation but follow a more practical approach by printing human-readable explanations of varying types. Therefore it is not straightforward to transform their explanations into formal constraints, as this is necessary for the envisaged applications.

Inconsistency management has also been studied for more specialized formalisms such as for multi-context systems (MCSs) [5], which are sets of knowledge-bases (implemented in possibly different formalisms and interconnected by so-called *bridge rules*), and DL-programs [6]; *inconsistency diagnoses* for MCSs use pairs of bridge rules which must be removed resp. added unconditionally to make the system consistent. Their notions are not directly comparable to ours as it refer to specific elements of the respective formalism (such as bridge rules of MCSs and the Abox of ontologies), which do not exist in general logic programs. Hence, our notion defines IRs more generically in terms of input facts. However, the main idea of explaining inconsistency is related.

Our complexity results relate to abductive reasoning using sceptical inference [7].

7   Conclusion and Outlook

We have introduced the novel notion of *inconsistency reasons* for explaining why an ASP- or HEX-program is inconsistent in terms of input facts. In contrast to previous notions from the area of answer set debugging, which usually explain why a certain interpretation is not an answer set of a concrete program, we consider programs with fixed proper rules, which are instantiated with various data parts. Moreover, while debugging approaches aim at explanations which support human users, ours was developed with respect to an envisaged application for optimizations in evaluation algorithms.

The next step is to realize this envisaged application. To this end, an efficient technique for computing IRs in the general case is needed. However, due to complexity this is not always possible, hence there is a tradeoff between the costs of computing inconsistency reasons and potential benefits. The usage of heuristics for estimating both factors appears to be an interesting starting point.

Acknowledgements. The author thanks Markus Bretterbauer, who developed a prototype implementation for computing inconsistency reasons as studied in this paper.

References

1. Brain, M., Gebser, M., Pührer, J., Schaub, T., Tompits, H., Woltran, S.: Debugging ASP programs by means of ASP. In: Baral, C., Brewka, G., Schlipf, J. (eds.) LPNMR 2007. LNCS (LNAI), vol. 4483, pp. 31–43. Springer, Heidelberg (2007). doi:10.1007/978-3-540-72200-7_5
2. Brain, M., Vos, M.D.: Debugging logic programs under the answer set semantics. In: Vos, M.D., Provetti, A. (eds.) Answer Set Programming, Advances in Theory and Implementation, Proceedings of the 3rd International ASP 2005 Workshop, Bath, UK, 27–29 September 2005, CEUR Workshop Proceedings, vol. 142. CEUR-WS.org (2005)
3. Eiter, T., Fink, M., Ianni, G., Krennwallner, T., Redl, C., Schüller, P.: A model building framework for answer set programming with external computations. Theor. Pract. Logic Program. **16**(4), 418–464 (2016)
4. Eiter, T., Fink, M., Krennwallner, T., Redl, C., Schüller, P.: Efficient HEX-program evaluation based on unfounded sets. J. Artif. Intell. Res. **49**, 269–321 (2014)
5. Eiter, T., Fink, M., Schüller, P., Weinzierl, A.: Finding explanations of inconsistency in multi-context systems. Artif. Intell. **216**, 233–274 (2014)

6. Eiter, T., Fink, M., Stepanova, D.: Inconsistency management for description logic programs and beyond. In: Faber, W., Lembo, D. (eds.) RR 2013. LNCS, vol. 7994, pp. 1–3. Springer, Heidelberg (2013). doi:10.1007/978-3-642-39666-3_1

7. Eiter, T., Gottlob, G., Leone, N.: Semantics and complexity of abduction from default theories. Artif. Intell. **90**(12), 177–223 (1997)

8. Eiter, T., Polleres, A.: Towards automated integration of guess and check programs in answer set programming: a meta-interpreter and applications. TPLP **6**(1–2), 23–60 (2006)

9. Faber, W.: Unfounded sets for disjunctive logic programs with arbitrary aggregates. In: Baral, C., Greco, G., Leone, N., Terracina, G. (eds.) LPNMR 2005. LNCS (LNAI), vol. 3662, pp. 40–52. Springer, Heidelberg (2005). doi:10.1007/11546207_4

10. Gebser, M., Puehrer, J., Schaub, T., Tompits, H.: A meta-programming technique for debugging answer-set programs. In: Fox, D., Gomes, C.P. (eds.) AAAI-08/IAAI-08 Proceedings, pp. 448–453 (2008). http://publik.tuwien.ac.at/files/PubDat_167810.pdf

11. Gelfond, M., Lifschitz, V.: Classical negation in logic programs and disjunctive databases. New Gener. Comput. **9**(3–4), 365–386 (1991)

12. Redl, C.: On equivalence and inconsistency of answer set programs with external sources. In: Proceedings of the Thirty-First AAAI Conference (AAAI 2017), San Francisco, California, USA. AAAI Press, February 2017

13. Reiter, R.: A theory of diagnosis from first principles. Artif. Intell. **32**(1), 57–95 (1987)

14. Syrjänen, T.: Debugging inconsistent answer set programs. In: Proceedings of the 11th International Workshop on Non-Monotonic Reasoning, pp. 77–84. Lake District, May 2006

Blending Lazy-Grounding and CDNL Search for Answer-Set Solving

Antonius Weinzierl[(✉)]

Institute of Information Systems, Knowledge-based Systems Group,
TU Wien, Favoritenstr. 9-11, 1040 Vienna, Austria
`weinzierl@kr.tuwien.ac.at`

Abstract. Efficient state-of-the-art answer-set solvers are two-phased: first grounding the input program, then applying search based on conflict-driven nogood learning (CDNL). The latter provides superior search performance but the former causes exponential memory requirements for many ASP programs. Lazy-grounding avoids this grounding bottleneck but exhibits poor search performance. The approach here aims for the best of both worlds: grounding and solving are interleaved, but there is a solving component distinct from the grounding component. The solving component works on (ground) nogoods, employs conflict-driven first-UIP learning and enables heuristics. Guessing is on atoms that represent applicable rules, atoms may be one of true, false, or must-be-true, and nogoods have a distinguished head literal. The lazy-grounding component is loosely coupled to the solver and may yield more ground instances than necessary, which avoids re-grounding whenever the solver moves from one search branch to another. The approach is implemented in the new ASP solver Alpha.

1 Introduction

This work presents an approach at blending lazy-grounding and search procedures based on conflict-driven nogood learning (CDNL) for Answer-Set Programming (ASP). The most efficient state-of-the-art ASP solvers (e.g., Clasp [7] or Wasp [1]) are two-phased: first, the input program is grounded, then the answer-sets of the resulting propositional ground program are searched using advanced search techniques like CDNL, showing superior search performance. Since the pre-grounding increases the program size exponentially, however, there are many real-world applications where ASP cannot be used due to this so-called *grounding bottleneck* (cf. [2]). Intelligent grounding [3] mitigates this issue by excluding those parts of the grounding that, in advance, can be determined to be unnecessary, constraint ASP approaches (cf. [10,16]) try escaping the bottleneck by adding constraint programming techniques, while incremental ASP [6] gradually expands certain parts of the program step-by-step. They are all inherently restricted, however, so users of ASP encounter the grounding bottleneck quickly.

This research has been funded by the Austrian Science Fund (FWF): P27730.

M. Balduccini and T. Janhunen (Eds.): LPNMR 2017, LNAI 10377, pp. 191–204, 2017.
DOI: 10.1007/978-3-319-61660-5_17

Lazy-grounding on the other hand promises to avoid the grounding bottle-neck in general by interleaving grounding and solving, i.e., rules of an ASP program are grounded during solving and only when needed. Indeed, lazy-grounding ASP systems like Gasp [15], Asperix [11–13], or Omiga [4,17] show little trouble with grounding size, but their search performance is quite bad compared to pre-grounding ASP systems. Lazy-grounding ASP systems do not profit from (up-to exponential) search-space reducing techniques like conflict-driven learning and backjumping, since lazy-grounding ASP solving differs significantly from CDNL-based ASP solving in the following ways.

Lazy-grounding is based on so-called computation sequences which correspond to sequences of firing rules. Consequently, guessing is on whether an applicable rule fires, while CDNL-based search guesses on whether an atom is *true* or *false*. Lazy-grounding solvers make use of *must-be-true* as a third truth value to distinguish whether an atom was derived by a rule firing or by a constraint. CDNL-based solvers, on the other hand, have no equivalent of *must-be-true* and propagation on nogoods must be adapted. Furthermore, ordinary nogoods (i.e., the input to CDNL-based ASP solvers) cannot distinguish between a rule and a constraint. Therefore, combining lazy-grounding with CDNL-based search is challenging and necessitates intricate adaptions in the search procedures as well as on the underlying data structures.

The contributions of this work are:

- an extension of nogoods to faithfully represent rules in the presence of *must-be-true*, allowing choices to correspond to the firing of applicable rules.
- lazy-grounding that is only loosely coupled to the search state in order to avoid re-grounding of the same rules in different parts of the search space.
- a CDNL-based algorithm, *AlphaASP*, employing lazy-grounding, conflict-driven learning following the first-UIP schema, backjumping, and heuristics.
- an implementation, the Alpha ASP system, combining the best of both worlds: lazy-grounding to avoid the grounding bottleneck and CDNL-based search procedures for good search performance.
- finally, by the above it also works out the differences between lazy-grounding and CDNL-based search and demonstrates their commonalities.

The remainder of this work is structured as follows. In Sect. 2 preliminary concepts are introduced and in Sect. 3 the novel nogood representation for rules is presented. Section 4 describes the loosely-coupled lazy-grounding and Sect. 5 presents the Alpha algorithm blending lazy-grounding and CDNL search. Discussion is in Sects. 6 and 7 concludes.

2   Preliminaries

Let $\mathcal{C}$ be a finite set of constants, $\mathcal{V}$ be a set of variables, and $\mathcal{P}$ be a finite set of predicates with associated arities, i.e., elements of $\mathcal{P}$ are of the form p/k where p is the predicate name and k its arity. We assume each predicate name occurs only with one arity. The set $\mathcal{A}$ of (non-ground) atoms is then given by

$\{p(t_1, \ldots, t_n) \mid p/n \in \mathcal{P}, t_1, \ldots, t_n \in \mathcal{C} \cup \mathcal{V}\}$. An atom $at \in \mathcal{A}$ is ground if no variable occurs in at and the set of variables occurring in at is denoted by $var(at)$. The set of all ground atoms is denoted by $\mathcal{A}_{grd}$. A (normal) rule is of the form:

$$at_0 \leftarrow at_1, \ldots, at_k, not\ at_{k+1}, \ldots, not\ at_n.$$

where each $at_i \in \mathcal{A}$ is an atom, for $0 \le i \le n$. For such a rule r the head, positive body, negative body, and body are defined as $H(r) = \{at_0\}$, $B^+(r) = \{at_1, \ldots, at_k\}$, $B^-(r) = \{at_{k+1}, \ldots, at_n\}$, and $B(r) = \{at_1, \ldots, at_n\}$, respectively. A rule r is a constraint if $H(r) = \emptyset$, a fact if $B(r) = \emptyset$, and ground if each $at \in B(r) \cup H(r)$ is ground. The variables occurring in r are given by $var(r) = \bigcup_{at \in H(r) \cup B(r)} var(at)$. A literal l is positive if $l \in \mathcal{A}$, otherwise it is negative. A rule r is *safe* if all variables occurring in r also occur in its positive body, i.e., $var(r) \subseteq \bigcup_{a \in B^+(r)} var(a)$.

A program P is a finite set of safe rules. P is ground if each $r \in P$ is. A (Herbrand) interpretation I is a subset of the Herbrand base wrt. P, i.e., $I \subseteq \mathcal{A}_{grd}$. An interpretation I satisfies a literal l, denoted $I \models l$ if $l \in I$ for positive l and $l \notin I$ for negative l. I satisfies a ground rule r, denoted $I \models r$ if $B^+(r) \subseteq I \wedge B^-(r) \cap I = \emptyset$ implies $H(r) \subseteq I$ and $H(r) \neq \emptyset$. Given an interpretation I and a ground program P, the FLP-reduct P^I of P wrt. I is the set of rules $r \in P$ whose body is satisfied by I, i.e., $P^I = \{r \in P \mid B^+(r) \subseteq I \wedge B^-(r) \cap I = \emptyset\}$. I is an *answer-set* of a ground program P if I is the subset-minimal model of P^I; the set of all answer-sets of P is denoted by $AS(P)$.

A substitution $\sigma : \mathcal{V} \to \mathcal{C}$ is a mapping of variables to constants. Given an atom at the result of applying a substitution σ to at is denoted by $at\sigma$; this is extended in the usual way to rules r, i.e., $r\sigma$ for a rule of the above form is $at_0\sigma \leftarrow at_1\sigma, \ldots, not\ at_n\sigma$. Then, the grounding of a rule is given by $grd(r) = \{r\sigma \mid \sigma$ is a substitution for all $v \in var(r)\}$ and the grounding $grd(P)$ of a program P is given by $grd(P) = \bigcup_{r \in P} grd(r)$. Elements of $grd(P)$ and $grd(r)$ are called ground instances of P and r, respectively. The answer-sets of a non-ground program P are given by the answer-sets of $grd(P)$.

CDNL-based ASP solving takes a ground program, translates it into nogoods and then runs a SAT-inspired (i.e., a DPLL-style) model building algorithm to find a solution for the set of nogoods. A nogood is comprised of Boolean signed literals, which are of the form $\mathbf{T}at$ and $\mathbf{F}at$ for $at \in \mathcal{A}$; a nogood $ng = \{s_1, \ldots, s_n\}$ is a set of Boolean signed literals s_i, $1 \le i \le n$, which intuitively states that a solution cannot satisfy all literals s_1 to s_n. Nogoods are interpreted over assignments, which are sets A of Boolean signed literals, i.e., an assignment is a (partial) interpretation where false atoms are represented explicitly. A solution for a set Δ of nogoods then is an assignment A such that $\{at \mid \mathbf{T}at \in A\} \cap \{at \mid \mathbf{F}at \in A\} = \emptyset$, $\{at \mid \mathbf{T}at \in A\} \cup \{at \mid \mathbf{F}at \in A\} = \mathcal{A}$, and no nogood $ng \in \Delta$ is violated, i.e., $ng \not\subseteq A$. A solution thus corresponds one-to-one to an interpretation that is a model of all nogoods. For more details and algorithms, see [7–9]. Also note that CDNL-based solvers for ASP employ additional checks to ensure that the constructed model is supported and unfounded-free, but these checks are not necessary in the approach presented.

Lazy-grounding, also called grounding on-the-fly, is built on the idea of a computation, which is a sequence $(A_0, \ldots, A_\infty)$ of assignments starting with the empty set and adding at each step the heads of applicable rules (cf. [4,11,15]). A ground rule r is *applicable* in a step A_k, if its positive body already has been derived and its negative body is not contradicted, i.e., $B^+(r) \subseteq A_k$ and $B^-(r) \cap A_k = \emptyset$. A computation $(A_0, \ldots, A_\infty)$ has to satisfy the following conditions besides $A_0 = \emptyset$, given the usual immediate-consequences operator T_P:

1. $\forall i \geq 1 : A_i \subseteq T_P(A_{i-1})$ (the computation contains only consequences),
2. $\forall i \geq 1 : A_{i-1} \subseteq A_i$ (the computation is monotonic),
3. $A_\infty = \bigcup_{i=0}^\infty A_i = T_P(A_\infty)$ (the computation converges), and
4. $\forall i \geq 1 : \forall at \in A_i \setminus A_{i-1}, \exists r \in P$ such that $H(r) = at$ and $\forall j \geq i - 1 :$ $B^+(r) \subseteq A_j \wedge B^-(r) \cap A_j = \emptyset$ (applicability of rules is persistent through the computation).

It has been shown that A is an *answer-set* of a normal logic program P iff there is a computation $(A_0, \ldots, A_\infty)$ for P such that $A = A_\infty$ [12]. Observe that A is finite, i.e., $A_\infty = A_n$ for some $n \in \mathbb{N}$, because $\mathcal{C}, \mathcal{P}$, and P are finite.

3   Nogood Representation of Rules

The goal of the Alpha approach is to combine lazy-grounding with CDNL-search in order to obtain an ASP solver that avoids the grounding bottleneck and shows very good search performance. The techniques of CDNL-based ASP solvers require a fully grounded input, i.e., their most basic ingredients are nogoods over ground literals. Alpha provides this by having a grounder component responsible for generating ground nogoods, and a solving component executing the CDNL-search. This separation is common for CDNL-based ASP solvers but differently from these, the grounding component in Alpha is not just used once but it is called every time after the solver has finished deriving new truth assignments. Hence, there is a cyclic interplay between the solving component and the grounding component. Different from CDNL-solving, however, the result of this interplay is a computation sequence in the style of lazy grounding. Most importantly, the solver does not guess on each atom whether it is *true* or *false*, but it guesses on ground instances of rules whether they fire or not.

Taking inspiration from other ASP solvers (e.g. DLV [14], Asperix, or Omiga), Alpha introduces *must-be-true* as a possible truth value for atoms. If an atom a is *must-be-true*, denoted $\mathbf{M}a$, then a must be true but there currently is no rule firing with a in its head. This allows to distinguish the case that a is true due to a constraint, from the case that a is derived because a rule fired. In Alpha an atom a therefore may be assigned one of *must-be-true*, *true*, or *false*, denoted by $\mathbf{M}$, $\mathbf{T}$, or $\mathbf{F}$, respectively. Nogoods however, are still over Boolean signed literals, i.e., only literals of form $\mathbf{T}a$ and $\mathbf{F}a$ occur in nogoods. For convenience, the complement of such a literal s is denoted by $\overline{s}$, formally $\overline{\mathbf{T}a} = \mathbf{F}a$ and $\overline{\mathbf{F}a} = \mathbf{T}a$. In order to capture propagation to $\mathbf{T}a$, nogoods may have a specifically indicated head. Usually a nogood propagates to $\mathbf{M}a$ or $\mathbf{F}a$, but a nogood with head may propagate to $\mathbf{T}a$ if certain conditions are met. Formally:

Definition 1. *A* nogood with head *is a nogood* $ng = \{s_1, \ldots, s_n\}$ *with one distinguished negative literal* s_i *such that* $s_i = \mathbf{F}a$ *for some* $a \in \mathcal{A}$ *and* $1 \leq i \leq n$. *A nogood with head is written as* $\{s_1, \ldots, s_n\}_i$ *where* i *is the index of the head; we write* $hd(ng)$ *to denote the head of* ng, *i.e.,* $hd(ng) = s_i$.

In the remainder of this work, the term nogood may denote a nogood with head or an ordinary nogood (without head). Next, we formally define assignments over $\mathbf{M}$, $\mathbf{T}$, and $\mathbf{F}$.

Definition 2. *An* Alpha assignment, *in the following just* assignment, *is a sequence* $A = (s_1, \ldots, s_n)$ *of signed literals* s_i, $1 \leq i \leq n$, *over* $\mathbf{T}$, $\mathbf{M}$, *and* $\mathbf{F}$.

In slight abuse of notation, we consider an assignment A to be a set, i.e., $s \in A$ for $A = (s_1, \ldots, s_n)$ holds if there exists $1 \leq i \leq n$ such that $s = s_i$. Furthermore, let $A^\mathcal{B}$ denote the *Boolean projection* of A defined as:

$$A^\mathcal{B} = \{\mathbf{T}a \mid \mathbf{T}a \in A \text{ or } \mathbf{M}a \in A\} \cup \{\mathbf{F}a \mid \mathbf{F}a \in A\}.$$

An assignment A is *consistent* if for every atom $a \in \mathcal{A}$ the following holds:

(i) $\mathbf{T}a \in A^\mathcal{B}$ implies $\mathbf{F}a \notin A$ and $\mathbf{F}a \in A$ implies $\mathbf{T}a \notin A^\mathcal{B}$ (*false* and *true* exclude each other),
(ii) if $\mathbf{M}a \in A$ and $\mathbf{T}a \in A$ with $\mathbf{M}a = s_i$ and $\mathbf{T}a = s_j$ for $1 \leq i, j \leq n$, then $i < j$ (*must-be-true* precedes *true*), and
(iii) if $\mathbf{X}a \in A$ for any $\mathbf{X} \in \{\mathbf{M}, \mathbf{T}, \mathbf{F}\}$ then there exists exactly one $1 \leq i \leq n$ with $\mathbf{X}a = s_i$ (every occurrence is unique).

Due to the nature of lazy-grounding, A needs not be complete with respect to $\mathcal{A}$, i.e., there may be $a \in \mathcal{A}$ such that $\mathbf{X}a \notin A$ holds for any $\mathbf{X} \in \{\mathbf{T}, \mathbf{F}, \mathbf{M}\}$. In the remainder of this work only consistent assignments are considered.

Example 1. Consider the assignment $A = (\mathbf{F}a, \mathbf{M}b, \mathbf{M}c, \mathbf{T}b)$. A is consistent and $A^\mathcal{B} = \{\mathbf{F}a, \mathbf{T}b, \mathbf{T}c\}$. The assignment $B = (\mathbf{M}b, \mathbf{F}b, \mathbf{T}a, \mathbf{M}a, \mathbf{F}b)$, however, is inconsistent, because it violates every single condition for consistency.

In order to establish unit-propagation on nogoods which propagates to *true*, *false*, or *must-be-true*, we use two different notions of being unit.

Definition 3. *Given a nogood* $ng = \{s_1, \ldots, s_n\}$ *and an assignment* A:

- ng *is* weakly-unit *under* A *for* s *if* $ng \setminus A^\mathcal{B} = \{s\}$ *and* $\overline{s} \notin A^\mathcal{B}$,
- ng *is* strongly-unit *under* A *for* s *if* ng *is a nogood with head,* $ng \setminus A = \{s\}$, $s = hd(ng)$, *and* $\overline{s} \notin A$.

If a nogood is strongly-unit, it also is weakly-unit but not the other way round.

Let ng be a nogood and A be an assignment, ng is *satisfied* by A if there exists $s \in ng$ with $\overline{s} \in A^\mathcal{B}$ and additionally for ng being a nogood with head $s = hd(ng)$, it holds that ng being strongly-unit under $A \setminus \{\overline{s}\}$ implies that $\overline{s} \in A$. The nogood ng is *violated* by the assignment A if $ng \subseteq A^\mathcal{B}$.

Definition 4. *An assignment A is a* solution *to a set Δ of nogoods if for every $ng \in \Delta$ it holds that ng is satisfied by A.*

Example 2. Let $n_1 = \{\mathbf{F}a, \mathbf{T}b, \mathbf{F}c\}_1$ and $n_2 = \{\mathbf{F}c, \mathbf{T}d\}$ be nogoods and $A = (\mathbf{M}b, \mathbf{F}c, \mathbf{F}d)$, $A' = (\mathbf{T}b, \mathbf{F}c, \mathbf{T}d)$ be assignments. Then, n_1 is weakly-unit under A but not strongly-unit, intuitively because b is assigned *must-be-true* and not *true*. However, n_1 is strongly-unit under A'. The nogood n_2 is satisfied by A while it is violated by A'.

In order to represent bodies of whole ground rules, Alpha introduces atoms representing such, similar as Clasp. Given a rule r and a variable substitution σ, the atom representing the body of $r\sigma$ is denoted by $\beta(r, \sigma)$. For convenience, we assume $\beta(r, \sigma) \in \mathcal{A}$ and β not occurring in any program, i.e., $\beta(r, \sigma)$ is a fresh and unique ground atom.

Definition 5. *Given a rule r and a substitution σ such that $r\sigma$ is ground, let $B^+(r\sigma) = \{a_1, \ldots, a_k\}$, $B^-(r\sigma) = \{a_{k+1}, \ldots, a_n\}$ and $H(r) = \{a_0\}$, the nogood representation $ng(r)$ of r is the following set of nogoods:*

$$ng(r) = \big\{\{\mathbf{F}\beta(r, \sigma), \mathbf{T}a_1, \ldots, \mathbf{T}a_k, \mathbf{F}a_{k+1}, \ldots, \mathbf{F}a_n\}_1, \{\mathbf{F}a_0, \mathbf{T}\beta(r, \sigma)\}_1,$$
$$\{\mathbf{T}\beta(r, \sigma), \mathbf{F}a_1\}, \ldots, \{\mathbf{T}\beta(r, \sigma), \mathbf{F}a_k\}$$
$$\{\mathbf{T}\beta(r, \sigma), \mathbf{T}a_{k+1}\}, \ldots, \{\mathbf{T}\beta(r, \sigma), \mathbf{T}a_n\}\big\}$$

Intuitively, the first nogood of the nogood representation $ng(r)$ expresses that whenever the body of $r\sigma$ holds, so does the atom representing the body, the second nogood expresses that whenever this atom is *true* then the head also is *true*. Note that both nogoods have heads, indicating that $\beta(r, \sigma)$ for the first and a_0 for the second nogood may be assigned to $\mathbf{T}$ upon unit-propagation. The nogoods of the last two rows express that $\beta(r, \sigma)$ is *true* only if the body of $r\sigma$ is satisfied. It is easy to see that $\beta(r, \sigma)$ is *true* iff the body of $r\sigma$ holds, formally:

Proposition 1. *Let r be a rule, σ be a substitution such that $r\sigma$ is ground, and let A be a consistent assignment such that A is a solution to $ng(r\sigma)$. Then, $\mathbf{T}\beta(r, \sigma) \in A$ iff $B^+(r\sigma) \subset A \wedge B^-(r\sigma) \cap A = \emptyset$.*

Note that $ng(r\sigma)$ contains no nogoods establishing support, because it is in general hard to determine the set of all ground rules having the same head; e.g. in $p(X) \leftarrow q(X, Y)$. the head $p(a)$ may be derived by many ground rules.

The following proposition shows that the nogood representation is correct, i.e., if an assignment, free of *must-be-true*, is a solution to $ng(r\sigma)$ then the corresponding Boolean interpretation satisfies the ground rule $r\sigma$.

Proposition 2. *Given a consistent assignment A with $A = A^{\mathcal{B}}$, a rule r, and a substitution σ s.t. $r\sigma$ is ground. Then, $A \models r\sigma$ iff A is a solution to $ng(r\sigma)$.*

Since the solver component of Alpha does not guess on every atom, but only on those representing applicable ground rules, i.e., it guesses on $\beta(r, \sigma)$ if $r\sigma$ is applicable under the current assignment, additional nogoods are added for the solver to detect this. Recall that a rule is applicable, if its positive body holds and its negative body is not contradicted, formally: $B^+(r\sigma) \subseteq A$ and $B^-(r\sigma) \cap A = \emptyset$. For a rule r' without a negative body, i.e., $B^-(r') = \emptyset$, no additional guessing is necessary, because r' is applicable only if r' fires, hence the rule cannot be guessed to not fire. Consequently, no choices need to be done for rules without negative body. For rules that may possibly serve as a choice point to guess on, the following nogoods are added.

Definition 6. *Given a ground rule r with $B^-(r) \neq \emptyset$, and a substitution σ such that $r\sigma$ is ground. Let $B^+(r\sigma) = \{a_1, \dots, a_k\}$, $B^-(r\sigma) = \{a_{k+1}, \dots, a_n\}$, and let $cOn(r, \sigma)$ and $cOff(r, \sigma)$ be fresh ground atoms not occurring in any input program. Then, the set $ng_{ch}(r\sigma)$ of choice nogoods is:*

$$ng_{ch}(r\sigma) = \{\{\mathbf{F}cOn(r, \sigma), \mathbf{T}a_1, \dots, \mathbf{T}a_k\}_1,$$
$$\{\mathbf{F}cOff(r, \sigma), \mathbf{T}a_{k+1}\}, \dots, \{\mathbf{F}cOff(r, \sigma), \mathbf{T}a_n\}\}$$

The choice nogoods encode applicability of the ground rule $r\sigma$ as follows: if the positive body holds, $cOn(r, \sigma)$ becomes *true* due to the first nogood, and if one of the negative body atoms becomes *true* or *must-be-true*, i.e., the rule is no longer applicable, then $cOff(r, \sigma)$ becomes *true* or *must-be-true*. The solver can then identify valid choices, i.e., applicable ground rules, by checking that for $\beta(r, \sigma)$, $cOn(r, \sigma)$ is *true* and $cOff(r, \sigma)$ does not hold (is *false* or unassigned).

Proposition 3. *Let r be a rule with $B^-(r) \neq \emptyset$, σ be a substitution such that $r\sigma$ is ground, and let A be an assignment such that A is a solution to $ng_{ch}(r, \sigma)$. Then, $r\sigma$ is applicable under $\{a \in A \mid \mathbf{T}a \in A\}$ iff $\mathbf{T}cOn(r, \sigma) \in A$, $\mathbf{M}cOff(r, \sigma) \notin A$, and $\mathbf{T}cOff(r, \sigma) \notin A$.*

Example 3. Consider the rule $r = p(X) \leftarrow q(X), not\, s(X)$. with substitution $\sigma : X \mapsto d$. Then, $ng_{ch}(r, \sigma) = \{\{\mathbf{F}cOn(r, \sigma), \mathbf{T}q(d)\}_1, \{\mathbf{F}cOff(r, \sigma), \mathbf{T}s(d)\}\}$. The assignment $A = (\mathbf{T}q(d))$ is no solution to $ng_{ch}(r, \sigma)$ but its extension $A' = (\mathbf{T}q(d), \mathbf{T}cOn(r, \sigma), \mathbf{F}s(d))$ is and $r\sigma$ is applicable in A', which is also reflected by $\mathbf{T}cOn(r, \sigma) \in A'$, $\mathbf{M}cOff(r, \sigma) \notin A'$, and $\mathbf{T}cOff(r, \sigma) \notin A'$ all holding.

4   Loosely Coupled Lazy-Grounding

How the lazy grounder generates the nogoods presented in the previous section is shown next. The main task of the lazy grounder is to construct substitutions σ for all rules $r \in P$. Necessary are not all possible substitutions, but only those that yield ground rules potentially applicable under the current assignment.

Definition 7. *Given an assignment A, a substitution σ for a rule $r \in P$ with $r\sigma$ being ground is of interest if $B^+(r\sigma) \subseteq A^\mathcal{B}$.*

It is easy to see that substitutions not of interest can be ignored for the computation of answer-sets, since they do not fire and do not lead to any guesses:

Algorithm 1. *lazyGround*

Input: A program P, an assignment A, and a state S.
Output: A set Δ of ground nogoods for substitutions of interest wrt. A.

$A_{new} \leftarrow \{a \mid \mathbf{T}a \in A^{\mathcal{B}}\} \setminus S$.
$S \leftarrow S \cup A_{new}$
$R_{subst} \leftarrow \{r\sigma \mid r \in P, r\sigma \text{ is ground}, B^+(r\sigma) \subseteq S, \text{ and } B^+(r\sigma) \cap A_{new} \neq \emptyset\}$ (a)
foreach $r\sigma \in R_{subst}$ **do**
$\quad \lfloor \quad \Delta \leftarrow \Delta \cup ng(r\sigma) \cup ng_{ch}(r\sigma)$ (b)
return Δ

Proposition 4. *Given an assignment A, a rule $r \in P$, and a substitution σ not of interest wrt. A. Then, $r\sigma$ is not applicable and it does not fire under $A^{\mathcal{B}}$.*

The lazy grounder therefore only needs to find substitutions of interest.

Definition 8. *Let P be a (non-ground) program and A be an assignment, the lazy-grounding wrt. A, denoted by $lgr_P(A)$, is the set of all ground rules that are of interest wrt. A. Formally, $lgr_P(A) = \{r\sigma \mid r \in P, B^+(r\sigma) \subseteq \{a \mid \mathbf{T}a \in A^{\mathcal{B}}\}\}$.*

Since each rule $r \in P$ is safe, $B^+(r\sigma) \subseteq \{a \mid \mathbf{T}a \in A^{\mathcal{B}}\}$ implies that $B^-(r\sigma)$ is ground. Consequently, all $r \in lgr_P(A)$ are ground.

Example 4. Consider the program $P = \{b \leftarrow not\,a. \quad p(X) \leftarrow q(X), not\,a.\}$ and the assignment $A = (\mathbf{M}q(c), \mathbf{F}q(d), \mathbf{T}q(e))$. The lazy-grounding wrt. A is $lgr_P(A) = \{b \leftarrow not\,a. \quad p(c) \leftarrow q(c), not\,a. \quad p(e) \leftarrow q(e), not\,a.\}$.

In Alpha the grounder is only loosely coupled with the state of the solver, i.e., the grounder returns all nogoods of interest, but it may report more nogoods. Although this looks inefficient at first, it can avoid a lot of re-grounding. In other lazy-grounding approaches the grounder is strongly bound to the current branch of the search and whenever the solver changes to another branch, all previously grounded rules are eliminated. In Alpha the lazy-grounder does not need to eliminate ground instances and therefore can re-use those ground instances for any future branch explored by the solver. Accumulating already derived ground substitutions therefore leads to an increasing number of substitutions of interest that need not be computed again. In the extreme case, the complete grounding is derived early on, leaving no work for the lazy-grounder.

In order to construct $lgr_P(A)$, the lazy-grounder maintains a state $S \subseteq \mathcal{A}_{grd}$, which is an accumulator of positive assignments previously seen. Substitutions of interest, which have not been computed before, yield at least one positive ground atom contained in $A \setminus S$, i.e., with increasing S the potentially relevant substitutions decrease. The lazy-grounding procedure is given by Algorithm 1:

- In (a) all relevant substitutions are computed. The restriction to substitutions $r\sigma$ with $B^+(r\sigma) \cap A_{new} \neq \emptyset$ enables the grounder to consider only rules that are relevant and this is a starting point for the construction of σ.

– In (b) the corresponding nogoods to be returned in the last step are generated. A semi-naive grounding algorithm is used here in the implementation.

Proposition 5. *For a program P and an assignment A, Algorithm 1 computes all ground rules of interest wrt. A, i.e., $lazyGround_P(A) = lgr_P(A)$ for a state $S = \emptyset$. Moreover, consecutive calls are such that: $\bigcup lazyGround_P(A) \subseteq lgr_P(A)$.*

Note that the accumulation of ground substitutions may exhaust all available memory. In such a situation, Alpha may simply forget all derived substitutions, except those in the current branch of the search space. Likewise, the solver can also forget previously obtained nogoods and thus free memory again.

5   The Alpha Approach

Based on the concepts above, Alpha blends lazy-grounding with CDNL search. At a glance, Alpha is a modified CDNL search algorithm using the lazy-grounder of the previous section and assignments with *must-be-true* in combination with nogoods with heads. The main algorithm is given by Algorithm 2, like for CDNL-based solvers it is comprised of one big loop where the search space is explored.

At a glance, the algorithm is as follows. After initialization, nogoods resulting from facts are requested from the grounder in (a) and the main loop is entered:

– First, unit propagation is applied to the known nogoods potentially extending the current assignment in (b) and then each iteration does one of:
– analyzing and learning from a conflict in (c),
– querying the lazy-grounding component for more nogoods in (d),
– introducing a choice in (e),
– assigning all unassigned atoms to false after the cycle of propagation, choice, and lazy-grounding reached a fixpoint in (f),
– reporting a found answer-set if it is free of *must-be-true* in (g), and finally
– backtracking if some assignment to *must-be-true* remained and the search branch thus yields no answer-set in (h).

For lazy-grounding, the function *lazyGround* of the previous section is used; for readability, however, the state S is assumed to be implicitly given. The *propagate* function applies unit-resolution to compute all inferences from the current assignment A and the set of nogoods Δ. Formally, let the immediate unit-propagation be given by:

$$\Gamma_\Delta(A) = A \cup \{\mathbf{T}a \mid \exists \delta \in \Delta, \delta \text{ is strongly-unit under } A \text{ for } s = \mathbf{F}a\}$$
$$\cup \{\mathbf{M}a \mid \exists \delta \in \Delta, \delta \text{ is weakly-unit under } A \text{ for } s = \mathbf{F}a\}$$
$$\cup \{\mathbf{F}a \mid \exists \delta \in \Delta, \delta \text{ is weakly-unit under } A \text{ for } s = \mathbf{T}a\}$$

Then, $propagate(\Delta, A) = (A', d)$ where $A' = lfp(\Gamma_\Delta(A))$ is the least fixpoint of the immediate unit-propagation and $d = 1$ if $A' \neq A$ and $d = 0$ otherwise indicates whether some new assignment has been derived by the propagation. For space reasons, the algorithm computing $propagate(\Delta, A)$ is omitted.

Algorithm 2. *AlphaASP*

Input: A (non-ground) normal logic program P.
Output: The set $AS(P)$ of all answer sets of P.

$AS \leftarrow \emptyset$	// Set of found answer-sets.
$A \leftarrow \emptyset$	// Current assignment.
$\Delta \leftarrow \emptyset$	// Set of nogoods.
$dl \leftarrow 0$	// Current decision level.
$exhausted \leftarrow 0$	// Stop condition.

$\Delta \leftarrow lazyGround_P(A)$ (a)
while $exhausted = 0$ **do**
   $(A', didPropagate) \leftarrow propagate(\Delta, A)$ (b)
   $A \leftarrow A'$
   if $\exists \delta \in \Delta : \delta \subseteq A$ **then**   (c)
      $(\delta_l, dl_{bj}) \leftarrow analyze(\delta, \Delta, A)$
      $A \leftarrow backjump(dl_{bj})$
      $dl \leftarrow dl_{bj}$
      $\Delta \leftarrow \Delta \cup \{\delta_l\}$
   else if $didPropagate = 1$ **then**   (d)
      $\Delta \leftarrow \Delta \cup lazyGround_P(A)$
   else if $acp(\Delta, A) \neq \emptyset$ **then**   (e)
      $dl \leftarrow dl + 1$
      $s \leftarrow select(acp(\Delta, A))$
      $A \leftarrow A \cup \{s\}$
   else if $atoms(\Delta) \setminus atoms(A) \neq \emptyset$ **then**   (f)
      $A \leftarrow A \cup \{\mathbf{F}a \mid a \in atoms(\Delta) \setminus atoms(A)\}$
   else if $\{\mathbf{M}a \in A\} = \emptyset$ **then**   (g)
      $AS \leftarrow AS \cup \{A\}$
      $\delta_{enum} \leftarrow enumNg(A)$
      $\Delta \leftarrow \Delta \cup \{\delta_{enum}\}$
      $A \leftarrow backtrack()$
      $dl \leftarrow dl - 1$
   else   (h)
      $A \leftarrow backtrack()$
      $dl \leftarrow dl - 1$
      if $dl = 0$ **then**
         $exhausted \leftarrow 1$

return AS

In (c) *AlphaASP* checks for a conflict caused by some nogood δ being violated in A. If so, $analyze(\delta, \Delta, A)$ analyzes the conflict employing ordinary conflict-driven nogood learning following the first-UIP schema. On an abstract level, the process applies resolution on certain nogoods from Δ, starting at δ and resolving them until a stop criterion is reached. The nogoods to resolve are obtained from the so-called implication graph, storing information about which assignments in A stem from which nogood by unit-resolution done during an earlier propagation.

The so-called first-UIP is used to stop, which is a certain graph dominator in the implication graph. First-UIP nogood learning is a well-established technique used here basically unaltered, for details see [9]. Backjumping is done by $backjump(dl_{bj})$ which removes from A all entries of a decision $dl > dl_{bj}$ higher than the decision level to backjump to and returns the resulting assignment; entries of the decision level dl_{bj} are kept.

In (d) new nogoods are requested from the grounder using *lazyGround*, given that the previous propagation derived new assignments. In the implementation, only the recently added assignments of A have to be transferred to the lazy-grounding component, since it has its internal state S and reports only those nogoods from new ground instances of a rule.

In (e) guessing takes place. For this, a list of active choice points is maintained, where each such choice point corresponds to an applicable rule. Formally, let the *atoms occurring* in a set Δ of nogoods be given by $atoms(\Delta) = \{a \mid \mathbf{X}a \in \delta, \delta \in \Delta, \mathbf{X} \in \{\mathbf{T}, \mathbf{F}\}\}$ and let the atoms occurring in an assignment A be $atoms(A) = \{a \mid \mathbf{X}a \in A, \mathbf{X} \in \{\mathbf{T}, \mathbf{F}, \mathbf{M}\}\}$. The set of *active choice points* wrt. a set of nogoods Δ and an assignment A is: $acp(\Delta, A) = \{\beta(r, \sigma) \in atoms(\Delta) \mid \mathbf{T}cOn(r, \sigma) \in A \wedge \mathbf{T}cOff(r, \sigma) \notin A \wedge \mathbf{M}cOff(r, \sigma) \notin A\}$. The *select* function simply takes an element from the set of active choice points. Heuristics can be employed to select a good guessing candidate.

When (f) is reached, the interplay between propagation, grounding, and guessing has reached a fixpoint, i.e., there are no more applicable ground rule instances and nothing can be derived by propagation or from further grounding. Since the guessing does not guess on all atoms, there may be atoms in A not having an assigned truth value. Due to the fixpoint, these atoms must be *false* and the algorithm assigns all to *false*. The propagation at the following iteration then guarantees that (potentially) violated nogoods are detected.

In (g) the solver checks whether there is an atom assigned to *must-be-true* but could not be derived to be *true*, i.e., no rule with the atom as its head fired. If there is an atom assigned *must-be-true*, then the assignment A is no answer set. If no such atom exists, then A is an answer-set and it is recorded. Furthermore, the function *enumNg(A)* generates a nogood that excludes the current assignment. This is necessary to avoid finding the same answer-set twice. Following [8], this enumeration nogood only needs to contain the decision literals of A, formally:

$$enumNg(A) = \{\mathbf{X}a \in A \mid a = \beta(r, \sigma) \text{ for some } r, \sigma, \text{ and } \mathbf{X} \in \{\mathbf{T}, \mathbf{F}\}\}$$

In (h) backtracking is done by *backtrack*, which removes from the assignment A all entries done in the current decision level and returns an assignment A' containing only elements of A that have not been assigned in the last decision level. For readability, the details of this are not made explicit (i.e., the assignment associating to each literal a decision level on which it was assigned). If the current decision level dl reaches 0, the search space exhausted.

The *AlphaASP* algorithm computes the answer-sets of normal logic programs.

Theorem 1. *Given a normal logic program P, then $AlphaASP(P) = AS(P)$.*

Proof. Ignoring branches not leading to an answer-set, *AlphaASP* constructs a computation sequence $A_s = (A_0, \ldots, A_\infty)$ where the assignment A in *AlphaASP* at the i-th iteration of the main loop corresponds to the Boolean assignment A_i in A_s by $A_i = A \setminus \{\mathbf{M}a \mid a \in \mathcal{A}\}$, i.e., assignments to *must-be-true* are removed. First observe that at the last step of *AlphaASP* the assignment A is free of *must-be-true* because otherwise the algorithm would have backtracked in (g), i.e., $A_\infty = A$. The sequence $(A_0, \ldots, A_\infty)$ is a computation, because: (1) by *propagate* and *acp* only containing consequences, propagation and guessing only add consequences, hence $\forall i \geq 1 : A_i \subseteq T_P(A_{i-1})$. (2) the current assignment only increases, i.e., $\forall i \geq 1 : A_{i-1} \subseteq A_i$. (3) due to the main loop only stopping after the search space is completely explored, the computation converges, i.e., $A_\infty = \bigcup_{i=0}^\infty A_i = T_P(A_\infty)$. (4) To show that $\forall i \geq 1 : \forall at \in A_i \setminus A_{i-1}, \exists r \in P$ such that $H(r) = at$ and $\forall j \geq i-1 : B^+(r) \subseteq A_j \wedge B^-(r) \cap A_j = \emptyset$ holds, observe that: (a) Guessing is solely on auxiliary atoms $\beta(r, \sigma)$ s.t. $\beta(r, \sigma) \in acp(\Delta, A)$ holds, i.e., $\mathbf{T}cOn(r, \sigma) \in A$ and by $ng_{ch}(r\sigma)$ it holds for each $a \in B^+(r\sigma)$ that $\mathbf{T}a \in A$, i.e., if $\beta(r, \sigma)$ is an active choice point then its positive body is wholly *true*. Additionally, nogoods in $ng(r\sigma)$ ensure that all atoms in $B^-(r\sigma)$ are *false* once $\beta(r, \sigma)$ is *true*. (b) Non-auxiliary atoms a become *true* only by *propagate* if there is a nogood δ strongly-unit, i.e., δ has head atom a and all atoms positively occurring in δ are *true* in A. For each ground $r\sigma$, there is exactly one nogood with head in $ng(r\sigma)$, thus $H(r\sigma)$ is *true* iff $r\sigma$ fires under A. In summary, *AlphaASP* is sound. Completeness follows from *AlphaASP* exploring the whole search space except for those parts excluded by learned nogoods; since *analyze* learns by resolution, completeness follows. Consequently, $AlphaASP(P) = AS(P)$. $\qed$

6   Discussion

The Alpha approach is implemented in the Alpha system, which is freely available at: https://github.com/alpha-asp. Early benchmarks showed a much improved performance on non-trivial search problems (e.g. graph 3-colorability) compared to other lazy-grounding ASP systems. Detailed benchmark results can be found at www.kr.tuwien.ac.at/research/systems/alpha. In Alpha, support nogoods (constraints) are not added in general, because they are not necessary and constructing them is hard for rules like $p(X) \leftarrow q(X, Y)$ which may have infinitely many ground instances all deriving the same ground atom $p(a)$. The solver may now guess that all rules with $p(a)$ in the head do not fire while also deriving that $p(a)$ *must-be-true*, which cannot yield an answer-set. The solver detects this but only after all atoms are assigned. Eventual unrelated guesses then make the search exponential where for traditional ASP systems it is not. Alpha mitigates this by adding support nogoods for a very limited class of atoms.

Similar to Alpha, in [5] a computation sequence is constructed using nogoods and conflict-driven learning. The ASP program, however, is grounded upfront, hence traditional techniques for conflict-driven ASP solving can be used directly. Integrating lazy-grounding into [5] very likely necessitates similar techniques as presented here, because the established CDNL techniques require support nogoods and loop nogoods, which are in general hard to construct (see above).

7   Conclusion

This work presented the Alpha approach that combines lazy-grounding with CDNL-based search techniques to create a novel ASP solver that avoids the grounding bottleneck and provides good search performance. For that, nogoods with heads and assignments with *must-be-true* are introduced such that the *AlphaASP* algorithm constructs answer-sets similar to other lazy-grounding systems, i.e., it guesses whether a rule fires and not whether an atom is *true*. *AlphaASP* makes use of a number of techniques for efficient answer-set solving: conflict-driven nogood learning following the first-UIP schema, backjumping, and heuristics. The Alpha approach is implemented in the freely available Alpha system. Early benchmarks show it avoids the grounding bottleneck and provides good search performance. Topics for future work are forgetting of nogoods and a lifting of learning to the first-order case akin to [17].

References

1. Alviano, M., Dodaro, C., Faber, W., Leone, N., Ricca, F.: WASP: a native ASP solver based on constraint learning. In: Cabalar, P., Son, T.C. (eds.) LPNMR 2013. LNCS (LNAI), vol. 8148, pp. 54–66. Springer, Heidelberg (2013). doi:10.1007/978-3-642-40564-8_6

2. Balduccini, M., Lierler, Y., Schüller, P.: Prolog and ASP inference under one roof. In: Cabalar, P., Son, T.C. (eds.) LPNMR 2013. LNCS (LNAI), vol. 8148, pp. 148–160. Springer, Heidelberg (2013). doi:10.1007/978-3-642-40564-8_15

3. Calimeri, F., Fuscà, D., Perri, S., Zangari, J.: I-DLV: the new intelligent grounder of DLV. In: AI*IA, pp. 192–207 (2016)

4. Dao-Tran, M., Eiter, T., Fink, M., Weidinger, G., Weinzierl, A.: OMiGA : an open minded grounding on-the-fly answer set solver. In: Cerro, L.F., Herzig, A., Mengin, J. (eds.) JELIA 2012. LNCS (LNAI), vol. 7519, pp. 480–483. Springer, Heidelberg (2012). doi:10.1007/978-3-642-33353-8_38

5. Dovier, A., Formisano, A., Pontelli, E., Vella, F.: A GPU implementation of the ASP computation. In: Gavanelli, M., Reppy, J. (eds.) PADL 2016. LNCS, vol. 9585, pp. 30–47. Springer, Cham (2016). doi:10.1007/978-3-319-28228-2_3

6. Gebser, M., Kaminski, R., Kaufmann, B., Ostrowski, M., Schaub, T., Thiele, S.: Engineering an incremental ASP solver. In: Garcia de la Banda, M., Pontelli, E. (eds.) ICLP 2008. LNCS, vol. 5366, pp. 190–205. Springer, Heidelberg (2008). doi:10.1007/978-3-540-89982-2_23

7. Gebser, M., Kaufmann, B., Neumann, A., Schaub, T.: *clasp*: a conflict-driven answer set solver. In: Baral, C., Brewka, G., Schlipf, J. (eds.) LPNMR 2007. LNCS, vol. 4483, pp. 260–265. Springer, Heidelberg (2007). doi:10.1007/978-3-540-72200-7_23

8. Gebser, M., Kaufmann, B., Neumann, A., Schaub, T.: Conflict-driven answer set enumeration. In: Baral, C., Brewka, G., Schlipf, J. (eds.) LPNMR 2007. LNCS (LNAI), vol. 4483, pp. 136–148. Springer, Heidelberg (2007). doi:10.1007/978-3-540-72200-7_13

9. Gebser, M., Kaufmann, B., Schaub, T.: Conflict-driven answer set solving: from theory to practice. Artif. Intell. **187**, 52–89 (2012)

10. Gebser, M., Ostrowski, M., Schaub, T.: Constraint answer set solving. In: Hill, P.M., Warren, D.S. (eds.) ICLP 2009. LNCS, vol. 5649, pp. 235–249. Springer, Heidelberg (2009). doi:10.1007/978-3-642-02846-5_22

11. Lefèvre, C., Beatrix, C., Stephan, I., Garcia, L.: Asperix, a first-order forward chaining approach for answer set computing. TPLP **17**(3), 266–310 (2017)

12. Lefèvre, C., Nicolas, P.: A first order forward chaining approach for answer set computing. In: Erdem, E., Lin, F., Schaub, T. (eds.) LPNMR 2009. LNCS (LNAI), vol. 5753, pp. 196–208. Springer, Heidelberg (2009). doi:10.1007/978-3-642-04238-6_18

13. Lefèvre, C., Nicolas, P.: The first version of a new ASP solver : ASPeRiX. In: Erdem, E., Lin, F., Schaub, T. (eds.) LPNMR 2009. LNCS (LNAI), vol. 5753, pp. 522–527. Springer, Heidelberg (2009). doi:10.1007/978-3-642-04238-6_52

14. Leone, N., Pfeifer, G., Faber, W., Eiter, T., Gottlob, G., Perri, S., Scarcello, F.: The DLV system for knowledge representation and reasoning. ACM Trans. Comput. Logic **7**, 499–562 (2002)

15. Palù, A.D., Dovier, A., Pontelli, E., Rossi, G.: Gasp: answer set programming with lazy grounding. Fundam. Inform. **96**(3), 297–322 (2009)

16. Teppan, E.C., Friedrich, G.: Heuristic constraint answer set programming. In: ECAI. FAIA, vol. 285, pp. 1692–1693. IOS Press (2016)

17. Weinzierl, A.: Learning non-ground rules for answer-set solving. In: Grounding and Transformation for Theories with Variables, pp. 25–37 (2013)

Answer Set Programming with Graded Modality

Zhizheng Zhang$^{(\boxtimes)}$

School of Computer Science and Engineering,
Southeast University, Nanjing, China
seu_zzz@seu.edu.cn

Abstract. Answer set programming with graded modality (ASP^{GM}) introduced in [6] provides an intuitive way for expressing modal concepts "at least as many as......" as well as "at most as many as......", and defaults. This paper studies the semantics of ASP^{GM} and investigates its connection to the most recent language of epistemic specification (ASP^{KM}) proposed in [3] and the answer set programming with epistemic negation (ASP^{NOT}) presented in [5]. Particularly, we define a new approach to evaluating graded modalities in ASP^{GM} such that ASP^{GM} is compatible with ASP^{KM} as well as ASP^{NOT} at the semantic level.

Keywords: Answer set · Graded modality · Introspection

1 Introduction

Several languages have been developed by extending the languages of answer set programming (ASP) using epistemic operators. The need for such extensions was early addressed by Gelfond in [1], where Gelfond proposed an extension of ASP with two modal operators K and M and their negations (ASP^{KM}). Informally, K p expresses p *is true in all belief sets of the agent*, M p means p *is true in some belief sets of the agent*. Recently, there is increasing research to address the long-standing problems of unintended world views due to recursion through modalities [3,4,7]. Very recently, Shen and Eiter [5] introduced general logic programs possible containing epistemic negation NOT (ASP^{NOT}), and defined its world views by minimizing the knowledge. ASP^{NOT} can not only express K p and M p formulas by *not* NOT p and NOT *not* p, but also offer a solution to the problems of unintended world views.

Zhang etc., [6] recently introduced a new graded modal operator M$_{[lb:ub]}$ for the representation of graded introspections: M$_{[lb:ub]}p$ intuitively means: it is known that the number of belief sets where p is true is between lb and ub, and then presented an extension of answer set programming using graded modalities (ASP^{GM}), that is capable of reasoning that combines nonmonotonic reasoning and graded epistemic reasoning. However, the approach to handling graded modalities proposed in [6] cannot address the problems of unintended world views.

In this paper, our purpose is to address the problems of unintended world views in ASP^{GM}. Inspired by the work in [5] and [4] where *minimizing the*

© Springer International Publishing AG 2017
M. Balduccini and T. Janhunen (Eds.): LPNMR 2017, LNAI 10377, pp. 205–211, 2017.
DOI: 10.1007/978-3-319-61660-5_18

knowledge is seen as a potent way to eliminate unintended world views, we consider the principle of *minimizing the knowledge* for ASP^{GM}. We present a new semantics for ASP^{GM} where a method of minimizing the knowledge is proposed. Then, we investigate the relationship between the new ASP^{GM} and ASP^{KM}, and the relationship between the new ASP^{GM} and ASP^{NOT}, and the implementation of computing the solution for the new ASP^{GM} programs.

2   The Original ASP^{GM}

In this section, we review the original ASP^{GM} introduced in [6]. ASP^{GM} is an extension of ASP using graded modalities. An ASP^{GM} program Π is a finite collection of rules of the form

$$l_1 \ or \ ... \ or \ l_k \leftarrow e_1, ..., e_m, s_1, ..., s_n.$$

where $k \geq 0$, $m \geq 0$, $n \geq 0$, l_i is a literal in first order logic language and is called an objective literals here, e_i is an extended literal which is an objective literal possibly preceded by a negation as failure operator *not*, s_i is a subjective literal of the form $\text{M}_\omega e$ where e is an extended literal and M_ω is a modal operator where ω is of the form $[lb : ub]$ or $[lb :]$ where lb and ub are natural numbers satisfying $lb \leq ub$. We use $head(r)$ to denote the set of objective literals in the head of a rule r and $body(r)$ to denote the set of extended literals and subjective literals in the body of r. Sometimes, we use $head(r) \leftarrow body(r)$ to denote a rule r. The positive body of a rule r is composed of the extended literals containing no *not* in its body. We use $body^+(r)$ to denote the positive body of r, $El(\Pi)$ to denote the set of all subjective literals appearing in Π. r is said to be *safe* if each variable in it appears in the positive body of the rule.

We restrict our definition of the semantics to ground programs. In the following definitions, l is used to denote a ground objective literal, e is used to denote a ground extended literal with or without one *not*.

Let W be a non-empty collection of consistent sets of ground objective literals, $< W, w >$ is a pointed structure of W where $w \in W$. We call w as a belief set in W. W is a model of a program Π if for each rule r in Π, r is satisfied by every pointed structure of W. The notion of satisfiability is defined below.

- $< W, w > \models l$ if $l \in w$
- $< W, w > \models not\ l$ if $l \notin w$
- $< W, w > \models \text{M}_{[lb:ub]}e$ if $lb \leq |\{w \in W : < W, w > \models e\}| \leq ub$
- $< W, w > \models \text{M}_{[lb:]}e$ if $|\{w \in W : < W, w > \models e\}| \geq lb$

Then, for a rule r in Π, $< W, w > \models r$ if $\exists l \in head(r)$: $< W, w > \models l$ or $\exists t \in body(r)$: $< W, w > \not\models t$. The satisfiability of a subjective literal does not depend on a specific belief set w in W, hence we can simply write $W \models \text{M}_\omega e$ if $< W, w > \models \text{M}_\omega e$. Let Π be an arbitrary ASP^{GM}, and W is a non-empty collection of consistent sets of ground objective literals in the language of Π, we use Π^W to denote the disjunctive logic program obtained by removing graded modalities using the following reduct laws:

1. removing from Π all rules containing subjective literals not satisfied by W.
2. replacing all other occurrences of subjective literals of the form $M_{[lb:ub]}$ l or $M_{[lb:]}$ l where $lb = |W|$ with l.
3. removing all occurrences of subjective literals of the form $M_{[lb:ub]}$ $not\ l$ or $M_{[lb:]}$ $not\ l$ where $lb = |W|$.
4. replacing other occurrences of subjective literals of the form $M_{[0:0]}$ e with e^{not}.
5. replacing other occurrences of subjective literals of the form M_ω e with e and e^{not} respectively.

where e^{not} is l if e is $not\ l$, and e^{not} is $not\ l$ if e is l. Then, W is a world view of Π if $W = AS(\Pi^W)$. Π^W is said to be the *introspective reduct* of Π with respect to W.

Here we use an example to illustrated the unintended world views problems in the original ASP^{GM}.

Example 1. Consider a program containing four rules: $p \leftarrow M_{[1:]}$ $q, not\ q$ and $p \leftarrow not\ not\ q, not\ q$ and $q \leftarrow M_{[1:]}$ $p, not\ p$, and $q \leftarrow not\ not\ p, not\ p^1$. Intuitively, the first two rules tell: *if q may be true in some belief sets, then p must be true in the belief sets where q is not proved to be true*, and the last two rules tell:*if p may be true in some belief sets, then q must be true in the belief sets where p is not proved to be true*. It seems that $\{\{p\}, \{q\}\}$ should be the only intended world view of the program. However, by the above semantics of ASP^{GM}, this program has another world view $\{\{\}\}$.

3   New Semantics of ASP^{GM}

Based on the notion of satisfiability defined in the original ASP^{GM}, new introspective reduct rules are given. Then, a way of minimizing the knowledge in ASP^{GM} is proposed.

Definition 1. *Let Π be an arbitrary ASP^{GM} program, and W is a non-empty collection of consistent sets of ground objective literals in the language of Π, Π^W is said to be the **introspective reduct** of Π with respect to W if Π^W is obtained by the following laws*

1. *removing from Π all rules containing subjective literals not satisfied by W.*
2. *replacing other occurrences of subjective literals of the form $M_{[0:0]}$ e with e^{not}.*
3. *replacing other occurrences of subjective literals of the form M_ω e with e and e^{not} respectively.*

We say W is a candidate world view of Π if $W = AS(\Pi^W)$.

---

[1] Here, we view *not not l* as a representation of *not l'* where we have $l' \leftarrow not\ l$ and l' is a fresh literal. It is worthwhile to note that CLINGO [8] is able to deal with *not not*.

Definition 2. *A non-empty collection W of consistent sets of ground objective literals is said to be a world view of a ASP^{GM} program Π if it satisfies all the following conditions*

- $W \in cwv(\Pi)$.
- $\nexists V \in cwv(\Pi)(\forall M_\omega \ e \in El(\Pi)(grade_V(e) > grade_W(e)))$

where $grade_X(e) = |\{x \in X : (X, x) \models e\}|$, $cwv(\Pi)$ denotes the set of all candidate world views of Π.

In order to distinguish between the new semantics of ASP^{GM} and the original one, we call the *introspective reduct/candidate world view/world view* defined in this section *GM-introspective reduct/GM-candidate world view/GM-world view*.

Consider the program given in the Example 1 again, it is easy to verify that the program has a unique world view $\{\{p\}, \{q\}\}$ as we expect.

4   Relation to ASP^{KM}

In [3] where the most recent version of ASP^{KM} is defined, an ASP^{KM} program is a set of rules of the form $h_1 \ or \ ... \ or \ h_k \leftarrow b_1, ..., b_m$ where $k \geq 0$, $m \geq 0$, h_i is an objective literal, and b_i is an objective literal possible preceded by a negation as failure operator *not*, a modal operator K or M, or a combination operator *not* K or *not* M. For distinguishment, we call the world view of the ASP^{KM} program **KM-world view**.

Definition 3. *Given an ASP^{KM} program Ω, an ASP^{GM} program is called a KM-GM-Image of Ω, denoted by $KM - GM - I(\Omega)$, if it is obtained by*

- *Replacing occurrences of literals of the form K l in Ω by $M_{[0:0]}$ not l.*
- *Replacing occurrences of literals of the form M l in Ω by $M_{[1:]}$ l and not not l respectively.*
- *Replacing occurrences of literals of the form not K l in Ω by $M_{[1:]}$ not l and not l respectively.*
- *Replacing occurrences of literals of the form not M l in Ω by $M_{[0:0]}$ l.*

Theorem 1. *Let Ω be an ASP^{KM} program, and Π be the ES-GM-Image of Ω, and W be a non-empty collection of consistent sets of ground objective literals, W is a GM-candidate world view of Π iff W is a KM-world view of Ω.*

5   Relation to ASP^{NOT}

Here, we consider the ASP^{NOT} program that is a set of the rules of the form $l_1 \ or \ ... \ or \ l_k \leftarrow e_1, ..., e_m, s_1, ..., s_n$ where $k \geq 0$, $m \geq 0$, $n \geq 0$, l_i is an objective literal, e_i is an extended literal, s_i is a subjective literal of the form NOT e or *not* NOT e. For distinguishment, we call the world view of an ASP^{NOT} program **NOT-world view**.

Definition 4. *Given an ASP^{NOT} program Ω, an ASP^{GM} program is called a NOT-GM-Image of Ω, denoted by NOT-GM-I(Ω), if it is obtained by*

- *Replacing all occurrences of literals of the form not NOT e in Ω by $M_{[0:0]} e^{not}$.*
- *Replacing all occurrences of literals of the form NOT e in Ω by $M_{[1:]} e^{not}$ and not e respectively.*

Theorem 2. *Let Ω be an ASP^{NOT} program, and Π be the NOT-GM-Image of Ω, and W be a non-empty collection of consistent sets of ground objective literals, W is a GM-world view of Π iff W is a NOT-world view of Ω.*

6   An Algorithm for Computing World Views

For a given ASP^{GM} program Π, we use $CWV_i(\Pi)$ ($i \in \mathbb{N}$ and $i \geq 1$) to denote the set of candidate world views of Π which contain exactly i belief sets. Obviously, we have $CWV(\Pi) = \bigcup_{i \geq 1} CWV_i(\Pi)$. We use $CWV_{>i}(\Pi)$ to denote the set of candidate world views of Π which contain strictly more than i belief sets. Then we have $CWV(\Pi) = (\bigcup_{1 \leq i \leq k} CWV_i(\Pi)) \cup CWV_{>k}(\Pi)$ for any natural number k and $k \geq 1$. Based on this idea, at a high level of abstraction, the algorithm for computing world views of an ASP^{GM} program Π composed of safe rules is as showed in Algorithm 1. In Algorithm 1, **ASP^{GM} Solver** first computes a dividing line n. Then, **ASP^{GM} Solver** adopts a *FOR-LOOP* to compute all GM-candidate world views of size less than or equal to n for Π by calling a function *CWViSolver* that computes all candidate world views of size i for Π. After that, *CWVgiSolver* is called to compute all GM-candidate world views of size strict greater than n for Π. At last, *MAX_GM* is called to compute all world views by the principle of minimizing the knowledge in ASP^{GM}.

Algorithm 1. ASP^{GM} Solver.

Input:
   Π: A ASP^{GM} program;
Output:
   All GM-world views of Π;
1: $n = max\{lb : M_{[lb:ub]}e$ or $M_{[lb:]}e$ in $\Pi\}$ {computes the maximal lb of subjective literals}
2: $CWV = \emptyset$
3: **for** every natural number $1 \leq k \leq n$ **do**
4:     $CWV_k = $ CWViSolver(Π, k) {computes all world views of size k}
5:     $CWV = CWV \cup CWV_k$
6: **end for**
7: $CWV_{>n} = $CWVgiSolver($\Pi, n$) {computes all world views of size strict greater than n}
8: $CWV = CWV \cup CWV_{>n}$
9: $WV = MAX\_GM(CWV)$
10: output WV

For ease of description of *CWViSolver* and *CWVgiSolver*, $m\_C\_lb\_U\_V\_p$ will be used to denote the fresh atom obtained from a subjective literals $M_\omega e$, where

p is the atom in e, and in the prefixes, V is t if e is p, V is f if e is $\neg p$, V is nt if e is $not\ p$, V is nf if e is $not\ \neg p$, C is 0 if ω is of the form $[lb:ub]$, C is 1 if ω is of the form $[lb:]$, and U is ub if ω is of the form $[lb:ub]$, and U is o if ω is of the form $[lb:]$. Thus, $m\_C\_lb\_U\_V\_p$ is called a *denoter* of $M_\omega e$ and also recorded as m_ω_l for convenience. We assume prefixes used here do not occur in Π. Other fresh atoms may be used to avoid conflicts.

$CWViSolver(\Pi, k)$ includes the following steps.

1. Create a disjunctive logic program Π' from Π.
   Rules without subjective literals are left unchanged. For each rule r containing a subjective literal $M_\omega e$
   (a) Eliminate $M_\omega e$ by the following laws:
      i. if ω is $[0:0]$, replace $M_\omega e$ with m_ω_e, e^{not}.
      ii. otherwise,
         A. Add a rule obtained from r by replacing $M_\omega e$ by m_ω_e, e, and
         B. Add a rule obtained from r by replacing $M_\omega e$ by m_ω_e, e^{not}
   (b) Add $m\_\omega\_e \leftarrow body^+(r), not\ \neg m\_\omega\_e$ and $\neg m\_\omega\_e \leftarrow body^+(r), not\ m\_\omega\_e$
2. Compute $AS(\Pi')$ of answer sets of Π' using ASP grounder-solver like DLV, CLINGO etc.
3. Generate a set $PWV(\Pi)$ of possible world views of k-size from $AS(\Pi')$.
   Group the answer sets in $AS(\Pi')$ by common $m\_$ and $\neg m\_$-literals. Each group is said to be a candidate world view.
4. Generate GM-candidate world views of Π by checking each possible world view in $PWV(\Pi)$.
   For each possible world view W, check that the following condition are met
   - $|W| = k$
   - if m_ω_e is a common literal in W, then $W \models M_\omega e$ is true.
   - if $\neg m\_\omega\_e$ is a common literal in W, then $W \models M_\omega e$ is false.
   Let W_S denote the set of literals with a prefixes $m\_$ or $\neg m\_$ in W. $\{A | \exists B \in W, A = B - W_S\}$ is a world view of Π if the above two conditions are met.

$CWVgiSolver(\Pi, n)$ can be obtained from $CWViSolver(\Pi, k)$ by replacing "k-size" in the step 3 with "size $> n$", and replacing "$|W| = k$" condition in the step 4 with "$|W| > n$".

Theorem 3. ASP^{GM} **Solver** *is sound and complete for computing world views.*

7 Conclusion and Future Work

Recently, *Minimizing the knowledge* is considered as an effective way to address the problems of unintended solutions for logic programs with strong introspections. In this paper, we present a new semantics for the answer set programming with graded modalities. The new semantics gives an approach to handling graded modalities and a way to minimize the knowledge. Our future work includes the investigation of the methodologies for modeling with ASP^{GM}, optimization of the implementation etc.

Acknowledgments. This work was supported by the National High Technology Research and Development Program of China (Grant No. 2015AA015406), and the National Science Foundation of China (Grant No. 60803061).

References

1. Gelfond, M.: Strong introspection. In: AAAI-1991, pp. 386–391 (1991)
2. Gelfond, M.: Logic programming and reasoning with incomplete information. Ann. Math. Artif. Intell. **12**(1–2), 89–116 (1994)
3. Kahl, P.T.: Refining the semantics for epistemic logic programs, Ph.D. thesis, Texas Tech University, USA (2014)
4. Farias del Cerro, L., Herzig, A., Su, E.I.: Epistemic equilibrium logic. In: IJCAI-2015, pp. 2964–2970 (2015)
5. Shen, Y.D., Eiter, T.: Evaluating epistemic negation in answer set programming. Artif. Intell. **237**, 115–135 (2016)
6. Zhang, Z., Zhang, S.: Logic programming with graded modality. In: Calimeri, F., Ianni, G., Truszczynski, M. (eds.) LPNMR 2015. LNCS (LNAI), vol. 9345, pp. 517–530. Springer, Cham (2015). doi:10.1007/978-3-319-23264-5_43
7. Gelfond, M.: New semantics for epistemic specifications. In: Delgrande, J.P., Faber, W. (eds.) LPNMR 2011. LNCS (LNAI), vol. 6645, pp. 260–265. Springer, Heidelberg (2011). doi:10.1007/978-3-642-20895-9_29
8. http://sourceforge.net/projects/potassco/files/guide/2.0/guide-2.0.pdf

LPNMR Systems

The ASP System DLV2

Mario Alviano[1], Francesco Calimeri[1], Carmine Dodaro[2(⊠)], Davide Fuscà[1],
Nicola Leone[1], Simona Perri[1], Francesco Ricca[1], Pierfrancesco Veltri[1],
and Jessica Zangari[1]

[1] DeMaCS, University of Calabria, Rende, Italy
{alviano,calimeri,fusca,leone,perri,ricca,veltri,zangari}@mat.unical.it
[2] DIBRIS, University of Genova, Genova, Italy
dodaro@dibris.unige.it

Abstract. We introduce DLV2, a new Answer Set Programming (ASP)
system. DLV2 combines $\mathcal{I}$-DLV, a fully-compliant ASP-Core-2 grounder,
with the well-assessed solver WASP. Input programs may be enriched by
annotations and directives that customize heuristics of the system and
extend its solving capabilities. An empirical analysis conducted on benchmarks from past ASP competitions shows that DLV2 outperforms the old
DLV system and is close to the state-of-the-art ASP system CLINGO.

Keywords: Answer Set Programming · Systems · Grounding · Solving

1 Introduction

Answer Set Programming (ASP) [1,2] is a declarative programming paradigm
proposed in the area of non-monotonic reasoning and logic programming. ASP
applications include product configuration, decision support systems for space
shuttle flight controllers, large-scale biological network repairs, data-integration
and scheduling systems (cfr. [3]). In ASP, computational problems are encoded
by logic programs whose answer sets, corresponding to solutions, are computed
by an ASP system [4]. Many ASP systems combine two modules, the grounder
and the solver [5]. The first module takes as input a program Π and instantiates
it by producing a program Π' semantically equivalent to Π, but not containing
variables. Then, the second module computes answer sets of Π' by adapting and
extending SAT solving techniques [6].

The inherent complexity of engineering a *monolithic* (i.e., including both
grounder and solver in the same implementation) ASP system favored the development of stand-alone grounders and solvers [6]. However, the requirements of
ASP applications are now renewing the interest in integrated solutions. Indeed,
monolithic systems offer both more control over the grounding and solving
process, and more flexibility in developing application-oriented features. The first
monolithic ASP system was called DLV [7], and more recently, also the grounder
GRINGO and the solver CLASP are released together in the CLINGO system [8].

In this paper we introduce DLV2, a new ASP system that updates DLV
with modern evaluation techniques and development platforms. In particular,

© Springer International Publishing AG 2017
M. Balduccini and T. Janhunen (Eds.): LPNMR 2017, LNAI 10377, pp. 215–221, 2017.
DOI: 10.1007/978-3-319-61660-5_19

DLV2 combines $\mathcal{I}$-DLV [9], a fully-compliant ASP-Core-2 grounder, with the well-assessed solver WASP [10]. These core modules are extended by application-oriented features, among them constructs such as annotations and directives that customize heuristics of the system and extend its solving capabilities. An empirical analysis conducted on benchmarks from past ASP competitions shows that DLV2 improves substantially over the old system and is comparable in performance with CLINGO.

2   System Description

In this section we first overview the core modules of DLV2, and then we illustrate its main application-oriented features in separate paragraphs.

Core Modules. The core modules of DLV2 are the grounder $\mathcal{I}$-DLV [9] and the solver WASP [10]. The grounder implements a bottom-up evaluation strategy based on the semi-naive algorithm and features enhanced indexing and other new techniques for increasing the system performance. Notably, the grounding module can be used as an effective deductive-database system, as it supports a number of ad-hoc strategies for query answering. The solver module implements a modern CDCL backtracking search algorithm, properly extended with custom propagation functions to handle the specific properties of ASP programs.

RDBMS Data Access. DLV2 can import relations from an RDBMS by means of an #import_sql directive. For example, #import_sql(DB, "user", "pass", "SELECT * FROM t", p) is used to access database DB and imports all tuples from table t into facts with predicate name p. Similarly, #export_sql directives are used to populate specific tables with the extension of a predicate.

Query Answering. DLV2 supports cautious reasoning over (non)ground queries. The computation of cautious consequences is done according to *anytime* algorithms [11], so that answers are produced during the computation even in computationally complex problems. Thus, the computation can be stopped either when a sufficient number of answers have been produced or when no new answer is produced after a specified amount of time, see [11] for more details. The magic-sets technique [12] can be used to further optimize the evaluation of queries.

Python Interface. DLV2 can be extended by means of a Python interface. On the grounding side, the input program can be enriched by external atoms [13] of the form $\&p(i_1,\ldots,i_n; o_1,\ldots,o_m)$, where p is the name of a Python function, $i_1,\ldots,i_n$ and $o_1,\ldots,o_m$ ($n, m \geq 0$) are input and output terms, respectively. For each instantiation $i'_1,\ldots,i'_n$ of the input terms, function p is called with arguments $i'_1,\ldots,i'_n$, and returns a set of instantiations for $o_1,\ldots,o_m$. For example, a single line of Python, def rev(s): s[::-1], is sufficient to define a function rev that reverse strings, and which can be used by a rule of the following form: revWord(Y) :- word(X), &rev(X;Y).

On the solving side, the input program can be enriched by external propagators. Communication with the Python modules follows a synchronous message-passing protocol implemented by means of method calls. Basically, an external module must comply with a specific interface, whose methods are associated to events occurring during the search for an answer set, e.g. a literal is inferred as true. Whenever a specific point of the computation is reached, the corresponding event is triggered, i.e., a method of the module is called. Some methods of the interface are allowed to return values that are subsequently interpreted by the solver. An example is reported below, and also used in experiments.

Example 1. Consider the following program, for a given positive n:

```
edb(1..n).    {val(X) : edb(X)} = 1.    {in(X) : edb(X)}.
:- val(X), X != #count{Y : in(Y)}.
```

The instantiation of the last constraint above results into n ground rules, each one comprising n instances of `in(Y)`. It turns out that the above program cannot be instantiated for large values of n due to memory consumption. Within DLV2, the expensive constraint can be replaced by the following propagator:

```
#propagator(@file="prop.py", @elements={X,v:val(X); X,i:in(X)}).
```

where `prop.py` is a Python script reacting on events involving instances of `val(X)` and `in(X)`, e.g. when the literal `val(1)` is inferred as true a notification is sent to the Python script. Without going into much details, the script checks whether candidate answer sets satisfy the expensive constraint, which however is not instantiated and therefore efficiently evaluated.   ◁

The Python interface also supports the definition of new heuristics [14], which are linked to input programs via Java-like annotations and therefore discussed in the next paragraph.

Java-like Annotations. Within DLV2, ASP programs can be enriched by global and local annotations, where each local annotation only affects the immediate subsequent rule. The system takes advantage of annotations to customize some of its heuristics. Customizations include *body ordering* and *indexing*, two of the crucial aspects of the grounding, and *solving heuristics* to tightly link encodings with specific domain knowledge. For example, the order of evaluation of body literals in a rule, say ≺, is constrained to satisfy `p(X)` ≺ `q(Y)` by adding the following local annotation: `%@rule_partial_order(@before={p(X)}, @after={q(Y)})`. Concerning solving heuristics, they are specified via the Python interface [14], and act on a set of literals of interests, where each literal is associated with a tuple of terms. For example, a heuristic saved in `heuristic.py` and acting on literals `p(X)` can be linked to a program by the following global annotation:

`%@global_heuristic(@file="heuristic.py", @elements={X : p(X)})`.

Note that annotations do not change the semantics of input programs. For this reason, their notation starts with %, which is used for comments in ASP-Core-2, so that other systems can simply ignore them.

3   Experiments

In order to assess the performance of DLV2, we compared it with CLINGO 5.1.0 [8] and DLV [7]. All systems were tested with their default configuration. CLINGO and DLV2 were tested on benchmarks taken from the latest ASP competition [15]. On the other hand, since DLV does not fully support the ASP-Core-2 standard language, the comparison between DLV and DLV2 is performed on benchmarks taken from the third competition [16]. An additional benchmark is obtained from the programs in Example 1. Experiments were performed on a NUMA machine equipped with two 2.8 GHz AMD Opteron 6320 processors. The time and memory were limited to 900 s and 15 GB, respectively.

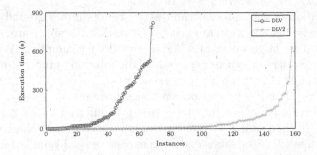

Fig. 1. DLV2 vs. DLV: cactus plot on benchmarks from ASP Competition 2011.

DLV *vs.* DLV2. Results are reported in the cactus plot in Fig. 1. A sensible improvement is obtained by the new version of the system. Indeed, the percentage gain of the solved instances is 128%, and it is even higher, namely 273% if the running time is bounded to 60 s.

CLINGO *vs.* DLV2. Results are reported in Table 1. On the overall, DLV2 solved 306 instances, 35 more than CLINGO. Such an advantage can be explained by the fact that DLV2 handles query answering, while CLINGO does not. If benchmarks with queries are ignored, the difference between the two systems is 5 instances in favor of CLINGO. Considering single benchmarks, CLINGO is sensibly faster than DLV2 on Combined Configuration, Complex Optimization Of Answer Sets and Graph Colouring, while the opposite happens for Crossing Minimization, Maximal Clique Problem and MaxSAT.

Table 1. DLV2 vs. CLINGO: solved instances and average running time (in seconds) on benchmarks from ASP Competition 2015 (20 instances per benchmark).

Benchmark	CLINGO		DLV2	
	Time	Solved	Time	Solved
Abstract Dialectical Frameworks WF Model (optimization)	8.89	20	137.31	15
Combined Configuration	286.73	10	150.62	1
Complex Optimization of Answer Sets	174.61	18	120.08	6
Connected Maximim-density Still Life (optimization)	193.63	6	73.40	9
Crossing Minimization (optimization)	65.63	6	2.90	19
Graceful Graphs	191.42	11	59.17	5
Graph Colouring	215.98	17	204.83	9
Incremental Scheduling	131.21	13	166.21	8
Knight Tour With Holes	15.00	10	41.83	10
Labyrinth	105.12	13	181.75	12
Maximal Clique Problem (optimization)	—	0	168.84	15
MaxSAT (optimization)	44.33	7	90.83	20
Minimal Diagnosis	7.74	20	38.39	20
Nomistery	163.46	8	118.42	9
Partner Units	35.40	14	375.99	9
Permutation Pattern Matching	180.30	12	153.64	20
Qualitative Spatial Reasoning	174.81	20	326.47	18
Ricochet Robots	130.42	8	267.87	9
Sokoban	86.52	10	174.69	10
Stable Marriage	430.49	5	459.31	9
System Synthesis (optimization)	—	0	—	0
Steiner Tree (optimization)	242.45	3	—	0
Valves Location (optimization)	42.51	16	68.40	16
Video Streaming (optimization)	56.96	13	0.10	9
Visit-all	248.40	11	68.94	8
Consistent Query Answering (query)	—	—	252.77	13
Reachability (query)	—	—	131.48	20
Strategic Companies (query)	—	—	30.07	7

Propagators. Results are reported in the plots in Fig. 2. As already observed in Example 1, the instantiation of the constraint is expensive in terms of memory consumption (right plot): the average memory consumption is around 5 GB on the 17 solved instances. Such an inefficiency also impacts on the execution time (left plot), which is on average 61 s on the solved instances. On the other hand, DLV2 takes a sensible advantage from the ad-hoc propagator, solving each tested instance in less than 1 s and 80 MB of memory.

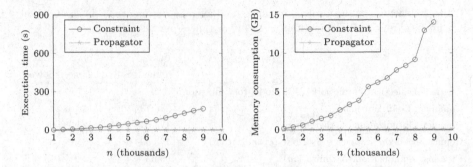

Fig. 2. Evaluation of DLV2 on programs from Example 1.

4 Conclusion

In this paper we presented DLV2, a new ASP system properly designed for combining the $\mathcal{I}$-DLV grounder [9] and the WASP solver [10]. We assessed the system performance against both DLV and CLINGO, and it turns out to be dramatically more efficient than DLV, and comparable with CLINGO. Moreover, the monolithic design of DLV2 eased the development of new features such as the Python interface for defining propagators and heuristics. Indeed, DLV2 automatically creates Python data structures based on the relevant information from the input program. In the Python interface of WASP, instead, such information had to be reconstructed by analyzing the table of symbols provided by the grounder, thus requiring an additional effort by the Python programmer. The system is available at the following URL: https://www.mat.unical.it/DLV2.

Acknowledgments. The work has been partially supported by the Italian ministry for economic development (MISE) under project "PIUCultura - Paradigmi Innovativi per l'Utilizzo della Cultura" (n. F/020016/01-02/X27), and under project "Smarter Solutions in the Big Data World (S2BDW)".

References

1. Brewka, G., Eiter, T., Truszczynski, M.: Answer set programming at a glance. Commun. ACM **54**(12), 92–103 (2011)
2. Gelfond, M., Lifschitz, V.: Classical negation in logic programs and disjunctive databases. New Gener. Comput. **9**(3/4), 365–386 (1991)
3. Erdem, E., Gelfond, M., Leone, N.: Applications of answer set programming. AI Mag. **37**(3), 53–68 (2016)
4. Lifschitz, V.: Answer set planning. In: ICLP, pp. 23–37. MIT Press (1999)
5. Kaufmann, B., Leone, N., Perri, S., Schaub, T.: Grounding and solving in answer set programming. AI Mag. **37**(3), 25–32 (2016). http://www.aaai.org/ojs/index.php/aimagazine/article/view/2672
6. Lierler, Y., Maratea, M., Ricca, F.: Systems, engineering environments, and competitions. AI Mag. **37**(3), 45–52 (2016). http://www.aaai.org/ojs/index.php/aimagazine/article/view/2675

7. Leone, N., Pfeifer, G., Faber, W., Eiter, T., Gottlob, G., Perri, S., Scarcello, F.: The DLV system for knowledge representation and reasoning. ACM Trans. Comput. Log. **7**(3), 499–562 (2006)

8. Gebser, M., Kaminski, R., Kaufmann, B., Ostrowski, M., Schaub, T., Wanko, P.: Theory solving made easy with clingo 5. In: ICLP TCs, pp. 2:1–2:15 (2016). http:// dx.doi.org/10.4230/OASIcs.ICLP.2016.2

9. Calimeri, F., Fuscà, D., Perri, S., Zangari, J.: I-DLV: the new Intelligent Grounder of DLV. Intelligenza Artificiale **11**(1), 5–20 (2017). doi:10.3233/IA-170104

10. Alviano, M., Dodaro, C., Leone, N., Ricca, F.: Advances in WASP. In: Calimeri, F., Ianni, G., Truszczynski, M. (eds.) LPNMR 2015. LNCS (LNAI), vol. 9345, pp. 40–54. Springer, Cham (2015). doi:10.1007/978-3-319-23264-5_5

11. Alviano, M., Dodaro, C., Ricca, F.: Anytime computation of cautious consequences in answer set programming. TPLP **14**(4–5), 755–770 (2014)

12. Alviano, M., Faber, W., Greco, G., Leone, N.: Magic sets for disjunctive datalog programs. Artif. Intell. **187**, 156–192 (2012)

13. Calimeri, F., Cozza, S., Ianni, G.: External sources of knowledge and value invention in logic programming. Ann. Math. Artif. Intell. **50**(3–4), 333–361 (2007)

14. Dodaro, C., Gasteiger, P., Leone, N., Musitsch, B., Ricca, F., Schekotihin, K.: Combining answer set programming and domain heuristics for solving hard industrial problems (application paper). TPLP **16**(5–6), 653–669 (2016)

15. Gebser, M., Maratea, M., Ricca, F.: The design of the sixth answer set programming competition – report. In: Calimeri, F., Ianni, G., Truszczynski, M. (eds.) LPNMR 2015. LNCS (LNAI), vol. 9345, pp. 531–544. Springer, Cham (2015). doi:10.1007/978-3-319-23264-5_44

16. Calimeri, F., Ianni, G., Ricca, F.: The third open answer set programming competition. TPLP **14**(1), 117–135 (2014). http://dx.doi.org/10.1017/S1471068412000105

LP2NORMAL — A Normalization Tool for Extended Logic Programs

Jori Bomanson[(✉)]

Department of Computer Science, Aalto University, Espoo, Finland
jori.bomanson@aalto.fi

Abstract. Answer set programming (ASP) features a rich rule-based modeling language for encoding search problems. While normal rules form the simplest rule type in the language, various forms of extended rules have been introduced in order to ease modeling of complex conditions and constraints. Normalization means replacing such extended rules with identically functioning sets of normal rules. In this system description, we present LP2NORMAL, which is a state-of-the-art normalizer that acts as a filter on ground logic programs produced by grounders, such as GRINGO. It provides options to translate away choice rules, cardinality rules, and weight rules, and to rewrite optimization statements using comparable techniques. The produced logic programs are suitable inputs to tools that lack support for extended rules, in particular. We give an overview of the normalization techniques currently supported by the tool and summarize its features. Moreover, we discuss the typical application scenarios of normalization, such as when implementing the search for answer sets using a back-end solver without direct support for cardinality constraints or pseudo-Boolean constraints.

1 Introduction

Answer set programming (ASP) [9,15] is a declarative programming paradigm that features a rich rule-based modeling language for encoding search problems. Normal rules form the base fragment of the language and their declarative interpretation is based on the notion of *answer sets* also known as *stable models* [12]. In order to ease modeling of complex conditions and constraints, different syntactic extensions have been introduced. In particular, the *extended rule types* of [16], i.e., *choice*, *cardinality*, and *weight* rules and *optimization* statements are central primitives for modeling in ASP [11].

When it comes to implementing language extensions, there are two basic strategies. The first is to extend the underlying answer-set solver to natively handle extended syntax. The second is to treat extensions as syntactic sugar and translate them away. The tool described in this paper is motivated by the latter *translation-based* strategy in a setting where the used ASP toolchain does not support extended rules. The *normalization* [14] of such rules means replacing them with sets of normal rules that are semantically indistinguishable from them in any context. For example, the weight rule $a :\text{-} 5 \leq \#\text{sum} \{2 : b; 3 : c; 5 : \text{not } d\}$

© Springer International Publishing AG 2017
M. Balduccini and T. Janhunen (Eds.): LPNMR 2017, LNAI 10377, pp. 222–228, 2017.
DOI: 10.1007/978-3-319-61660-5_20

can be normalized into a :- b, c and a :- not d. While the normalization of choice rules is straightforward, sophisticated normalization techniques for other extended rules have been developed in a trilogy of papers [6–8].

In this system description, we present a tool called LP2NORMAL $(2.27)^1$, which is a state-of-the-art normalizer that can be used to filter ground logic programs produced by a grounder before forwarding the programs to a solver for the computation of answer sets. It ships with the functionality required to translate away extended rules, by which we exclusively refer to choice, cardinality, and weight rules in the sequel, as well as to *rewrite* optimization statements in an analogous way.

The rest of this paper is organized as follows. In Sect. 2, we give a brief overview of translation techniques that can be exploited in the process of normalizing extended rule types. Specific features of LP2NORMAL available through numerous command-line options are summarized in Sect. 3. Some existing application scenarios of normalization are discussed in Sect. 4. Finally, we conclude the paper in Sect. 5.

2   Overview of Normalization Techniques

In this section, we look into techniques for normalizing cardinality and weight rules. We also describe methods for simplifying the rules before normalization and the idea of a *cone of influence* for pruning the normalized output. Finally, we turn to the task of rewriting optimization statements. The underlying translation schemes to be described can be interpreted as specifications of counting and comparison operations between numbers expressed in unary, binary, or even mixed-radix bases.

In ASP, to say that at least k out of n literals are satisfied one typically uses a single *cardinality rule*. This condition is also expressible in a number of normal rules, such as in $O(n(n-k))$ rules and auxiliary atoms that form a Binary Decision Diagram (BDD) [10] or a *counting grid* [14,16]. The rows of this grid are built in a dynamic programming fashion based on the previous rows, and each row encodes a partial count of satisfied input literals as a unary number. Translation schemes identical or close to this involve the Sinz counter [17] and sequential counters (SEQ) [13]. More concise encodings are obtainable via merge sorting, where again partial counts are built up of smaller counts, but this time by recursively merging halves of the input. We obtain an ASP encoding of an odd-even sorting network of size $O(n(\log n)^2)$ by using odd-even merging networks [5], and expressing their building blocks, i.e., comparators, in three normal rules each. On the other hand, we obtain a *totalizer* of size $O(n^2)$ by using direct mergers that require no additional auxiliary variables [4].

The translations mentioned so far can be encoded without introducing negated literals beyond those in the input. With the use of additional negation, even more concise schemes are attainable. For example, an encoding of a

---

[1] Available at http://research.ics.aalto.fi/software/asp.

binary adder yields no more than $O(n \log k)$ rules and atoms. However, analogous encodings have proven difficult in SAT solving due to their poor propagation properties. Moreover, they may lead to soundness issues in ASP, where negation and positive recursion require extra care.

Lower bounds in cardinality rules, such as k above, can be replaced or complemented by upper bounds in typical ASP systems, which convert the latter to the former in the grounding phase [16]. Hence, only lower bounds remain for normalization. Similarly, syntactic constructions that combine choice and cardinality rules can be interpreted as short hands for the two types of rules, which can be normalized separately.

When it comes to *weight rules*, in which literals are generalized to have weights w_i, we may apply simplifications prior to normalization, potentially reducing numbers of literals or their weights. Both of these outcomes generally lower the size of the subsequent normalization. Among these simplifications, we have some basic ones such as removal of literals with large weights $w_i \geq k$ that can be compensated with simple normal rules. Also, we may factor out any common divisor d of the weights and divide the bound, rounding it up. Furthermore, this division is applicable even when one of the weights w_i results in a remainder, as long as the corresponding quotient is afterward incremented by one in the case that $k \leq w_i \pmod{d}$. Other scenarios providing simplification opportunities include cases where a number of the largest weights in a rule are always required in order to reach the bound; where a pair of weights together satisfy the bound; where a weight is too small to ever make a difference; and certain cases where analysis of the residues resulting from division of the bound and weights with a heuristically chosen divisor reveals that the division can be done given minor adjustments to some of the numbers.

The actual normalization techniques in the tool for weight rules mainly revolve around two types of translations. On the one hand, we have sequential weight counters (SWC) [13] and Reduced-Ordered Binary Decision Diagrams (ROBDDs) [1], both of the size $O(nk)$. They are particularly compact for small rules, for which the asymptotic size is not relevant in practice. On the other hand, we have sorting and merging based normalizations, which are $O(n(\log n)^2 \log w_{\max})$ in size, e.g., when odd-even mergers are used, where $w_{\max}$ denotes the largest input weight [6,10]. In the constructions, there are $c \leq \log_2 w_{\max}$ sorters, each of which intuitively counts digits of a certain significance, which are followed by $c-1$ mergers that perform deferred carry propagation. These normalizations can be compressed with *structure sharing* [6]. This sharing method stems from the observation that the sets of inputs to the sorters overlap significantly in general. When merge sorters are used, the overlap leads to duplication in their structure, which may be maximized via optimization and then eliminated.

In normalizations of cardinality and weight rules, there is only a single output atom per rule that we are interested in. Yet, in their basic form, the outlined normalization strategies, with the exception of those based on (RO)BDDs, define sequences of outputs that are not all needed. Namely, counting grids, mergers, sorters, and SWCs produce entire vectors of atoms with sorted truth values. Now one may imagine that we mark a single output, and propagate this information of what is wanted and what is not backward through the rules in the translation, so as to compute what we call a *cone of influence*. Then, the actual normalization may be produced in a forward phase, where only the rules defining atoms that fell inside the cone are included. From our practical experience, this pruning technique is important for sorters, and sorter based translations as well as SWCs. Moreover, it also brings down the asymptotic size of odd-even merge sorting programs, which it prunes down from a size of $O(n(\log n)^2)$ to *selection programs* of size $O(n(\log m)^2)$, where m is the lesser of the bound k and $n - k + 1$. The resulting size rivals that of selection network designs used in SAT [2].

Whereas the above options concern normalization resulting in purely normal rules, LP2NORMAL also supports *rewriting of optimization statements* using techniques similar to those used in normalization, but which generate modified optimization statements in adddition to normal rules. The rewritings generally define auxiliary atoms using normal rules so as to encode the sum, or some partial sums of the weighted literals that make up an optimization statement. Such a statement is then replaced by one or more statements specified in terms of the new auxiliary atoms. These rewritings are solely aimed at boosting solving performance by offering new atoms to branch on and to use in learnt nogoods. However, these rewritings may grow impractically large in terms of the generated atoms and rules, especially when applied to optimization statements that carry substantial amounts of information. To alleviate this issue, we have developed ways to limit the size increase via refined control over how much rewriting is done.

3   Implementation

In this section, we cover the usage and highlight some implementation details of the normalizer tool LP2NORMAL. A summary of the discussed options is shown in Table 1.

By default, all extended rules are translated into normal rules, while leaving other rules and statements intact. This behaviour is configurable via command line options, which are prefixed with -c, -w, and -o for cardinality, weight, and optimization statements, respectively. Rules can be kept as they are with the option -k for choice rules and -ck, -wk, -ok for the rest.

Cardinality rules are by default normalized using an automatic scheme -cc that generates merge sorting programs built recursively of direct mergers and odd-even mergers. The choice between them is based on trial runs that reveal which one introduces fewer atoms and rules. The decision can be fixed to direct mergers with -ct and to odd-even mergers with -ch. Moreover, an option -cs is available for basing the normalization on selection networks instead.

For weight rules, one may pick the schemes based on SWCs with -wc and ROBDDs with -wb. Weight rule translations constructed from sorters and mergers are primarily controlled with the option -wq. Due to a design choice in

Table 1. Command-line options of translations in LP2NORMAL by extended rule type and with asymptotic sizes after cone-of-influence simplification. Here $m = \min\{k, n - k + 1\}$.

	Options	Rules (Atoms)	
Choice		$O(n)$	
Cardinality	-cn	$O(n(n - k))$	Counting grid [10,14,16]
Cardinality	-ct	$O(nm)\ (O(n \log m))$	Totalizer [4]
Cardinality	-ch	$O(n(\log m)^2)$	Odd-even merge sort [5,8,10]
Cardinality	-cc	$O(n(\log m)^2)$	Automatic merge sort
Weight	-wc	$O(nk)$	SWC [13]
Weight	-wb	$O(nk)$	ROBDD [1]
Weight	-wq -cc	$O(n(\log n)^2 \log w_{\max})$	Network of merge sorters [3,6,10]
Optimization	-oqn -cc	$O(n(\log n)^2 \log w_{\max})$	Network of merge sorters [7]
Optimization	-oKpq -oqn -cc	$O(nq(\log n)^2)$	Rewrite q most significant digits [7]

LP2NORMAL aiming for simplicity, the choice of sorters and mergers used here are inferred from any active cardinality rule options. That is, if a user requests cardinality rules to be translated using odd-even mergers and sorters with -ch and weight rules with -wq, then those types of mergers and sorters are used in the weight rule normalizations as well.

For optimization statements, the option -oqn instructs LP2NORMAL to proceed with optimization statements in the same way as when it applies -wq to weight rules, but with the following difference. The translation is cut short before a bound check is encoded, and instead the atoms that the bound check would have depended on are printed in an optimization statement. Moreover, there is an option -ox to use certain weight rule translation techniques, primarily those based on SWCs or ROBDDs, to produce a single sorted and weighted sequence of atoms that encodes all subset sums of the contents of an optimization statement. This option is not always feasible, but it serves as a proof-of-concept for how to use these weight rule techniques to rewrite optimization statements in a natural way. Finally, we highlight a collection of options prefixed by -oK or -ok that select parts of optimization statements to be rewritten or kept from being rewritten. For example, with -oKp3 one may instruct LP2NORMAL to first split every weight after three of its most significant digits and then to apply any specified rewriting, such as -oqn, to the more significant part only.

4   Applications

Our main driving motivation for normalization in ASP has been to add support for extended rules to solvers that would otherwise accept only normal rules.

To this end, LP2NORMAL took part in several systems submitted by the Aalto team to the Sixth Answer Set Programming Competition[2]. The following is an example of a pipeline relying on the SAT solver LINGELING for solving ASP problems. In the pipeline, LP2NORMAL translates away extended rules, while the tools in the middle take care of translating the resulting normal rules to SAT.

<code>lp2normal | lp2acyc | lp2sat -b | lingeling</code>

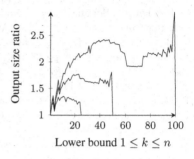

In the above, we could alternatively use normalization capabilities of the state-of-the-art ASP solver CLASP (3.2.2) via its options `--pre` and `--trans-ext=all`. However, the techniques implemented therein generally yield larger output. Figure 1 depicts the case for cardinality rules on $n \in \{25, 50, 100\}$ atoms. As another use case, one of the systems in the competition combines LP2NORMAL and CLASP in a configuration where the normalizer translates cardinality and weight rules of only modest size, and rewrites parts of optimization statements. In this case, the role of the normalizer is to alter the solving performance of the solver,

Fig. 1. Ratios of the numbers of integers in the normalizations produced by CLASP in comparison to LP2NORMAL.

which can handle extended rules natively as well. The impact on performance varies benchmark by benchmark, which frequently improves but also sometimes degrades.

5   Conclusion

We summarized the most important capabilities of the tool LP2NORMAL concerning the normalization of cardinality and weight rules and rewriting of optimization statements in answer set programs. In future development, we plan to incorporate optimal sorting networks into the tool together with other compact networks generated offline for small, fixed ranges of input parameters. Finally, we continue to explore partial rewriting, which has proven to be an effective way to limit translation size and enhance optimization performance [7].

References

1. Abío, I., Nieuwenhuis, R., Oliveras, A., Rodríguez-Carbonell, E., Mayer-Eichberger, V.: A new look at BDDs for pseudo-Boolean constraints. J. Artif. Intell. Res. **45**, 443–480 (2012)

---

[2] Participant systems are available at http://aspcomp2015.dibris.unige.it/participants.

2. Asín, R., Nieuwenhuis, R., Oliveras, A., Rodríguez-Carbonell, E.: Cardinality networks: a theoretical and empirical study. Constraints **16**(2), 195–221 (2011)
3. Bailleux, O., Boufkhad, Y., Roussel, O.: New encodings of pseudo-Boolean constraints into CNF. In: Kullmann, O. (ed.) SAT 2009. LNCS, vol. 5584, pp. 181–194. Springer, Heidelberg (2009). doi:10.1007/978-3-642-02777-2_19
4. Bailleux, O., Boufkhad, Y.: Efficient CNF encoding of Boolean cardinality constraints. In: Rossi, F. (ed.) CP 2003. LNCS, vol. 2833, pp. 108–122. Springer, Heidelberg (2003). doi:10.1007/978-3-540-45193-8_8
5. Batcher, K.: Sorting networks and their applications. In: AFIPS Spring Joint Computer Conference, pp. 307–314. ACM (1968)
6. Bomanson, J., Gebser, M., Janhunen, T.: Improving the normalization of weight rules in answer set programs. In: Fermé, E., Leite, J. (eds.) JELIA 2014. LNCS (LNAI), vol. 8761, pp. 166–180. Springer, Cham (2014). doi:10.1007/978-3-319-11558-0_12
7. Bomanson, J., Gebser, M., Janhunen, T.: Rewriting optimization statements in answer-set programs. In: Technical Communications of ICLP 2016, vol. 52, OASIcs, pp. 5:1–5:15 (2016)
8. Bomanson, J., Janhunen, T.: Normalizing cardinality rules using merging and sorting constructions. In: Cabalar, P., Son, T.C. (eds.) LPNMR 2013. LNCS (LNAI), vol. 8148, pp. 187–199. Springer, Heidelberg (2013). doi:10.1007/978-3-642-40564-8_19
9. Brewka, G., Eiter, T., Truszczyński, M.: Answer set programming at a glance. Commun. ACM **54**(12), 92–103 (2011)
10. Eén, N., Sörensson, N.: Translating pseudo-Boolean constraints into SAT. J. Satisfiability Boolean Model. Comput. **2**, 1–26 (2006)
11. Gebser, M., Schaub, T.: Modeling and language extensions. AI Mag. **37**(3), 33–44 (2016)
12. Gelfond, M., Lifschitz, V.: The stable model semantics for logic programming. In: Proceedings of ICLP 1988, pp. 1070–1080. MIT Press (1988)
13. Hölldobler, S., Manthey, N., Steinke, P.: A compact encoding of pseudo-Boolean constraints into SAT. In: Glimm, B., Krüger, A. (eds.) KI 2012. LNCS (LNAI), vol. 7526, pp. 107–118. Springer, Heidelberg (2012). doi:10.1007/978-3-642-33347-7_10
14. Janhunen, T., Niemelä, I.: Compact translations of non-disjunctive answer set programs to propositional clauses. In: Balduccini, M., Son, T.C. (eds.) Logic Programming, Knowledge Representation, and Nonmonotonic Reasoning. LNCS (LNAI), vol. 6565, pp. 111–130. Springer, Heidelberg (2011). doi:10.1007/978-3-642-20832-4_8
15. Janhunen, T., Niemelä, I.: The answer set programming paradigm. AI Mag. **37**(3), 13–24 (2016)
16. Simons, P., Niemelä, I., Soininen, T.: Extending and implementing the stable model semantics. Artif. Intell. **138**(1–2), 181–234 (2002)
17. Sinz, C.: Towards an optimal CNF encoding of Boolean cardinality constraints. In: Beek, P. (ed.) CP 2005. LNCS, vol. 3709, pp. 827–831. Springer, Heidelberg (2005). doi:10.1007/11564751_73

Harvey: A System for Random Testing in ASP

Alexander Greßler[1], Johannes Oetsch[2]([✉]), and Hans Tompits[1]

[1] Institute of Information Systems 184/3, Vienna University of Technology,
Favoritenstraße 9-11, 1040 Vienna, Austria
{gressler,tompits}@kr.tuwien.ac.at
[2] Institute of Computer Languages 185/2, Vienna University of Technology,
Favoritenstraße 9-11, 1040 Vienna, Austria
johannes.oetsch@tuwien.ac.at

Abstract. We present Harvey, a tool for random testing in answer-set programming (ASP) that allows to incorporate constraints to guide the generation of test inputs. Due to the declarative nature of ASP, it can be argued that there is less need for testing than in conventional software development. However, it is shown in practice that testing is still needed when more sophisticated methods are not viable. Random testing is recognised as a simple yet effective method in this regard. The approach described in this paper allows for random testing of answer-set programs in which both test-input generation and determining test verdicts is facilitated using ASP itself: The test-input space is defined using ASP rules and uniformity of test-input selection is achieved by using XOR sampling. This allows to go beyond simple random testing by adding further ASP constraints in the process.

1 Introduction

Testing is an essential part of every software development process. No system for software quality management can spare testing completely, and there exists a vast body of literature on this subject [1,8]. Although answer-set programming (ASP) plays an important role in the field of declarative programming with numerous practical applications, no dedicated tools for testing answer-set programs have been introduced so far. In this paper, we address this issue and present the system Harvey for realising random testing in ASP.[1] As a matter of fact, we go beyond simple random testing by allowing to incorporate constraints in the process to guide the input generation towards especially interesting areas of the test-input space.

Due to the declarative nature of ASP, it can be argued that there is less need for testing than in conventional imperative software development. Still, testing is often required when more sophisticated methods like correctness proofs or equivalence checks are not viable. This can be due to a lack of skills of an

---

[1] The name of our tool refers to Harvey Dent, one of the infamous adversaries of Batman, who, under his pseudonym Two-Face, makes all important decisions by flipping a coin.

© Springer International Publishing AG 2017
M. Balduccini and T. Janhunen (Eds.): LPNMR 2017, LNAI 10377, pp. 229–235, 2017.
DOI: 10.1007/978-3-319-61660-5_21

ASP engineer or simply because of limited time resources. Also, the ongoing development of solver languages makes testing—as a light-weight verification approach—especially appealing. Accordingly, several works on testing methods for ASP have been discussed in the literature so far, including investigations on white-box testing (also referred to as *structure-based testing*) [5,6] and mutation testing along with bounded-exhaustive testing [9].

We realise random testing for ASP where both test-input generation and determining test verdicts is done employing ASP itself by means of our tool Harvey, which is comprised of two scripts, implemented in Python: the first one generates random inputs by exploiting ASP rules and achieves uniformity of the test-input selection by using XOR streamlining. This is a method to obtain near-uniform samples of combinatorial search spaces by adding parity constraints, adopted from similar approaches used in the area of SAT solving [2,4], and which is basically a re-implementation of the system xorro [11] taking current ASP language features into account. The second script of Harvey takes care for the actual testing procedure, where answer sets of the tested program joined with the generated input are compared with those of a test oracle under the respective inputs. For the answer-set generation, the solver clasp [11] is used.

2   Testing Methodology

We are concerned with testing *uniform problem encodings*, i.e., we distinguish between a *fixed program* that specifies solutions to a problem at hand and *problem instances* that are represented as sets of facts. Consider the following reviewer-assignment problem: We have reviewers and papers and the objective is to find an assignment subject to the following constraints: (i) every paper is assigned to three reviewers, (ii) no reviewer gets more than four papers, and (iii) no paper gets assigned to a reviewer who has declared a conflict of interest. We consider the following encoding in gringo syntax [11]:

```
{ assigned(R,P) : reviewer(R) } = 3 :- paper(P).
:- { assigned(R,P) : paper(P) } > 4, reviewer(R).
:- assigned(R,P), conflict(R,P).
#show assigned/2.
```

Each rule directly encodes one of the three conditions from the above. Now, a problem instance is described using the predicates paper/1, reviewer/1, and conflict/2, serving as input for the problem encoding, while the output is given by the answer sets (projected to the output predicate assigned/2) of the encoding joined with the facts. Testing simply means to select different problem instances (inputs) and verify that the computed answer sets indeed correspond to all the solutions that we would expect. If testing is to be automated, verifying that the actual output matches the intended output requires a test oracle, i.e., some computational means to determine the correct outcome.

In ASP, it is often sensible to assume that such an oracle is given by an answer-set program itself: it is usually rather simple to come up with an initial

ASP encoding that is easily seen to be correct but lacks other properties like scalability. In the course of refining and rewriting this first design towards a more efficient and thus useful program, different versions will emerge that have to be tested against the original.

Different strategies for selecting test inputs in ASP have been looked at. One strategy is to use inputs that "cover" different elements of the program in structure-based testing [5, 6]. Another strategy is to exhaust the test-input space within a fixed scope in bounded-exhaustive testing [9]. Random testing is a selection strategy were inputs are sampled uniformly at random. Like in bounded-exhaustive testing, some size limit on a usually otherwise infinitely large input space is required.

Assume now the program to be tested, referred to as the *program under test*, is some encoding of the reviewer-assignment problem as described above. The first thing that is needed for testing is a program that is known to be correct. We will use the encoding given above for this purpose and refer to it as the *test oracle*. The only assumption is that the program under test and the test oracle have the same output predicates.

Writing a test-input generator. The next step is to define the test input space. This is conveniently done using ASP itself. Following previous work [9], we use a program whose answer sets represent all valid inputs:

```
#const n=4.
{ noR(R) : R=0..n } = 1.   { noP(P) : P=0..n } = 1.
reviewer(R)  :- noR(M), R <= M, R=1..n.
paper(P)     :- noP(M), P <= M, P=1..n.
{ conflict(R,P) } :- reviewer(R), paper(P).
#show reviewer/1.  #show paper/1.  #show conflict/2.
```

We use a constant n as a size limit for the inputs. The actual number of reviewers is nondeterministically set to a value between 0 and n. The same holds for the number of papers. Facts `reviewer/1` and `paper/1` are then derived to represent the actual input instance. Also, for any pair of reviewers and papers, we non-deterministically choose if there should be a conflict of interest. Finally, #show statements are used to define the *input signature*, i.e., the predicates that are used to describe a problem instance. Smaller values for n result in a smaller size of the input space and inputs can be generated faster; larger values yield more inputs for a more complete testing. Adequate values in practice are the smallest values such that all interesting structural variations of input instances are included while the generation of inputs is still feasible [9].

Note that the inputs are not arbitrary sets of facts over the input alphabet but structures that adhere to certain constraints: in our example, `conflict/2` is a relation between existing reviewers and existing papers. In general, there will be different assumptions and restrictions regarding what we consider as valid or *admissible input* for some ASP encoding. They can be regarded as preconditions and are often implicit. Defining an input generator makes them explicit.

Different approaches to random testing. There are different options to implement random testing of the program under test. Following the concept of running a test case from previous work [5], we would first compute one answer set of the input generator, then we would verify that *all* answer sets of the input joined with the test oracle and the program under test are matching, i.e., coincide on the output predicates. Sometimes, this can be realised using equivalence checkers for ASP [7,10], but they come with their own limitations like scalability or certain syntactical restrictions on the programs. A more pragmatic approach is to compute only one answer set of the program under test joined with the input. If there is no answer set, we check if the test oracle yields no answer set as well on the test input. Otherwise, we check if this resulting answer set, projected to the output predicates, is an answer-set of the test oracle as well. This is the approach that is used in Harvey. In testing, practicality beats purity!

This method resembles random testing for conventional programming paradigms. Since our ASP test oracle can be used not only to check if a given solution is correct but also to compute solutions itself, we can inverse the roles of the program under test and the test oracle: we would first compute an answer set of the test oracle and check if it is also an answer set of the program under test. This means to check whether expected solutions are indeed produced by the program under test. While the conventional method relates to the correctness of a program, this method is rather concerned with its completeness. In Harvey, this can be done by simply swapping the test oracle and the program under test.

For some problems, the distribution of instances that have solutions is very sparse. This means that for some programs, there are no answer sets for most input instances. Random testing would thus become uninteresting and any trivial inconsistent encoding would probably pass all tests. A simple remedy would be to guide the generation of test inputs to areas of the test-input space where we have solutions. This can be done by simply adding the program under test (or the test oracle) entirely to the input generator. Generalising this idea, we can of course add arbitrary further constraints to the input generator if we want to focus on inputs with special properties. In our running example, we could add constraints to get test inputs with many papers and few reviewers if we are expecting that such scenarios are particularly relevant.

3   Architecture and Implementation

Harvey has been implemented in Python 3.4 and consists of two separate scripts, `xorsample.py` and `harvey.py`, where `xorsample.py` generates random inputs and `harvey.py` runs the tests using the inputs and the test oracle.

Generating random inputs. The task of generating a random sample of test inputs is reduced to computing answer sets of an ASP input generator. The simplest way would be to enumerate all answer sets and pick the required number of answer sets at random. Unfortunately, a complete enumeration is seldom an option as the number of answer sets will often be prohibitively large. However, to achieve a near-uniform distribution, an approach based on XOR streamlining [4] can be utilised. It has been used for SAT but is applicable to general combinatorial spaces. An XOR constraint is basically a parity constraint on a set S of atoms that expresses that an odd number of atoms from S has to be true. We can add such constraints to a program, each one cuts the solution space in half. We keep adding constraints until a complete enumeration is possible. Basically, the script `xorsample.py` re-implements the tool `xorro` [11] taking current ASP language features into account as well as utilising certain optimisations that are relevant for test-input generation.

Assume U is a set of propositional atoms. An *XOR constraint over U* is an expression of form $a_1 \oplus \cdots \oplus a_n$, where all a_i, $1 \leq i \leq n$, are from U. Such an XOR constraint is *satisfied* in an interpretation I iff an odd number of atoms from $\{a_1, \ldots a_n\}$ is true in I. We translate an XOR constraint into ASP rules as follows:

$$x_1 \leftarrow a_1. \quad \cdots \quad x_n \leftarrow a_n, \text{not } x_{n-1}. \quad x_n \leftarrow \text{not } a_n, x_{n-1}.$$

Then, we add the constraint $\bot \leftarrow \text{not } x_n$ to enforce that the XOR constraint is satisfied, or we add $\bot \leftarrow x_n$ to express that it must not be satisfied.

The process of computing random answer sets is as follows: First, we add a number of XOR constraints over the ground atoms to an encoding. For each atom and each XOR constraint, it is decided with a fixed probability p (default is 0.5) whether it should be in that constraint. Also, it is decided with probability 0.5 whether this constraint has to be satisfied or not. The probability p can be changed by the user; smaller values yield shorter constraints and help to boost performance in practice without changing the distribution too much from an ideal uniform one [3].

Computing a given number of random test inputs proceeds basically in three steps:

1. The input generator is grounded and s XOR constraints are added to the program. By default, $s = \log(n)$, where n is the number of atoms.
2. We try to enumerate all answer sets, if no answer set is found, the number of XOR constraints is decreased. If the enumeration cannot be completed within a given time limit, the number is increased.
3. If the number of resulting answer sets is greater than the number of inputs that we still need, we select n answer sets randomly and discard the rest. Otherwise, we keep all of them and proceed with Step 2, where we use a fresh set of XOR constraints.

Testing with random inputs. After test inputs have been generated, we next use them for actual testing. For each test input, we proceed as follows: We ground the program under test joined with the test input. We then use `clasp` to compute one answer set. We distinguish between two cases: First, if no answer set is found, we check whether we get no answer set with the test oracle as well. If so, the test case passed, otherwise it failed. For the second case, assume that $A = \{a_1, \ldots a_m\}$ is an answer set of the program under test. We need to verify that it is an answer set of the test oracle on the test input as well. To this end, we temporally add the following rules to the test oracle:

$$passed \leftarrow a_1, \ldots, a_m, \text{not } a_{m+1}, \ldots, \text{not } a_n. \quad \bot \leftarrow \text{not } passed.$$

Here, $\{a_{m+1}, \ldots, a_n\}$ is the set of atoms of the grounding of the program under test minus $\{a_1, \ldots, a_m\}$. This guarantees that the solution of the program under test is included in the solutions of the test oracle for a particular input instance. Accordingly, if the test oracle joined with the input has no answer sets, the test failed, otherwise it passed. Note that we could replace this part of Harvey with an external call to some other computational method to verify the correctness of a solution, therefore this testing framework can also be used when an ASP test oracle is not available.

Usage of Harvey. Assume that the file `oracle.lp` contains the encoding of our reviewer-assignment problem, `inp-gen.lp` contains the input generator as discussed above, and `program.lp` is some ASP encoding of the reviewer-assignment problem that we want to test. To simply check if solutions produced using `program.lp` are correct for 100 random inputs, we could run Harvey as follows:

```
python3.4 harvey.py --n=100 --g="-c n=5" --cf=inp-gen.lp
    --rf=oracle.lp --tf=program.lp
```

Option `--n` specifies the number of test inputs and option `--g` can be used to pass command-line arguments to `gringo`. In this example, we use it to set the constant `n` that is used in the input-generator to restrict the size of input instances. Likewise, `-c` can be used to pass command-line arguments to `clasp`. Options `--rf` and `--tf` are used to define the test oracle and the program under test, respectively. If we would like to run tests that check whether expected solutions are produced by the program under test, we can simply swap the two programs in the invocation. If run-time performance is poor, we could make use of option `--q`, e.g., by writing `--q=0.1`. This option defines the probability an atom is contained in an XOR constraint. Better to have a less uniform distribution than to have no test inputs at all. Another option is to set a time limit in seconds for `clasp` with option `--t` (the default is no time limit). Also, sometimes the initial choice of the number of XOR constraints that are added is not optimal. If the number of solutions can be estimated to be 2^s, then s would be a good number of constraints to start with and can be set with option `-s`.

4   Concluding Remarks

We presented Harvey, a tool for random testing ASP programs that relies on XOR streamlining [4] to achieve a near-uniform distribution of test inputs. An ASP engineer has more options than pure random testing, e.g., adding constraints or testing not only for correctness but also for completeness of solutions. Harvey provides support for an ASP engineer when a program is modified and one needs a simple method to check if emerging versions are still correct. Also, it can serve as a baseline for comparisons with other testing approaches. A detailed empirical evaluation of the tool is a goal for future work. Harvey can be downloaded from http://www.kr.tuwien.ac.at/research/systems/harvey/.

References

1. Beizer, B.: Black-Box Testing: Techniques for Functional Testing of Software and Systems. Wiley, New York (1999)
2. Chakraborty, S., Meel, K.S., Vardi, M.Y.: A scalable and nearly uniform generator of SAT witnesses. In: Sharygina, N., Veith, H. (eds.) CAV 2013. LNCS, vol. 8044, pp. 608–623. Springer, Heidelberg (2013). doi:10.1007/978-3-642-39799-8_40
3. Gomes, C.P., Hoffmann, J., Sabharwal, A., Selman, B.: Short XORs for model counting: from theory to practice. In: Marques-Silva, J., Sakallah, K.A. (eds.) SAT 2007. LNCS, vol. 4501, pp. 100–106. Springer, Heidelberg (2007). doi:10.1007/978-3-540-72788-0_13
4. Gomes, C.P., Sabharwal, A., Selman, B.: Near-uniform sampling of combinatorial spaces using XOR constraints. In: Proceedings of NIPS 2006, pp. 481–488. MIT Press (2006)
5. Janhunen, T., Niemelä, I., Oetsch, J., Pührer, J., Tompits, H.: On testing answer-set programs. In: Proceedings of ECAI 2010, pp. 951–956. IOS Press (2010)
6. Janhunen, T., Niemelä, I., Oetsch, J., Pührer, J., Tompits, H.: Random vs. structure-based testing of answer-set programs: An experimental comparison. In: LPNMR 2011, pp. 242–247. Springer (2011)
7. Janhunen, T., Oikarinen, E.: LPEQ and DLPEQ – translators for automated equivalence testing of logic programs. In: Lifschitz, V., Niemelä, I. (eds.) LPNMR 2004. LNCS, vol. 2923, pp. 336–340. Springer, Heidelberg (2003). doi:10.1007/978-3-540-24609-1_30
8. Myers, G.J.: Art of Software Testing. Wiley, Hoboken (1979)
9. Oetsch, J., Prischink, M., Pührer, J., Schwengerer, M., Tompits, H.: On the small-scope hypothesis for testing answer-set programs. In: Proceedings of KR 2012, pp. 43–53. AAAI Press (2012)
10. Oetsch, J., Seidl, M., Tompits, H., Woltran, S.: Testing relativised uniform equivalence under answer-set projection in the system ccT. In: Seipel, D., Hanus, M., Wolf, A. (eds.) INAP/WLP -2007. LNCS, vol. 5437, pp. 241–246. Springer, Heidelberg (2009). doi:10.1007/978-3-642-00675-3_16
11. Potassco – The Potsdam answer set solving collection. https://potassco.org/

NoHR: Integrating XSB Prolog with the OWL 2 Profiles and Beyond

Carlos Lopes, Matthias Knorr$^{(\boxtimes)}$, and João Leite

NOVA LINCS, Departamento de Informática, Faculdade de Ciências e Tecnologia,
Universidade Nova de Lisboa, 2829-516 Caparica, Portugal
`mkn@fct.unl.pt`

Abstract. We present the latest, substantially improved, version of
NoHR, a reasoner designed to answer queries over hybrid theories composed of an OWL ontology in Description Logics and a set of non-
monotonic rules in Logic Programming. Whereas the need to combine
the distinctive features of these two knowledge representation and reasoning approaches stems from real world applications, their integration is
nevertheless theoretically challenging due to their substantial semantical
differences. NoHR has been developed as a plug-in for the widely used
ontology editor Protégé - in fact, the first hybrid reasoner of its kind
for Protégé, building on a combination of reasoners dedicated to OWL
and rules - but it is also available as a library, allowing for its integration within other environments and applications. Compared to previous
versions of NoHR, this is the first that supports all polynomial OWL
profiles, and even beyond, allowing for its usage with real-world ontologies that do not fit within a single profile. In addition, NoHR has now
an enhanced integration with its rule engine, which provides support for
a vast number of standard built-in Prolog predicates that considerably
extend its usability.

Keywords: Query answering · Logic programming · Description logic
ontologies

1 Introduction

Ontology languages based on Description Logics (DLs) [4] and non-monotonic
rule languages as known from Logic Programming (LP) [6] are both well-known
formalisms in knowledge representation and reasoning (KRR) each with its own
distinct benefits and features. This is also witnessed by the emergence of the
Web Ontology Language (OWL) [13] and the Rule Interchange Format (RIF)
[17] in the ongoing standardization of the Semantic Web driven by the W3C.[1]

On the one hand, ontology languages have become widely used to represent and reason over taxonomic knowledge. Since DLs are (usually) decidable
fragments of first-order logic, they are monotonic by nature, which means that

---

[1] http://www.w3.org.

© Springer International Publishing AG 2017
M. Balduccini and T. Janhunen (Eds.): LPNMR 2017, LNAI 10377, pp. 236–249, 2017.
DOI: 10.1007/978-3-319-61660-5_22

drawn conclusions persist when adopting new additional information. Furthermore, they allow reasoning on abstract information, such as relations between classes of objects, even without knowing any concrete instances. The balance between expressiveness and complexity of reasoning with ontology languages, inherited from DLs, is witnessed by the fact that the very expressive general language OWL 2, with its high worst-case complexity, includes three tractable (polynomial) profiles [20] each with a different application purpose in mind.

On the other hand, non-monotonic rules explicitly represent inference, from premises to conclusions, focusing on reasoning over instances. They commonly employ the Closed World Assumption (CWA), i.e., the absence of a piece of information suffices to derive it being false, until new information to the contrary is provided, hence being non-monotonic. This permits to declaratively model defaults and exceptions, in the sense that the absence of an exceptional feature can be used to derive that the (more) common case applies, and also integrity constraints, which can be used to ensure that the data under consideration is conform to the desired specifications.

Combining both formalisms, though a non-trivial problem due to the mismatch between semantic assumptions of the two formalisms, and the considerable differences as to how decidability is ensured in each of them, has been frequently requested by applications [1,22,23]. For example, in clinical health care, large ontologies such as SNOMED CT,[2] that are captured by the OWL 2 profile OWL 2 EL and its underlying description logic (DL) $\mathcal{EL}^{++}$ [5], are used for electronic health record systems, clinical decision support systems, or remote intensive care monitoring, to name only a few. Yet, expressing conditions such as dextrocardia, i.e., that the heart is exceptionally on the right side of the body, is not possible and requires non-monotonic rules. Another example can be found in [22], where modeling pharmacy data of patients with the closed-world assumption would have been preferred in the study to match patient records with clinical trials criteria, because usually it can be assumed that a patient is not under a specific medication unless explicitly known. Also, in [1] it is shown that in Legal Reasoning, besides the well known need for default reasoning afforded by non-monotonic rules, it is also necessary to reason in the absence of concrete known individuals (instances), hence requiring features found in ontology languages such as DL. Moreover, another application scenario can be found in the risk assessment of cargo shipment, which we will describe in more detail in Sect. 2.3. Notably, ontologies developed for such applications, are often covered by the constructors the polynomial OWL 2 profiles provide, but do not necessarily fall precisely into one of those profiles.

In this paper, we describe the latest version of NoHR[3] (Nova Hybrid Reasoner), a plug-in for the ontology editor Protégé 5.X,[4] that allows the user to query combinations of ontologies and non-monotonic rules in a top-down manner, which substantially extends the usability of NoHR w.r.t. both the ontology and the rule part.

---

[2] http://www.ihtsdo.org/snomed-ct/.
[3] http://nohr.di.fct.unl.pt.
[4] http://protege.stanford.edu.

NoHR is theoretically founded on the formalism of Hybrid MKNF under the well-founded semantics [18] which comes with two main arguments in its favor (cf. the related work in [9,21] on combining DLs with non-monotonic rules). First, the overall approach, introduced in [21] and based on the logic of minimal knowledge and negation as failure (MKNF) [19], provides a very general and flexible framework for combining DL ontologies and non-monotonic rules (see [21]). Second, [18], which is a variant of [21] based on the well-founded semantics [10] for logic programs, has a lower data complexity than the former – it is polynomial for polynomial DLs – and is amenable for applying top-down query procedures, such as $\mathbf{SLG}(\mathcal{O})$ [2], to answer queries based only on the information relevant for the query, and without computing the entire model – no doubt a crucial feature when dealing with large ontologies and huge amounts of data.

NoHR – the first Protégé plug-in to integrate non-monotonic rules and top-down queries – is implemented in a way that combines the capabilities of the DL reasoners ELK [16], HermiT [11], and Konclude [24] with the rule engine XSB Prolog,[5] exhibiting the following additional features:[6]

- Support for ontologies written in any of the three tractable OWL 2 Profiles, and even beyond for those combining all the permitted constructors;
- Support for a vast number of standard built-in Prolog predicates;
- Rule editor within Protégé accompanied with a completely overhauled rule parser to match the novel extensions;
- Possibility to define predicates with arbitrary arity in Protégé;
- Guaranteed termination of query answering;
- Robustness w.r.t. inconsistencies between the ontology and the rules;
- Scalable fast interactive response times.

2   Hybrid MKNF Knowledge Bases

We start with illustrating the kind of hybrid knowledge bases considered by NoHR. First, we present an overview on the kind of ontologies in description logics permitted here, referring to [4] for a more general and thorough introduction to DLs. Then we recall hybrid MKNF knowledge bases, followed by an example scenario and some general considerations on how to perform query answering.

2.1   Description Logics

Description logics (DLs) are usually decidable fragments of first-order logic, commonly defined over disjoint countably infinite sets of *concept names* N_C, *role names* N_R, and *individual names* N_I, matching unary predicates, binary predicates and constants resp. Building on these, *complex concepts* (and sometimes also *complex roles*) are introduced based on the logical constructors a concrete

---

[5] http://xsb.sourceforge.net.
[6] NoHR 3.0 Beta can be downloaded for testing from http://nohr.di.fct.unl.pt/.

DL admits to be used. An ontology $\mathcal{O}$ then is a finite set of *inclusion axioms* of the form $C \sqsubseteq D$ where C and D are both (complex) concepts (or roles) and *assertions* of the form $C(a)$ or $R(a, b)$ for concepts C, roles R, and individuals a, b. The semantics of such ontologies is defined in terms on *interpretation* $\mathcal{I} = (\Delta^{\mathcal{I}}, \cdot^{\mathcal{I}})$ consisting of a non-empty domain $\Delta^{\mathcal{I}}$ and an *interpretation function* $\cdot^{\mathcal{I}}$ in a standard way for first-order logic. Typical reasoning tasks are model-checking or consistency or classification which requires computing all concept inclusions between atomic concepts entailed by $\mathcal{O}$.

While the description logic $\mathcal{SROIQ}$ underlying the W3C standard OWL 2 is very general and highly expressive as it admits many different constructors, reasoning with it is highly complex, which is why the profiles OWL 2 EL, OWL 2 QL and OWL 2 RL have been defined [20] for which reasoning is tractable. Since a detailed account on the full technical specification of these profiles would be beyond the scope of this paper, we rather give a brief overview on important features as supported in NoHR.

First, $\mathcal{EL}_{\perp}^{+}$, a large fragment of $\mathcal{EL}^{++}$ [5], the DL underlying the tractable profile OWL 2 EL [20], only allows conjunction of concepts, existential restriction of concepts ($\exists R.C$ for roles R and concepts C basically corresponding to $\forall x \exists y R(x, y) \wedge C(y)$), hierarchies of roles, and disjoint concepts. $\mathcal{EL}_{\perp}^{+}$ is tailored towards reasoning with large conceptual models, i.e., large TBoxes.

DL-Lite$_R$, one language of the *DL-Lite* family [3] and underlying OWL 2 QL, admits in addition role inverses and disjoint roles, but in exchange only simple role hierarchies, no conjunction, and limitations on the use of existential restrictions in particular on the left hand side of inclusion axioms. This profile focuses on answering queries over huge amounts of data, and is amenable to the usage of relational database technology.

Finally, Description Logic Programs [12], which underly OWL 2 RL, have been introduced with the aim to find a fragment that can be directly translated into rules (and vice-versa), and therefore can also be implemented using a rule reasoner. Due to that, this profile allows more constructors than the other two profiles, but usually restricts their usage to one side of the inclusion axioms to ensure that the translation to rules is possible.

Many ontologies matching one of these profiles exist and are used in applications, but there are also those that do use features from more than one of those, such as the bio-ontology Galen[7] or the famous benchmark ontology LUBM.[8] For these, reasoning is no longer polynomial, but highly efficient general purpose OWL reasoners nevertheless make their usage feasible in practice.

2.2 MKNF Knowledge Bases

MKNF knowledge bases (KBs) build on the logic of minimal knowledge and negation as failure (MKNF) [19]. Two main different semantics have been defined

---

[7] https://bioportal.bioontology.org/ontologies/GALEN.

[8] http://swat.cse.lehigh.edu/projects/lubm/.

[18,21], and we focus on the well-founded version [18], due to its lower computational complexity and amenability to top-down querying without computing the entire model. Here, we only point out important notions following [14], and refer to [18] and [2] for the details.

MKNF knowledge bases as presented in [2] combine an ontology and a set of non-monotonic rules (similar to a normal logic program).

A *rule* r is of the form $H \leftarrow A_1, \ldots, A_n, \textbf{not } B_1, \ldots, \textbf{not } B_m$ where the *head* of r, H, and all A_i with $1 \leq i \leq n$ and B_j with $1 \leq j \leq m$ in the *body* of r are atoms. A *program* $\mathcal{P}$ is a finite set of rules, $\mathcal{O}$ is an ontology, and an *MKNF knowledge base* $\mathcal{K}$ is a pair $(\mathcal{O}, \mathcal{P})$. A rule r is *safe* if all its variables occur in at least one A_i with $1 \leq i \leq n$, and $\mathcal{K}$ is *safe* if all its rules are safe.[9]

The semantics of MKNF knowledge bases $\mathcal{K}$ is usually given by a translation π into an MKNF formula $\pi(\mathcal{K})$, i.e., a formula over first-order logic extended with two modal operators $\textbf{K}$ and $\textbf{not}$. It is shown in [18], that if $\mathcal{K}$ is *MKNF-consistent*, then a unique model of $\mathcal{K}$ can be determined. Here, $\mathcal{K}$ may indeed not be MKNF-consistent if the ontology alone is unsatisfiable, or by the combination of appropriate axioms in $\mathcal{O}$ and rules in $\mathcal{P}$, e.g., axiom $A \sqsubseteq \neg B$ in $\mathcal{O}$, and facts $A(a)$ and $B(a)$ in $\mathcal{P}$. In the former case, we argue that the ontology alone should be consistent and be repaired if necessary before combining it with non-monotonic rules. Thus, we assume that $\mathcal{O}$ occurring in $\mathcal{K}$ is consistent, which does not truly constitute a restriction as we can always make $\mathcal{O}$ by appropriately turning assertions in $\mathcal{O}$ into rules without any effect on the semantics of $\mathcal{K}$.

2.3 Use Case

The customs service for any developed country assesses imported cargo for a variety of risk factors including terrorism, narcotics, food and consumer safety, pest infestation, tariff violations, and intellectual property rights.[10] Assessing this risk, even at a preliminary level, involves extensive knowledge about commodities, business entities, trade patterns, government policies and trade agreements. Some of this knowledge may be external to a given customs agency: for instance the broad classification of commodities according to the international Harmonized Tariff System (HTS), or international trade agreements. Other knowledge may be internal to a customs agency, such as lists of suspected violators or of importers who have a history of good compliance with regulations.

Figure 1 shows a simplified fragment $\mathcal{K} = (\mathcal{O}, \mathcal{P})$ of such a knowledge base. In this fragment, a shipment has several attributes: the country of its origination, the commodity it contains, its importer and producer. The ontology contains a geographic classification, along with information about producers who are located in various countries. It also contains (partial) information about three

---

[9] In general, the notion of DL-safety is used in this context which requires that these variables occur in atoms that do themselves not occur in the ontology, but due to the particular reasoning method employed here, we can relax that restriction.

[10] The system described here, originally presented in [23], is not intended to reflect the policies of any country or agency.

* * * $\mathcal{O}$ * * *

Commodity ≡ (∃HTSCode.⊤)   Tomato ⊑ EdibleVegetable
CherryTomato ⊑ Tomato   GrapeTomato ⊑ Tomato
CherryTomato ⊓ GrapeTomato ⊑ ⊥   Bulk ⊓ Prepackaged ⊑ ⊥
EURegisteredProducer ≡ (∃RegisteredProducer.EUCountry)
LowRiskEUCommodity ≡ (∃ExpeditableImporter.⊤) ⊓ (∃CommodCountry.EUCountry)
Inspection ⊓ NoInspection ⊑ ⊥

ShpmtCommod(s_1, c_1)   ShpmtDeclHTSCode(s_1, h7022)
ShpmtImporter(s_1, i_1)   CherryTomato(c_1)   Bulk(c_1)
ShpmtCommod(s_2, c_2)   ShpmtDeclHTSCode(s_2, h7022)
ShpmtImporter(s_2, i_2)   GrapeTomato(c_2)   Prepackaged(c_2)
ShpmtCountry($s_2, turkey$)
ShpmtCommod(s_3, c_3)   ShpmtDeclHTSCode(s_3, h7021)
ShpmtImporter(s_3, i_3)   GrapeTomato(c_3)   Bulk(c_3)
ShpmtCountry($s_3, portugal$)   ShpmtProducer(s_3, p_1)
RegisteredProducer($p_1, portugal$)   EUCountry($portugal$)
RegisteredProducer($p_2, slovakia$)   EUCountry($slovakia$)

* * * $\mathcal{P}$ * * *

AdmissibleImporter($\mathbf{x}$) ← ShpmtImporter($\mathbf{y}, \mathbf{x}$), **not** SuspectedBadGuy($\mathbf{x}$).
SuspectedBadGuy(i_1).
ApprovedImporterOf($i_2, \mathbf{x}$) ← EdibleVegetable($\mathbf{x}$).
ApprovedImporterOf($i_3, \mathbf{x}$) ← GrapeTomato($\mathbf{x}$).
CommodCountry($\mathbf{x}, \mathbf{y}$) ← ShpmtCommod($\mathbf{z}, \mathbf{x}$), ShpmtCountry($\mathbf{z}, \mathbf{y}$).
ExpeditableImporter($\mathbf{x}, \mathbf{y}$) ← ShpmtCommod($\mathbf{z}, \mathbf{x}$), ShpmtImporter($\mathbf{z}, \mathbf{y}$),
                              AdmissibleImporter($\mathbf{y}$), ApprovedImporterOf($\mathbf{y}, \mathbf{x}$).
CompliantShpmt($\mathbf{x}$) ← ShpmtCommod($\mathbf{x}, \mathbf{y}$), HTSCode($\mathbf{y}, \mathbf{z}$), ShpmtDeclHTSCode($\mathbf{x}, \mathbf{z}$).
Random($\mathbf{x}$) ← ShpmtCommod($\mathbf{x}, \mathbf{y}$), **not** Random($\mathbf{x}$).
NoInspection($\mathbf{x}$) ← ShpmtCommod($\mathbf{x}, \mathbf{y}$), CommodCountry($\mathbf{y}, \mathbf{z}$), EUCountry($\mathbf{z}$).
Inspection($\mathbf{x}$) ← ShpmtCommod($\mathbf{x}, \mathbf{y}$), **not** NoInspection($\mathbf{x}$), Random($\mathbf{x}$).
Inspection($\mathbf{x}$) ← ShpmtCommod($\mathbf{x}, \mathbf{y}$), **not** CompliantShpmt($\mathbf{x}$).
Inspection($\mathbf{x}$) ← ShpmtCommod($\mathbf{x}, \mathbf{y}$), Tomato($\mathbf{y}$), ShpmtCountry($\mathbf{x}, slovakia$).
HTSChapter($\mathbf{x}, 7$) ← EdibleVegetable($\mathbf{x}$).
HTSHeading($\mathbf{x}, 702$) ← Tomato($\mathbf{x}$).
HTSCode($\mathbf{x}, $h7022) ← CherryTomato($\mathbf{x}$).
HTSCode($\mathbf{x}, $h7021) ← GrapeTomato($\mathbf{x}$).
TariffCharge($\mathbf{x}, 0$) ← CherryTomato($\mathbf{x}$), Bulk($\mathbf{x}$).
TariffCharge($\mathbf{x}, 40$) ← GrapeTomato($\mathbf{x}$), Bulk($\mathbf{x}$).
TariffCharge($\mathbf{x}, 50$) ← CherryTomato($\mathbf{x}$), Prepackaged($\mathbf{x}$).
TariffCharge($\mathbf{x}, 100$) ← GrapeTomato($\mathbf{x}$), Prepackaged($\mathbf{x}$).

Fig. 1. MKNF knowledge base for Cargo Imports

shipments: s_1, s_2 and s_3. There is also a set of rules indicating information about importers, and about whether to inspect a shipment either to check for compliance of tariff information or for food safety issues. For that purpose, the set of rules also includes a classification of commodities based on their harmonized tariff information (HTS chapters, headings and codes, cf. http://www.usitc.gov/tata/hts), and tariff information, based on the classification of commodities as given by the ontology.

The overall task then is to access all the information and assess whether some shipment should be inspected in full detail, under certain conditions randomly, or not at all. In fact, an inspection is considered if either a random inspection is indicated, or some shipment is not compliant, i.e., there is a mismatch between the filed cargo codes and the actually carried commodities, or some suspicious cargo is observed, in this case tomatoes from slovakia. In the first case, a potential random inspection is indicated whenever certain exclusion conditions do not hold. To ensure that one can distinguish between strictly required and random inspections, a random inspection is assigned the truth value undefined based on the rule Random$(\mathbf{x})$ ← ShpmtCommod$(\mathbf{x}, \mathbf{y})$, **not** Random$(\mathbf{x})$. Note that the example indeed utilizes the features of rules and ontologies: for example exceptions to the potential random inspections can be expressed, but at the same time, taxonomic and non-closed knowledge is used, e.g., some shipment may in fact originate from the EU, this information is just not available.

Querying MKNF knowledge bases is based on $\mathbf{SLG}(\mathcal{O})$, as defined in [2]. This procedure extends SLG resolution with tabling [7] with an *oracle* to $\mathcal{O}$ that handles ground queries to the DL-part of $\mathcal{K}$ by returning (possibly empty) sets of atoms that, together with $\mathcal{O}$ and information already proven true, allows us to derive the queried atom. We refer to [2] for the full account of $\mathbf{SLG}(\mathcal{O})$. E.g., the result of querying the example knowledge base for Inspection$(\mathbf{x})$ reveals that of the three shipments, s_2 requires an inspection (due to mislabeling), s_1 may be subject to a random inspection as it does not knowingly originate from the EU, while s_3 is indicated as inconsistent, due to the fact that Inspection and NoInspection hold, which are required to be disjoint.

3   NoHR Plug-in

In this section, we describe the architecture of the plug-in for Protégé as shown in Fig. 2 and discuss some features of the implementation, in particular the new ones that considerably extend the applicability of the tool.

3.1   Architecture

The input for the plug-in consists of an OWL file written in a description logic as described in Sect. 2.1, which can be loaded and manipulated as usual in Protégé, and a rule file. For the latter, we provide a tab called NoHR Rules that allows us to load, save and edit rule files in a dedicated panel.

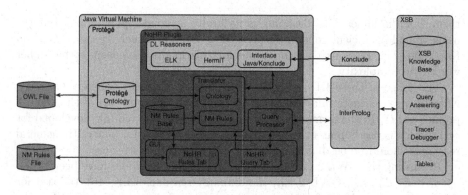

Fig. 2. System architecture of NoHR

The NoHR Query tab also allows for the visualization of the rules, but its main purpose is to provide an interface for querying the combined KB. Whenever the first query is posed by pushing "Execute", the translator is started, which with the help of one of the integrated OWL ontology reasoners ELK [16], HermiT [11], and Konclude [24] transforms the ontology axioms into a set of rules that are equivalent to the ontology axioms in terms of answers to ground queries (see details in Sect. 3.2). The translation result is joined with the given non-monotonic rules in $\mathcal{P}$, which is further transformed if inconsistency detection is required (in the presence of certain DL constructs in the ontology, such as DisjointWith axioms).

The result is used as input for the top-down query engine XSB Prolog which realizes the well-founded semantics for logic programs [10]. The transfer to XSB is realized via InterProlog,[11] which is an open-source Java front-end allowing the communication between Java and a Prolog engine.

Next, the query is sent via InterProlog to XSB, and answers are returned to the query processor, which collects them and sets up a table showing for which variable substitutions we obtain true, undefined, or inconsistent valuations (or just shows the truth value for a ground query). XSB itself not only answers queries very efficiently in a top-down manner, with tabling, it also avoids infinite loops, thus ensuring termination of the query process.

Once the query has been answered, the user may pose other queries, and the system will simply send them directly without any repeated preprocessing. If the user changes data in the ontology or in the rules, then the system recompiles, but always restricted to the part that actually changed.

3.2 Support for OWL 2 Profiles

The new version of NoHR supports the usage of ontologies written in any of the OWL 2 profiles and even those that combine expressive features from several of

---

[11] http://interprolog.com/java-bridge/.

them. For the latter, DL reasoning is naturally no longer polynomial, but the usage of highly efficient general purpose DL reasoners, namely HermiT [11], and Konclude [24], allow us to compensate for this in practice despite the higher worst-case complexity.

The usage of these DL reasoners depends in fact on the concrete ontology. Namely, a switch determines whether the ontology falls into one of the OWL 2 profiles or not. In the former case, the specific translation module developed for each profile is used, unless the user indicates specifically that one of the general reasoners should be used, which are used anyway in the latter case.

Regarding the OWL 2 profiles, specific translation modules have been developed. For $\mathcal{EL}_{\perp}^{+}$, i.e., OWL 2 EL, the ontology reasoner ELK [16], tailored for $\mathcal{EL}_{\perp}^{+}$ and considerably faster than other reasoners when comparing classification time, is used to classify the ontology resulting from normalizing the given ontology $\mathcal{O}$ to ensure that no possible derivations w.r.t. answering ground queries are lost during the subsequent translation into rules.[12] The inferred axioms together with $\mathcal{O}$ are translated discarding certain axioms which are irrelevant for answering ground queries. For $DL\text{-}Lite_R$, i.e., OWL 2 QL, a dedicated direct translation without prior classification is used, introducing some auxiliary predicates instead to compensate for the missing inferred axioms (see [8,14] for the respective details on both approaches). The new module for DLP, i.e., OWL 2 RL, makes use of a direct translation that does not require any auxiliary predicates as the profile supports the direct translation into rules by design.

Concerning an ontology that does not fall into one single profile, a new general translation module is used that follows the methodology employed for $\mathcal{EL}_{\perp}^{+}$, i.e., (partial) normalization, classification, and translation of the (relevant majority of the) inferred axioms, where the translation function itself results from a merge of those for the single profiles. The reason why we apply two different general purpose DL reasoners is as follows. It has been shown that Konclude is in general the most efficient reasoner currently available [24], unfortunately, unlike Protégé, it is not programmed in Java, but in C++, and the necessary effort to interface with Konclude constitutes a considerable drain on efficiency of reasoning, because no native Java interface compatible with the current version of the OWL API exists. HermiT, on the other side, does not suffer from that problem, but has some limitations when reasoning with very large ontologies [11]. Therefore, both reasoners are integrated and the user can choose in the preference panel which of the two should be employed.

Note that if we consider the hybrid knowledge base as presented in Fig. 1 whose ontology is in $\mathcal{EL}_{\perp}^{+}$, and query for Inspection(x), then we would obtain that $s1$ is undefined, s_2 is true, and s_3 is inconsistent. However, if we add an additional property ImportedBy and declare it to be the inverse of ShpmtImporter, then the resulting ontology would no longer fit any profile and the previous version of NoHR would simply cease to work. In our new version, the ontology change is detected and the general translation module is used instead, that is,

---

[12] Similar techniques have been used independently that allow the usage of OWL 2 RL reasoners for answering ground queries for ontologies outside of OWL 2 RL [25].

NoHR is capable to interactively adjust to changes that alter the DL fragment of the considered ontology.

Finally, note that, while HermiT and Konclude are in fact applicable to arbitrary DL ontologies, the method applied in NoHR is not: it is well known that constructors such as first-order disjunction cannot be captured in non-monotonic rules in an equivalent way, which is why the approach is focused on the constructors occurring in the OWL 2 profiles.

3.3 Extended Prolog Support

The rule editor and parser have also been significantly improved, both in terms of user-friendliness and applicability. To begin with, the rule syntax has been slightly changed so that variables are now always written with a leading "?". While this deviates from standard prolog syntax, it permits the user and the system to differentiate between variables and constant names from the ontology where, in general, there is no restriction to only have names in lowercase. This is also aligned with common rule notation in SWRL,[13] a monotonic first-order rule language that has been used in Protégé.

With the new version, NoHR was also extended with the ability to use a large set of inbuilt XSB prolog predicates. This extension allows the user to specify rules that make use of arithmetic, comparison and list-based predicates for enriching the knowledge base. The novel features include:

– Built-in XSB Prolog predicates prefixed with #. For example:

$$\texttt{A(?X) :- B(?X), C(?Y), \#compare("<",?X,?Y).}$$

– Inline form of built-in XSB Prolog predicates. For example:

$$\texttt{A(?X) :- B(?Y), C(?Z), ?X is (?Y + ?Z) * 2.}$$

– Numeric and list expressions within rule syntax and integration with numeric datatypes from OWL 2.

In addition, the rule editor allows switching between the labels of ontology names and their Internationalized Resource Identifiers (IRIs), which can be useful in case the labels used in the ontology are not unique.

4   Evaluation

Previous tests on the $\mathcal{EL}^+_\bot$ component have shown that a) different $\mathcal{EL}$ ontologies can be preprocessed for querying in a short period of time (around one minute for SNOMED CT with over 300,000 concepts), b) adding rules increases the time of the translation only linearly, and c) querying time is in general neglectable, in comparison to a) and b) [14]. In subsequent tests on improved versions of

---

[13] http://www.w3.org/Submission/SWRL/.

	Axioms	EL	ELK	HermiT	Konclude
Fly Anatomy	19,211	yes	0.75	1.54	1.63
Full-Galen	37,696	no	–	–	11.21
LUBM1	93	no	–	1.81	3.82
Snomed Anatomy	40,485	yes	1.76	3.08	12.39
Snomed CT	294,479	yes	13.89	–	63.71

Fig. 3. Reasoners average classification run times (in seconds) when called from Java.

both components (for $\mathcal{EL}^+_\perp$ and $DL\text{-}Lite_R$) [8], we have shown that i) NoHR scales reasonably well for OWL QL query answering without non-monotonic rules (only slowing down for memory-intensive cases), ii) preprocessing is even faster when compared to NoHR's previous version using a classifier (for EL), iii) querying scales well, even for over a million facts/assertions in the ABox, despite being slightly slower on average in comparison to EL, and iv) adding rules scales linearly for pre-processing and querying, even for an ontology with many negative inclusions (for $DL\text{-}Lite_R$).

Here, we are additionally interested as to how using the general reasoners compares to applying the dedicated translation module. For that purpose, we first compared how the preprocessing period is affected by the usage of the different reasoners using standard ontologies of different expressiveness (in Fig. 3, we mention the number of inclusion axioms and whether the ontology fits the OWL 2 EL profile). All tests were performed on a 4 GHz Intel i7 processor with 16 GB under Windows 7 (64-bit) with 12 GB of maximum memory for Java. The results are shown in Fig. 3. We can observe that ELK is indeed always fastest, which confirms that by default, whenever the ontology fits $\mathcal{EL}^+_\perp$, the dedicated translation module should be used. Additionally, we observe that HermiT is faster than Konclude in all instances where it does not time out, which justifies using HermiT whenever it is capable of classifying the given ontology. Still, Konclude turns out to be useful, in the cases where HermiT fails to classify an ontology that does not fit the OWL 2 EL profile.

In addition, we also compared HermiT and the translation module for OWL 2 QL that does not use any classifier. This test is actually quite similar to the one in [8] comparing the modules for QL and EL. The difference is that we can use a more expressive ontology that only fits the QL profile, and not EL. Figure 4 shows the results for ii) where we considered LUBM[14], a standard benchmark for evaluating queries over a large data set, which also includes a given set of standard queries. We created instances of $LUBM_n$ with $n = 1, 5, 10$ using the provided generator, and a restricted version of LUBM which fits OWL QL (thus rendering only one standard query meaningless), with the number of assertions ranging from roughly 100,000 to over 1,400,000. Note that "Initialization" includes loading the ontology and, for HermiT, also classifying it, "Ontology Processing" includes the actual translation, and "XSB Processing" the writing of the rule file and loading it in XSB. We observe that QL is considerably faster,

---

[14] http://swat.cse.lehigh.edu/projects/lubm/.

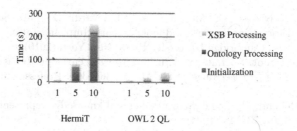

Fig. 4. Preprocessing time for LUBM for the two translation modes

indeed up to factor 35 for $LUBM_{10}$, which is mainly due to the absence of a classification step. When querying, we observed a slight compensation as running queries using the HermiT translation is slightly faster (up to factor 2.5), but faster preprocessing clearly favors the dedicated OWL 2 QL module here.

5 Conclusions

The Protégé plugin NoHR – also distributed as an API – affords us the possibility to query knowledge bases composed of both an ontology in one of the OWL 2 profiles and even a union of their language features and a set of non-monotonic rules, using a top-down reasoning approach, which means that only the part of the ontology and rules that is relevant for the query is actually evaluated. Its sound theoretical foundation together with the fast interactive response times make NoHR a truly one-of-a-kind reasoner and the novel extensions have considerably widened its applicability.

Ongoing work focuses on supporting different sets of rules and allowing the import of other rule file formats such as Prolog. In terms of future work, the integration of database access is of interest as this could considerably reduce the time it takes to load prolog files with huge amounts of data in XSB. Also, adjusting NoHR to the paraconsistent semantics for MKNF knowledge bases of [15] would provide better support to the already observed paraconsistent behavior.

Acknowledgments. We would like to acknowledge the valuable contribution of both Nuno Costa and Vadim Ivanov to the development of NoHR. This work was partially supported by Fundação para a Ciência e a Tecnologia (FCT) under UID/CEC/04516/2013, and grant SFRH/BPD/86970/2012 (M. Knorr).

References

1. Alberti, M., Knorr, M., Gomes, A.S., Leite, J., Gonçalves, R., Slota, M.: Normative systems require hybrid knowledge bases. In: AAMAS, pp. 1425–1426 (2012)
2. Alferes, J.J., Knorr, M., Swift, T.: Query-driven procedures for hybrid MKNF knowledge bases. ACM Trans. Comput. Log. **14**(2), 1–43 (2013)
3. Artale, A., Calvanese, D., Kontchakov, R., Zakharyaschev, M.: The *DL-Lite* family and relations. J. Artif. Intell. Res. (JAIR) **36**, 1–69 (2009)

4. Baader, F., Calvanese, D., McGuinness, D.L., Nardi, D., Patel-Schneider, P.F. (eds.): The Description Logic Handbook: Theory, Implementation, and Applicationsm, 3rd edn. Cambridge University Press, New York (2010)

5. Baader, F., Brandt, S., Lutz, C.: Pushing the $\mathcal{EL}$ envelope. In: Kaelbling, L.P., Saffiotti, A. (eds.) Proceedings of IJCAI, pp. 364–369. Professional Book Center (2005)

6. Baral, C., Gelfond, M.: Logic programming and knowledge representation. J. Log. Program. **19**(20), 73–148 (1994)

7. Chen, W., Warren, D.S.: Tabled evaluation with delaying for general logic programs. J. ACM **43**(1), 20–74 (1996)

8. Costa, N., Knorr, M., Leite, J.: Next step for NoHR: OWL 2 QL. In: Arenas, M., et al. (eds.) ISWC 2015. LNCS, vol. 9366, pp. 569–586. Springer, Cham (2015). doi:10.1007/978-3-319-25007-6_33

9. Eiter, T., Ianni, G., Lukasiewicz, T., Schindlauer, R., Tompits, H.: Combining answer set programming with description logics for the semantic web. Artif. Intell. **172**(12–13), 1495–1539 (2008)

10. Gelder, A.V., Ross, K.A., Schlipf, J.S.: The well-founded semantics for general logic programs. J. ACM **38**(3), 620–650 (1991)

11. Glimm, B., Horrocks, I., Motik, B., Stoilos, G., Wang, Z.: Hermit: an OWL 2 reasoner. J. Autom. Reason. **53**(3), 245–269 (2014)

12. Grosof, B.N., Horrocks, I., Volz, R., Decker, S.: Description logic programs: combining logic programs with description logic. In: Hencsey, G., White, B., Chen, Y.R., Kovács, L., Lawrence, S. (eds.) Proceedings of WWW, pp. 48–57. ACM (2003)

13. Hitzler, P., Krötzsch, M., Parsia, B., Patel-Schneider, P.F., Rudolph, S. (eds.): OWL 2 Web Ontology Language: Primer, 2nd edn. W3C Recommendation, Cambridge (2012). http://www.w3.org/TR/owl2-primer/

14. Ivanov, V., Knorr, M., Leite, J.: A query tool for $\mathcal{EL}$ with non-monotonic rules. In: Alani, H., et al. (eds.) ISWC 2013. LNCS, vol. 8218, pp. 216–231. Springer, Heidelberg (2013). doi:10.1007/978-3-642-41335-3_14

15. Kaminski, T., Knorr, M., Leite, J.: Efficient paraconsistent reasoning with ontologies and rules. In: Yang, Q., Wooldridge, M. (eds.) Proceedings of IJCAI (2015)

16. Kazakov, Y., Krötzsch, M., Simančík, F.: The incredible ELK: from polynomial procedures to efficient reasoning with $\mathcal{EL}$ ontologies. J. Autom. Reason. **53**, 1–61 (2013)

17. Kifer, M., Boley, H. (eds.): RIF Overview, 2nd edn. W3C Working Group Note, Cambridge (2013). http://www.w3.org/TR/rif-overview/

18. Knorr, M., Alferes, J.J., Hitzler, P.: Local closed world reasoning with description logics under the well-founded semantics. Artif. Intell. **175**(9–10), 1528–1554 (2011)

19. Lifschitz, V.: Nonmonotonic databases and epistemic queries. In: Mylopoulos, J., Reiter, R. (eds.) Proceedings of IJCAI, pp. 381–386. Morgan Kaufmann (1991)

20. Motik, B., Grau, B.C., Horrocks, I., Wu, Z., Fokoue, A., Lutz, C. (eds.): OWL 2 Web Ontology Language: Profiles, 2nd edn. W3C Recommendation, Cambridge (2012). http://www.w3.org/TR/owl2-profiles/

21. Motik, B., Rosati, R.: Reconciling description logics and rules. J. ACM **57**(5), 93–154 (2010)

22. Patel, C., Cimino, J., Dolby, J., Fokoue, A., Kalyanpur, A., Kershenbaum, A., Ma, L., Schonberg, E., Srinivas, K.: Matching patient records to clinical trials using ontologies. In: Aberer, K., et al. (eds.) ASWC/ISWC -2007. LNCS, vol. 4825, pp. 816–829. Springer, Heidelberg (2007). doi:10.1007/978-3-540-76298-0_59

23. Slota, M., Leite, J., Swift, T.: On updates of hybrid knowledge bases composed of ontologies and rules. Artif. Intell. **229**, 33–104 (2015)

24. Steigmiller, A., Liebig, T., Glimm, B.: Konclude: system description. J. Web Sem. **27**, 78–85 (2014)

25. Stoilos, G., Cuenca Grau, B., Motik, B., Horrocks, I.: Repairing ontologies for incomplete reasoners. In: Aroyo, L., Welty, C., Alani, H., Taylor, J., Bernstein, A., Kagal, L., Noy, N., Blomqvist, E. (eds.) ISWC 2011. LNCS, vol. 7031, pp. 681–696. Springer, Heidelberg (2011). doi:10.1007/978-3-642-25073-6_43

ArgueApply: A Mobile App for Argumentation

Jörg Pührer[(✉)]

Institute of Computer Science, Leipzig University, Leipzig, Germany
puehrer@informatik.uni-leipzig.de

Abstract. Formal models developed in the field of argumentation allow for analysing and evaluating problems that have previously been studied by philosophers on an informal level only. Importantly, they also give rise to the development of computational tools for argumentation. In this paper we report on `ArgueApply`, a mobile app for argumentation that is based on the Grappa framework, an extension of, e.g., abstract argumentation in the sense of Dung. With `ArgueApply` users can engage in online discussions and evaluate their semantics. Each of the resulting interpretations can be seen as a coherent view on a discussion in which some of the discussions statements are accepted and others rejected. Being a mobile tool, `ArgueApply` is intended to be more accessible than existing systems for computing argumentation semantics allowing, e.g., for spontaneous analysis of an ongoing discussion or collective preparation for an important debate. While having a practical system for these applications is the final goal of our work, an immediate objective is using the system for exploring which type of Grappa frameworks under which semantics are best suited for such applications.

Keywords: Argumentation · Grappa · Mobile application · Android · Answer-set programming

1   Introduction

Argumentation is an area at the intersection of Philosophy and several sub-fields of Artificial Intelligence (AI), in particular knowledge representation, non-monotonic reasoning, and multi-agent systems. It has seen a steady rise of interest over the last two decades, mainly due to new AI techniques which allow for a formal investigation of problems that have been studied informally only by philosophers, and which also allow for the development of computational tools for argumentation. In this work, we present `ArgueApply` a mobile application that makes use of recent advancements in formal argumentation. It allows users to participate in online debates and to analyse these discussions as well as their inherent viewpoints revealed by argumentation semantics. To this end, `ArgueApply` uses the Grappa framework [1], a recent extension of argumentation frameworks (AFs) by Dung [2]. Grappa was introduced to make the

---

This work was supported by the German Research Foundation (DFG) under grant BR 1817/7-2.

M. Balduccini and T. Janhunen (Eds.): LPNMR 2017, LNAI 10377, pp. 250–262, 2017.
DOI: 10.1007/978-3-319-61660-5_23

powerful argumentation semantics of abstract dialectical frameworks (ADFs) [3] more accessible to users by applying them on simple structures in the form of labelled trees and a pattern language for defining statement acceptance. With ArgueApply we want to go one step further in usability and offer argumentation technology for the lay user. In a first application phase, ArgueApply should serve as a prototype for exploring the requirements of social argumentation tools. Thus, the first goal of using ArgueApply is to evaluate different fragments and semantics of Grappa for their suitability for three usage scenarios presented in this paper: Providing a neutral evaluation of a dispute, preparing for a debate against an opponent, and live analysis of a discussion. After the initial phase of application we want to release ArgueApply with optimised Grappa settings on prominent mobile distribution platforms to attract users. Arguably, an automated evaluation of an argumentative structure does not guarantee that everybody agrees on its results. It is often the one who shouts the loudest or the one with best rhetorical skills who wins a real debate. Moreover, there might be diverging opinions about the importance of statements and different judgements of the mutual influence of statements. Nevertheless, we hope that the availability of tools like ArgueApply can promote and support discussions based on reason and good arguments.

The paper is outlined as follows. In Sect. 2 we provide background on the Grappa framework. Section 3 describes the core functionality of our mobile application. Then, in Sect. 4 we discuss intended usage scenario and how we want to exploit ArgueApply for testing which types of Grappa instances are best suited to address certain applications in practice. We also detail which Grappa configurations we currently use. Section 5 covers the user interface and design decisions we took. In Sect. 6 we provide an overview of the technology we used for the implementation. We report on related work in Sect. 7 and then conclude the paper with an outlook on future work.

2 Grappa

ArgueApply is based on the Grappa framework by Brewka and Woltran [1] which we quickly recall next. For more intuition on the concepts presented here, we refer to their paper. Grappa allows to describe argumentation scenarios using arbitrary directed edge-labelled graphs. A Grappa *instance* is a tuple $G = \langle S, E, L, \lambda, \pi \rangle$ where S is a set of *statements*, $E \subseteq S \times S$ a set of *links*, L a set of *labels*, λ an assignment of labels to links, and π an assignment of *acceptance patterns* over L (defined next) to nodes. Whether a statement is accepted or rejected depends on its acceptance pattern that describes how the influence of the incoming links determine the acceptance status. *Acceptance patterns* over a set of labels L are defined as follows:

- A *term* over L is of the form $\#(l)$, $\#_t(l)$ (with $l \in L$), or *min*, min_t, *max*, max_t, *sum*, sum_t, *count*, $count_t$.
- A *basic acceptance pattern* (over L) is of the form $a_1 t_1 + \cdots + a_n t_n \, R \, a$, where the t_i are terms over L, the a_is and a are integers and $R \in \{<, \leq, =, \neq, \geq, >\}$.

– An *acceptance pattern* (over L) is a basic acceptance pattern or a Boolean combination of acceptance patterns.

For example, the term $\#(l)$ refers to the number of incoming links with label l whose source nodes are accepted, while *count* refers to the number of different labels of incoming links whose source nodes are accepted. The terms with subscript t take all incoming links into account, independent of their acceptance.

The semantics of a graph is defined in terms of (three-valued) interpretations, assigning each statement in S one of the truth values **true** (**t**), **false** (**f**), or **undecided** (**u**). For interpretations v and w, w *extends* v if for all $s \in S$, $v(s) \neq \mathbf{u}$ implies $v(s) = w(s)$. We denote by $[v]_2$ the set of all completions of an interpretation v, i.e. 2-valued interpretations that extend v.

To evaluate the acceptance pattern of a statement s we need the fixed value function val_s^m that maps terms to numbers, depending on a multiset of labels $m : L \to \mathbb{N}$.

$$
\begin{aligned}
val_s^m(\#(l)) &= m(l) \\
val_s^m(\#_t(l)) &= |\{(e,s) \in E \mid \lambda((e,s)) = l\}| \\
val_s^m(min) &= \min\{l \in L \mid m(l) > 0\} \\
val_s^m(min_t) &= \min\{\lambda((e,s)) \mid (e,s) \in E\} \\
val_s^m(max) &= \max\{l \in L \mid m(l) > 0\} \\
val_s^m(max_t) &= \max\{\lambda((e,s)) \mid (e,s) \in E\} \\
val_s^m(sum) &= \sum_{l \in L} m(l) \\
val_s^m(sum_t) &= \sum_{(e,s) \in E} \lambda((e,s)) \\
val_s^m(count) &= |\{l \mid m(l) > 0\}| \\
val_s^m(count_t) &= |\{\lambda((e,s)) \mid (e,s) \in E\}|
\end{aligned}
$$

The function is undefined for $min_{(t)}$, $max_{(t)}$, $sum_{(t)}$ in case of non-numerical labels. For $\emptyset$ they yield the neutral element of the corresponding operation, i.e. $val_s^m(sum) = val_s^m(sum_t) = 0$, $val_s^m(min) = val_s^m(min_t) = \infty$, as well as $val_s^m(max) = val_s^m(max_t) = -\infty$.

Let m and s be as before. For basic acceptance patterns the *satisfaction relation* $\models$ is defined by

$$
(m,s) \models a_1 t_1 + \cdots + a_n t_n \, R \, a \text{ iff } \sum_{i=1}^{n} (a_i \, val_s^m(t_i)) \, R \, a.
$$

The extension to Boolean combinations is as usual.

We need to define the *characteristic operator* for a Grappa instance $G = \langle S, E, L, \lambda, \pi \rangle$. It is a function that maps one three-valued interpretation v to another:

$$
\Gamma_G(v) = P_G^\gamma(v) \cup N_G^\gamma(v)
$$

where

$$
\begin{aligned}
P_G^\gamma(v) &= \{s \mid (m,s) \models \pi(s) \text{ for each } m \in \{m_s^{v_2} \mid v_2 \in [v]_2\}\} \\
N_G^\gamma(v) &= \{\neg s \mid (m,s) \not\models \pi(s) \text{ for each } m \in \{m_s^{v_2} \mid v_2 \in [v]_2\}\}
\end{aligned}
$$

and

$$m_s^v(l) = [(e, s) \in E \mid v(e) = \mathbf{t}, \lambda((e, s)) = l]$$

denotes the multiset of active labels of s generated by v.

With this characteristic function different semantics of G can be defined: An interpretation v is *admissible* w.r.t. G if $\Gamma_G(v)$ extends v; it is *complete* w.r.t. G if $v = \Gamma_G(v)$; it is *grounded* w.r.t. G if it is complete and extends no other interpretation that is complete w.r.t. G; it is *preferred* w.r.t. G if v is admissible and no other admissible interpretation w.r.t. $\leq_i$ extends v. Even more semantics for Grappa instances have been defined (see [1]).

Example 1. Consider the Grappa instance G with $S = \{a, b\}$, $E = \{(b, a), (b, b)\}$, $L = \{\mathbf{+}, \mathbf{-}\}$, (b, a) being labelled with $\mathbf{-}$ and (b, b) with $\mathbf{+}$, a has the acceptance pattern $\#(\mathbf{+}) - \#(\mathbf{-}) \geq 0$ and b the pattern $\#(\mathbf{+}) - \#(\mathbf{-}) > 0$. The following interpretations are admissible w.r.t. G: $v_1 = \{a{\rightarrow}\mathbf{u}, b{\rightarrow}\mathbf{u}\}$, $v_2 = \{a{\rightarrow}\mathbf{u}, b{\rightarrow}\mathbf{t}\}$, $v_3 = \{a{\rightarrow}\mathbf{u}, b{\rightarrow}\mathbf{f}\}$, $v_4 = \{a{\rightarrow}\mathbf{t}, b{\rightarrow}\mathbf{f}\}$, $v_5 = \{a{\rightarrow}\mathbf{f}, b{\rightarrow}\mathbf{t}\}$. Moreover, v_1, v_4, and v_5 are complete and v_4, and v_5 are preferred w.r.t. G. The unique grounded interpretation is v_1.

3   Principle Functionality

Users of `ArgueApply` can participate in online discussions, each of which corresponds to a Grappa instance. Every participant can post new statements in a similar way as sending messages in a chat application. A statement is permanently associated to the user who authored it (or to a philosopher avatar—philosopher avatars are discussed in the end of the section). This way we can measure how many statements by a user are accepted, respectively rejected, with respect to a solution of the Grappa instance. Links between statements however are not owned by users. Participants of the discussion can express which influence they see between two statements but this information is not automatically turned into a link but counts as a vote for a (labelled) link. The rationale behind this design decision is that different opinions on how statements influence each other could otherwise destabilise the discussion: a single user unhappy with the evaluation results could add unquestionable links that significantly influence the outcome. In contrast to links which carry meta-information, it makes sense that everybody can add new statements without a voting, as they contribute content to the discussion and should only be questioned by further arguments.

A user can trigger the evaluation of the Grappa instance at all times and then browse through the resulting interpretations such that for each statement it is highlighted whether it is accepted or rejected with respected to the currently selected solution.

In addition to ordinary users we added ten pre-defined avatars which can be used to post statements that do not reflect one's own opinion (see Fig. 3). In a number of use cases (see next section) it makes sense to model also the point of view of someone else. The avatars are named after philosopher that are allegedly depicted in the 'The School of Athens', the famous fresco by Raphael that serves as theme of our application.

4   Usage Scenarios and Testbed Objective

The purpose of ArgueApply is to clearly structure and analyse a debate and its inherent viewpoints. This includes fully understanding not only one's own point of view but also that of the opponent. Another goal is helping to reveal inconsistencies in the lines of argumentation of its users. We see different usage scenarios that we want to exemplify in the following.

Who is right? A group of friends has a dispute and a neutral point of view is required. Each person enters her contributing statements in an ArgueApply discussion, and collectively they determine the link structure (as described in Sect. 3). The evaluation reveals that only the arguments of one opinion group are acceptable and thus, this subgroup has 'won' the debate. In this type of application scenario, ArgueApply acts as a neutral instance, e.g., deciding who won a bet, or, which decision should be taken by the overall group. In the latter application, simplifying group decisions, ArgueApply offers the advantage over simple voting for a decision that the group acts on the basis of clearly structured arguments rather than the average of intuitions. While it is not clear that this approach leads to better overall decisions, it makes at least the motivation for group decisions more transparent, documentable, and verifiable.

Preparation for a Debate. A non-governmental organization (NGO) fighting for some cause is invited to send a representative to a media debate. In the run-up to the event, members of the organisation use ArgueApply to find out an optimal line of argumentation as preparation for the representative. In a brainstorming process, arguments for and against the cause of the NGO are collected as statements of an ArgueApply discussion. Here, philosopher avatars as described in Sect. 3 are used for statements of opponents. When modelling links, the team tries to judge the influences between statements in the way most members of their target group would. If the evaluation reveals that there is a line of argumentation of the opponent that is consistent, the team can search for new arguments that break this line.

Live Analysis of a Debate. A group of journalists is responsible for a live ticker of a news portal in which they comment on an ongoing political discussion. As an extra service to their readers they provide an analysis of the argumentative structure of the debate using ArgueApply. For each of the debating parties, one of the reporters is responsible for summarising their respective arguments and for entering them as statements into the system. Other journalists judge how these statements influence each other and create respective links. Having the structure of the discussion formalised is already of benefit to commentators, as it allows to pinpoint to the weaknesses of the argumentations. With the help of the evaluation (that can be updated during the debate as new arguments come in) the journalists comment on which party has a consistent view or which arguments need to be accepted for sure (according to the chosen Grappa semantics).

ArgueApply has been developed in a research project that focuses on bridging the gap between theory and application of formal argumentation models. While the sketched usage scenarios illustrate the intended final application area of the system, we follow a further objective with ArgueApply, i.e., determining which type of Grappa frameworks are best suited for the target applications. This includes the choice of link labels and acceptance patterns, and also the decision on which semantics to use. Therefore, we intend to use ArgueApply in a first phase of application as a testbed for the practicality of different Grappa configurations. To find good settings that lead to intuitive results is also important for a later second phase of application when we want to make the application available on big distribution platforms. For usability considerations we want the application to offer non-expert users only a maximally restricted choice of Grappa parameters. That is, the default settings should lead to an intuitive user experience and further options should only be available in a dedicated expert mode. While argumentation semantics [4] have been extensively investigated formally, even for standard AFs, there is relatively little work on their suitability for different application scenarios and their intuition [5,6]. For the more recent frameworks of ADFs and Grappa, there have not been any systematic studies in this direction yet, although first applications of ADFs have been presented [7]. The analysis of suitability of argumentation semantics and Grappa parameters is future work and beyond the scope of this system description, nevertheless, we want to explain in the following which configuration we used in the initial setup of ArgueApply and why.

Currently ArgueApply uses four different link labels, ++,+, -, and -- that express strong support, support, attacks, and strong attacks, respectively. This choice was taken simply because this labels are used in the standard examples of Grappa literature. The voting mechanism for links described in Sect. 3 is handled outside of Grappa and thus need not be considered in acceptance patterns, i.e., different version of the same link entered by different users are not part of the Grappa instance but only the link resulting from the vote is. Handling voting within Grappa would require an extension of the framework as, by definition, multiple links with the same start and destination statements are not allowed. Moreover, the voting arithmetic is hard to express using the currently available pattern language. At the moment ArgueApply uses two closely related acceptance patterns:

$$2 * \#(++) + \#(+) - 2 * \#(--) - \#(-) \geq 0$$

and

$$2 * \#(++) + \#(+) - 2 * \#(--) - \#(-) > 0.$$

The patterns express that the statement should be accepted if there are more positive links than negative links from acceptance statements, where labels with doubled symbols count twice as positive, respectively negative, than their single counterparts. The first pattern accepts statements with a neutral balance. By default, we use this pattern for all arguments. Our rationale is that a new statement in a debate that is not known to be influenced by any other statement, i.e., being unquestioned, should be accepted. As a first lesson learned from using ArgueApply, we noticed that there are often situations when we want to express a doubt in a statement for which we do not know whether it is true but we do not have any appropriate counterargument. Hence, we implemented a feature do express general doubt in a statement. If one of the participants of the discussion doubts a statement, two modifications happen:

1. the statement is evaluated with the second acceptance pattern and
2. a link labelled + is added from the statement to itself.

As a consequence, when no further link ends at the statement, it is not by default accepted, but leads to resulting interpretations where it is accepted and others where it is rejected under the ADF semantics we implemented so far (see below), similar like a choice atom $\{s\}$ in answer-set programming. The self-link is not shown as a further link to the user (as this might be confusing), however users get some indication whether someone doubted the statement.

We implemented three Grappa semantics in ArgueApply using an translational approach that was introduced in a recent paper [8] (for implementation details see Sect. 6): admissible, complete, and preferred semantics. We chose to start with these three for pragmatic reasons, because we could use existing translations. The user can change the used semantics in the settings section of ArgueApply.

As mentioned, we do not yet release the system on a larger scale, however, it can be downloaded from our website under
http://www.informatik.uni-leipzig.de/~puehrer/ArgueApply/

5   Interface

After starting ArgueApply (when logged in), the user is shown the list of discussions to which she has subscribed (Fig. 1). When a user starts a new discussion she is considered the moderator who has, e.g., the right to modify properties or delete the discussion. A discussion can be public or private, i.e., joined on invitation only. When selecting a discussion from the list it is opened for participation. Clearly, the most natural way to depict Grappa instances is drawing its graph. In a mobile app, however, a graph-based user interface is not the best choice due to the small size of common smartphone displays. We decided to display the discussion in an expandable list of statements as shown in Fig. 2. All statements form the top-level elements of the list. New statements are added by entering them in the text area on the bottom of the screen. Note that due to the integration in the Android platform we can make use of the built-in speech recognition feature for vocal input of statements. This alternative input method works quite well in practice. On the left-hand side of the new statement area we can select whether the new statement is sent in the name of the user or one of the philosopher avatars (Fig. 3).

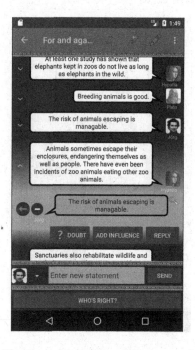

Fig. 1. The home activity of `ArgueApply` with the list of discussions the user is subscribed to. Every user can start new discussions or join any public discussion from here.

Fig. 2. A discussion on the pros and cons of zoos in `ArgueApply`. All statements are shown in an expandable list. On expansion of a statement item its links are displayed. Here, the fourth statement is expanded, showing one link. The blue arrow indicates that the link is incoming and the red minus symbol depicts the label of the link. (Color figure online)

When a statement item is expanded, the links of the statements are shown, by listing the influencing statements. The user may decide whether to show incoming, outgoing or both type of links. Link labels are depicted by small symbols. Note that the label displayed is the one which has currently the most votes (cf. Sect. 3). New links between two existing statements are proposed using the 'Add influence' button, whereas the 'Quote' button creates a link to or from a new statement. For existing links, one can add a vote for a label and see all existing votes by participants using the edit button (pencil) visible in Fig. 3. It is shown in edit mode only which may be activated in the top bar menu. Doubts (see Sect. 4) are expressed using the 'Doubt' button. If a statement has been doubted it is annotated with a question mark such as the second statement in Fig. 4.

The evaluation of a discussion is triggered by the 'Who's right?' button. The user can browse between all resulting interpretations in a new area that appears at the bottom of the screen as depicted in Fig. 4. Accepted statements

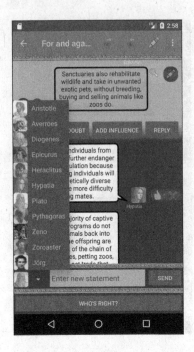

Fig. 3. Besides the user account, ten philosophers can be selected as avatars for publishing statements. These are used for modelling opinions that no human participant of the discussion shares.

Fig. 4. After evaluating the Grappa instance, the user can select which resulting interpretation to view in the bottom area. The statements in the list are coloured and annotated according to their acceptance under the chosen solution. The Grappa instance is a variant of that in Example 1. (Color figure online)

are marked green while rejected ones get a red border in addition to a label stating the acceptance state. Undecided statements remain unchanged.

6   Technology and Implementation

An overview of the technology used in `ArgueApply` is given in Fig. 5. The system is an Android application written primarily in Java and requires a minimum SDK level of 23 (Android 4.4). The evaluation of the semantics in `ArgueApply` is based on answer-set programming (ASP) [9–11]. We implemented transformations from Grappa to ASP for admissible, complete, and preferred semantics presented in a recent paper [8]. The system is open to extensions by further semantics, given the availability of a transformation to ASP (e.g., [12]). As a solving backend we rely on `clingo` [13,14]. The solver is executed locally on the mobile device. We implemented an ASP service that may run independent of

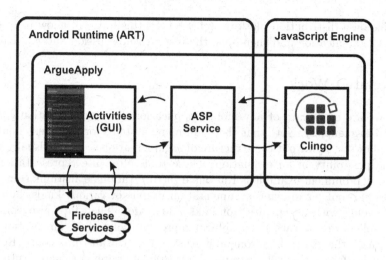

Fig. 5. Technology overview for `ArgueApply`.

the Android application and encapsulates a JavaScript version of `clingo` 5.2 generated by the `emscripten` source-to-source compiler. The availability of a JavaScript interpreter by default on the Android platform guarantees that we can use the solver without the need to compile it natively on every device. A drawback of local computation is that a solver invocation is computationally expensive and that it has to be carried out on each device independently. To tackle this problem, we implemented online caching of solving results such that only the device of one participant of a discussion needs to execute the solver after a modification. We also consider a backend solution for solving while keeping the local computation capability as a fallback mechanism. Adding additional server side computation would be relatively easy because the cloud services we use (and discuss next) offer good interfaces for a reasoning server to plugin as yet another client.

`ArgueApply` builds on the web application platform Firebase [15]. We exploit its user authentication service that allows for e-mail based accounts or user identification via social login providers. Moreover, we use the realtime database of Firebase to store all online data, including discussions and subscription information. It is a non-relational database that represents data as a JSON tree. The API of the Firebase framework allows to listen to changes of nodes of this tree. Exploiting this feature, modifications of a discussion by one user can immediately be followed on other users' devices. We also use a Firebase feature for inviting people to join a discussion. Users can send a link by different means (e.g., e-mail, chat or text message) from within the application. If the link is opened on an Android device where `ArgueApply` is installed, the discussion will be automatically opened in the application. Otherwise, the link leads to a website with instructions how to obtain the system. In a later phase, when the app will be available on a distribution platform (cf. Sect. 4), this link will allow to directly install the application and then open the respective discussion automatically.

7 Related Work

First, we are not aware of any mobile applications for argumentation comparable to ArgueApply. Note that there are many systems for argument mapping which allow for laying out the structure of argumentation scenario visually without the possibility of an evaluation. These tools often use representations as graphs. A prominent example is Rationale (http://rationale.austhink.com/), an educational tool for drawing argumentation graphs of various kinds, but they do not come with a semantics of links and nodes. There are also commercial suppliers of software for graphical representations of argumentation scenarios, e.g., the Australian company Austhink (www.austhink.com). Related internet platforms are Debategraph (http://debategraph.org) and TruthMapping (http://truthmapping.com). The web service argüman (en.arguman.org) allows for building tree-shaped argument mappings that are public and structured by three types of premises ('because', 'but', and 'however'). Schneider, Groza, and Passant provide a comprehensive overview of approaches to modelling argumentation for the Social Semantic Web [16] including several online tools. We now focus on systems that allow for both, evaluating and editing instances of frameworks for argumentation.

Most related to ArgueApply is the GrappaVis environment that is also based on Grappa and allows for creating Grappa instances in a graphical environment [17] for desktop computers. The system is written in Java and also uses ASP for evaluating Grappa semantics. In addition to Grappa instances it can also handle ADFs. While our tool is written for lay end users offering a deliberately restricted choice of options, GrappaVis is a tool for knowledge engineers that is not targeted towards specific use cases and allows for designing arbitrary frameworks.

The graphical browser-based OVA system (Online Visualisation of Argument) supports analysis of textual arguments in the Argument Web [18]. In order to evaluate a debate, models are currently fed into an ASPIC solver [19] which itself relies on instantiating the ASPIC specification in Dung AFs. One drawback of this approach is that the result on the AF level is not easily translatable back to the debate model. In future work we want to allow for interchange between OVA and ArgueApply instances.

The VUE software tool [20] is based on the early Issue Based Information System (IBIS) approach [21], a graph-based technique for formalising the decisions made during a design process and their reasons. Thus, the system is focused on design debates. The goal is to support engineers by an automated evaluation of alternative design solutions given the underlying graph structure developed during the design process.

8   Conclusion and Outlook

We presented **ArgueApply**, a mobile app for argumentation based on the Grappa framework. To the best of our knowledge, it is the first smartphone application based on techniques from formal argumentation theory and publicly available from our website. The system allows users to formalise and evaluate debates provided that they collaboratively enter statements and their mutual argumentative influences. In future work, we want to explore enforcing features (as, e.g., studied for AFs [22]) to improve the system's capabilities for scenario like the one for preparing a debate: What type of arguments do I need to add or alter in order to improve the evaluation results for my party? We also want to further integrate **ArgueApply** with other mobile applications in order to directly import statements from their data. As mentioned earlier, we plan to achieve interoperability with the OVA platform [18]. It would be nice to have a tablet computer version or a web version for bigger screens, where we could display Grappa instances as labelled graphs. This would be helpful in the preparation scenario, where individual users enter information using their smartphone but there is at the same time also a good overview in a bigger visual representation. Finally, once we finished the testbed phase, we want to release **ArgueApply** on a bigger scale on a popular distribution platform for mobile applications.

References

1. Brewka, G., Woltran, S.: GRAPPA: a semantical framework for graph-based argument processing. In: Proceedings of the 21st European Conference on Artificial Intelligence (ECAI 2014), Including Prestigious Applications of Intelligent Systems (PAIS 2014), Prague, Czech Republic, pp. 153–158, 18–22 August 2014
2. Dung, P.M.: On the acceptability of arguments and its fundamental role in non-monotonic reasoning, logic programming and n-Person games. Artif. Intell. **77**(2), 321–358 (1995)
3. Brewka, G., Strass, H., Ellmauthaler, S., Wallner, J.P., Woltran, S.: Abstract dialectical frameworks revisited. In: Proceedings of the 23rd International Joint Conference on Artificial Intelligence (IJCAI 2013), Beijing, China, pp. 803–809, 3–9 August 2013
4. Baroni, P., Caminada, M., Giacomin, M.: An introduction to argumentation semantics. Knowl. Eng. Rev. **26**(4), 365–410 (2011)
5. Rahwan, I., Madakkatel, M.I., Bonnefon, J., Awan, R.N., Abdallah, S.: Behavioral experiments for assessing the abstract argumentation semantics of reinstatement. Cogn. Sci. **34**(8), 1483–1502 (2010)
6. Gaggl, S.A., Rudolph, S., Thomazo, M.: What is a reasonable argumentation semantics? In: Eiter, T., Strass, H., Truszczyński, M., Woltran, S. (eds.) Advances in Knowledge Representation, Logic Programming, and Abstract Argumentation. LNCS, vol. 9060, pp. 309–324. Springer, Cham (2015). doi:10.1007/978-3-319-14726-0_21
7. Al-Abdulkarim, L., Atkinson, K., Bench-Capon, T.J.M.: A methodology for designing systems to reason with legal cases using abstract dialectical frameworks. Artif. Intell. Law **24**(1), 1–49 (2016)

8. Brewka, G., Diller, M., Heissenberger, G., Linsbichler, T., Woltran, S.: Solving advanced argumentation problems with answer-set programming. In: Proceedings of the 31st AAAI Conference on Artificial Intelligence, AAAI 2017 (2017)

9. Marek, V.W., Truszczyński, M.: Stable models and an alternative logic programming paradigm. In: Apt, K.R., Marek, V.W., Truszczynski, M., Warren, D.S. (eds.) The Logic Programming Paradigm. Artificial Intelligence, pp. 375–398. Springer, Heidelberg (1999)

10. Niemelä, I.: Logic programs with stable model semantics as a constraint programming paradigm. Ann. Math. Artif. Intell. 25(3–4), 241–273 (1999)

11. Gebser, M., Kaminski, R., Kaufmann, B., Schaub, T.: Answer Set Solving in Practice. Synthesis Lectures on Artificial Intelligence and Machine Learning. Morgan and Claypool Publishers (2012)

12. Berthold, M.: Extending the DIAMOND system to work with GRAPPA. In: Proceedings of the 1st International Workshop on Systems and Algorithms for Formal Argumentation (SAFA), pp. 52–62 (2016)

13. Gebser, M., Kaminski, R., Kaufmann, B., Ostrowski, M., Schaub, T., Schneider, M.: Potassco: the potsdam answer set solving collection. AI Commun. 24(2), 107–124 (2011)

14. Gebser, M., Kaminski, R., Kaufmann, B., Romero, J., Schaub, T.: Progress in *clasp* series 3. In: Calimeri, F., Ianni, G., Truszczynski, M. (eds.) LPNMR 2015. LNCS, vol. 9345, pp. 368–383. Springer, Cham (2015). doi:10.1007/978-3-319-23264-5_31

15. Firebase Platform (2017). https://firebase.google.com/

16. Schneider, J., Groza, T., Passant, A.: A review of argumentation for the social semantic web. Semant. Web 4(2), 159–218 (2013)

17. Heissenberger, G., Woltran, S.: GrappaVis - a system for advanced graph-based argumentation. In: Baroni, P., Gordon, T.F., Scheffler, T., Stede, M. (eds.) Computational Models of Argument - Proceedings of COMMA 2016. Frontiers in Artificial Intelligence and Applications, vol. 287, Potsdam, Germany, pp. 473–474. IOS Press, 12–16 September 2016

18. Janier, M., Lawrence, J., Reed, C.: OVA+: an argument analysis interface. In: Parsons, S., Oren, N., Reed, C., Cerutti, F. (eds.) Computational Models of Argument - Proceedings of COMMA 2014. Frontiers in Artificial Intelligence and Applications, vol. 266, Atholl Palace Hotel, Scottish Highlands, UK, pp. 463–464. IOS Press, 9–12 September 2014

19. Prakken, H.: An abstract framework for argumentation with structured arguments. Argument Comput. 1(2), 93–124 (2010)

20. Baroni, P., Romano, M., Toni, F., Aurisicchio, M., Bertanza, G.: Automatic evaluation of design alternatives with quantitative argumentation. Argument Comput. 6(1), 24–49 (2015)

21. Kunz, W., Rittel, H.: Issues as elements of information systems. Working Paper 131, Institute of Urban and Regional Development. University of California, Berkeley, California (1970)

22. Baumann, R.: What does it take to enforce an argument? Minimal change in abstract argumentation. In: Raedt, L.D., Bessière, C., Dubois, D., Doherty, P., Frasconi, P., Heintz, F., Lucas, P.J.F. (eds.) Proceedings of the 20th European Conference on Artificial Intelligence (ECAI 2012). Including Prestigious Applications of Artificial Intelligence (PAIS 2012) System Demonstrations Track. Frontiers in Artificial Intelligence and Applications, vol. 242, Montpellier, France, pp. 127–132. IOS Press, 27–31 August 2012

LPNMR Applications

catnap: Generating Test Suites of Constrained Combinatorial Testing with Answer Set Programming

Mutsunori Banbara[1], Katsumi Inoue[2], Hiromasa Kaneyuki[1], Tenda Okimoto[1], Torsten Schaub[3(✉)], Takehide Soh[1], and Naoyuki Tamura[1]

[1] Kobe University, 1-1 Rokko-dai Nada-ku, Kobe, Hyogo 657-8501, Japan
[2] NII, 2-1-2 Hitotsubashi, Chiyoda-ku, Tokyo 101-8430, Japan
[3] Universität Potsdam, August-Bebel-Strasse 89, 14482 Potsdam, Germany
torsten@cs.uni-potsdam.de

Abstract. We develop an approach to test suite generation for Constrained Combinatorial Testing (CCT), one of the most widely studied combinatorial testing techniques, based on Answer Set Programming (ASP). The resulting *catnap* system accepts a CCT instance in fact format and combines it with a first-order encoding for generating test suites, which can subsequently be solved by any off-the-shelf ASP systems. We evaluate the effectiveness of our approach by empirically contrasting it to the best known bounds obtained via dedicated implementations.

1 Introduction

Software testing can generally be defined as the task of analyzing software systems to detect failures. Recently, software testing has become an area of increasing interest in several Software Engineering communities involving both researchers and practitioners, such as the international series of ICST/IWCT conferences. The typical topics of this area include *model-based testing, combinatorial testing, security testing, domain specific testing*, etc. In this paper, we consider a problem of generating test suites (viz. sets of test cases) for Combinatorial Testing (CT; [7,15,19]) and its extensions.

CT is an effective black-box testing technique to detect elusive failures of software. Modern software systems are highly configurable. It is often impractical to test all combinations of configuration options. CT techniques have been developed especially for such systems to avoid falling into combinatorial explosion. CT relies on the observation that most failures are caused by interactions between a small number of configuration options. For example, *strength-t CT* tests all t-tuples of configuration options in a systematic way. Such testing requires much smaller test suites than exhaustive testing, and is more effective than random testing.

This work was partially funded by JSPS (KAKENHI 15K00099) and DFG (SCHA 550/11).

ⓒ Springer International Publishing AG 2017
M. Balduccini and T. Janhunen (Eds.): LPNMR 2017, LNAI 10377, pp. 265–278, 2017.
DOI: 10.1007/978-3-319-61660-5_24

Constrained CT (*CCT*; [8]) is an extension of CT with constraint handling; it is one of the most widely studied CT techniques in recent years. In many configurable systems, there exist constraints between specific configuration options that make certain combinations invalid. CCT provides a combinatorial approach to testing, involving hard and soft constraints on configuration options. Hard constraints must be strictly satisfied. Soft constraints must not necessarily be satisfied but the overall number of violations should be minimal. Therefore, CCT cannot only exclude *invalid* tuples that cannot be executed, but also minimize the ones that might be undesirable.

The CCT problem of generating optimal (smallest) CCT test suites is known to be difficult. Several methods have been proposed such as greedy algorithms [8,10,21,22], metaheuristics-based algorithms [12,14], and Satisfiability Testing (SAT) [16,20]. However, each method has strengths and weaknesses. Greedy algorithms can quickly generate test suites. Metaheuristics-based implementations can give smaller ones by spending more time. On the other hand, both of them cannot guarantee their optimality. Although complete methods like SAT can guarantee optimality, it is costly to implement a dedicated encoder from CCT problems into SAT. For constraint solving in CCT, most methods utilize off-the-shelf constraint solvers as back-ends[1]. They are used to check whether each test case satisfies a given set of constrains during test suite generation as well as to calculate valid tuples at prepossessing. However, there are few implementations that provide CCT with soft constraints and limiting resources (the number of test cases, time, and etc.). It is therefore challenging to develop a universal CCT solver which can efficiently generate optimal test suites as well as suboptimal ones for CCT instances having a wide range of hard and soft constraints, even if the resources are limited.

In this paper, we describe an approach to solving the CCT problems based on Answer Set Programming (ASP; [4]). The resulting *catnap* system accepts a CCT instance in fact format and combines it with a first-order ASP encoding for CCT solving, which is subsequently solved by an off-the-shelf ASP system, in our case *clingo*. Our approach draws upon distinct advantages of the high-level approach of ASP, such as expressive language, extensible encodings, flexible multi-criteria optimization, etc. However, the question is whether *catnap*'s high-level approach matches the performance of state-of-the-art CCT solving techniques. We address this question by empirically contrasting *catnap* with dedicated implementations. From an ASP perspective, we gain insights into advanced modeling techniques for CCT. *catnap*'s encoding for CCT solving has the following features: (a) a series of compact ASP encodings, (b) easy extension for CCT under limiting resources, and (c) easy extension of CCT with soft constraints.

In the sequel, we assume some familiarity with ASP, its semantics as well as its basic language constructs. Our encodings are given in the language of *gringo* series 4. Although we provide a brief introduction to CCT in the next section, we refer the reader to the literature [17] for a broader perspective.

---

[1] SAT solvers in [8,12,14,16,20], PB (Pseudo Boolean) solver in [22], and CSP (Constraint Satisfaction Problem) solver in [21].

	Product Line Options				
	Display	Email Viewer (Email)	Camera	Video Camera (Video)	Video Ringtones (Ringtones)
Possible Values	16 Million Colors (16MC)	Graphical (GV)	2 Megapixels (2MP)	Yes	Yes
	8 Million Colors (8MC)	Text (TV)	1 Megapixel (1MP)	No	No
	Black and White (BW)	None	None		

Constraints on valid configurations:

(1) Graphical email viewer **requires** color display.
(2) 2 Megapixel camera **requires** a color display.
(3) Graphical email viewer **not supported** with the 2 Megapixel camera.
(4) 8 Million color display **does not support** a 2 Megapixel camera.
(5) Video camera **requires** a camera **and** a color display.
(6) Video ringtones **cannot occur** with No video camera.
(7) The combination of 16 Million colors, Text and 2 Megapixel camera is **not supported**.

Fig. 1. Mobile phone product line in [8]

2   Background

Generating a CCT test suite is to find a *Constrained Mixed-level Covering Array* (*CMCA*; [8]), which is an extension of a *Covering Array* [9] with constraints. A *CMCA* of size N is a $N \times k$ array $A = (a_{ij})$, written as $CMCA(t, k, (v_1, \ldots, v_k), \mathcal{C})$. The integer constant t is the *strength* of the coverage of interactions, k is the number of parameters, and v_j is the number of values for each parameter j $(1 \leq j \leq k)$. The constraint $\mathcal{C}$ is given as a Boolean formula in Conjunctive Normal Form (CNF):

$$\mathcal{C} = \bigwedge_{1 \leq \ell \leq h} \left(\bigvee_{1 \leq m \leq m_\ell} pos(j_m^\ell, d_m^\ell) \quad \vee \bigvee_{m_\ell+1 \leq n \leq n_\ell} neg(j_n^\ell, d_n^\ell) \right).$$

Each constraint clause identified by ℓ $(1 \leq \ell \leq h)$ is a disjunction of predicates $pos/2$ and $neg/2$. The atom $pos(j, d)$ expresses that the parameter j has a value d, and the atom $neg(j, d)$ expresses its negation. The constants n_ℓ and m_ℓ $(n_\ell \geq m_\ell \geq 1)$ are the number of both type of atoms and *pos* atoms, respectively, for each clause ℓ. Then *CMCA* has the following properties:

- $a_{ij} \in \{0, 1, 2, \ldots, v_j - 1\}$,
- every row r $(1 \leq r \leq N)$ satisfies constraint $\mathcal{C}$ (called *domain constraints*).

$$\bigwedge_{1 \leq r \leq N, 1 \leq \ell \leq h} \left(\bigvee_{1 \leq m \leq m_\ell} (a_{r,j_m^\ell} = d_m^\ell) \quad \vee \bigvee_{m_\ell+1 \leq n \leq n_\ell} \neg(a_{r,j_n^\ell} = d_n^\ell) \right)$$

- in every $N \times t$ sub-array, all possible t-tuples (pairs when $t = 2$) that satisfy the constraint $\mathcal{C}$ occur at least once (also called *coverage constraints*).

	Display	Email	Camera	Video	Ringtones
1	8MC	**TV**	**None**	No	No
2	BW	**None**	**None**	No	No
3	16MC	**None**	**2MP**	Yes	Yes
4	16MC	None	2MP	No	No
5	16MC	**GV**	**None**	No	No
6	16MC	**TV**	**1MP**	Yes	Yes
7	BW	TV	1MP	No	No
8	8MC	**None**	**1MP**	Yes	No
9	8MC	**GV**	**1MP**	Yes	Yes

Constraints on valid configurations:

C_1: ¬(Email = GV) ∨ (Display = 16MC) ∨ (Display = 8MC)
C_2: ¬(Camera = 2MP) ∨ (Display = 16MC) ∨ (Display = 8MC)
C_3: ¬(Email = GV) ∨ ¬(Camera = 2MP)
C_4: ¬(Display = 8MC) ∨ ¬(Camera = 2MP)
C_5: ¬(Video = Yes) ∨ (Camera = 2MP) ∨ (Camera = 1MP)
C_6: ¬(Video = Yes) ∨ (Display = 16MC) ∨ (Display = 8MC)
C_7: ¬(Ringtones = Yes) ∨ ¬(Video = No)
C_8: ¬(Display = 16MC) ∨ ¬(Email = TV) ∨ ¬(Camera = 2MP)

Fig. 2. An optimal test suite of strength-2 CCT

A *CMCA* of size N is *optimal* if N is equal to the smallest n such that a *CMCA* of size n exists. We refer to a t-tuple as a *valid* t-tuple if it satisfies the constraint C, otherwise we call it *invalid*. Note that the constraint C is defined as hard constraint, and we have no soft constraints in this definition.

As an illustration, we use a simplified software product line of mobile phones proposed in [8]. The product line of Fig. 1 has five configuration options, three of which have three values, while others have two choices of values (Yes and No). "Display" has exactly one value among 16MC, 8MC, and BW. The product line has seven constraints on valid configurations. The constraint (3) forbids a pair (GV, 2MP) in the interaction of (Email, Camera). Exhaustive testing requires $2^2 \times 3^3 = 108$ test cases (or configurations) for all different phones produced by instantiating this product line. The constraints reduce the number of test cases to 31. However, in general, such testing fails to scale to large product lines. Instead, strength-t CCT is able to provide effective testing while avoiding the combinatorial explosion. The question is what is the smallest number of test cases for the product phone line. An optimal test suite of strength-2 CCT is shown in Fig. 2. It consists of 9 test cases, and gives an answer to the question. Each row represents an individual test case, which satisfies the domain constraints in (1–7). Each column represents a configuration option. In the interaction of (Email, Camera), we highlight the different pairs to show that all valid pairs (7 combinations) occur at least once. Note that invalid pairs (GV, 2MP) and (TV, 2MP) do not occur. This property holds for all interactions of two

(Display, Email)	(Display, Camera)	(Display, Video)	(Display, Ringtones)	(Email, Camera)
(16MC, GV)	(16MC, 2MP)	(16MC, Yes)	(16MC, Yes)	(GV, 2MP) ♠
(16MC, TV)	(16MC, 1MP)	(16MC, No)	(16MC, No)	(GV, 1MP)
(16MC, None)	(16MC, None)	(8MC, Yes)	(8MC, Yes)	(GV, None)
(8MC, GV)	(8MC, 2MP) ♠	(8MC, No)	(8MC, No)	(TV, 2MP) ♠
(8MC, TV)	(8MC, 1MP)	(BW, Yes) ♠	(BW, Yes) ♠	(TV, 1MP)
(8MC, None)	(8MC, None)	(BW, No)	(BW, No)	(TV, None)
(BW, GV) ♠	(BW, 2MP) ♠			(None, 2MP)
(BW, TV)	(BW, 1MP)			(None, 1MP)
(BW, None)	(BW, None)			(None, None)
(Email, Video)	(Email, Ringtones)	(Camera, Video)	(Camera, Ringtones)	(Video, Ringtones)
(GV, Yes)	(GV, Yes)	(2MP, Yes)	(2MP, Yes)	(Yes, Yes)
(GV, No)	(GV, No)	(2MP, No)	(2MP, No)	(Yes, No)
(TV, Yes)	(TV, Yes)	(1MP, Yes)	(1MP, Yes)	(No, Yes) ♠
(TV, No)	(TV, No)	(1MP, No)	(1MP, No)	(No, No)
(None, Yes)	(None, Yes)	(None, Yes) ♠	(None, Yes) ♠	
(None, No)	(None, No)	(None, No)	(None, No)	

Fig. 3. All pairs of configuration options

configuration options and thus satisfies the coverage constraints. This test suite is an instance of an optimal $CMCA(2, 5, 3^3 2^2, \mathcal{C})$ of size 9, where the notation $3^3 2^2$ is an abbreviation of $(3, 3, 3, 2, 2)$. The CNF formula $\mathcal{C}$ consists of eight constraint clauses shown at the bottom of Fig. 2. The notation $(j = d)$ and $\neg(j = d)$ is used for convenience, instead of $pos(j, d)$ and $neg(j, d)$. The clause C_ℓ represents the constraint (ℓ) for $1 \leq \ell \leq 4$ in Fig. 1. The conjunction of C_5 and C_6 represents constraint (5). The clauses C_7 and C_8 represent the constraints (6) and (7) respectively. As can be seen in Fig. 3, this array has 67 pairs in total for all interactions of parameter values, 57 of which are valid, while other 10 (followed by ♠) are invalid. For example, (GV, 2MP) in the interaction of (Email, Camera) is an invalid pair which is directly derived from C_3. (BW, Yes) in the interaction of (Display, Ringtones) is an implicit invalid pair which is derived from C_6, C_7, and the constraint that each parameter must have exactly one value. Such implicit invalid pairs make it difficult to find all valid (or invalid) pairs manually. Thus, current existing implementations calculate them at preprocessing.

3   The *catnap* Approach

We begin with describing *catnap*'s fact format for $CMCA$ instances and then present ASP encodings for $CMCA$ finding. Due to lack of space, *catnap* encodings presented here are restricted to strength $t = 2$ and stripped off capacities for handling $t \geq 3$. We also omit the explanation of calculating valid tuples at prepossessing.

Fact Format. Facts express the parameter values and constraints of a $CMCA$ instance in the syntax of ASP grounders, in our case *gringo*. Their format can be easily explained via the phone product line in Sect. 2. Its fact representation is shown in Listing 1. The facts of the predicate p/2 provide parameter

```
1   p("Display",("16MC"; "8MC"; "BW")). p("Email",("GV"; "TV"; "None")).
2   p("Camera",("2MP"; "1MP"; "None")). p("Video",("Yes"; "No")). p("Ringtones",("Yes"; "No")).

4   c(1,(neg("Email","GV"); pos("Display","16MC"); pos("Display","8MC"))).
5   c(2,(neg("Camera","2MP"); pos("Display","16MC"); pos("Display","8MC"))).
6   c(3,(neg("Email","GV"); neg("Camera","2MP"))).
7   c(4,(neg("Display","8MC"); neg("Camera","2MP"))).
8   c(5,(neg("Video","Yes"); pos("Camera","2MP"); pos("Camera","1MP"))).
9   c(6,(neg("Video","Yes"); pos("Display","16MC"); pos("Display","8MC"))).
10  c(7,(neg("Ringtones","Yes"); neg("Video","No"))).
11  c(8,(neg("Display","16MC"); neg("Email","TV"); neg("Camera","2MP"))).
```

Listing 1. Facts representing the phone product line of Fig. 1

```
1   row(1..n). col(I) :- p(I,_). c(ID) :- c(ID,_).

3   1 { assigned(R,I,A) : p(I,A) } 1 :- row(R); col(I).

5   % domain constraints
6   :- not assigned(R,I,A) : c(ID,pos(I,A)); assigned(R,J,B) : c(ID,neg(J,B)); c(ID); row(R).

8   % coverage constraints
9   covered(I,J,A,B) :- assigned(R,I,A); assigned(R,J,B); I<J.
10  :- not covered(I,J,A,B); pair(I,J,A,B).
```

Listing 2. ASP encoding for strength-2 *CMCA* finding

values in Line 1–2. The fact p(j,d) expresses that a parameter j can have a value d. Note that the ';' in the second argument is syntactic sugar, and the first fact in Line 1 is expanded into three facts p("Display","16MC"), p("Display","8MC"), and p("Display","BW"). The facts of the predicate c/2 provide constraints in Line 4–11. The fact c(ℓ, *lit*) expresses that a constraint clause identified by ℓ has literal *lit*. Again, the fact in Line 4 is expanded into three facts c(1,neg("Email","GV")), c(1,pos("Display","16MC")), and c(1,pos("Display","8MC")).

First-Order Encoding. We introduce the predicate assigned/3 to provide the assignments of parameter values. Note that a solution is composed of a set of these assignments. The atom assigned(R, I, A) expresses that a value A is assigned to the (R, I)-entry of the array. We also use the predicate pair/4 to provide pre-calculated valid pairs. The atom pair(I, J, A, B) expresses a valid pair (A, B) in the interaction of parameters (I, J).

Our encoding for strength-2 *CMCA* finding is shown in Listing 2. Given an instance of fact format with size n, the rules in Line 1 generate row(R), col(I), and c(ID) for each row R, column I, and constraint ID. The rule in Line 3, for every row R and column I, generates a candidate of assignments at first and then constrains that there is exactly one value A such that assigned(R,I,A) holds.

For domain constraints, the rule in Line 6, for every row R and constraint ID, ensures that a value A is assigned to the (R,I)-entry if c(ID,pos(I,A)) holds and value B is not if c(ID,neg(J,B)) holds. As an example, for the first row and the

```
1   row(1..n). col(I) :- p(I,_). c(ID) :- c(ID,_).

3   % activation atoms
4   { activated(R) : row(R) }.
5   #minimize{ 1,R : activated(R) }.
6   :- not activated(R); activated(R+1); R>0. % can be omitted

8   1 { assigned(R,I,A) : p(I,A) } 1 :- activated(R); col(I).

10  % domain constraints
11  :- not assigned(R,I,A) : c(ID,pos(I,A)); assigned(R,J,B) : c(ID,neg(J,B)); c(ID); row(R).

13  % coverage constraints
14  covered(I,J,A,B) :- assigned(R,I,A); assigned(R,J,B); I<J.
15  :- not covered(I,J,A,B); pair(I,J,A,B).
```

Listing 3. ASP encoding for optimal strength-2 *CMCA* finding

constraint "c(3,(neg("Email","GV");neg("Camera","2MP")))", this rule is grounded to ":- assigned(1,"Email","GV"),assigned(1,"Camera","2MP")."

For coverage constraints, the rule in Line 9, for every row R, different columns I and J (I $\leq$ J), and values A and B, generates an atom covered(I,J,A,B) if assigned(R,I,A) and assigned(R,J,B) hold. The atom covered(I,J,A,B) expresses that a pair (A,B) is covered in the interaction of parameter (I,J). Then, the rule in Line 10 ensures that a pair (A,B) in (I,J) is covered if pair(I,J,A,B) holds for every different columns I and J, and values A and B.

For optimal *CMCA* finding, we use the idea of *blocking variables*, in our case the predicate activated/1. The atom activated(R) expresses that a row R is used in a resulting array. Listing 3 shows our encoding of optimal strength-2 *CMCA* finding for a given instance with initial bound n. The differences from Listing 3 are Line 4–6 and 8. The rule in Line 4 generates an atom activated(R) for each row R. An optimal array can be found by minimizing the number of these atoms in Line 5. The rule in Line 8 is adjusted to generate the assignments of parameter values only for activated rows. Although the rule in Line 6 can be omitted, we keep it as an additional rule for performance improvement. We refer to the encoding of Listing 3 as *basic encoding*.

4   Extensions

We next extend the basic *catnap* encoding in view of enhancing the flexibility of multi-criteria optimization.

Easy extension for CCT under limiting resources. The basic encoding can concisely implement CCT solving based on *CMCA* as defined in Sect. 2. However, it does not provide CCT solving under a limited number of test cases, which happens in the real world. More precisely, the basic encoding cannot generate any test suite if the initial bound n is less than the minimal size. This is because the coverage constraints are not satisfied. To solve this practical issue, we weaken the coverage constraints by switching them from hard to soft (and refer to them as *weak coverage constraints*).

```
1  row(1..n). col(I) :- p(I,_). c(ID) :- c(ID,_).

3  % activation atoms
4  { activated(R) : row(R) }.
5  #minimize{ 1@size,R : activated(R) }.
6  :- not activated(R); activated(R+1); R>0. % can be omitted

8  1 { assigned(R,I,A) : p(I,A) } 1 :- activated(R); col(I).

10 % domain constraints
11 :- not assigned(R,I,A) : c(ID,pos(I,A)); assigned(R,J,B) : c(ID,neg(J,B)); c(ID); row(R).

13 % coverage constraints (soft)
14 covered(I,J,A,B) :- assigned(R,I,A); assigned(R,J,B); I<J.
15 #maximize{ 1@coverage,I,J,A,B : covered(I,J,A,B) }.
```

Listing 4. ASP encoding for strength-2 CCT solving with weak coverage constraints

Listing 4 shows our encoding of strength-2 CCT solving with weak coverage constraints. The main difference from the basic encoding is that the number of covered pairs is maximized in Line 15. This encoding has two criteria, the minimality of size and the maximality of coverage. Their priority levels are defined by integer constants `size` and `coverage` on the right-hand side of @ (`size`<`coverage` by default). An optimal solution can be found by a well-known multi-criteria optimization strategy called lexicographic optimization in *clingo*. It enables us to optimize criteria in a lexicographic order based on their priorities.

We refer to the encoding of Listing 4 as *weakened encoding*. The idea of weak coverage constraints allows for flexible CCT solving. The weakened encoding can generate test suites of maximal coverage under limiting resources if an initial bound n is less than the minimal size, otherwise it can find optimal *CMCA*s. Moreover, it does not require the pre-calculation of valid pairs which existing implementations rely upon. From a viewpoint of hybridization, the weakened encoding proposes a complementary approach to *prioritized CT* [6,18], which is an extension of CT with ordering (or re-ordering) strategies between test cases for detecting failures as early as possible.

Easy extension of CCT with soft constraints. Soft constraints are useful to express preferences and costs in a wide range of combinatorial optimization problems. However, there are few implementations that provide CCT solving with soft constraints. We here extend *catnap*'s fact format and weakened encoding with soft constraints. For this, we utilize the idea of *constraint atoms* used for timetabling [2]. The basic encoding can be extended in the same way.

Constraint atoms are instances of two predicates: `hard_constraint`/1 and `soft_constraint`/2. The atom `hard_constraint`(C) expresses that a constraint clause C is a hard constraint. The atom `soft_constraint`(C,W) expresses that C is a soft constraint and its weight is W. Listing 5 shows an extension of the fact representation of the phone product line with constraint atoms. In this case, new constraint clauses in Line 3–4 are added as soft constraints.

```
1  hard_constraint(1..8).
2  soft_constraint(9..10,1).
3  c(9,(neg("Display","BW"); pos("Camera","None"))).
4  c(10,(neg("Camera","None"); neg("Display","BW"); neg("Email","None"))).
```

Listing 5. An extended fact representation with constraint atoms

```
1  % domain constraints (hard)
2  :- not assigned(R,I,A) : c(ID,pos(I,A)); assigned(R,J,B) : c(ID,neg(J,B));
3     hard_constraint(ID); row(R).
4
5  % domain constraints (soft)
6  penalty(ID,R,W) :- not assigned(R,I,A) : c(ID,pos(I,A));
7                         assigned(R,J,B) : c(ID,neg(J,B));
8                     soft_constraint(ID,W); row(R).
9  #minimize{ W@soft,ID,R : penalty(ID,R,W) }.
```

Listing 6. An extended domain constraints with constraint atoms

Extending the weakened encoding with constraint atoms can be done by replacing the rule in Line 11 of Listing 4 with the rules shown in Listing 6. For hard domain constraints, the rule in Line 2–3 is the same as before except `hard_constraint(ID)`. For soft domain constraints, the rule in Line 6–8 generates an atom `penalty(ID,R,W)` if the assignments in R violate a constraint ID for every row R and soft constraint identified by ID of weight W. That is, for each violation in R for ID, the atom `penalty(ID,R,W)` is generated. Then, the number of these atoms is minimized in Line 9, where its priority level is defined by an integer constant `soft`.

The resulting encoding has three criteria, the minimality of size, the maximality of coverage, and the minimality of penalty costs. The default ordering of priority levels is `soft`<`size`<`coverage`, but can be changed by the command line option of *clingo*. As an example, in a case of `soft`=`size`, the encoding can generate optimal test suites of minimal sum of size and penalty costs.

CCT solving with *catnap* can be promising, since it allows for flexible CCT solving of generating optimal test suites by varying a set of hard and soft constraints, switching them between hard and soft, and varying the priority levels of criteria, even if the number of test cases is limited.

5   Experiments

As we have mentioned, *catnap* accepts a CCT instance in fact format and combines it with a first-order ASP encoding for CCT solving, which is subsequently solved by the ASP system *clingo* that returns an assignment representing a solution to the original CCT instance. The *catnap* system also accepts the *CASA* format [12]. For this, we implemented a converter that provides us with the resulting CCT instance in *catnap*'s fact format.

Table 1. Comparison between the basic and weakened encodings

instance	t	k	$(v_1, \ldots, v_k)$	$|\mathcal{C}|$	Weakened	Basic
benchmark_apache	2	172	$2^{158}3^84^45^16^1$	7	**30**	**30**
benchmark_bugzilla	2	52	$2^{49}3^14^2$	5	**16**	**16**
benchmark_gcc	2	199	$2^{189}3^{10}$	40	**15**	**15**
benchmark_spins	2	18	$2^{13}4^5$	13	**19**	**19**
benchmark_spinv	2	55	$2^{42}3^24^{11}$	49	**31**	**31**
benchmark_1	2	97	$2^{86}3^34^15^56^2$	24	**38**	**38**
benchmark_2	2	94	$2^{86}3^34^35^16^1$	22	**30**	**30**
benchmark_3	2	29	$2^{27}4^2$	10	**18**	**18**
benchmark_4	2	58	$2^{51}3^44^25^1$	17	**20**	$t.o$
benchmark_5	2	174	$2^{155}3^74^35^56^4$	39	**46**	**46**
benchmark_6	2	77	$2^{73}4^36^1$	30	**24**	$t.o$
benchmark_7	2	30	$2^{29}3^1$	15	**9**	**9**
benchmark_8	2	119	$2^{109}3^24^25^36^3$	37	**37**	38
benchmark_9	2	61	$2^{57}3^14^15^16^1$	37	**20**	$t.o$
benchmark_10	2	147	$2^{130}3^64^55^26^4$	47	**41**	**41**
benchmark_11	2	96	$2^{84}3^44^25^26^4$	32	**40**	41
benchmark_12	2	147	$2^{136}3^44^35^16^3$	27	**36**	**36**
benchmark_13	2	133	$2^{124}3^44^15^26^2$	26	**36**	**36**
benchmark_14	2	92	$2^{81}3^54^36^3$	15	**36**	**36**
benchmark_15	2	58	$2^{50}3^44^15^26^1$	22	**30**	**30**
benchmark_16	2	87	$2^{81}3^34^26^1$	34	**24**	$t.o$
benchmark_17	2	137	$2^{128}3^34^25^16^3$	29	**36**	**36**
benchmark_18	2	141	$2^{127}3^24^45^66^2$	28	**41**	**41**
benchmark_19	2	197	$2^{172}3^94^95^36^4$	43	**44**	$t.o$
benchmark_20	2	158	$2^{138}3^44^55^46^7$	48	59	**54**
benchmark_21	2	85	$2^{76}3^34^25^16^3$	46	**36**	**36**
benchmark_22	2	79	$2^{72}3^44^16^2$	22	**36**	**36**
benchmark_23	2	27	$2^{25}3^16^1$	15	**12***	**12***
benchmark_24	2	119	$2^{110}3^25^36^4$	29	43	**42**
benchmark_25	2	134	$2^{118}3^64^25^26^6$	27	**48**	50
benchmark_26	2	95	$2^{87}3^14^35^4$	32	**27**	**27**
benchmark_27	2	62	$2^{55}3^24^25^16^2$	20	**36**	**36**
benchmark_28	2	194	$2^{167}3^{16}4^25^36^6$	37	**50**	51
benchmark_29	2	144	$2^{134}3^75^3$	22	**25**	**25**
benchmark_30	2	79	$2^{73}3^34^3$	35	**16**	**16**
#best					33	26

Table 2. Comparison of *catnap* with other approaches

Instance	*catnap*	*TCA*	*CASA*	*Cascade*	*ACTS*
benchmark_apache	**30**	**30**	32	n.a	33
benchmark_bugzilla	**16**	**16**	**16**	20	19
benchmark_gcc	**15**	16	19	n.a	23
benchmark_spins	**19**	**19**	**19**	27	26
benchmark_spinv	**31**	**31**	36	41	45
benchmark_1	38	**36**	38	n.a	48
benchmark_2	**30**	**30**	**30**	n.a	32
benchmark_3	**18**	**18**	**18**	19	19
benchmark_4	**20**	**20**	**20**	24	22
benchmark_5	46	**43**	45	n.a	54
benchmark_6	**24**	**24**	**24**	30	25
benchmark_7	**9**	**9**	**9**	12	12
benchmark_8	**37**	**37**	38	n.a	47
benchmark_9	**20**	**20**	**20**	23	22
benchmark_10	41	**40**	42	n.a	47
benchmark_11	40	**39**	41	n.a	47
benchmark_12	**36**	**36**	39	n.a	43
benchmark_13	**36**	**36**	**36**	n.a	40
benchmark_14	**36**	**36**	37	n.a	39
benchmark_15	**30**	**30**	**30**	37	32
benchmark_16	**24**	**24**	**24**	n.a	25
benchmark_17	**36**	**36**	38	n.a	41
benchmark_18	41	**39**	41	n.a	52
benchmark_19	44	**43**	47	n.a	51
benchmark_20	59	**49**	52	n.a	60
benchmark_21	**36**	**36**	**36**	n.a	39
benchmark_22	**36**	**36**	**36**	n.a	37
benchmark_23	**12**	**12**	**12**	14	14
benchmark_24	43	**40**	42	n.a	48
benchmark_25	48	**45**	47	n.a	52
benchmark_26	**27**	**27**	30	n.a	34
benchmark_27	**36**	**36**	**36**	45	37
benchmark_28	50	**47**	50	n.a	57
benchmark_29	**25**	**25**	29	n.a	29
benchmark_30	**16**	**16**	19	n.a	22
#best	25	34	15	0	0

For our experiments, we use all 35 instances[2] proposed in [8], five of which are from highly configurable software systems such as *SPIN*, *GCC*, *Apache*, and *Bugzilla*. Note that these are *CMCA* instances in *CASA* format which have no soft constraints. We ran them on a multi-core Linux machine equipped with Xeon 3.16 GHz and 32GB RAM. We imposed a time-limit (*t.o*) of 1 hour for each run. We used the ASP system *clingo* (version 4.5.3) with the multi-threaded portfolio search of four configurations.[3]

At first, we analyze the difference between the basic and weakened encodings. Table 1 contrasts the bounds obtained from both encodings. The information of the $CMCA(t, k, (v_1, \ldots, v_k), \mathcal{C})$ instances is given in the first five columns. We highlight the best bound of different encodings for each instance. The #best row gives the number of best bounds for each encoding. The symbol '*' indicates that *catnap* proved the optimality of the obtained bound. The weakened encoding was able to find the best bounds for 33 instances compared with 26 obtained with the basic encoding. The basic encoding found no solution to 5 instances in the time limit. Both encodings were able to prove that the previous known bound (12) for `benchmark_23` is optimal. Because of these observations, we adopt the weakened encoding as the default setting of *catnap*.

Next, we compare the performance of *catnap* with other approaches. Table 2 contrasts the bounds obtained by *catnap* with the best known ones in [14] obtained from dedicated implementations: *CASA* [12], *TCA* [14], *ACTS* [21], and *Cascade* [22]. *CASA* and *TCA* are metaheuristics-based dedicated implementations. *ACTS* and *Cascade* are based on greedy algorithms. The symbol '*n.a*' indicates that a solver found no solution in [14]. The *catnap* system was able to find the best bounds for 25 instances, compared with 34 of *TCA*, 15 of *CASA*, and 0 of *Cascade* and *ACTS*. For the large instance `benchmark_gcc`, *catnap* was able to find a better bound (15) than the others. Although it does not fully match the performance of *TCA*, *catnap* can be competitive to *CASA* and can outperform *Cascade* and *ACTS*.

Finally, we discuss some more details of our experimental results. We used a simple configuration for multi-threaded portfolio search of *clingo*, but it took a longer time to find the best bounds of many instances than the others. This is a limitation of our approach at present. To overcome this issue, we will investigate the best configuration of *clingo*, since it offers several optimization strategies. Evaluating the scalability of *catnap* for higher strength $t \geq 3$ is also an important future work.

6   Conclusion

From an ASP perspective, the most relevant related works are ASP encodings of *event-sequence testing* [3,5,11]. Event-sequence testing is a testing technique especially for event-driven systems and is different from CCT that focuses on

---

[2]   http://cse.unl.edu/~citportal/public/tools/casa/benchmarks.zip.
[3]   The combination of `--config={trendy, jumpy}` and `--opt-strat={bb,1,usc,1}`.

highly configurable systems. The ASP encodings in [5,11] use #maximize statements including **covered** atoms for two purposes. One is to generate single test cases of maximal coverage in each iteration of ASP-based greedy algorithms. *Cascade* adopts a similar technique. Another is to generate strength-t test suites of maximal $(t + 1)$-coverage for early failure detection. This technique is closely related to prioritized CT [18]. Note that the purpose of the weak coverage constraints in Sect. 4 is to provide CCT under limiting resources.

We presented an ASP-based approach to solving the CCT problems. The resulting system *catnap* consists of first-order encodings and delegates solving tasks to general-purpose ASP systems. We showed that ASP is an ideal modeling language for combinatorial testing, as demonstrated by *catnap*'s compact encodings for CCT solving. We contrasted the performance of *catnap* to the best known bounds obtained via dedicated implementations. The *catnap* system demonstrated that ASP's general-purpose technology allows us to compete with state-of-the-art CCT solving techniques. All source code of *catnap* is available from https://potassco.org/doc/apps/.

Closely related to constraint solving in CCT, recent advances in Constraint ASP (CASP; [1,13]) open up a successful direction to extend ASP to be more expressive. CASP solvers such as *clingcon* can solve finite linear Constraint Satisfaction Problems in a declarative way. We will investigate the possibilities of CCT solving with CASP to extend CCT with richer constraints.

References

1. Balduccini, M.: Representing constraint satisfaction problems in answer set programming. In: Faber, W., Lee, J. (eds.) Proceedings of the Second Workshop on Answer Set Programming and Other Computing Paradigms (ASPOCP 2009), pp. 16–30 (2009)
2. Banbara, M., Inoue, K., Kaufmann, B., Schaub, T., Soh, T., Tamura, N., Wanko, P.: teaspoon: Solving the curriculum-based course timetabling problems with answer set programming. In: Proceedings of the Eleventh International Conference of the Practice and Theory of Automated Timetabling (PATAT 2016), pp. 13–32 (2016)
3. Banbara, M., Tamura, N., Inoue, K.: Generating event-sequence test cases by answer set programming with the incidence matrix. In: Technical Communications of the 28th International Conference on Logic Programming (ICLP 2012), LIPIcs, vol. 17, pp. 86–97. Schloss Dagstuhl (2012)
4. Baral, C.: Knowledge Representation, Reasoning and Declarative Problem Solving. Cambridge University Press, Cambridge (2003)
5. Brain, M., Erdem, E., Inoue, K., Oetsch, J., Pührer, J., Tompits, H., Yilmaz, C.: Event-sequence testing using answer-set programming. Int. J. Adv. Softw. **5**(3–4), 237–251 (2012)
6. Bryce, R.C., Colbourn, C.J.: Prioritized interaction testing for pair-wise coverage with seeding and constraints. Inf. Softw. Technol. **48**(10), 960–970 (2006)
7. Cohen, D.M., Dalal, S.R., Fredman, M.L., Patton, G.C.: The AETG system: an approach to testing based on combinatiorial design. IEEE Trans. Softw. Eng. **23**(7), 437–444 (1997)

8. Cohen, M.B., Dwyer, M.B., Shi, J.: Constructing interaction test suites for highly-configurable systems in the presence of constraints: a greedy approach. IEEE Trans. Softw. Eng. **34**(5), 633–650 (2008)
9. Colbourn, C.J., Dinitz, J.H.: Handbook of Combinatorial Designs, 2nd edn. Chapman & Hall/CRC, Boca Raton (2006)
10. Czerwonka, J.: Pairwise testing in real world. Practical extensions to test case generators. In: Proceedings of the 24th Annual Pacific Northwest Software Quality Conference, pp. 419–430 (2006)
11. Erdem, E., Inoue, K., Oetsch, J., Pührer, J., Tompits, H., Yilmaz, C.: Answer-set programming as a new approach to event-sequence testing. In: Proceedings of the 3rd International Conference on Advances in System Testing and Validation Lifecycle (VALID 2011), pp. 25–34. Xpert Publishing Services (2011)
12. Garvin, B.J., Cohen, M.B., Dwyer, M.B.: An improved meta-heuristic search for constrained interaction testing. In: Proceedings of the first International Symposium on Search Based Software Engineering (SSBSE 2009), pp. 13–22. IEEE (2009)
13. Gebser, M., Ostrowski, M., Schaub, T.: Constraint answer set solving. In: Hill, P.M., Warren, D.S. (eds.) ICLP 2009. LNCS, vol. 5649, pp. 235–249. Springer, Heidelberg (2009). doi:10.1007/978-3-642-02846-5_22
14. Lin, J., Luo, C., Cai, S., Su, K., Hao, D., Zhang, L.: TCA: An efficient two-mode meta-heuristic algorithm for combinatorial test generation. In: Proceedings of 30th IEEE/ACM International Conference on Automated Software Engineering (ASE 2015), pp. 494–505. IEEE (2015)
15. Mandl, R.: Orthogonal latin squares: an application of experiment design to compiler testing. Commun. ACM **28**(10), 1054–1058 (1985)
16. Nanba, T., Tsuchiya, T., Kikuno, T.: Constructing test sets for pairwise testing: a SAT-based approach. In: Proceedings of the Second International Conference on Networking and Computing (ICNC 2011), pp. 271–274. IEEE (2011)
17. Nie, C., Leung, H.: A survey of combinatorial testing. ACM Comput. Surv. **43**(2), 11:1–11:29 (2011). http://doi.acm.org/10.1145/1883612.1883618
18. Petke, J., Yoo, S., Cohen, M.B., Harman, M.: Efficiency and early fault detection with lower and higher strength combinatorial interaction testing. In: Proceedings of ESEC/FSE 2013, pp. 26–36. ACM (2013)
19. Tatsumi, K.: Test case design support system. In: Proceedings of the International Conference on Quality Control (ICQC 1987), pp. 615–620 (1987)
20. Yamada, A., Kitamura, T., Artho, C., Choi, E., Oiwa, Y., Biere, A.: Optimization of combinatorial testing by incremental SAT solving. In: Proceedings of 8th IEEE International Conference on Software Testing, Verification and Validation (ICST 2015), pp. 1–10. IEEE (2015)
21. Yu, L., Lei, Y., Nourozborazjany, M., Kacker, R., Kuhn, D.R.: An efficient algorithm for constraint handling in combinatorial test generation. In: Proceedings of the Sixth IEEE International Conference on Software Testing, Verification and Validation (ICST 2013), pp. 242–251. IEEE (2013)
22. Zhang, Z., Yan, J., Zhao, Y., Zhang, J.: Generating combinatorial test suite using combinatorial optimization. J. Syst. Softw. **98**, 191–207 (2014)

Automatic Synthesis of Optimal-Size Concentrators by Answer Set Programming

Marc Dahlem[1]([✉]), Tripti Jain[2], Klaus Schneider[2], and Michael Gillmann[1]

[1] Insiders Technologies GmbH, Kaiserslautern, Germany
m_dahlem@cs.uni-kl.de
[2] Department of Computer Science,
University of Kaiserslautern, Kaiserslautern, Germany
https://insiders-technologies.de,
http://es.cs.uni-kl.de

Abstract. A concentrator is a circuit with N inputs and $M \leq N$ outputs that can route any given subset of $K \leq M$ valid inputs to K of its M outputs. Concentrator circuits are important building blocks of many parallel algorithms. The design of optimal concentrator circuits is however a challenging task that has already been considered in many research papers. In this paper, we show how answer set programming can be used to automatically generate concentrator circuits of provably optimal size.

1 Introduction

A (N, M)-concentrator [17] is a circuit with N inputs and $M \leq N$ outputs that can route any given number $K \leq M$ of valid inputs to K of its M outputs. The selection of K of the M outputs and the mapping of the inputs to these is thereby not of interest, but it must be possible to connect any $K \leq M$ inputs with some K of the M outputs.

Pinsker [17] proved that there exist (N, M)-concentrators with no more than $29N$ edges for sufficiently large N. Chung [5] improved this to $27N$ edges but still with an unknown depth. Gabber and Galil presented then a concentrator of approximately size $273N$ and depth $O(\log(N))$ [9]. Schöning defined in [19] super-concentrators of approximate size $28N$. However, all of these linear-size constructions are based on expander graphs which are notoriously difficult to implement, and while asymptotically superior, the constants are too large to be of practical interest.

More practical concentrators were implemented by permutation networks (see references in [13]) where concentrators of size $O(N \log(N))$ and depth $O(\log(N))$ can be obtained. It is also known that some of the switches of the permutation networks are not needed [18]. Sorting networks [1] can also be used as concentrators. However, the practically best known sorting networks [1,16] require $O(\log(N)^2)$ depth and $O(N \log(N)^2)$ size (in terms of comparators) and therefore do not lead to competitive concentrator designs.

© Springer International Publishing AG 2017
M. Balduccini and T. Janhunen (Eds.): LPNMR 2017, LNAI 10377, pp. 279–285, 2017.
DOI: 10.1007/978-3-319-61660-5_25

Automatic synthesis became more and more successful during the past decade. Starting with early work of Church, Pnueli and Rosner, there are now many powerful tools that are based on the success of recent (quantified) SAT solvers. Work in that area that is close to ours is the automatic synthesis of sorting networks that has been initiated in [15] and has now lead recently to new optimality results in a couple of papers (see e.g. [2,3,6,7]).

Answer Set Programming (ASP) [8] is an attractive approach to combine verification techniques with programming. While SAT solvers compute one satisfying assignment for a given propositional logic formula, ASP solvers compute *all* satisfying models (the 'answer sets'). ASP is NP-complete and has its root in logical programming and non-monotonic reasoning [12]. Therefore, it is not a surprise that the syntax of ASP shares some similarities with the logical programming language Prolog. Programming in ASP is declarative, i.e., the solutions to a problem are specified with constraints and boundary conditions. Thus, the main idea is to write precise specifications for the solution instead of encoding a way how to compute it. Besides the usual logic operators, ASP solvers offer minimize and maximize statements, cardinality constraints, helper function definitions, a lot of syntactic sugar, etc. There exists a standardized core language for ASP [4] which is used in most ASP frameworks. The most recent and modern frameworks are clasp [10,11] and dlv [14], and in this paper, the clasp framework has been used.

In this paper, we show how ASP [8] can be used to automatically generate concentrator circuits of provably minimal size which is in the spirit of previous work where SAT solvers have been used to generate optimal size and optimal depth sorting networks. We show that minimal networks can be generated which are smaller in size than optimal sorting networks.

2   Representing Concentrators as ASP Problem

The general idea of describing the network minimizing problem for concentrators as a specification in ASP is as follows: A network has three kinds of nodes, namely input nodes, output nodes, and internal switches. Input nodes have no input ports

Fig. 1. Switch configurations

and one output port, output nodes have one input port, but no output ports, and switches have two input ports and two output ports. Every output port of a node can have a connection to an input port of another node.

Beside the different kinds of nodes, there are obvious constraints on the connections which must be defined: An input port can only be connected to one output port, every input port must be connected to the network, etc.

The goal of the shown answer set program is to find a suitable connection for such a network as an answer set. Hereby, every combination of choosing M out of the N inputs must be able to be routed to the last M out of the N outputs. The routing is given by a switch configuration: a switch can either route its two inputs through to the corresponding two outputs or it may swap these as shown in Fig. 1.

The ASP program can be configured by three constants: the number of inputs N, the number of outputs M, and the upper bound *num_switches* of the switches to be used. The latter is used to reduce the search space, as it is assumed that the solution to be found is below *max_switches*.

```
1  #minimize{N: num_switches(N)}
2  #const N = 3.
3  #const M = 2.
4  #const max_switches = 5.
```

Given these parameters, the basic nodes of the network can be configured.

```
1  in(1..N).
2  out(1..N).
3  used_output(Out) :- out(Out), Out>(N-M).
```

The first two lines define exactly N input and output ports of the network. The last line defines all last M of the outputs as used for concentrating. These used outputs should be reachable from all possible $K \leq M$ out of N inputs. It is enough to find a concentrator for $K = M$, because this concentrator retains the concentrator property for all $K \leq M$. Concerning the switches, first of all it is defined how many switches a network can/must have:

```
1  possible_switch_amount(0..max_switches).
2  1 {num_switches(N): possible_switch_amount(N)} 1.
3  switch(1..N) :- num_switches(N), N>0.
```

It is specified above that it is possible to have 0 up to *max_switches* as number of switches in the concentrator. For the result, ASP should choose exactly one of these numbers of switches and create as many switches as chosen. Without further restrictions, this network would minimize according to the overall goal to size 0. Therefore, a clear way of defining the constraints on the network must be found. The first step is to define the goals the network must fulfill:

```
1  1 {goal(C,In,Out):used_output(Out)} 1 :- contains(In,C).
2  :- goal(C,In,Out), goal(C,In2,Out), In != In2.
```

Considering all combinations of picking M out of the N inputs in a predicate C, for every input In of such a combination C, exactly one $used\_output(Out)$ must be assigned to that input. Furthermore, for two different inputs In and $In2$ in the same combination C, the output cannot be the same.

Next, a predicate *goal* is defined which holds for all combinations of inputs a unique assignment of the chosen inputs to the used outputs of our network. More precisely, the amount of goals per combination of the inputs is the same as M. Furthermore, there are $\binom{N}{M}$ combinations possible which sums up to $\binom{N}{M} \cdot M$ goals. Each of these $\binom{N}{M} \cdot M$ goals must be fulfilled by the network, whereby for every combination C another switch configuration can be chosen. Therefore, $\binom{N}{M}$ different switch configurations for the network must be found. The defined goals must be reachable in the network, given a switch configuration for the combination C.

```
1  :- goal(C,In,O), #count{1,Out:reachable(C,in(In),out(Out))}=N, N!=1.
2  :- goal(C,In,Out), reachable(C,in(In),out(Out2)), Out2!=Out.
```

The first line states that the input of our goal must reach exactly one output of the network for one combination. The second line states that this output is the same as the output defined in our goal. The recursive predicate *reachable* is thereby dependent on the combination, as for every input combination, other switch positions of the network must be chosen.

```
1  reachable(combination(ID),in(In),D) :- connection(in(In),D), →
       combination(ID).
2  reachable(C,In,D) :- connection(switch(S),SV,D), →
       switchConfig(C,S,IV,SV), reachable(C,In,switch(S,IV)).
```

A node D is reachable from an input In for a combination C, if

- either D is directly connected to the input In
- or the predecessor of D is a switch S which is already reachable: the input port IV of S is already reachable from In, and IV routes for the combination C (with switch position SV) to the node D

To conclude with, some more constraints can be added on the reachability, as e.g.:

```
1  :- reachable(C,In,S), reachable(C,In2,S), In!=In2.
```

First of all, it is not allowed, that the input port of a switch S can be reached from different inputs. Second, it is also not allowed that the same output S can be reached from different inputs.

3   Experimental Results

The above ASP encoding has been executed with clingo 4.5.4 on a machine with a i5-6600@3.30 GHz running Ubuntu 16.04.1 LTS. The machine had 16 GB memory available, and the process used approximately 2 GB. The ASP solver has been called multiple times for different numbers of N and M. Figure 1 shows the generated concentrators for different N and M, and Table. 1 shows the required runtimes (total time, time needed to prove the optimality, and time needed to generate the first concentrator of that size).

Table 1. Runtimes in seconds required for generating optimal concentrators (runtimes for $N < 5$ are negligible).

N/M	total	optimality	model
5/1	0.26	0.198	0.015
5/2	145.79	145.245	0.354
5/3	303.70	302.949	0.415
5/4	0.88	0.394	0.365
6/1	24.11	23.850	0.087
6/2	33042.85	33031.465	8.439
6/3	317913.72	317906.224	5.799
6/4	60621.53	60594.077	11.865
6/5	113.61	112.970	0.208
7/1	5618.63	5618.438	0.038
7/2	1870617.88 (aborted)	N/A	3.392

The ASP solver returns almost immediately the optimal concentrators up to $N = 4$ inputs. Starting from $N = 5$, the time required to prove that there is no

Table 2. Optimal-size concentrators generated by ASP (the outputs with a circle are those where the valid inputs will be routed to).

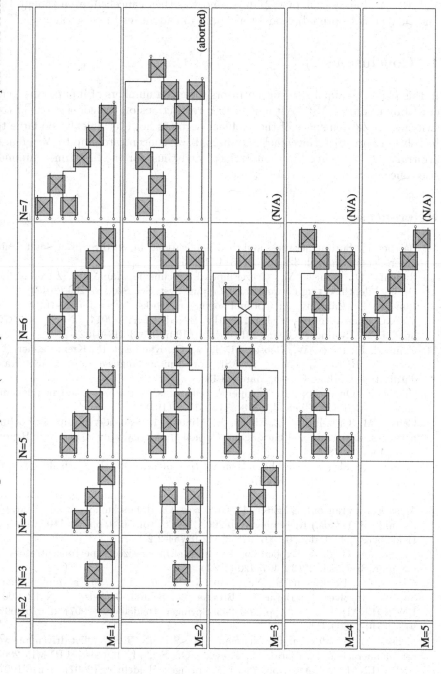

smaller concentrator grows enormously: The computation of a minimal network for $N = 6$ and $M = 3$ took more than three days; the computation for $N = 7$ and $M = 2$ already took over 21 days, and was then cancelled, even though the first (maybe not optimal) model was found already after three seconds.

4 Conclusions

In this paper, minimal size concentrators for given numbers of inputs were generated by means of ASP. Without further restrictions on the connections of the switches, the search space of the problem becomes however quickly too large to be solved in practice. Nevertheless, minimal-size concentrators up to $N = 6$ and arbitrary $M \leq N$ have been determined, and thus, for the first time, minimal size concentrators are proved this way.

References

1. Batcher, K.: Sorting networks and their applications. In: AFIPS Spring Joint Computer Conference, vol. 32, pp. 307–314 (1968)
2. Bundala, D., Codish, M., Cruz-Filipe, L., Schneider-Kamp, P., Závodný, J.: Optimal-depth sorting networks. J. Comput. Syst. Sci. **84**, 185–204 (2017)
3. Bundala, D., Závodný, J.: Optimal sorting networks. In: Dediu, A.-H., Martín-Vide, C., Sierra-Rodríguez, J.-L., Truthe, B. (eds.) LATA 2014. LNCS, vol. 8370, pp. 236–247. Springer, Cham (2014). doi:10.1007/978-3-319-04921-2_19
4. Calimeri, F., Faber, W., Gebser, M., Ianni, G., Kaminski, R., Krennwallner, T., Leone, N., Ricca, F., Schaub, T.: ASP-Core-2 input language format, ASP Standardization Working Group, March 2013
5. Chung, F.: On concentrators, superconcentrators, generalizers, and nonblocking networks. Bell Syst. Tech. J. **58**(8), 1765–1777 (1978)
6. Codish, M., Cruz-Filipe, L., Ehlers, T., Müller, M., Schneider-Kamp, P.: Sorting networks: to the end and back again. Cornell University, July 2015. arXiv Report arXiv:1507.01428v1
7. Ehlers, T., Müller, M.: New bounds on optimal sorting networks. Cornell University Library, January 2015. arXiv Report arXiv:1501.06946v1
8. Eiter, T., Ianni, G., Krennwallner, T.: Answer set programming: a primer. In: Tessaris, S., Franconi, E., Eiter, T., Gutierrez, C., Handschuh, S., Rousset, M.-C., Schmidt, R.A. (eds.) Reasoning Web 2009. LNCS, vol. 5689, pp. 40–110. Springer, Heidelberg (2009). doi:10.1007/978-3-642-03754-2_2
9. Gabber, O., Galil, Z.: Explicit constructions of linear-sized superconcentrators. J. Comput. Syst. Sci. **22**(3), 407–420 (1981)
10. Gebser, M., Kaufmann, B., Neumann, A., Schaub, T.: *clasp*: a conflict-driven answer set solver. In: Baral, C., Brewka, G., Schlipf, J. (eds.) LPNMR 2007. LNCS (LNAI), vol. 4483, pp. 260–265. Springer, Heidelberg (2007). doi:10.1007/978-3-540-72200-7_23
11. Gebser, M., Kaufmann, B., Neumann, A., Schaub, T.: Conflict-driven answer set enumeration. In: Baral, C., Brewka, G., Schlipf, J. (eds.) LPNMR 2007. LNCS (LNAI), vol. 4483, pp. 136–148. Springer, Heidelberg (2007). doi:10.1007/978-3-540-72200-7_13

12. Gelfond, M., Lifschitz, V.: The stable model semantics for logic programming. In: Kowalski, R., Bowen, K. (eds.) Logic Programming, Seattle, Washington, USA, pp. 1070–1080. MIT Press (1988)

13. Jain, T., Schneider, K.: Verifying the concentration property of permutation networks by BDDs. In: Leonard, E., Schneider, K. (eds.) Formal Methods and Models for Codesign (MEMOCODE), Kanpur, India, pp. 43–53. IEEE Computer Society (2016)

14. Leone, N., Pfeifer, G., Faber, W., Calimeri, F., Dell'Armi, T., Eiter, T., Gottlob, G., Ianni, G., Ielpa, G., Koch, C., Perri, S., Polleres, A.: The DLV system. In: Flesca, S., Greco, S., Ianni, G., Leone, N. (eds.) JELIA 2002. LNCS (LNAI), vol. 2424, pp. 537–540. Springer, Heidelberg (2002). doi:10.1007/3-540-45757-7_50

15. Morgenstern, A., Schneider, K.: Synthesis of parallel sorting networks using SAT solvers. In: Methoden und Beschreibungssprachen zur Modellierung und Verifikation von Schaltungen und Systemen (MBMV), Oldenburg, Germany, pp. 71–80 (2011)

16. Parberry, I.: The pairwise sorting network. Par. Proc. Lett. (PPL) 2(2–3), 205–211 (1992)

17. Pinsker, M.: On the complexity of a concentrator. In: International Teletraffic Conference (ITC), Stockholm, Sweden, pp. 318:1–318:4 (1973)

18. Quinton, B., Wilton, S.: Concentrator access networks for programmable logic cores on SoCs. In: International Symposium on Circuits and Systems (ISCAS), Kobe, Japan, vol. 1, pp. 45–48. IEEE Computer Society (2005)

19. Schöning, U.: Smaller superconcentrators of density 28. Inf. Process. Lett. (IPL) 98(4), 127–129 (2006)

plasp 3: Towards Effective ASP Planning

Yannis Dimopoulos[1], Martin Gebser[2], Patrick Lühne[2], Javier Romero[2], and Torsten Schaub[2(✉)]

[1] University of Cyprus, Nicosia, Cyprus
[2] University of Potsdam, Potsdam, Germany
torsten@cs.uni-potsdam.de

Abstract. We describe the new version of the PDDL-to-ASP translator *plasp*. First, it widens the range of accepted PDDL features. Second, it contains novel planning encodings, some inspired by SAT planning and others exploiting ASP features such as well-foundedness. All of them are designed for handling multi-valued fluents in order to capture both PDDL as well as SAS planning formats. Third, enabled by multi-shot ASP solving, it offers advanced planning algorithms also borrowed from SAT planning. As a result, *plasp* provides us with an ASP-based framework for studying a variety of planning techniques in a uniform setting. Finally, we demonstrate in an empirical analysis that these techniques have a significant impact on the performance of ASP planning.

1 Introduction

Reasoning about actions and change constitutes a major challenge to any formalism for knowledge representation and reasoning. It therefore comes as no surprise that Automated Planning [4] was among the first substantial application of Answer Set Programming (ASP [12]). Meanwhile this has led to manifold action languages [9], various applications in dynamic domains [1], but only few adaptions of Automated Planning techniques [16]. Although this has provided us with diverse insights into how relevant concepts are expressed in ASP, almost no attention has been paid to making reasoning about actions and change effective. This is insofar surprising as a lot of work has been dedicated to planning with techniques from the area of Satisfiability Testing (SAT [2]), a field often serving as a role model for ASP.

We address this shortcoming with the third series of the *plasp* system. From its inception, the purpose of *plasp* was to provide an elaboration-tolerant platform to planning by using ASP. Already its original design [7] foresaw to compile planning problems formulated in the Planning Domain Definition Language (PDDL [13]) into ASP facts and to use ASP meta-encodings for modeling alternative planning techniques. These could then be solved with fixed horizons (and optimization) or in an incremental fashion. The redesigned *plasp* 3 system features optional preprocessing by the state-of-the-art planning system *Fast Downward* [10] (via the intermediate SAS format), a homogeneous factual representation capturing both PDDL and SAS input (with multi-valued fluents), and

M. Balduccini and T. Janhunen (Eds.): LPNMR 2017, LNAI 10377, pp. 286–300, 2017.
DOI: 10.1007/978-3-319-61660-5_26

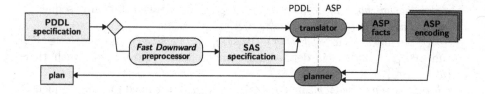

Fig. 1. Solving PDDL inputs with *plasp*'s workflow (highlighted in blue)

a normalization step to support advanced PDDL features. Moreover, *plasp* 3 provides a spectrum of ASP encodings ranging from adaptions of known SAT encodings [15] to novel encodings taking advantage of ASP-specific concepts. Finally, *plasp* 3 offers sophisticated planning algorithms, also stemming from SAT planning [15], by taking advantage of multi-shot ASP solving. The common structure of various incremental ASP encodings makes *plasp*'s planning framework also applicable to dynamic domains beyond PDDL. The usual workflow of *plasp* 3, though, is summarized in Fig. 1.

2   ASP Encodings for Planning

We consider STRIPS-like (multi-valued) *planning tasks* according to [10], given by a 4-tuple $\langle \mathcal{F}, s_0, s_\star, \mathcal{O} \rangle$, in which

- $\mathcal{F}$ is a finite set of state variables, also called *fluents*, where each $x \in \mathcal{F}$ has an associated finite domain x^d of possible values for x,
- s_0 is a *state*, i.e., a (total) function such that $s_0(x) \in x^d$ for each $x \in \mathcal{F}$,
- $s_\star$ is a *partial state* (listing goal conditions), i.e., a function such that $s_\star(x) \in x^d$ for each $x \in \tilde{s}_\star$, where $\tilde{s}_\star$ denotes the set of all $x \in \mathcal{F}$ such that $s_\star(x)$ is defined, and
- $\mathcal{O}$ is a finite set of operators, also called *actions*, where a^c and a^e in $a = \langle a^c, a^e \rangle$ are partial states denoting the *precondition* and *postcondition* of a for each $a \in \mathcal{O}$.

Given a state s and an action $a \in \mathcal{O}$, the *successor state* $o(a, s)$ obtained by applying $a = \langle a^c, a^e \rangle$ in s is defined if $a^c(x) = s(x)$ for each $x \in \tilde{a}^c$, and undefined otherwise. Provided that $s' = o(a, s)$ is defined, $s'(x) = a^e(x)$ for each $x \in \tilde{a}^e$, and $s'(x) = s(x)$ for each $x \in \mathcal{F} \setminus \tilde{a}^e$. That is, if the successor state $o(a, s)$ is defined, it includes the postcondition of a and keeps any other fluents unchanged from s. We extend the notion of a successor state to sequences $\langle a_1, \ldots, a_n \rangle$ of actions by letting $o(\langle a_1, \ldots, a_n \rangle, s) = o(a_n, o(\ldots, o(a_1, s) \ldots))$, provided that $o(a_i, o(\ldots, o(a_1, s) \ldots))$ is defined for all $1 \leq i \leq n$. Given this, a *sequential plan* is a sequence $\langle a_1, \ldots, a_n \rangle$ of actions such that $s' = o(\langle a_1, \ldots, a_n \rangle, s_0)$ is defined and $s'(x) = s_\star(x)$ for each $x \in \tilde{s}_\star$.

Several *parallel* representations of sequential plans have been investigated in the literature [4,15,17]. We call a set $\{a_1, \ldots, a_k\} \subseteq \mathcal{O}$ of actions *confluent* if $a_i^e(x) = a_j^e(x)$ for all $1 \leq i < j \leq k$ and each $x \in \tilde{a}_i^e \cap \tilde{a}_j^e$. Given a state s and a confluent set $A = \{a_1, \ldots, a_k\}$ of actions, A is

- $\forall$-*step serializable* in s if $o(\langle a'_1, \ldots, a'_k \rangle, s)$ is defined for any sequence $\langle a'_1, \ldots, a'_k \rangle$ such that $\{a'_1, \ldots, a'_k\} = A$;
- $\exists$-*step serializable* in s if $a^e(x) = s(x)$, for each $a \in A$ and $x \in \tilde{a}^c$, and $o(\langle a'_1, \ldots, a'_k \rangle, s)$ is defined for some sequence $\langle a'_1, \ldots, a'_k \rangle$ such that $\{a'_1, \ldots, a'_k\} = A$;
- *relaxed* $\exists$ -*step serializable* in s if $o(\langle a'_1, \ldots, a'_k \rangle, s)$ is defined for some sequence $\langle a'_1, \ldots, a'_k \rangle$ such that $\{a'_1, \ldots, a'_k\} = A$.

Note that any $\forall$-step serializable set A of actions is likewise $\exists$-step serializable, and similarly any $\exists$-step serializable A is relaxed $\exists$-step serializable. In particular, the condition that any sequence built from a $\forall$-step serializable A leads to a (defined) successor state implies that the precondition of each action in A must already be established, which is also required for $\exists$-step serializable sets, but not for relaxed $\exists$-step serializable sets. We extend the three serialization concepts to plans by calling a sequence $\langle A_1, \ldots, A_m \rangle$ a $\forall$-*step*, $\exists$-*step*, or *relaxed* $\exists$-*step plan* if $s_m(x) = s_\star(x)$, for each $x \in \tilde{s}_\star$, and each set A_i of actions is $\forall$-step, $\exists$-step, or relaxed $\exists$-step serializable, respectively, in s_{i-1} for $1 \leq i \leq m$, where $s_i(x) = a^e(x)$ for each $a \in A_i$ and $x \in \tilde{a}^e$, and $s_i(x) = s_{i-1}(x)$ for each $x \in \mathcal{F} \backslash \bigcup_{a \in A_i} \tilde{a}^e$. That is, parallel representations partition some sequential plan such that each part A_i is $\forall$-step, $\exists$-step, or relaxed $\exists$-step serializable in the state obtained by applying the actions preceding A_i.

Example 1. Consider a planning task $\langle \mathcal{F}, s_0, s_\star, \mathcal{O} \rangle$ with $\mathcal{F} = \{x_1, x_2, x_3, x_4, x_5\}$ such that $x_1^d = x_2^d = x_3^d = x_4^d = x_5^d = \{0,1\}$, $s_0 = \{x_1 = 0, x_2 = 0, x_3 = 0, x_4 = 0, x_5 = 0\}$, $s_\star = \{x_4 = 1, x_5 = 1\}$, and $\mathcal{O} = \{a_1, a_2, a_3, a_4\}$, where $a_1 = \langle \{x_1 = 0\}, \{x_1 = 1, x_2 = 1\} \rangle$, $a_2 = \langle \{x_3 = 0\}, \{x_1 = 1, x_3 = 1\} \rangle$, $a_3 = \langle \{x_2 = 1, x_3 = 1\}, \{x_4 = 1\} \rangle$, and $a_4 = \langle \{x_2 = 1, x_3 = 1\}, \{x_5 = 1\} \rangle$. One can check that $\langle a_1, a_2, a_3, a_4 \rangle$ and $\langle a_1, a_2, a_4, a_3 \rangle$ are the two sequential plans consisting of four actions. The $\forall$-step plan with fewest sets of actions is given by $\langle \{a_1\}, \{a_2\}, \{a_3, a_4\} \rangle$. Similarly, $\langle \{a_1, a_2\}, \{a_3, a_4\} \rangle$ is the $\exists$-step plan with fewest sets of actions. Finally, the relaxed $\exists$-step plan $\langle \{a_1, a_2, a_3, a_4\} \rangle$ consists of one set of actions only. ∎

In ASP, we represent a planning task like the one in Example 1 by facts as follows:

```
fluent(x1).   fluent(x2).   fluent(x3).   fluent(x4).   fluent(x5).
value(x1,0). value(x2,0). value(x3,0). value(x4,0). value(x5,0).
value(x1,1). value(x2,1). value(x3,1). value(x4,1). value(x5,1).

init(x1,0). init(x2,0). init(x3,0). init(x4,0). init(x5,0).
                              goal(x4,1). goal(x5,1).

action(a1).     action(a2).     action(a3).     action(a4).
prec(a1,x1,0). prec(a2,x3,0). prec(a3,x2,1). prec(a4,x2,1).
post(a1,x1,1). post(a2,x1,1). prec(a3,x3,1). prec(a4,x3,1).
post(a1,x2,1). post(a2,x3,1). post(a3,x4,1). post(a4,x5,1).
```

Listing 1. Common part of sequential and parallel encodings for STRIPS-like planning

```
1  holds(X,V,0) :- init(X,V).

3  #program check(t).

5  :- query(t), goal(X,V), not holds(X,V,t).

7  #program step(t).

9  {holds(X,V,t) : value(X,V)} = 1 :- fluent(X).

11  {occurs(A,t)} :- action(A).

13  :- occurs(A,t), post(A,X,V), not holds(X,V,t).

15  change(X,t) :- holds(X,V,t-1), not holds(X,V,t).
16  effect(X,t) :- occurs(A,t), post(A,X,V).
17  :- change(X,t), not effect(X,t).
```

The facts can then be combined with encodings such that stable models correspond to sequential, ∀-step, ∃-step, or relaxed ∃-step plans. The rules as well as integrity constraints in Listing 1 form the common core of respective incremental encodings [6] and are grouped into three parts: a subprogram base, including the rule in Line 1, which is not preceded by any **#program** directive; a parameterized subprogram check(t), containing the integrity constraint in Line 5, in which the parameter t serves as placeholder for successive integers starting from 0; and a parameterized subprogram step(t), comprising the rules and integrity constraints below the **#program** directive in Line 7, whose parameter t stands for successive integers starting from 1. By first instantiating the base subprogram along with check(t), where t is replaced by 0, and then proceeding with integers from 1 for t in check(t) and step(t), an incremental encoding can be gradually unrolled. We take advantage of this to capture plans of increasing length, expressed by the latest integer used to replace t with.

In more detail, the rule in Line 1 of Listing 1 maps facts specifying s_0 to atoms over the predicate holds/3, in which the third argument 0 refers to the given state. Starting from 0 for the parameter t, the integrity constraint in Line 5 then tests whether the conditions of $s_\star$ are established, where the dedicated atom query(t) is set to true only for the latest integer taken for t. This allows for increasing the plan length by successively instantiating the subprograms check(t) and step(t) with further integers. The latter subprogram includes the choice rule in Line 9 to generate a successor state such that each fluent $x \in \mathcal{F}$ is mapped to some value in its domain x^d. The other choice rule in Line 11 permits to unconditionally pick actions to apply, expressed by atoms over

Listing 2. Extension of Listing 1 for encoding sequential plans

```
19   :- occurs(A,t), prec(A,X,V), not holds(X,V,t-1).

21   :- #count{A : occurs(A,t)} > 1.
```

Listing 3. Extension of Listing 1 for encoding ∀-step plans

```
19   :- occurs(A,t), prec(A,X,V), not holds(X,V,t-1).

21   :- occurs(A,t), prec(A,X,V), not post(A,X,_), not holds(X,V,t).

23   single(X,t) :- occurs(A,t), prec(A,X,V1), post(A,X,V2), V1 != V2.
24   :- single(X,t), #count{A : occurs(A,t), post(A,X,V)} > 1.
```

occurs/2, in order to obtain a corresponding successor state. Given that both sequential and parallel plans are such that the postcondition of an applied action holds in the successor state, the integrity constraint in Line 13 asserts respective postcondition(s). On the other hand, fluents unaffected by applied actions must remain unchanged, which is reflected by the rules in Lines 15 and 16 along with the integrity constraint in Line 17, restricting changed fluents to postconditions of applied actions.

The common encoding part described so far takes care of matching successor states to postconditions of applied actions, while requirements regarding preconditions are subject to the kind of plan under consideration and expressed by dedicated additions to the step(t) subprogram. To begin with, the two integrity constraints added in Listing 2 address sequential plans by, in Line 19, asserting the precondition of an applied action to hold at the state referred to by t-1 and, in Line 21, denying multiple actions to be applied in parallel. Note that, if the plan length or the latest integer taken for t, respectively, exceeds the minimum number of actions required to establish the conditions of s_*, the encoding of sequential plans given by Listings 1 and 2 permits idle states in which no action is applied. While idle states cannot emerge when using the basic *iclingo* control loop [6] of *clingo* to compute shortest plans, they are essential for the planner presented in Sect. 3 in order to increase the plan length in more flexible ways. Turning to parallel representations, Listing 3 shows additions dedicated to ∀-step plans, where the integrity constraint in Line 19 is the same as in Listing 2 before. This guarantees the preconditions of applied actions to hold, while their confluence is already taken care of by means of the integrity constraint in Line 13. It thus remains to make sure that applied actions do not interfere in a way that would disable any serialization, which essentially means that the precondition of an applied action a must not be invalidated by another action applied in parallel. For a fluent $x \in \tilde{a}^c$ that is not changed by a itself, i.e., $x \notin \tilde{a}^e$ or $a^e(x) = a^c(x)$, the integrity constraint in Line 21, which applies in case of

Listing 4. Extension of Listing 1 for encoding ∃-step plans

```
19  :- occurs(A,t), prec(A,X,V), not holds(X,V,t-1).

21  apply(A1,t) :- action(A1),
22     ready(A2,t) : post(A1,X,V1), prec(A2,X,V2), A1 != A2, V1 != V2.

24  ready(A,t) :- action(A), not occurs(A,t).
25  ready(A,t) :- apply(A,t).
26  :- action(A), not ready(A,t).
```

$x \notin \tilde{a}^e$, suppresses a parallel application of actions a' such that $x \in \tilde{a}'^e$ and $a'^e(x) \neq a^c(x)$. (If $a^e(x) = a^c(x)$, the integrity constraint in Line 13 already requires x to remain unchanged.) On the other hand, the situation becomes slightly more involved when $x \in \tilde{a}^e$ and $a^e(x) \neq a^c(x)$, i.e., the application of a invalidates its own precondition. In this case, no other action a' such that $x \in \tilde{a}'^e$ can be applied in parallel, either because $a'^e(x) \neq a^e(x)$ undermines confluence or since $a'^e(x) = a^e(x)$ disrespects the precondition of a. To account for such situations and address all actions invalidating their precondition regarding x at once, the rule in Line 23 derives an atom over `single/2` to indicate that at most (and effectively exactly) one action affecting x can be applied, as asserted by the integrity constraint in Line 24. As a consequence, no action applied in parallel can invalidate the precondition of another action, so that any serialization leads to the same successor state as obtained in the parallel case.

Example 2. The two sequential plans from Example 1 correspond to two stable models, obtained with the encoding of sequential plans given by Listings 1 and 2, both including the atoms $occurs(a_1,1)$ and $occurs(a_2,2)$. In addition, one stable model contains $occurs(a_3,3)$ along with $occurs(a_4,4)$, and the other $occurs(a_4,3)$ as well as $occurs(a_3,4)$, thus exchanging the order of applying a_3 and a_4. Given that a_3 and a_4 are confluent, the independence of their application order is expressed by a single stable model, obtained with the encoding part for ∀-step plans in Listing 3 instead of the one in Listing 2, comprising $occurs(a_3,3)$ as well as $occurs(a_4,3)$ in addition to $occurs(a_1,1)$ and $occurs(a_2,2)$. Note that, even though the set $\{a_1, a_2\}$ is confluent, it is not ∀-step serializable (in s_0), and a parallel application is suppressed in view of the atom $single(x_1,1)$, derived since a_1 invalidates its precondition regarding x_1. Moreover, the requirement that the precondition of an applied action must be established in the state before permits only $\langle\{a_1\}, \{a_2\}, \{a_3, a_4\}\rangle$ as ∀-step plan or its corresponding stable model, respectively, with three sets of actions.   ■

Additions to Listing 1 addressing ∃-step plans are given in Listing 4. As before, the integrity constraint in Line 19 is included to assert the precondition of an applied action to hold at the state referred to by `t-1`. Unlike with ∀-step plans, however, an applied action may invalidate the precondition of another action, in which case the other action must come first in a serialization, and the

Listing 5. Replacement of Lines 21–26 in Listing 4 by **#edge** statement

```
21  #edge((A1,t),(A2,t)) : occurs(A1,t),
22                         post(A1,X,V1), prec(A2,X,V2), A1 != A2, V1 != V2.
```

aim is to make sure that there is some compatible serialization. To this end, the rule in Lines 21–22 expresses that an action can be safely applied, as indicated by a respective instance of the head atom `apply(A1,t)`, once *all* other actions whose preconditions it invalidates are captured by corresponding instances of `ready(A2,t)`. The latter provide actions that are not applied or whose application is safe, i.e., no yet pending action's precondition gets invalidated, and are derived by means of the rules in Lines 24 and 25. In fact, the least fixpoint obtained via the rules in Lines 21–25 covers all actions exactly if the applied actions do not circularly invalidate their preconditions, and the integrity constraint in Line 26 prohibits any such circularity, which in turn means that there is a compatible serialization. Excluding circular interference also lends itself to an alternative implementation by means of the **#edge** directive [5] of *clingo*, in which case built-in acyclicity checking [3] is used. A respective replacement of Lines 21–26 is shown in Listing 5, where the **#edge** directive in Lines 21–22 asserts edges from an applied action to all other actions whose preconditions it invalidates, and acyclicity checking makes sure that the graph induced by applied actions remains acyclic.

The encoding part for relaxed ∃-step plans in Listing 6 deviates from those given so far by not necessitating the precondition of an applied action to hold in the state before. Rather, the preconditions of actions applied in parallel may be established successively, where confluence along with the condition that an action is applicable only after other actions whose preconditions it invalidates have been processed guarantee the existence of a compatible serialization. In fact, the rules in Lines 22–26 are almost identical to their counterparts in Listing 4, and the difference amounts to the additional prerequisite '`reach(X,V,t) : prec(A1,X,V)`' in Line 22. Instances of `reach(X,V,t)` are derived by means of the rules in Lines 19 and 20 to indicate fluent values from the state referred to by `t-1` along with postconditions of actions whose application has been determined to be safe. The prerequisites of the rule in Lines 22–23 thus express that an action can be safely applied once its precondition is established, possibly by means of other actions preceding it in a compatible serialization, *and* if it does not invalidate any pending action's precondition. Similar to its counterpart in Listing 4, the integrity constraint in Line 27 then makes sure that actions are not applied unless their application is safe in the sense of a relaxed ∃-step serializable set.

Example 3. The ∀-step plan $\langle\{a_1\},\{a_2\},\{a_3,a_4\}\rangle$ from Example 1 can be condensed into $\langle\{a_1,a_2\},\{a_3,a_4\}\rangle$ when switching to ∃-step serializable sets. Corresponding stable models obtained with the encodings given by Listing 1 along with Listing 4 or 5 include $occurs(a_1,1)$, $occurs(a_2,1)$, $occurs(a_3,2)$,

Listing 6. Extension of Listing 1 for encoding relaxed ∃-step plans

```
19   reach(X,V,t) :- holds(X,V,t-1).
20   reach(X,V,t) :- occurs(A,t), apply(A,t), post(A,X,V).

22   apply(A1,t) :- action(A1), reach(X,V,t) : prec(A1,X,V);
23       ready(A2,t) : post(A1,X,V1), prec(A2,X,V2), A1 != A2, V1 != V2.

25   ready(A,t) :- action(A), not occurs(A,t).
26   ready(A,t) :- apply(A,t).
27   :- action(A), not ready(A,t).
```

and $\texttt{occurs}(a_4,2)$. Regarding the **#edge** directive in Listing 5, these atoms induce the graph $(\{\langle a_1,1\rangle, \langle a_2,1\rangle\}, \{\langle\langle a_2,1\rangle, \langle a_1,1\rangle\rangle\})$, which is clearly acyclic. Its single edge tells us that a_1 must precede a_2 in a compatible serialization, while the absence of a cycle means that the application of a_1 does not invalidate the precondition of a_2. In terms of the encoding part in Listing 4, $\texttt{apply}(a_1,1)$ and $\texttt{ready}(a_1,1)$ are derived first, which in turn allows for deriving $\texttt{apply}(a_2,1)$ and $\texttt{ready}(a_2,1)$. The requirement that the precondition of an applied action must be established in the state before, which is shared by Listings 4 and 5, however, necessitates at least two sets of actions for an ∃-step plan or a corresponding stable model, respectively. Unlike that, the encoding of relaxed ∃-step plans given by Listings 1 and 6 yields a stable model containing $\texttt{occurs}(a_1,1)$, $\texttt{occurs}(a_2,1)$, $\texttt{occurs}(a_3,1)$, and $\texttt{occurs}(a_4,1)$, corresponding to the relaxed ∃-step plan $\langle\{a_1,a_2,a_3,a_4\}\rangle$. The existence of a compatible serialization is witnessed by first deriving, amongst other atoms, $\texttt{reach}(x_1,0,1)$ and $\texttt{reach}(x_3,0,1)$ in view of s_0. These atoms express that the preconditions of a_1 and a_2 are readily established, so that $\texttt{apply}(a_1,1)$ along with $\texttt{reach}(x_2,1,1)$ and $\texttt{ready}(a_1,1)$ are derived next. The latter atom indicates that a_1 can be safely applied before a_2, which then leads to $\texttt{apply}(a_2,1)$ along with $\texttt{reach}(x_3,1,1)$. Together $\texttt{reach}(x_2,1,1)$ and $\texttt{reach}(x_3,1,1)$ reflect that the precondition of a_3 as well as a_4 can be established by means of a_1 and a_2 applied in parallel, so that $\texttt{apply}(a_3,1)$ and $\texttt{apply}(a_4,1)$ are derived in turn.   ■

In order to formalize the soundness and completeness of the presented encodings, let B stand for the rule in Line 1 of Listing 1, $Q(i)$ for the integrity constraint in Line 5 with the parameter t replaced by some integer i, and $S(i)$ for the rules and integrity constraints below the **#program** directive in Line 7 with i taken for t. Moreover, we refer to specific encoding parts extending $S(i)$, where the parameter t is likewise replaced by i, by $S^s(i)$ for Listing 2, $S^\forall(i)$ for Listing 3, $S^\exists(i)$ for Listing 4 $S^E(i)$ for Line 19 of Listing 4 along with Listing 5, and $S^R(i)$ for Listing 6. Given that $S^E(i)$ includes an **#edge** directive subject to acyclicity checking, we understand stable models in the sense of [3], i.e., the graph induced by a (regular) stable model, which is empty in case of no **#edge** directives, must be acyclic.

Theorem 1. *Let I be the set of facts representing planning task $\langle \mathcal{F}, s_0, s_\star, \mathcal{O} \rangle$, $\langle a_1, \ldots, a_n \rangle$ be a sequence of actions, and $\langle A_1, \ldots, A_m \rangle$ be a sequence of sets of actions. Then,*

- *$\langle a_1, \ldots, a_n \rangle$ is a sequential plan iff $I \cup B \cup Q(0) \cup \bigcup_{i=1}^{n} (Q(i) \cup S(i) \cup S^s(i)) \cup \{\text{query}(n).\}$ has a stable model M such that $\{\langle a, i \rangle \mid \text{occurs}(a, i) \in M\} = \{\langle a_i, i \rangle \mid 1 \le i \le n\}$;*
- *$\langle A_1, \ldots, A_m \rangle$ is a $\forall$-step (resp., $\exists$-step or relaxed $\exists$-step) plan iff $I \cup B \cup Q(0) \cup \bigcup_{i=1}^{m} (Q(i) \cup S(i) \cup S^p(i)) \cup \{\text{query}(m).\}$, where $S^p(i) = S^{\forall}(i)$ (resp., $S^p(i) \in \{S^{\exists}(i), S^E(i)\}$ or $S^p(i) = S^R(i)$) for $1 \le i \le m$, has a stable model M such that $\{\langle a, i \rangle \mid \text{occurs}(a, i) \in M\} = \{\langle a, i \rangle \mid 1 \le i \le m, a \in A_i\}$.*

Let us note that, with each of the considered encodings, any plan corresponds to a unique stable model, as the latter is fully determined by atoms over occurs/2, i.e., corresponding (successor) states as well as auxiliary predicates functionally depend on the applied actions. Regarding the encoding part for relaxed $\exists$-step plans in Listing 6, we mention that acyclicity checking cannot (in an obvious way) be used instead of rules dealing with the safe application of actions. To see this, consider $\langle \mathcal{F}, s_0, s_\star, \mathcal{O} \rangle$ with $\mathcal{F} = \{x_1, x_2, x_3\}$ such that $x_1^d = x_2^d = x_3^d = \{0, 1\}$, $s_0 = \{x_1 = 0, x_2 = 0, x_3 = 0\}$, $s_\star = \{x_3 = 1\}$, and $\mathcal{O} = \{a_1, a_2\}$, where $a_1 = \langle \emptyset, \{x_1 = 1, x_2 = 1\} \rangle$ and $a_2 = \langle \{x_1 = 1, x_2 = 0\}, \{x_3 = 1\} \rangle$. There is no sequential plan for this task since only a_1 is applicable in s_0, but its application invalidates the precondition of a_2. Concerning the (confluent) set $\{a_1, a_2\}$, the graph $(\{(a_1, 1), (a_2, 1)\}, \{\langle (a_1, 1), (a_2, 1) \rangle\})$ is acyclic and actually includes the information that a_2 should precede a_1 in any compatible serialization. However, if the prerequisite in Line 23 of Listing 6 were dropped to "simplify" the encompassing rule, the application of a_1 would be regarded as safe, and then the precondition of a_2 would seem established as well. That is, it would be unsound to consider the establishment and invalidation of preconditions in separation, no matter the respective implementation techniques.

As regards encoding techniques, common ASP-based approaches, e.g., [12], define successor states, i.e., the predicate holds/3, in terms of actions given by atoms over occurs/2. Listing 1, however, includes a respective choice rule, which puts it inline with SAT planning, where our intention is to avoid asymmetries between fluents and actions, as either of them would in principle be sufficient to indicate plans [11]. Concerning (relaxed) $\exists$-step plans, the encoding parts in Listings 4 and 6 make use of the built-in well-foundedness requirement in ASP and do, unlike [15], not unfold the order of actions applied in parallel. In contrast to the SAT approach to relaxed $\exists$-step plans in [17], we do not rely on a fixed (static) order of actions, and to our knowledge, no encoding similar to the one in Listing 6 has been proposed so far.

3   A Multi-shot ASP Planner

Planning encodings must be used with a strategy for fixing the plan length. For example, the first approaches to planning in SAT and ASP follow a sequential algorithm starting from 0 and successively incrementing the length by 1 until a plan is found.

For parallel planning in SAT, more flexible strategies were proposed in [15], based on the following ideas. First, minimal parallel plans do not coincide with shortest sequential plans. Hence, it is unclear whether parallel plans should be minimal. Second, solving times for different plan lengths follow a certain pattern, which can be exploited. To illustrate this, consider the solving times of a typical instance in Fig. 2. For lengths 0 to 4, in gray, the instance is unsatisfiable, and time

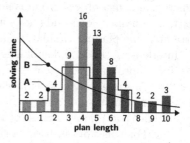

Fig. 2. Exemplary solving times

grows exponentially. Then, the first satisfiable instances, in light green, are still hard, but they become easier for greater plan lengths. However, for even greater plan lengths, the solving time increases again because the size becomes larger. Accordingly, [15] suggests not to minimize parallel plan length but rather avoid costly unsatisfiable parts by moving early to easier satisfiable lengths.

The sequential algorithm (S) solves the instance in Fig. 2 in 46 time units, viz. $2+2+4+9+16+13$, by trying plan lengths 0 to 4 until it finds a plan at 5. The idea of algorithm A [15] is to simultaneously consider n plan lengths. In our example, fixing n to 5, A starts with lengths 0 to 4. After 2 time units, lengths 0 and 1 are finished, and 5 and 6 are added. Another 2 units later, length 2 is finished, and 7 is started. Finally, at time 8, length 7 yields a plan. The times spent by A for each length are indicated in Fig. 2, and summing them up amounts to 40 time units in total. Algorithm B [15] distributes time non-uniformly over plan lengths: if length n is run for t time units, then lengths $n + i$ are run for $t * \gamma^i$ units, where $i \geq 1$ and γ lies between 0 and 1. In our example, we set γ to 0.8. When length 3 has been run for 6 units, previous lengths are already finished, and the times for the following lengths are given by curve B in Fig. 2. At this point, length 8 is assigned 2 units ($6 * 0.8^5$) and yields a plan, leading to a total time of 38 units: 8 units for lengths 0 to 2, and 30 for the rest. (The 30 units correspond to the area under the curve from length 3 on.) Note that both A and B find a plan before finishing the hardest instances and, in practice, often save significant time over S.

We adopted algorithms A (S if $n = 1$) and B, and implemented them as planning strategies of *plasp* via multi-shot ASP solving. In general, they can be applied to any incremental encoding complying with the threefold structure of `base`, `step(t)`, and `check(t)` subprograms. Assuming that the subprograms adhere to *clingo*'s modularity condition [6], they are assembled to ASP programs of the form

$$P(n) = \text{base} \cup \bigcup_{i=0}^{n} \text{check}(i) \cup \bigcup_{i=1}^{n} \text{step}(i)$$

where n gives the length of the unrolled encoding. The planner then looks for an integer n such that $P(n) \cup \{\text{query}(n).\}$ is satisfiable, and algorithms S, A, and B provide different strategies to approach such an integer.

The planner is implemented using *clingo*'s multi-shot solving capacities, where one *clingo* object grounds and solves incrementally. This approach avoids extra grounding efforts and allows for taking advantage of previously learned constraints. The planner simulates the parallel processing of different plan lengths by interleaving sequential subtasks. To this end, the *clingo* object is used to successively unroll an incremental encoding up to integer(s) n. In order to solve a subtask for some $m < n$, the unrolled part $P(n)$ is kept intact, while query(m) is set to true instead of query(n). That is, the search component of *clingo* has to establish conditions in check(m), even though the encoding is unrolled up to $n \geq m$. For this approach to work, we require that $P(m) \cup \{\text{query}(m).\}$ is satisfiable iff $P(n) \cup \{\text{query}(m).\}$ is satisfiable for $0 \leq m \leq n$. An easy way to guarantee this property is to tolerate idle states in-between m and n, as is the case with the encodings given in Sect. 2.

4   System and Experiments

Like its predecessor versions, the third series of *plasp*[1] provides a translator from PDDL specifications to ASP facts. Going beyond the STRIPS-like fragment, it incorporates a normalization step to support advanced PDDL features such as nested expressions in preconditions, conditional effects, axiom rules, as well as existential and universal quantifiers. Moreover, *plasp* allows for optional preprocessing by *Fast Downward*, leading to an intermediate representation in the SAS planning format. This format encompasses multi-valued fluents, mutex groups, conditional effects, and axiom rules, which permit a compact (propositional) specification of planning tasks and are (partially) inferred by *Fast Downward* from PDDL inputs. Supplied with PDDL or SAS inputs, *plasp* produces a homogeneous factual representation, so that ASP encodings remain independent of the specific input format.[2]

To empirically contrast the different encodings and planning algorithms presented in Sects. 2 and 3, we ran *plasp* on PDDL specifications from the International Planning Competition. For comparison, we also include two variants of the state-of-the-art SAT planning system *Madagascar* [14], where M stands for the standard version and M_p for the use of a specific planning heuristic. All experiments were performed on a Linux machine equipped with Intel Core

---

[1] Available at: https://github.com/potassco/plasp.

[2] The encodings given in Sect. 2 focus on STRIPS-like planning tasks with multi-valued fluents as well as mutex groups, where the latter have been omitted for brevity. In contrast to the ease of incorporating mutex groups, extending parallel encodings to conditional effects or axiom rules is not straightforward [15], while sequential encodings for them are shipped with *plasp*.

i7-2600 processor at 3.8 GHz and 16 GB RAM, limiting time and memory per run to 900 s and 8 GB, while charging 900 s per aborted run in the tables below.

Regarding *plasp*, we indicate the encoding of a particular kind of plan by a superscript to the planning algorithm (denoted by its letter), where ^{s} stands for sequential, $^\forall$ for $\forall$-step, $^\exists$ for $\exists$-step, ^{E} for $\exists$-step by means of acyclicity checking, and ^{R} for relaxed $\exists$-step plans; e.g., $B^\exists$ refers to algorithm B applied to the encoding of $\exists$-step plans given by Listings 1 and 4. Moreover, each combination of algorithm and encoding can optionally be augmented with a planning heuristic [8], which has been inspired by M_p and is denoted by an additional subscript $_p$, like in $B_p^\exists$. (The parameters of A and B are set to $n = 16$ or $\gamma = 0.9$, respectively, as suggested in [15].)

Tables 1 and 2 show total numbers of solved instances and average runtimes, in total and for individual domains of PDDL specifications, for the two *Madagascar* variants and different *plasp* settings. In case of Table 1, all systems take PDDL inputs directly, while preprocessing by *Fast Downward* is used for *plasp* in Table 2. Note that the tables refer to different subsets of instances, as we omit instances solved by some *plasp* setting in less than 5 s or unsolved by all of them within the given resource limits. To give an account of the impact of preprocessing, we list the best-performing *plasp* setting with or without preprocessing in the last row of the respective other table. As the $B^\exists$ setting of *plasp*, relying on algorithm B along with the (pure) ASP encoding of $\exists$-step plans, turns out to perform generally robust as well as comparable to the alternative implementation provided by B^E, we choose it as the baseline for varying the planning algorithm, encoding, or heuristic.

Table 1. Solved instances and average runtimes without preprocessing by *Fast Downward*

	solved	Ø time	gripper	logistics	blocks	elevator	depots	driverlog
M_p	76/76	0.97	0.03	0.02	0.54	0.02	0.07	4.73
M	76/76	1.18	0.08	0.02	0.15	0.02	0.15	6.44
$B_p^\exists$	72/76	98.06	416.21	46.15	79.34	0.77	129.58	15.66
$B^\exists$	60/76	238.64	39.90	46.16	637.86	10.90	305.45	146.29
$A^\exists$	59/76	257.40	51.85	46.22	654.63	10.62	392.52	149.02
B^E	56/76	255.12	34.30	46.10	657.20	2.37	413.98	139.97
B^R	54/76	348.67	134.96	44.49	741.69	66.72	547.06	274.72
B°	49/76	412.90	622.39	62.83	517.26	89.37	669.86	460.33
$B^\forall$	48/76	402.64	516.33	46.12	537.78	434.63	283.20	251.81
$S^\exists$	47/76	389.78	720.20	46.18	590.74	317.76	261.66	140.42
$B_p^\exists$	75/76	15.48	1.73	0.47	3.10	0.89	10.44	74.31

In Table 1, where *plasp* takes PDDL inputs directly, we first observe that the *Madagascar* variants solve the respective instances rather easily. Unlike that, we checked that the size of ground instantiations often becomes large and constitutes a bottleneck with all *plasp* settings, which means that the original PDDL specifications are not directly suitable for ASP-based planning. The sequential algorithm or encoding, respectively, of $S^\exists$ and B^s is responsible for increased search efforts or plan lengths, so that these settings solve considerably fewer instances than the baseline provided by $B^\exists$, which is closely followed by the $A^\exists$ setting that uses algorithm A instead of B. While encodings of (relaxed) $\exists$-step plans help to reduce plan lengths, the more restrictive $\forall$-step plans aimed at by $B^\forall$ are less effective, especially in the *gripper* and *elevator* domains. Apart from a few outliers, the alternative implementations of $\exists$-step plans in $B^\exists$ and B^E perform comparable, while further reductions by means of relaxed $\exists$-step plans turn out to be minor and cannot compensate for the more sophisticated encoding of B^R. With the exception of the *gripper* domain, the planning heuristic applied by $B_p^\exists$ significantly boosts search performance, and the last row of Table 1 indicates that this *plasp* setting even comes close to *Madagascar* once preprocessing by *Fast Downward* is used.

Table 2. Solved instances and average runtimes with preprocessing by *Fast Downward*

	solved	∅ time	grid	gripper	logistics	mystery	blocks	elevator	freecell	depots	driverlog
M_p	136/136	2.76	0.78	0.04	13.09	4.11	1.92	0.02	0.59	0.22	0.21
M	135/136	12.89	1.40	0.34	13.18	71.07	1.12	0.03	18.91	7.38	3.16
$B_p^\exists$	121/136	131.03	72.14	5.89	76.82	282.99	27.52	2.22	656.76	40.22	27.91
B^E	116/136	159.52	460.51	19.29	32.12	345.20	7.47	1.92	655.19	135.48	206.84
$B^\exists$	114/136	168.44	63.01	19.42	32.80	351.92	21.48	2.00	656.47	223.28	208.59
$A^\exists$	114/136	174.68	297.15	15.62	32.93	362.73	21.50	2.01	656.49	234.51	218.26
$B^\forall$	107/136	231.49	459.96	277.92	48.00	387.43	4.92	4.95	652.87	344.64	289.63
B^R	105/136	248.73	481.89	35.34	206.25	474.35	58.31	9.15	660.35	374.21	316.21
$S^\exists$	103/136	255.83	71.41	771.20	32.89	292.88	21.49	82.91	656.95	130.63	203.49
B^s	88/136	367.90	451.49	699.70	754.42	89.69	3.13	9.23	104.12	755.56	597.61
$B_p^\exists$	51/136	584.52	900.00	668.33	900.00	900.00	625.17	2.10	900.00	429.19	118.06

Table 2 turns to *plasp* settings run on SAS inputs provided by *Fast Downward*. Beyond information about mutex groups, which is not explicitly available in PDDL, the preprocessing yields multi-valued fluents that conflate several Booleans from a PDDL specification. This makes ground instantiations much more compact than before and significantly increases the number of instances *plasp* can solve. However, a sequential algorithm or encoding as in $S^\exists$ and B^s remains less effective than the other settings, also with SAS format. Interestingly, the encoding of $\forall$-step plans used in $B^\forall$ leads to slightly better overall performance than the relaxed $\exists$-step plans of B^R, although variations are domain-specific, as becomes apparent when comparing *gripper* and *logistics*. Given that the parallel plans permitted by B^R are most general, this tells us that the overhead of encoding them does not pay off in the domains at hand, while yet lacking optimizations, e.g., based on disabling–enabling-graphs [17], still leave room for future improvements. As with PDDL inputs, $\exists$-step plans again constitute the best trade-off between benefits and efforts of a parallel encoding, where the planning algorithm of $B^\exists$ is comparable to $A^\exists$. The alternative implementation by means of acyclicity checking in B^E improves the performance on the instances in Table 2 to some extent but not dramatically. Unlike that, the planning heuristic of $B_p^{\exists}$ leads to substantial performance gains, now also in the *gripper* domain, whose PDDL specification had been problematic for the heuristic before. The large gap in comparison to $B_p^{\exists}$ on direct PDDL inputs in the last row confirms the high capacity of preprocessing by *Fast Downward*. Finally, we note that the two *Madagascar* variants remain significantly ahead of the $B_p^{\exists}$ setting of *plasp*, which is related to their streamlined yet planning-specific implementation of grounding, while *plasp* brings the advantage that first-order ASP encodings can be used to prototype and experiment with different features.

5 Summary

We presented the key features of the new *plasp* system. Although it addresses PDDL-based planning, *plasp*'s major components, such as the encodings as well as its planner, can be applied to dynamic domains at large. While our general-purpose approach cannot compete with planning-specific systems at eye level, we have shown how careful encodings and well-engineered solving processes can boost performance. This is reflected, e.g., by an increase of 25 additional instances solved by $B_p^{\exists}$ over $S^{\exists}$ in Table 1. A further significant impact is obtained by extracting planning-specific constraints, as demonstrated by the increase of the performance of $B_p^{\exists}$ from 51 to 121 solved instances in Table 2.

Acknowledgments. This work was partially funded by DFG grant SCHA 550/9.

References

1. Baral, C., Gelfond, M.: Reasoning agents in dynamic domains. In: Logic-Based Artificial Intelligence, pp. 257–279. Kluwer (2000)
2. Biere, A., Heule, M., van Maaren, H., Walsh, T.: Handbook of Satisfiability. IOS (2009)
3. Bomanson, J., Gebser, M., Janhunen, T., Kaufmann, B., Schaub, T.: Answer set programming modulo acyclicity. Fundamenta Informaticae **147**(1), 63–91 (2016)
4. Dimopoulos, Y., Nebel, B., Koehler, J.: Encoding planning problems in non-monotonic logic programs. In: Steel, S., Alami, R. (eds.) ECP 1997. LNCS, vol. 1348, pp. 169–181. Springer, Heidelberg (1997). doi:10.1007/3-540-63912-8_84
5. Gebser, M., Kaminski, R., Kaufmann, B., Ostrowski, M., Schaub, T., Wanko, P.: Theory solving made easy with clingo 5. In: Technical Communications, ICLP, pp. 2:1–2:15. OASIcs (2016)
6. Gebser, M., Kaminski, R., Kaufmann, B., Schaub, T.: Clingo = ASP + control: Preliminary report. In: Technical Communications, ICLP (2014). arXiv:1405.3694
7. Gebser, M., Kaminski, R., Knecht, M., Schaub, T.: plasp: a prototype for PDDL-Based planning in ASP. In: Delgrande, J.P., Faber, W. (eds.) LPNMR 2011. LNCS (LNAI), vol. 6645, pp. 358–363. Springer, Heidelberg (2011). doi:10.1007/978-3-642-20895-9_41

8. Gebser, M., Kaufmann, B., Otero, R., Romero, J., Schaub, T., Wanko, P.: Domain-specific heuristics in answer set programming. In: Proceedings AAAI, pp. 350–356. AAAI (2013)
9. Gelfond, M., Lifschitz, V.: Action languages. Electron. Trans. Artif. Intell. **3**(6), 193–210 (1998)
10. Helmert, M.: The fast downward planning system. J. Artif. Intell. Res. **26**, 191–246 (2006)
11. Kautz, H., McAllester, D., Selman, B.: Encoding plans in propositional logic. In: Proceedings of KR, pp. 374–384. Morgan Kaufmann (1996)
12. Lifschitz, V.: Answer set programming and plan generation. Artif. Intell. **138**(1–2), 39–54 (2002)
13. McDermott, D.: PDDL – the planning domain definition language. TR Yale (1998)
14. Rintanen, J.: Madagascar: scalable planning with SAT. In: Proceedings of IPC, pp. 66–70 (2014)
15. Rintanen, J., Heljanko, K., Niemelä, I.: Planning as satisfiability: parallel plans and algorithms for plan search. Artif. Intell. **170**(12–13), 1031–1080 (2006)
16. Son, T., Baral, C., Nam, T., McIlraith, S.: Domain-dependent knowledge in answer set planning. ACM Trans. Comput. Logic **7**(4), 613–657 (2006)
17. Wehrle, M., Rintanen, J.: Planning as satisfiability with relaxed ∃-step plans. In: Proceedings of AI, pp. 244–253. Springer, Heidelberg (2007)

Nurse Scheduling via Answer Set Programming

Carmine Dodaro[✉] and Marco Maratea

DIBRIS, University of Genova, Genoa, Italy
{dodaro,marco}@dibris.unige.it

Abstract. The Nurse Scheduling problem (NSP) is a combinatorial problem that consists of assigning nurses to shifts according to given practical constraints. In previous years, several approaches have been proposed to solve different variants of the NSP. In this paper, an ASP encoding for one of these variants is presented, whose requirements have been provided by an Italian hospital. We also design a second encoding for the computation of "optimal" schedules. Finally, an experimental analysis has been conducted on real data provided by the Italian hospital using both encodings. Results are very positive: the state-of-the-art ASP system CLINGO is able to compute one year schedules in few minutes, and it scales well even when more than one hundred nurses are considered.

Keywords: Answer Set Programming · Scheduling · Nurse Scheduling Problem

1 Introduction

The Nurse Scheduling problem (NSP) consists of generating a schedule of working and rest days for nurses working in hospital units. The schedule should determine the shift assignments of nurses for a predetermined window of time, and must satisfy requirements imposed by the Rules of Procedure of hospitals. A proper solution to the NSP is crucial to guarantee the high level of quality of health care, to improve the degree of satisfaction of nurses and the recruitment of qualified personnel. For these reasons, several approaches to solve the NSP are reported in the literature, including those based on integer programming [1,2], genetic algorithms [3], fuzzy approaches [4], and ant colony optimization algorithms [5], to mention a few. However, such approaches are not directly comparable with each other, since the requirements usually depend on the policy of the specific hospitals [6]. Detailed surveys on the NSP can be found in [7,8].

Complex combinatorial problems, such as the NSP, are usually the target for the application of logic formalisms such as Answer Set Programming (ASP). Indeed, the simple syntax [9] and the intuitive semantics [10], combined with the availability of robust implementations (see, e.g. [11,12]), make ASP an ideal candidate for addressing such problems. As a matter of fact, ASP has been successfully used in several research areas, including Artificial Intelligence [13], Bio-informatics [14,15], and Databases [16]; more recently ASP has been applied

© Springer International Publishing AG 2017
M. Balduccini and T. Janhunen (Eds.): LPNMR 2017, LNAI 10377, pp. 301–307, 2017.
DOI: 10.1007/978-3-319-61660-5_27

to solve industrial applications [17–19]. However, to the best of our knowledge, no ASP encoding has presented to solve the NSP.

In this paper, we report an ASP encoding to address a variant of the NSP (see Sect. 3), whose requirements, presented in Sect. 2, have been provided by an Italian hospital. We also present a variant of the encoding for the computation of "optimal" schedules. The encoding is natural and intuitive, in the sense that it was obtained by applying the standard modeling methodology, and it is easy to understand. Moreover, an experimental analysis has been conducted on real data provided by an Italian hospital using both encodings. Results are very positive: the state-of-the-art ASP system CLINGO [12] is able to compute one year schedules in few minutes. Moreover, scalability analysis shows that CLINGO scales well even when more than one hundred nurses are considered (see Sect. 4).

2   Nurse Scheduling Problem

The NSP involves the generation of schedules for nurses consisting of working and rest days over a predetermined period of time, which is fixed to one year in this paper. Moreover, the schedules must satisfy a set of requirements. In this section, we informally describe the ones used in the paper as specified by an Italian hospital.

Hospital requirements. We consider three different shifts: *morning* (7 A.M.–2 P.M.), *afternoon* (2 P.M.–9 P.M.), and *night* (9 P.M.–7 A.M.). In order to ensure the best assistance program for patients, each shift is associated with a minimum and a maximum number of nurses that must be present in the hospital.

Nurses requirements. Concerning nurses, the schedules have to guarantee a fair workload. Thus, a limit on the minimum and maximum number of working hours per year is imposed. Moreover, additional requirements are imposed to ensure an adequate rest period to each nurse: *(a)* nurses are legally guaranteed 30 days of paid vacation, *(b)* the starting time of a shift must be at least 24 h later than the starting time of the previous shift, and *(c)* each nurse has at least two rest days each fourteen days window. In addition, after two consecutive working nights there is one special rest day which is not included in the rest days of *(c)*.

Balance requirements. Finally, the number of times a nurse can be assigned to morning, afternoon and night shifts is fixed. However, schedules where this number ranges over a set of acceptable values are also valid. Thus, we identify two variants of the encoding that will be discussed in the next section.

3   Answer Set Programming Encoding

The NSP described in the previous section has been solved by means of an ASP encoding. Answer sets of the logic program presented in this section correspond to the solutions of the NSP. In the following, we assume that the reader is familiar with Answer Set Programming and ASP-CORE-2 input language specification [9].

The encoding has been created following the *Guess&Check* programming methodology. In particular, the following choice rule guesses an assignment to exactly one shift for each nurse and for each day.

$$1 \leq \{assign(N, S, D) : shift(S, H)\} \leq 1 :\!- day(D), \; nurse(N). \qquad (1)$$

Note that the rule also filters out assignments where a nurse works twice during the same day. Instances of the predicate *assign(N, S, D)* are used to store the shift assignment S for a nurse N in a specific day D. Instances of the predicate *shift(S, H)* are used to represent the shifts, where S is a shift and H is the number of working hours associated to the shift. In our setting, we consider the following instances of the predicate: *shift("1−mor", 7)*, *shift("2−aft", 7)*, *shift("3−nig", 10)*, *shift("4−specres", 0)*, *shift("5−rest", 0)*, and *shift("6−vac", 0)*. Actually, the latest three instances are used in our encoding to model the nurses days off. Note that the number used in the name of the shift will be used in the following to compactly encoding constraints in (5).

Hospital requirements. Hospitals need to guarantee that a minimum and a maximum number of nurses are present in the hospital during a specific shift:

$$:\!- day(D), \; \#count\{N : assign(N, "1{-}mor", D)\} > K, \; maxNurseMorning(K).$$
$$:\!- day(D), \; \#count\{N : assign(N, "1 - mor", D)\} < K, \; minNurseMorning(K). \quad (2)$$

In particular, constraints reported in (2) refers to the morning shift. The ones related to afternoon and the night shifts are similar and are not reported here for space constraints.

Nurse requirements. First of all, requirements to guarantee a fair workload of nurses are accomplished as follows:

$$:\!- nurse(N), \; maxHours(M), \; \#sum\{H, D : assign(N, S, D), \; shift(S, H)\} > M.$$
$$:\!- nurse(N), \; minHours(M), \; \#sum\{H, D : assign(N, S, D), \; shift(S, H)\} < M. \quad (3)$$

In particular, for each nurse the number of working hours is bounded by a minimum and a maximum number provided as input by the hospital. Moreover, nurses are guaranteed exactly 30 days of vacation. Thus, the following constraint filters out assignments where the number of vacation days is different from 30:

$$:\!- nurse(N), \; \#count\{D : assign(N, "6{-}vac", D)\} \neq 30. \qquad (4)$$

Then, the starting time of a shift must be at least 24 h later than the starting time of the previous shift. In other words, a nurse assigned to the afternoon shift cannot be assigned to the morning shift of the day after; and a nurse assigned to the night shift cannot be assigned to the morning and to the afternoon shifts of the day after. These requirements are compactly expressed by means of the following constraint:

$$:\!- nurse(N), \; assign(N, S1, D), \; assign(N, S2, D+1), \; S2 < S1 \leq "3 - nig". \qquad (5)$$

The correctness of this constraint is guaranteed by the following ordering on shifts: $"1-mor" < "2-aft" < "3-nig"$. As implementation note, the lexicographic order of strings is not always guaranteed by current systems, thus in the tested encoding only integers are used, e.g. $shift("1-mor", 7$ is replaced by $shift(1,7)$. Moreover, each nurse is guaranteed at least two rest days for each fourteen days. Assignments violating such requirement are filtered out by the following constraint (where YD is the number of days in a year):

$$:- nurse(N),\ day(D),\ days(YD),\ D \leq YD - 13,$$
$$\#count\{D1 : assign(N,"5-rest",D1),\ D \leq D1 \leq D + 13\} < 2. (6)$$

Finally, after two consecutive working nights one special rest day is guaranteed.

$$:- assign(N,"3-nig",D-2),\ assign(N,"3-nig",D-1),$$
$$not\ assign(N,"4-specrest",D).$$
$$:- assign(N,"4-specrest",D),\ not\ assign(N,"3-nig",D-2).$$
$$:- assign(N,"4-specrest",D),\ not\ assign(N,"3-nig",D-1). \quad (7)$$

Balance requirements. We report here the two variants considered in the paper. The first variant is expressed by means of the following constraints:

$$:- nurse(N),\ \#count\{D : assign(N,"1-mor",D)\} > M,\ maxMorning(M).$$
$$:- nurse(N),\ \#count\{D : assign(N,"1-mor",D)\} < M,\ minMorning(M). \quad (8)$$

The idea of (8) is to filter out only assignments where shifts assigned to nurses are out of a fixed range of acceptable values.

The second variant of the encoding considers also the following weak constraint:

$$:\sim nurse(N),\ nbMorning(M),\ \#count\{D : assign(N,"1-mor",D)\} = X,$$
$$minMorning(M_1),\ maxMorning(M_2),\ M_1 \leq X \leq M_2,\ Y =| X - M |. [Y@1, N] (9)$$

In this case the weak constraint is used to assign a cost for each assignment measuring the distance between the assignment and the target M. The optimum assignment is the one with the minimum cost. Note that in (8) and in (9) we reported only the constraints related to the morning shift since the ones related to other shifts are similar.

4   Empirical Evaluation

In this section we report about the results of an empirical analysis conducted on real data provided by an Italian hospital used as reference in this paper. The scheduling has been created for a one year time window and the maximum (resp. minimum) number of hours per year is set to 1692 (resp. 1687). The number of nurses working to the considered hospital unit is 41, and the number of nurses working during the morning and the afternoon shifts ranges from 6

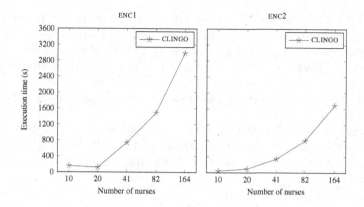

Fig. 1. Scalability analysis with 10, 20, 41, 82 and 164 nurses.

to 9, whereas the number of nurses working during the night shift ranges from 4 to 7. Concerning the holidays, we considered 15 days of vacation chosen by the nurses, according to the vacations selected in the year 2015, whereas the other 15 days of vacation are assigned according to the needs of the hospital. The desired number of working mornings and afternoons is equal to 78, whereas the desired number of working nights is equal to 60. In addition, the hospital can accept schedules where the number of working mornings and afternoons per year ranges from 74 to 82, whereas the number of working nights per year ranges from 58 to 61. We tested two variants of the encoding. The first one, reported as ENC1, considers constraints from (1) to (8), whereas the second variant, reported as ENC2, considers also the weak constraint (9). The experiments were run on an Intel Xeon 2.4 GHz. Time and memory were limited to 1 h and 15 GB, respectively. Concerning the ASP systems, we used CLINGO [12] and WASP [20]. However, the performance of the latter are not reported since it is slower than CLINGO in all tested instances.

Concerning encoding ENC1, CLINGO computes a schedule in 12 min with a peak of memory usage of 300 MB. The performance of CLINGO executed on encoding ENC2 are even better. In fact, CLINGO configured with the option --opt-strategy=usc is able to find the optimum solution in 6 min with a peak of memory usage of 224 MB. The difference in performance can be explained by the strategy employed by CLINGO in presence of weak constraints. In particular, weak constraints are first considered as hard constraint and then relaxed when they cannot be satisfied.

Scalability. We also performed an analysis about the scalability of the encoding, considering different numbers of nurses. In particular, we considered different test cases containing 10, 20, 41, 82 and 164 nurses, respectively. For each test case, we proportionally scaled the number of working nurses during each shift, whereas other requirements are not modified. Results are reported in Fig. 1.

Concerning the encoding ENC1, we note that all instances have been solved within the allotted time. In particular, less than 3 min are needed to compute the schedule with 10 nurses and approximately 100 MB of memory consumption. Similar results have been obtained for computing the schedule with 20 nurses. Concerning the scheduling with 82 and 164 nurses, CLINGO computes the solution within 25 and 50 min, respectively, whereas its memory usage is 500 and 990 MB, respectively.

Results obtained with the encoding ENC2 are also very positive: CLINGO finds the optimal assignment within 60 min and 1024 MB for all considered test cases, with a peak of 28 min and 838 MB of memory usage when 164 nurses are considered.

5 Conclusion

In this paper we described two ASP encodings for addressing a variant of the NSP. Since ASP programs are executable specifications, we obtained a practical tool for supporting the head nurse in producing the schedule for the next year. We experimented with our implementation on real data provided by an Italian hospital, and all instances are solved within one hour with both encodings, even with more that 100 nurses.

Acknowledgments. We would like to thank Nextage srl for providing partial funding for this work. The funding has been provided in the framework of a research grant by the Liguria POR-FESR 2014–2020 programme.

References

1. Azaiez, M.N., Sharif, S.S.A.: A 0–1 goal programming model for nurse scheduling. Comput. OR **32**, 491–507 (2005)
2. Bard, J.F., Purnomo, H.W.: Preference scheduling for nurses using column generation. Eur. J. Oper. Res. **164**(2), 510–534 (2005)
3. Aickelin, U., Dowsland, K.A.: An indirect genetic algorithm for a nurse-scheduling problem. Comput. OR **31**(5), 761–778 (2004)
4. Topaloglu, S., Selim, H.: Nurse scheduling using fuzzy modeling approach. Fuzzy Sets Syst. **161**(11), 1543–1563 (2010). http://dx.doi.org/10.1016/j.fss.2009.10.003
5. Gutjahr, W.J., Rauner, M.S.: An ACO algorithm for a dynamic regional nurse-scheduling problem in Austria. Comput. OR **34**(3), 642–666 (2007)
6. Millar, H.H., Kiragu, M.: Cyclic and non-cyclic scheduling of 12 h shift nurses by network programming. Eur. J. Oper. Res. **104**(3), 582–592 (1998)
7. Burke, E.K., Causmaecker, P.D., Berghe, G.V., Landeghem, H.V.: The state of the art of nurse rostering. J. Sched. **7**(6), 441–499 (2004)
8. Cheang, B., Li, H., Lim, A., Rodrigues, B.: Nurse rostering problems - a bibliographic survey. Eur. J. Oper. Res. **151**(3), 447–460 (2003)
9. Calimeri, F., Faber, W., Gebser, M., Ianni, G., Kaminski, R., Krennwallner, T., Leone, N., Ricca, F., Schaub, T.: ASP-Core-2 Input Language Format (2013). https://www.mat.unical.it/aspcomp.2013/files/ASP-CORE-2.01c.pdf

10. Gelfond, M., Lifschitz, V.: Classical negation in logic programs and disjunctive databases. New Gener. Comput. **9**(3/4), 365–386 (1991)
11. Alviano, M., Dodaro, C.: Anytime answer set optimization via unsatisfiable core shrinking. TPLP **16**(5–6), 533–551 (2016)
12. Gebser, M., Kaufmann, B., Kaminski, R., Ostrowski, M., Schaub, T., Schneider, M.T.: Potassco: the potsdam answer set solving collection. AI Commun. **24**(2), 107–124 (2011)
13. Balduccini, M., Gelfond, M., Watson, R., Nogueira, M.: The USA-Advisor: a case study in answer set planning. In: Eiter, T., Faber, W., Truszczyński, M. (eds.) LPNMR 2001. LNCS (LNAI), vol. 2173, pp. 439–442. Springer, Heidelberg (2001). doi:10.1007/3-540-45402-0_39
14. Erdem, E., Öztok, U.: Generating explanations for biomedical queries. TPLP **15**(1), 35–78 (2015)
15. Koponen, L., Oikarinen, E., Janhunen, T., Säilä, L.: Optimizing phylogenetic supertrees using answer set programming. TPLP **15**(4–5), 604–619 (2015)
16. Marileo, M.C., Bertossi, L.E.: The consistency extractor system: Answer set programs for consistent query answering in databases. Data Knowl. Eng. **69**(6), 545–572 (2010)
17. Abseher, M., Gebser, M., Musliu, N., Schaub, T., Woltran, S.: Shift design with answer set programming. Fundam. Inform. **147**(1), 1–25 (2016)
18. Dodaro, C., Gasteiger, P., Leone, N., Musitsch, B., Ricca, F., Schekotihin, K.: Combining answer set programming and domain heuristics for solving hard industrial problems (application paper). TPLP **16**(5–6), 653–669 (2016)
19. Dodaro, C., Leone, N., Nardi, B., Ricca, F.: Allotment problem in travel industry: a solution based on ASP. In: Cate, B., Mileo, A. (eds.) RR 2015. LNCS, vol. 9209, pp. 77–92. Springer, Cham (2015). doi:10.1007/978-3-319-22002-4_7
20. Alviano, M., Dodaro, C., Leone, N., Ricca, F.: Advances in WASP. In: Calimeri, F., Ianni, G., Truszczynski, M. (eds.) LPNMR 2015. LNCS (LNAI), vol. 9345, pp. 40–54. Springer, Cham (2015). doi:10.1007/978-3-319-23264-5_5

Hybrid Metabolic Network Completion

Clémence Frioux[1,2], Torsten Schaub[1,3(✉)], Sebastian Schellhorn[3],
Anne Siegel[2], and Philipp Wanko[3]

[1] Inria, Rennes, France
[2] IRISA, Université de Rennes 1, Rennes, France
[3] Universität Potsdam, Potsdam, Germany
torsten@cs.uni-potsdam.de

Abstract. Metabolic networks play a crucial role in biology since they capture all chemical reactions in an organism. While there are networks of high quality for many model organisms, networks for less studied organisms are often of poor quality and suffer from incompleteness. To this end, we introduced in previous work an ASP-based approach to metabolic network completion. Although this qualitative approach allows for restoring moderately degraded networks, it fails to restore highly degraded ones. This is because it ignores quantitative constraints capturing reaction rates. To address this problem, we propose a hybrid approach to metabolic network completion that integrates our qualitative ASP approach with quantitative means for capturing reaction rates. We begin by formally reconciling existing stoichiometric and topological approaches to network completion in a unified formalism. With it, we develop a hybrid ASP encoding and rely upon the theory reasoning capacities of the ASP system *clingo* for solving the resulting logic program with linear constraints over reals. We empirically evaluate our approach by means of the metabolic network of *Escherichia coli*. Our analysis shows that our novel approach yields greatly superior results than obtainable from purely qualitative or quantitative approaches.

1 Introduction

Among all biological processes occurring in a cell, metabolic networks are in charge of transforming input nutrients into both energy and output nutrients necessary for the functioning of other cells. In other words, they capture all chemical reactions occurring in an organism. In biology, such networks are crucial from a fundamental and technological point of view to estimate and control the capability of organisms to produce certain products. Metabolic networks of high quality exist for many model organisms. In addition, recent technological advances enable their semi-automatic generation for many less studied organisms. However, the resulting metabolic networks are of poor quality, due to error-prone, genome-based construction processes and a lack of (human) resources. As a consequence, they usually suffer from substantial incompleteness. The common fix is to fill the gaps by completing a draft network by borrowing chemical pathways from reference networks of well studied organisms until the augmented network provides the measured functionality.

© Springer International Publishing AG 2017
M. Balduccini and T. Janhunen (Eds.): LPNMR 2017, LNAI 10377, pp. 308–321, 2017.
DOI: 10.1007/978-3-319-61660-5_28

In previous work [19], we introduced a logical approach to *metabolic network completion* by drawing on the work in [10]. We formulated the problem as a qualitative combinatorial (optimization) problem and solved it with Answer Set Programming (ASP [2]). The basic idea is that reactions apply only if all their reactants are available, either as nutrients or provided by other metabolic reactions. Starting from given nutrients, referred to as *seeds*, this allows for extending a metabolic network by successively adding operable reactions and their products. The set of metabolites in the resulting network is called the *scope* of the seeds and represents all metabolites that can principally be synthesized from the seeds. In metabolic network completion, we query a database of metabolic reactions looking for (minimal) sets of reactions that can restore an observed bio-synthetic behavior. This is usually expressed by requiring that certain *target* metabolites are in the scope of some given seeds. For instance, in the follow-up work in [4,15], we successfully applied our ASP-based approach to the reconstruction of the metabolic network of the macro-algae *Ectocarpus siliculosus*, using the collection of reference networks at http://metacyc.org.

Although we evidenced in [16] that our ASP-based method effectively restores the bio-synthetic capabilities of moderately degraded networks, it fails to restore the ones of highly degraded metabolic networks. The main reason for this is that our purely qualitative approach misses quantitative constraints accounting for the law of mass conservation, a major hypothesis about metabolic networks. This law stipulates that each internal metabolite of a network must balance its production rate with its consumption rate. Such rates are given by the weighted sums of all reaction rates consuming or producing a metabolite, respectively. This calculation is captured by the *stoichiometry*[1] of the involved reactions. Hence, the qualitative ASP-based approach fails to tell apart solution candidates with correct and incorrect stoichiometry and therefore reports inaccurate results for highly degraded networks.

We address this by proposing a hybrid approach to metabolic network completion that integrates our qualitative ASP approach with quantitative techniques from *Flux Balance Analysis* (FBA[2] [12]), the dominating quantitative approach for capturing reaction rates in metabolic networks. We accomplish this by taking advantage of recently developed theory reasoning capacities for the ASP system *clingo* [7]. More precisely, we use an extension of *clingo* with linear constraints over reals, as dealt with in Linear Programming (LP [5]). This extension provides us with an extended ASP modeling language as well as a generic interface to alternative LP solvers, viz. *cplex* and *lpsolve*, for dealing with linear constraints. We empirically evaluate our approach by means of the metabolic network of *Escherichia coli*. Our analysis shows that our novel approach yields superior results than obtainable from purely qualitative or quantitative approaches. Moreover, our hybrid application provides a first evaluation of the theory extensions of the ASP system *clingo* with linear constraints over reals in a non-trivial setting.

---

[1] See also https://en.wikipedia.org/wiki/Stoichiometry.
[2] See also https://en.wikipedia.org/wiki/Flux_balance_analysis.

2   Metabolic Network Completion

We represent a *metabolic network* as a labeled directed bipartite graph $G = (R \cup M, E, s)$, where R and M are sets of nodes standing for *reactions* and *metabolites*, respectively. When $(m, r) \in E$ or $(r, m) \in E$ for $m \in M$ and $r \in R$, the metabolite m is called a *reactant* or *product* of reaction r, respectively. More formally, for any $r \in R$, define $rcts(r) = \{m \in M \mid (m, r) \in E\}$ and $prds(r) = \{m \in M \mid (r, m) \in E\}$. The *edge labeling* $s : E \to \mathbb{R}$ gives the stoichiometric coefficients of a reaction's reactants and products, respectively. Finally, the activity rate of reactions is bound by lower and upper bounds, denoted by $lb_r \in \mathbb{R}_0^+$ and $ub_r \in \mathbb{R}_0^+$ for $r \in R$, respectively. Whenever clear from the context, we refer to metabolic networks with G (or G', etc.) and denote the associated reactions and metabolites with M and R (or M', R' etc.), respectively.

We distinguish a set $S \subseteq M$ of metabolites as initiation *seeds*, that is, compounds initially present due to experimental evidence. Another set of metabolites is assumed to be activated by default. These *boundary metabolites* are defined as: $S_b(G) = \{m \in M \mid r \in R, m \in prds(r), rcts(r) = \emptyset\}$. For simplicity, we assume that all boundary compounds are seeds: $S_b(G) \subseteq S$. Note that concepts like reachability and activity in network completion are independent of this assumption.

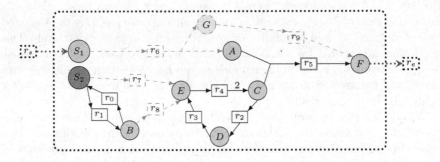

Fig. 1. Example of a metabolic network

For illustration, consider the metabolic network in Fig. 1 and ignore the shaded part. The network consists of 8 reactions, r_s, r_e and r_0 to r_5, and 8 metabolites, $A, \ldots, F$, S_1, S_2. Here, $S = \{S_1, S_2\}$, S_1 being the only boundary compound of the network. Consider reaction $r_4 : E \to 2C$ transforming one unit of E into two units of C. We have $rcts(r_4) = \{E\}$, $prds(r_4) = \{C\}$, along with $s(E, r_4) = 1$ and $s(r_4, C) = 2$.

In biology, several concepts have been introduced to model the activation of reaction fluxes in metabolic networks, or to synthesize metabolic compounds. To model this, we introduce a function *active* that given a metabolic network G takes a set of seeds $S \subseteq M$ and returns a set of activated reactions $active_G(S) \subseteq R$. With it, *metabolic network completion* is about ensuring that a set of target reactions is activated from seed compounds in S by possibly extending the metabolic network with reactions from a reference network (cf. shaded part in Fig. 1).

Formally, given a metabolic network $G = (R \cup M, E, s)$, a set $S \subseteq M$ of seed metabolites such that $S_b(G) \subseteq S$, a set $R_T \subseteq R$ of target reactions, and a reference network $(R' \cup M', E', s')$, the *metabolic network completion problem* is to find a set $R'' \subseteq R' \setminus R$ of reactions of minimal size such that $R_T \subseteq active_{G''}(S)$ where[3]

$$G'' = ((R \cup R'') \cup (M \cup M''), E \cup E'', s''), \tag{1}$$

$$M'' = \{m \in M' \mid r \in R'', m \in rcts(r) \cup prds(r)\}, \tag{2}$$

$$E'' = E' \cap ((M'' \times R'') \cup (R'' \times M'')), \text{ and} \tag{3}$$

$$s'' = s \cup s'. \tag{4}$$

We call R'' a *completion* of $(R \cup M, E, s)$ from $(R' \cup M', E', s')$ wrt S and R_T.

Our concept of activation allows us to capture different biological paradigms. Accordingly, different formulations of metabolic network completion can be characterized: the stoichiometric, the relaxed stoichiometric, the topological, and the hybrid one. We elaborate upon their formal characterizations in the following sections.

Stoichiometric Metabolic Network Completion. The first activation semantics has been introduced in the context of Flux Balance Analysis capturing reaction flux distributions of metabolic networks at steady state. In this paradigm, each reaction r is associated with a *metabolic flux value*, expressed as a real variable v_r confined by the minimum and maximum rates:

$$lb_r \leq v_r \leq ub_r \qquad \text{for } r \in R \tag{5}$$

Flux distributions are formalized in terms of a system of equations relying on the stoichiometric coefficients of reactions. Reaction rates are governed by the *law of mass conservation* under a steady state assumption, that is, the input and output rates of reactions consuming and producing a metabolite are balanced:

$$\sum_{r \in R} s(r, m) \cdot v_r + \sum_{r \in R} -s(m, r) \cdot v_r = 0 \qquad \text{for } m \in M \tag{6}$$

Given a target reaction $r_T \in R_T$, a metabolic network $G = (R \cup M, E, s)$ and a set of seeds S, *stoichiometric activation* is defined as follows:

$$r_T \in active_G^s(S) \text{ iff } v_{r_T} > 0 \text{ and } (5) \text{ and } (6) \text{ hold for } M \text{ and } R.$$

Note that the condition $v_{r_T} > 0$ strengthens the flux condition for $r_T \in R$ in the second part. More generally, observe that activated target reactions are not directly related to the network's seeds S. However, the activation of targets highly depends on the boundary metabolites in $S_b(G)$ for which (6) is always satisfied and thus initiates the fluxes.

---

[3] Since s, s' have disjoint domains we view them as relations and compose them by union.

To solve metabolic network completion with flux-balance activated reactions, Linear Programming can be used to maximize the flux rate v_{r_T} provided that the linear constraints are satisfied. This problem turns out to be hard to solve in practice and existing approaches scale poorly to real-life applications (cf. [13]).

This motivated the use of approximate methods. The relaxed problem is obtained by weakening the mass-balance equation (6) as follows:

$$\sum_{r \in R} s(r, m) \cdot v_r + \sum_{r \in R} -s(m, r) \cdot v_r \geq 0 \qquad \text{for } m \in M \qquad (7)$$

This lets us define the concept of *relaxed stoichiometric activation*:

$$r_T \in active_G^r(S) \quad \text{iff} \quad v_{r_T} > 0 \text{ and (5) and (7) hold for } M \text{ and } R.$$

The resulting problem can now be efficiently solved with Linear Programming [18]. Note however that for strict steady-state modeling an *a posteriori* verification of solutions is needed to warrant the exact mass-balance equation (6).

In our draft network G, consisting of all bold nodes and edges depicted in Fig. 1 (viz. reactions r_s, r_e and r_0 to r_5 and metabolites $A, \ldots, F$, S_1 and S_2 and r_5 the single target reaction) and the reference network G', consisting of the shaded part of Fig. 1, (viz. reactions r_6 to r_9 and metabolite G) a strict stoichiometry-based completion aims to obtain a solution with $r_5 \in active_{G''}^s(\{S_1, S_2\})$ where v_{r_5} is maximal. This can be achieved by adding the completion $R_1'' = \{r_6, r_9\}$. The cycle made of compounds E, C, D is already balanced and notably self-activated. Such self-activation of cyclic pathways is an inherent problem of purely stoichiometric approaches to network completion. This is a drawback of the semantics since the effective activation of the cycle requires the additional (and unchecked) condition that at least one of the compounds was present as the initial state of the system [16]. The instance of Eq. (6) controlling the reaction rates related to metabolite C is $2 \cdot v_{r_4} - v_{r_2} - v_{r_5} = 0$.

Existing systems addressing strict stoichiometric network completion either cannot guarantee optimal solutions [11] or do not support a focus on specific target reactions [21]. Other approaches either partially relax the problem [22] or solve the relaxed problem based on Eq. (7), like the popular system *gapfill* [18].

Topological Metabolic Network Completion. A qualitative approach to metabolic network completion relies on the topology of networks for capturing the activation of reactions. Given a metabolic network G, a reaction $r \in R$ is *activated* from a set of seeds S if all reactants in $rcts(r)$ are reachable from S. Moreover, a metabolite $m \in M$ is *reachable* from S if $m \in S$ or if $m \in prds(r)$ for some reaction $r \in R$ where all $m' \in rcts(r)$ are reachable from S. The *scope* of S, written $\Sigma_G(S)$, is the closure of metabolites reachable from S. In this setting, *topological activation* of reactions from a set of seeds S is defined as follows:

$$r_T \in active_G^t(S) \quad \text{iff} \quad rcts(r_T) \subseteq \Sigma_G(S).$$

Note that this semantics avoids self-activated cycles by imposing an external entry to all cycles. The resulting network completion problem can be expressed as a combinatorial optimization problem and effectively solved with ASP [19].

For illustration, consider again the draft and reference networks G and G' in Fig. 1. We get $\Sigma_G(\{S_1, S_2\}) = \{S_1, S_2, B\}$, indicating that target reaction r_5 is not activated from the seeds with the draft network because A and C are not reachable. This changes once the network is completed. Valid minimal completions are $R_2'' = \{r_6, r_7\}$ and $R_3'' = \{r_6, r_8\}$ because $r_5 \in active_{G_i''}^t(\{S_1, S_2\})$ since $\{A, C\} \subseteq \Sigma_{G_i''}(\{S_1, S_2\})$ for all extended networks G_i'' obtained from completions R_i'' of G for $i \in \{2, 3\}$. Relevant elements from the reference network are given in dashed gray.

Hybrid Metabolic Network Completion. The idea of hybrid metabolic network completion is to combine the two previous activation semantics: the topological one accounts for a well-founded initiation of the system from the seeds and the stoichiometric one warrants its mass-balance. We thus aim at network completions that are both topologically functional and flux balanced (without suffering from self-activated cycles). More precisely, a reaction $r_T \in R_T$ is *hybridly activated* from a set S of seeds in a network G, if both criteria apply:

$$r_T \in active_G^h(S) \quad \text{iff} \quad r_T \in active_G^s(S) \text{ and } r_T \in active_G^t(S)$$

Applying this to our example in Fig. 1, we get the (minimal) hybrid solutions $R_4'' = \{r_6, r_7, r_9\}$ and $R_5'' = \{r_6, r_8, r_9\}$. Both (topologically) initiate paths of reactions from the seeds to the target, ie. $r_5 \in active_{G_i''}^t(\{S_1, S_2\})$ since $\{A, C\} \subseteq \Sigma_{G_i''}(\{S_1, S_2\})$ for both extended networks G_i'' obtained from completions R_i'' of G for $i \in \{4, 5\}$. Both solutions are as well stoichiometrically valid and balance the amount of every metabolite, hence we also have $r_5 \in active_{G_i''}^s(\{S_1, S_2\})$.

3   Answer Set Programming with Linear Constraints

For encoding our hybrid problem, we rely upon the theory reasoning capacities of the ASP system *clingo* that allows us to extend ASP with linear constraints over reals (as addressed in Linear Programming). We confine ourselves below to features relevant to our application and refer the interested reader for details to [7].

As usual, a *logic program* consists of *rules* of the form

```
a0 :- a1,...,am,not am+1,...,not an
```

where each a_i is either a *(regular) atom* of form $p(t_1, \ldots, t_k)$ where all t_i are terms or a *linear constraint atom* of form[4] '&sum{a₁*x₁;...;a_l*x_l} <= k' that stands for the linear constraint $a_1 \cdot x_1 + \cdots + a_l \cdot x_l \leq k$. All a_i and k are finite sequences of digits with at most one dot[5] and represent real-valued coefficients a_i and k. Similarly all x_i stand for the real-valued variables x_i. As usual, **not** denotes (default) *negation*. A rule is called a *fact* if $n = 0$.

Semantically, a logic program induces a set of *stable models*, being distinguished models of the program determined by stable models semantics [9].

---

[4] In *clingo*, theory atoms are preceded by '&'.

[5] In the input language of *clingo*, such sequences must be quoted to avoid clashes.

Such a stable model X is an *LC-stable model* of a logic program P,[6] if there is an assignment of reals to all real-valued variables occurring in P that (i) satisfies all linear constraints associated with linear constraint atoms in P being in X and (ii) falsifies all linear constraints associated with linear constraint atoms in P being not in X. For instance, the (non-ground) logic program containing the fact `'a("1.5").'` along with the rule `'&sum{R*x} <= 7 :- a(R).'` has the stable model

　　　`{a("1.5"), &sum{"1.5"*x}<=7}`.

This model is LC-stable since there is an assignment, e.g. $\{x \mapsto 4.2\}$, that satisfies the associated linear constraint '$1.5*x \leq 7$'. We regard the stable model along with a satisfying real-valued assignment as a solution to a logic program containing linear constraint atoms.

To ease the use of ASP in practice, several extensions have been developed. First of all, rules with variables are viewed as shorthands for the set of their ground instances. Further language constructs include *conditional literals* and *cardinality constraints* [20]. The former are of the form `a:b`$_1$`,...,b`$_m$`,` the latter can be written as `s{d`$_1$`;...;d`$_n$`}t`, where `a` and `b`$_i$ are possibly default-negated (regular) literals and each `d`$_j$ is a conditional literal; `s` and `t` provide optional lower and upper bounds on the number of satisfied literals in the cardinality constraint. We refer to `b`$_1$`,...,b`$_m$ as a *condition*. The practical value of both constructs becomes apparent when used with variables. For instance, a conditional literal like `a(X):b(X)` in a rule's antecedent expands to the conjunction of all instances of `a(X)` for which the corresponding instance of `b(X)` holds. Similarly, `2{a(X):b(X)}4` is true whenever at least two and at most four instances of `a(X)` (subject to `b(X)`) are true. Finally, objective functions minimizing the sum of weights w_i subject to condition c_i are expressed as `#minimize{`w_1`:`c_1`;...;`w_n`:`c_n`}`.

In the same way, the syntax of linear constraints offers several convenience features. As above, elements in linear constraint atoms can be conditioned, viz.

　　　`'&sum{a`$_1$`*x`$_1$`:c`$_1$`;...;a`$_l$`*x`$_l$`:c`$_n$`} <= y'`

where each c_i is a condition. Moreover, the theory language for linear constraints offers a domain declaration for real variables, `'&dom{lb..ub} = x'` expressing that all values of `x` must lie between `lb` and `ub`. And finally the maximization (or minimization) of an objective function can be expressed with `&maximize{a`$_1$`*x`$_1$`:c`$_1$`;...;a`$_l$`*x`$_l$`:c`$_n$`}` (by minimize). The full theory grammar for linear constraints over reals is available at https://potassco.org.

4　Solving Hybrid Metabolic Network Completion

In this section, we present our hybrid approach to metabolic network completion. We start with a factual representation of problem instances. A metabolic network G with a typing function $t : M \cup R \rightarrow \{\texttt{d},\texttt{r},\texttt{s},\texttt{t}\}$, indicating the origin of the respective entities, is represented as follows:

---

[6] This corresponds to the definition of T-stable models using a *strict* interpretation of theory atoms [7], and letting T be the theory of linear constraints over reals.

$$F(G,t) = \{\texttt{metabolite}(m,t(m)) \mid m \in M\}$$
$$\cup \{\texttt{reaction}(r,t(r)) \mid r \in R\}$$
$$\cup \{\texttt{bounds}(r,lb_r,ub_r) \mid r \in R\} \cup \{\texttt{objective}(r,t(r)) \mid r \in R\}$$
$$\cup \{\texttt{reversible(r)} \mid r \in R, rcts(r) \cap prds(r) \neq \emptyset\}$$
$$\cup \{\texttt{rct}(m,s(m,r),r,t(r)) \mid r \in R, m \in rcts(r)\}$$
$$\cup \{\texttt{prd}(m,s(r,m),r,t(r)) \mid r \in R, m \in prds(r)\}$$

While most predicates should be self-explanatory, we mention that `reversible` identifies bidirectional reactions. Only one direction is explicitly represented in our fact format. The four types d, r, s, and t tell us whether an entity stems from the draft or reference network, or belongs to the seeds or targets.

In a metabolic network completion problem, we consider a draft network $G = (R \cup M, E, s)$, a set S of seed metabolites, a set R_T of target reactions, and a reference network $G' = (R' \cup M', E', s')$. An instance of this problem is represented by the set of facts $F(G,t) \cup F(G',t')$. In it, a key role is played by the typing functions that differentiate the various components:

$$t(n) = \begin{cases} \texttt{d}, \text{ if } n \in (M \setminus (T \cup S)) \cup (R \setminus (R_{S_b} \cup R_T)) \\ \texttt{s}, \text{ if } n \in S \cup R_{S_b} \\ \texttt{t}, \text{ if } n \in T \cup R_T \end{cases} \quad \text{and} \quad t'(n) = \texttt{r},$$

where $T = \{m \in rcts(r) \mid r \in R_T\}$ is the set of target metabolites and $R_{S_b} = \{r \in R \mid m \in S_b(G), m \in prds(r)\}$ is the set of reactions related to boundary seeds.

Our encoding of hybrid metabolic network completion is given in Listing 1. Roughly, the first 10 lines lead to a set of candidate reactions for completing the draft network. Their topological validity is checked in lines 12–16 with regular ASP, the stoichiometric one in lines 18–24 in terms of linear constraints. (Lines 1–16 constitute a revision of the encoding in [19].) The last two lines pose a hybrid optimization problem, first minimizing the size of the completion and then maximizing the flux of the target reactions.

In more detail, we begin by defining the auxiliary predicate edge/4 representing directed edges between metabolites connected by a reaction. With it, we calculate in Line 4 and 5 the scope $\Sigma_G(S)$ of the draft network G from the seed metabolites in S; it is captured by all instances of scope(M,d). This scope is then extended in Line 7/8 via the reference network G' to delineate all possibly producible metabolites. We draw on this in Line 10 when choosing the reactions R'' of the completion (cf. Sect. 2) by restricting their choice to reactions from the reference network whose reactants are producible. This amounts to a topological search space reduction.

The reactions in R'' are then used in lines 12–14 to compute the scope $\Sigma_{G''}(S)$ of the completed network. And R'' constitutes a topologically valid completion if all targets in T are producible by the expanded draft network G'': Line 16 checks whether $T \subseteq \Sigma_{G''}(S)$ holds, which is equivalent to $R_T \subseteq active^t_{G''}(S)$. Similarly, R'' is checked for stoichiometric validity in lines 18–24. For simplicity,

```
1    edge(R,M,N,T) :- reaction(R,T), rct(M,_,R,T), prd(N,_,R,T).
2    edge(R,M,N,T) :- reaction(R,T), rct(N,_,R,T), prd(M,_,R,T), reversible(R).

4    scope(M,d) :- metabolite(M,s).
5    scope(M,d) :- edge(R,_,M,T), T!=r, scope(N,d):edge(R,N,_,T'), N!=M, T'!=r.

7    scope(M,x) :- scope(M,d).
8    scope(M,x) :- edge(R,_,M,_), scope(N,x):edge(R,N,_,_), N!=M.

10   { completion(R) : edge(R,M,N,r), scope(N,x), scope(M,x) }.

12   scope(M,c) :- scope(M,d).
13   scope(M,c) :- edge(R,_,M,T), T!=r, scope(N,c):edge(R,N,_,T'), T'!=r, N!=M.
14   scope(M,c) :- completion(R), edge(R,_,M,r), scope(N,c):edge(R,N,_,r), N!=M.

16   :- metabolite(M,t), not scope(M,c).

18   &dom{L..U} = R :- bounds(R,L,U).

20   &sum{ IS*IR : prd(M,IS,IR,T), T!=r;  IS'*IR' : prd(M,IS',IR',r), completion(IR');
21        -OS*OR : rct(M,OS,OR,T), T!=r; -OS'*OR' : rct(M,OS',OR',r), completion(OR')
22        } = "0" :- metabolite(M,_).

24   &sum{ R } > "0" :- reaction(R,t).

26   &maximize{   R : objective(R,t) }.
27   #minimize{ 1,R : completion(R)  }.
```

Listing 1. Encoding of hybrid metabolic network completion

we associate reactions with their rate and let their identifiers take real values. Accordingly, Line 18 accounts for (5) by imposing lower and upper bounds on each reaction rate. The mass-balance equation (6) is enforced for each metabolite M in lines 20–22; it checks whether the sum of products of stoichiometric coefficients and reaction rates equals zero, viz. IS*IR, -OS*OR, IS'*IR', and -OS'*OR'. Reactions IR, OR and IR', OR' belong to the draft and reference network, respectively, and correspond to $R \cup R''$. Finally, by enforcing $r_T > 0$ for $r_T \in R_T$ in Line 24, we make sure that $R_T \subseteq active^s_{G''}(S)$.

In all, our encoding ensures that the set R'' of reactions chosen in Line 10 induces an augmented network G'' in which all targets are activated both topologically as well as stoichiometrically, and is optimal wrt the hybrid optimization criteria.

5 System and Experiments

In this section, we introduce *fluto*, our new system for hybrid metabolic network completion, and empirically evaluate its performance. The system relies on the hybrid encoding described in Sect. 4 along with the hybrid solving capacities of *clingo* [7] for implementing the combination of ASP and LP. We use *clingo* 5.2.0 incorporating as LP solvers either *cplex* 12.7.0.0 or *lpsolve* 5.5.2.5 via their respective Python interfaces. We describe the details of the underlying solving techniques in a separate paper and focus below on application-specific aspects.

Table 1. Comparison of qualitative results.

Degradation	F(BB)		F(USC)		F(BB+USC)		Verified		
	#SOLS	#OPTS	#SOLS	#OPTS	#SOLS	#OPTS	F(BB+USC)	M	G
10% (900)	**900**	**900**	892	892	900	900	**900**	660	56
20% (900)	**830**	669	793	**769**	867	814	**867**	225	52
30% (900)	**718**	88	461	**344**	780	382	**780**	61	0
all (2700)	**2448**	1657	2146	**2005**	2547	2096	**2547**	946	108

Table 2. Comparison of system options.

Configuration	F(BB)		F(USC)	
	T	TO	T	TO
DEFAULT	377	121	**190**	46
CORE-50	358	109	230	75
CORE-0	**350**	**96**	233	76
PROP-50	363	112	226	69
PROP-100	386	105	360	139
HEURISTIC	542	178	252	**28**

Table 3. Results using best system options.

Degradation	F(VBS)		Verified		
	#SOLS	#OPTS	F(VBS)	M	G
10% (900)	900	900	**900**	660	56
20% (900)	896	855	**896**	225	52
30% (900)	848	575	**848**	61	0
40% (900)	681	68	**681**	29	0
All (3600)	3325	2398	**3325**	975	108

The output of *fluto* consists of two parts. First, the completion R'', given by instances of predicate `completion`, and second, an assignment of floats to (metabolic flux variables v_r for) all $r \in R \cup R''$. In our example, we get

$R'' = \{\text{completion}(r_6), \text{completion}(r_8), \text{completion}(r_9)\}$ and $\{r_s = 49999.5, r_9 = 49999.5, r_3 = 49999.5, r_2 = 49999.5, r_e = 99999.0, r_6 = 49999.5, r_5 = 49999.5, r_4 = 49999.5\}$. Variables assigned 0 are omitted. Note the flux value $r_8 = 0$ even though $r_8 \in R''$. This is to avoid the self-activation of cycle C, D and E. By choosing r_8, we ensure that the cycle has been externally initiated at some point but activation of r_8 is not necessary at the current steady state.

We analyze (i) the quality of *fluto*'s approach to metabolic network completion and (ii) the impact of different system configurations. To have a realistic setting, we use degradations of a functioning metabolic network of *Escherichia coli* [17] comprising 1075 reactions. The network was randomly degraded by 10, 20, and 30%, creating 10 networks for each degradation by removing reactions until the target reactions were inactive according to *Flux Variability Analysis* [3]. 90 target reactions with varied reactants were randomly chosen for each network, yielding 2700 problem instances in total. The reference network consists of reactions of the original metabolic network.

We ran each benchmark on a Xeon E5520 2.4 GHz processor under Linux limiting RAM to 20 GB. At first, we investigate two alternative optimization strategies for computing completions of minimum size. The first one, *branch-and-bound* (BB), iteratively produces solutions of better quality until the optimum is found and the other, *unsatisfiable core* (USC), relies on successively identifying and relaxing unsatisfiable cores until an optimal solution is obtained. Note that we are not only interested in optimal solutions but if unavailable also solutions activating target reactions without trivially restoring the whole reference network. In *clingo*, BB naturally produces these solutions in contrast to USC. Therefore, we use USC with stratification [1], which provides at least some suboptimal solutions. Each obtained best solution was checked with *cobrapy* 0.3.2 [6], a renowned system implementing an FBA-based gold standard (for verification only).

Table 1 gives the number of solutions (#SOLS) and optima (#OPTS) obtained by *fluto* (F) in its default setting within 20 min for BB, USC and the best of both (BB+USC), individually for each DEGRADATION and over**all**. For 94.3% of the instances *fluto*(BB+USC) found a solution within the time limit and 82.3% of them were optimal. We observe that BB provides overall more useful solutions but USC acquires more optima, which was to be expected by the nature of the optimization techniques. Additionally, each technique finds solutions to problem instances where the other exceeds the time limit, underlining the merit of using both in tandem. Column VERIFIED compares the quality of solutions provided by *fluto*, *meneco* 1.4.3 (M) [16] and *gapfill*[7] (G) [18]. Both *meneco* and *gapfill* are systems for metabolic network completion. While *meneco* pursues the topological approach, *gapfill* applies the relaxed stoichiometric variant using Eq. (7). The numbers represent how many problem instances had verified solutions for each system.[8] All solutions found by *fluto* could be verified by *cobrapy*. In detail, *fluto* found a smallest set of reactions completing the draft network for 77.6%, a suboptimal solution for 16.7%, and no solution for 5.6% of the problem instances. In comparison, for *meneco* 35.0%, and for *gapfill* merely 4.0% of its solutions passed verification. The ignorance of *meneco* regarding stoichiometry leads to possibly unbalanced networks, which particularly outcrops for higher degradation. The simplified view of *fluto* in terms of stoichiometry misguides the search for possible completions and eventually leads to unbalanced networks. Moreover, *gapfill*'s ignorance of network topology results in self-activated cycles. By exploiting both topology and stoichiometry, *fluto* avoids such cycles and scales much better with increasing reference network size and degradation of the network.

The configuration space of *fluto* is huge. In addition to its own parameters, the ones of *clingo* and the respective LP solver amplify the number of options. We thus focus on distinguished features revealing an impact in our experiments. First, the *fluto* option CORE-n invokes the irreducible inconsistent set algorithm [14] whenever n% of atoms are decided. This algorithm extracts a minimal set of conflicting linear constraints for a given conflict. Second, PROP-n controls the frequency of LP propagation: the consistency of linear constraints is only checked if n% of atoms are decided. Finally, HEURISTIC allows for using *clingo*'s domain-specific heuristics. Such heuristics are expressed in the input language with the directive #heuristic. In *fluto*, we use the statement

---

[7] Update of 2011-09-23 see http://www.maranasgroup.com/software.htm.

[8] The results for *meneco* and *gapfill* are taken from previous work [16], where they were run to completion with *no* time limit.

```
#heuristic completion(R) : interesting(R). [1,true]
```
to make the solver first decide interesting reactions and assign them true. A reaction of G' is **interesting** if it is on a direct path in $G' \cup G$ from a seed metabolite to a target metabolite. The DEFAULT is to use CORE-100, PROP-0, disable HEURISTIC, and use LP solver *cplex*. This allows us to detect conflicts among the linear constraints as soon as possible and only perform expensive conflict analysis on the full assignment.

For our experiments, we selected at random three networks with at least 20 instances for which BB and USC could find the optimum in 100 to 600 s. With the resulting 270 medium to hard instances, we compared DEFAULT as baseline, $n \in \{0, 50, 100\}$ for CORE-n and PROP-n, respectively, and HEURISTIC, limiting time to 600 s. Table 2 gives the overall average time in seconds (T) and number of timeouts (TO). The first column reflects the CONFIGURATIONs. We focus on the impact of distinguished parameters wrt the default setting and leave a more exhaustive exploration to future work.[9] Overall, USC performs best as regards average time, and USC and HEURISTIC yield the least number of timeouts. BB works well with frequent conflict analysis (CORE-0), while it weakens USC's performance. On the other hand, unlike BB, USC favors frequent theory propagation (DEFAULT). BB learns weaker constraints while optimizing only pertaining to the best known bound, thus the improvement step is less constraint compared to USC. Due to this, conflicts are more likely to appear later on and be of less quality, enhancing the potential of conflict analysis and hampering the usefulness of frequent LP propagation. USC on the other hand, aims at quickly identifying unsatisfiable partial assignments and learning structural constraints building upon each other, which is enhanced by frequent conflict detection. Thus, higher quality conflicts are likely detected earlier where conflict minimization has less potential and produces overhead. HEURISTIC reduces performance for BB. Even though the bound of the initial solution might be lower, the solver derives no additional information from this bound, and the heuristic hurts the unsatisfiability proof at the end. USC works surprisingly well with HEURISTIC. Since all heuristically modified variables are part of the optimization, the first USC optimization step disregards the heuristics entirely because no reactions from the reference network are selected. Afterward, the learned unsatisfiable core is relaxed by choosing heuristically modified reactions first. This might lead to unsatisfiable cores with higher quality since they arguably include relevant reactions. Iterating this process, the solver appears to learn shortcuts, fixing sets of important reactions that have to be included in solutions, thus reducing the complexity of the remaining search. Note that instead of modifying all atoms in the optimization statement which was shown to be unsuccessful in [8], we specifically select topologically relevant reactions.

Finally, we take the best configurations and examine how *fluto* scales on harder instances. To this end, we use configurations with bold rows in Table 2. We rerun the first experiment after adding 900 instances degraded by 40% (Table 3). F(VBS) denotes the virtual best results, meaning for each problem

---

[9] Also, we do not present results of *lpsolve* since it produced inferior results.

instance the best known solution among the three configurations was verified. For 20% and 30% degradation, we obtain additional 29 and 68 solutions and 41 and 193 optima, respectively. Overall, we find solutions for 92.4% out of the 3600 instances and 72.1% of them are optimal. The number of solutions decreases slightly and the number of optima more drastically with higher degradation. While again 100% of *fluto*'s solutions could be verified, only 27.1% and 3% are obtained for *meneco* and *gapfill*, respectively.

6   Discussion

We presented the first hybrid approach to metabolic network completion by combining topological and stoichiometric constraints in a uniform setting. To this end, we elaborated a formal framework capturing different semantics for the activation of reactions. Based upon these formal foundations, we developed a hybrid ASP encoding reconciling disparate approaches to network completion. The resulting system, *fluto*, thus combines the advantages of both approaches and yields greatly superior results compared to purely quantitative or qualitative existing systems. Our experiments show that *fluto* scales to more highly degraded networks and produces useful solutions in reasonable time. In fact, all of *fluto*'s solutions passed the biological gold standard. The exploitation of the network's topology guides the solver to more likely completion candidates, and furthermore avoids self-activated cycles, as obtained in FBA-based approaches. Also, unlike other systems, *fluto* allows for establishing optimality and address the strict stoichiometric completion problem without approximation.

fluto takes advantage of the hybrid reasoning capacities of the ASP system *clingo* for extending logic programs with linear constraints over reals. This provides us with a practically relevant application scenario for evaluating this hybrid form of ASP. To us, the most surprising empirical result was the observation that domain-specific heuristic allow for boosting unsatisfiable core based optimization. So far, such heuristics have only been known to improve satisfiability-oriented reasoning modes, and usually hampered unsatisfiability-oriented ones (cf. [8]).

Acknowledgments. This work was partially funded by DFG grant SCHA 550/9 and 11.

References

1. Ansótegui, C., Bonet, M., Levy, J.: SAT-based MaxSAT algorithms. Artif. Intell. **196**, 77–105 (2013)
2. Baral, C.: Knowledge Representation, Reasoning and Declarative Problem Solving. Cambridge University Press, New York (2003)
3. Becker, S., Feist, A., Mo, M., Hannum, G., Palsson, B., Herrgard, M.: Quantitative prediction of cellular metabolism with constraint-based models: the COBRA toolbox. Nat. Protoc. **2**(3), 727–738 (2007)

4. Collet, G., Eveillard, D., Gebser, M., Prigent, S., Schaub, T., Siegel, A., Thiele, S.: Extending the metabolic network of *Ectocarpus Siliculosus* using answer set programming. In: Cabalar, P., Son, T.C. (eds.) LPNMR 2013. LNCS, vol. 8148, pp. 245–256. Springer, Heidelberg (2013). doi:10.1007/978-3-642-40564-8_25

5. Dantzig, G.: Linear Programming and Extensions. Princeton University Press, Princeton (1963)

6. Ebrahim, A., Lerman, J., Palsson, B., Hyduke, D.: COBRApy: COnstraints-Based Reconstruction and Analysis for Python. BMC Syst. Biol. **7**, 74 (2013)

7. Gebser, M., Kaminski, R., Kaufmann, B., Ostrowski, M., Schaub, T., Wanko, P.: Theory solving made easy with clingo 5. In: Technical Communication of ICLP, pp. 2:1–2:15. OASIcs (2016)

8. Gebser, M., Kaminski, R., Kaufmann, B., Romero, J., Schaub, T.: Progress in *clasp* series 3. In: Calimeri, F., Ianni, G., Truszczynski, M. (eds.) LPNMR 2015. LNCS (LNAI), vol. 9345, pp. 368–383. Springer, Cham (2015). doi:10.1007/978-3-319-23264-5_31

9. Gelfond, M., Lifschitz, V.: Classical negation in logic programs and disjunctive databases. New Gener. Comput. **9**, 365–385 (1991)

10. Handorf, T., Ebenhöh, O., Heinrich, R.: Expanding metabolic networks: scopes of compounds, robustness, and evolution. J. Mol. Evol. **61**(4), 498–512 (2005)

11. Latendresse, M.: Efficiently gap-filling reaction networks. BMC Bioinform. **15**(1), 225 (2014)

12. Maranas, C., Zomorrodi, A.: Optimization Methods in Metabolic Networks. Wiley, Hoboken (2016)

13. Orth, J., Palsson, B.: Systematizing the generation of missing metabolic knowledge. Biotechnol. Bioeng. **107**(3), 403–412 (2010)

14. Ostrowski, M., Schaub, T.: ASP modulo CSP: the clingcon system. Theory Pract. Logic Program. **12**(4–5), 485–503 (2012)

15. Prigent, S., Collet, G., Dittami, S., Delage, L., Ethis de Corny, F., Dameron, O., Eveillard, D., Thiele, S., Cambefort, J., Boyen, C., Siegel, A., Tonon, T.: The genome-scale metabolic network of ectocarpus siliculosus (ectogem): a resource to study brown algal physiology and beyond. Plant J. **80**(2), 367–381 (2014)

16. Prigent, S., Frioux, C., Dittami, S., Thiele, S., Larhlimi, A., Collet, G., Gutknecht, F., Got, J., Eveillard, D., Bourdon, J., Plewniak, F., Tonon, T., Siegel, A.: Meneco, a topology-based gap-filling tool applicable to degraded genome-wide metabolic networks. PLOS Comput. Biol. **13**(1), e1005276 (2017)

17. Reed, J., Vo, T., Schilling, C., Palsson, B.: An expanded genome-scale model of Escherichia coli K-12 (iJR904 GSM/GPR). Genome Biol. **4**(9), R54 (2003)

18. Satish Kumar, V., Dasika, M., Maranas, C.: Optimization based automated curation of metabolic reconstructions. BMC Bioinform. **8**(1), 212 (2007)

19. Schaub, T., Thiele, S.: Metabolic network expansion with ASP. In: Proceedings ICLP, pp. 312–326. Springer (2009)

20. Simons, P., Niemelä, I., Soininen, T.: Extending and implementing the stable model semantics. Artif. Intell. **138**(1–2), 181–234 (2002)

21. Thiele, I., Vlassis, N., Fleming, R.: fastGapFill: efficient gap filling in metabolic networks. Bioinformatics **30**(17), 2529–2531 (2014)

22. Vitkin, E., Shlomi, T.: MIRAGE: a functional genomics-based approach for metabolic network model reconstruction and its application to cyanobacteria networks. Genome Biol. **13**(11), R111 (2012)

Action Language Hybrid AL

Alex Brik[1(✉)] and Jeffrey Remmel[2(✉)]

[1] Google Inc, Mountain View, USA
abrik@google.com
[2] Department of Mathematics, UC San Diego, San Diego, USA
jremmel@ucsd.edu

Abstract. This paper introduces an extension of the action language $\mathcal{AL}$ to Hybrid $\mathcal{AL}$. A program in Hybrid $\mathcal{AL}$ specifies both a transition diagram and associated computations for observing fluents and executing actions. The semantics of $\mathcal{AL}$ is defined in terms of Answer Set Programming (ASP). Similarly, the semantics of Hybrid $\mathcal{AL}$ is defined using Hybrid ASP which is an extension of ASP that allows rules to control sequential execution of arbitrary algorithms.

Constructing a mathematical model of an agent and its environment based on the theory of action languages has been studied and has applications to planning and diagnostic problems, see [10] for an overview. In the realm of diagnostic problems, the goal is to find explanations of unexpected observations. We are interested in solving diagnostic problems such as those that arise diagnosing malfunctions of a large distributed software system, as described in [14].

The approach to solving a diagnostic problems described in [1] is based on the idea of using a mathematical model of the agent's domain, created using a description in the action language $\mathcal{AL}$ [2] to find explanations for unexpected observations. Central to this approach is the notion of the **agent loop** [10] which we modify to underline the relevance to the diagnostic problem.

1. Observe the world, check that observations are consistent with expectations, and update the knowledge base.
2. Select an appropriate goal G.
3. Explain unexpected observations and search for a plan (a sequence of actions) to achieve G.
4. Execute an initial part of the plan, update the knowledge base, go back to step 1.

The description and the facts from the knowledge base are translated into a logic program in a language of answer set programming (ASP) [11]. An ASP solver is then used to find stable models of the program, which are descriptions of possible trajectories of the underlying domain. These can be used to carry out steps 1 and 3 of the agent loop.

The two assumptions for the applicability of the agent loop are: (1) the agent is capable of making correct observations, performing actions, and recording

© Springer International Publishing AG 2017
M. Balduccini and T. Janhunen (Eds.): LPNMR 2017, LNAI 10377, pp. 322–335, 2017.
DOI: 10.1007/978-3-319-61660-5_29

these observations and actions (not defeasible), and (2) normally the agent is capable of observing all relevant exogenous actions occurring in its environment (defeasible). Hybrid $\mathcal{AL}$ is introduced to help solve a diagnostic problem where both (1) and (2) are defeasible, and where to decrease the size of the search space the following are assumed: (A1) the agent is normally capable of determining a small set of possible actions occurring in its environment by computationally simulating its environment for a number of steps starting from a known state in the past, or by performing other relevant external computations (here, by a small set we mean a set that can practically be represented by the enumeration), (A2) the agent has a description of at least one past state, and that description is sufficient to satisfy the assumption A1.

Under these assumptions, the agent may need to perform sequential computations (where the choice of the computations at step j may depend on the output of computations at step $j - 1$) in order to determine sets of possible actions and states of the domain. Our hypothesis is that under these assumptions, computational efficiency can be improved if ASP-like processing and the external computations are merged. Such a merging can reduce the number of possible actions and states that need to determine the next action.

One way to address such issues is to extend $\mathcal{AL}$ to a richer action language that provides mechanism for performing computations and passing input and output parameters between the computation steps. Hybrid $\mathcal{AL}$, introduced in this paper is one such extension. While descriptions in $\mathcal{AL}$ are translated into ASP, the descriptions in Hybrid $\mathcal{AL}$ are translated into Hybrid ASP (H-ASP) which is an extension of ASP, introduced by the authors in [3] that allows rules to control sequential execution of arbitrary algorithms. This functionality makes H-ASP well suited for solving a diagnostic problem under our assumptions.

The outline of this paper is as follows. In Sect. 1, we will discuss an example to motivate our need to extend $\mathcal{AL}$ to Hybrid $\mathcal{AL}$, and we will briefly describe $\mathcal{AL}$ and H-ASP. In Sect. 2 we will introduce Hybrid $\mathcal{AL}$. In Sect. 3 we will revisit the example from Sect. 1 and show how it can be described in Hybrid $\mathcal{AL}$. Section 4 contains discussion of related work and conclusions.

1 Motivation for Hybrid $\mathcal{AL}$ and Preliminaries

To motivate the introduction of Hybrid $\mathcal{AL}$, we will consider an example of a hypothetical video processing system (Fig. 1). The system selects a video from a video library. The choice depends on the initial state and time. It then checks the quality of the video. If the check fails, then a system is said to malfunction, which is unexpected.

We are interested in creating a diagnostic agent of this system. We will assume that the agent is only able to determine a subset of the videos $v_1, ..., v_k$ dependent on the initial state, one of which was chosen by the system. The agent will then investigate k possible trajectories of the system - one for each possible video. For each such trajectory, the agent will check quality of the video and determine whether it could have caused unexpected behavior.

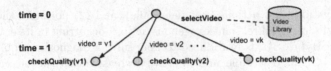

Fig. 1. Video processing system

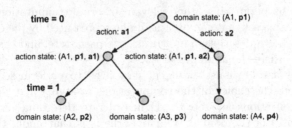

Fig. 2. Hybrid transition diagram

Since the subset of the videos selected by the agent is time dependent, it may not be practical to pre-compute the values of properties of the videos. Instead, if the system malfunctions, then the agent will access the video library, select a set of videos and run the quality check algorithm on each of the selected videos to explain the unexpected behavior of the system.

A key concept related to action languages is that of a transition diagram, which is a labeled directed graph where vertices are states of a dynamic domain, and edges are labeled with subsets of actions. In Hybrid $\mathcal{AL}$, one considers **hybrid transition diagrams**, which are directed graphs with two types of vertices: action states and domain states. A **domain state** is a pair $(A, \mathbf{p})$ where A is a set of propositional atoms and $\mathbf{p}$ is a vector of sequences of 0s and 1s. We can think of A as a set of values of the properties of a system, and $\mathbf{p}$ as the description of the parameters used by external computations. An **action state** is a tuple $(A, \mathbf{p}, a)$ where A and $\mathbf{p}$ are as in the domain state, and a is a set of actions. An out edge from a domain state must have an action state as its destination. An out edge from an action state must have a domain state as its destination. Moreover, if $(A, \mathbf{p})$ is a domain state that has an out-edge to an action state $(B, \mathbf{r}, a)$, then $A = B$ and $\mathbf{p} = \mathbf{r}$. There is a simple bijection between the set of transition diagrams and the set of hybrid transition diagrams.

An example of a hybrid transition diagram is in Fig. 2. Two actions $a1$ and $a2$ can be performed in the state $(A1, \mathbf{p}1)$. Thus, the action states are $(A1, \mathbf{p}1, a1)$ and $(A1, \mathbf{p}1, a2)$. The consequents of applying action $a1$ at the state $(A1, \mathbf{p}1)$ are two domain states $(A2, \mathbf{p}2)$ and $(A3, \mathbf{p}3)$, for action $a2$ at $(A1, \mathbf{p}1)$ it is the domain state $(A4, \mathbf{p}4)$.

We will now briefly review action language $\mathcal{AL}$. Our review is based on Chap. 8 of [10] where one can find more details. $\mathcal{AL}$ has three special sorts: **statics, fluents,** and **actions**. The fluents are partitioned into two sorts: **inertial** and **defined**. Statics and fluents are referred to as **domain properties**. Intuitively, statics are properties of the system that don't change with time. Inertial fluents are properties that are subject to the law of inertia. Their values can be directly influenced by actions, and in the absence of such a change the values remain unchanged. Defined fluents are properties defined in terms of other fluents and cannot be directly influenced by actions. A **domain literal** is a domain property p or its negation $\neg p$. If a domain literal l is formed by a fluent, it is referred to as **fluent literal**; otherwise it is a **static literal**. A set S of domain literals is called **complete** if for any domain property p either p or $\neg p$ is in S. S is called **consistent** if there is no p such that $p \in S$ and $\neg p \in S$.

$\mathcal{AL}$ allows the following types of statements:

1. **Causal Laws:** a causes l_{in} if $p_0, ..., p_m$,
2. **State constraints:** l if $p_0, ..., p_m$, and
3. **Executability conditions:** impossible $a_0, ..., a_k$ if $p_0, ..., p_m$
   where a is an action, l is an arbitrary domain literal, l_{in} is a literal formed by an inertial fluent, $p_0, ..., p_m$ are domain literals, $k \geq 0$ and $m \geq -1$.

A system description $\mathcal{SD}$ in $\mathcal{AL}$ specifies a transition diagram. Intuitively, state constraints specify sets of allowed states. Given a state S, the set of executability conditions specifies concurrent actions executable at S, i.e. sets of actions that can decorate out edges of S. Given a state S and a set of actions A, causal laws together with state constraints determine the set of possible consequent states that result from executing A at S, i.e. neighbor states connected to S by out edges decorated by A.

Formally, a complete and consistent set σ of domain literals is a state of a transition diagram defined by $\mathcal{SD}$ if σ is the unique answer set of program $\Pi_c(\mathcal{SD}) \cup \sigma_{nd}$, where σ_{nd} is the collection of all domain literals of σ formed by inertial fluents and statics (the definition of $\Pi_c(\mathcal{SD})$ is omitted for brevity).

A system description $\mathcal{SD}$ of $\mathcal{AL}$ is called **well founded** if for any complete and consistent set of fluent literals σ satisfying the state constraints of $\mathcal{SD}$, the program $\Pi_c(\mathcal{SD}) \cup \sigma_{nd}$ has at most one answer set. Sufficient conditions for well-foundedness are expressed in terms of the **fluent dependency graph**, which is a directed graph such that its vertices are arbitrary domain literals and where it has an edge: **(a)** from l to l' if l is formed not by a defined fluent and $\mathcal{SD}$ contains a state constraint with the head l and the body contains l', **(b)** from f to l' if f is a defined fluent and $\mathcal{SD}$ contains a state constraint with head f and body containing l' and not containing f, **(c)** from $\neg f$ to f for every defined fluent f. A fluent dependency graph is said to be **weakly acyclic** if it does not contain any paths from defined fluents to their negations.

Proposition (Proposition 1 in [9]). If a system description $\mathcal{SD}$ of $\mathcal{AL}$ is weakly acyclic, then $\mathcal{SD}$ is well-founded.

A transition $\langle \sigma_0, a, \sigma_1 \rangle$ is described in terms of a program $\Pi(\mathcal{SD}, \sigma_0, a)$ (the definition of $\Pi(\mathcal{SD}, \sigma_0, a)$ is omitted for brevity). A state-action-state triple

$\langle \sigma_0, a, \sigma_1 \rangle$ is a **transition** of $T(\mathcal{SD})$ iff $\Pi(\mathcal{SD}, \sigma_0, a)$ has an answer set A such that $\sigma_1 = \{l : h(l, 1) \in A\}$.

We now give a brief overview of H-ASP restricted to the relevant rules. A H-ASP program P has an underlying parameter space S and a set of atoms At. Elements of S, called *generalized positions*, are of the form $\mathbf{p} = (t, x_1, \ldots, x_m)$ where t is time and x_i are parameter values. We let $t(\mathbf{p})$ denote t and $x_i(\mathbf{p})$ denote x_i for $i = 1, \ldots, m$. The universe of P is $At \times S$. A pair $(Z, \mathbf{p})$ where $Z \subseteq At$ and $\mathbf{p} \in S$ will be referred to as a *hybrid state*. For $M \subseteq At \times S$, we write $\mathbb{GP}(M) = \{\mathbf{p} \in S : (\exists a \in At)((a, \mathbf{p}) \in M)\}$, $W_M(\mathbf{p}) = \{a \in At : (a, \mathbf{p}) \in M\}$, and $(Z, \mathbf{p}) \in M$ if $\mathbf{p} \in \mathbb{GP}(M)$ and $W_M(\mathbf{p}) = Z$. A *block* B is an object of the form $B = a_1, \ldots, a_n, not\ b_1, \ldots, not\ b_m$ where $a_1, \ldots, a_n, b_1, \ldots, b_m \in At$. We let $B^- = not\ b_1, \ldots, not\ b_m$, and $B^+ = a_1, \ldots, a_n$. We write $M \models (B, \mathbf{p})$, if $(a_i, \mathbf{p}) \in M$ for $i = 1, \ldots, n$ and $(b_j, \mathbf{p}) \notin M$ for $j = 1, \ldots, m$.

Advancing rules are of the form: $a \leftarrow B : A, O$. Here B is a block, $O \subseteq S$, for all $\mathbf{p} \in O$ $A(\mathbf{p}) \subseteq S$, and for all $\mathbf{q} \in A(\mathbf{p})$, $t(\mathbf{q}) > t(\mathbf{p})$. A represents a partial function $S \rightarrow 2^S$, and is called an *advancing algorithm*. The idea is that if $\mathbf{p} \in O$ and B is satisfied at $\mathbf{p}$, then A can be applied to $\mathbf{p}$ to produce a set of generalized positions O' such that if $\mathbf{q} \in O'$, then $t(\mathbf{q}) > t(\mathbf{p})$ and $(a, \mathbf{q})$ holds.

Stationary-i rules are of the form: $a \leftarrow B_1; B_i : H, O$ (where for $i = 1$ we mean $a \leftarrow B_1 : H, O$). Here B_i are blocks and H is a Boolean algorithm defined on O. The idea is that if $(\mathbf{p}_1, \mathbf{p}_i) \in O$ (where for $i = 1$ we mean $\mathbf{p}_1 \in O$), B_k is satisfied at $\mathbf{p}_k$ for $k = 1, i$, and $H(\mathbf{p}_1, \mathbf{p}_i)$ is true (where for $i = 1$ we mean $H(\mathbf{p}_1)$), then $(a, \mathbf{p}_i)$ holds.

A *H-ASP Horn program* P is a H-ASP program which does not contain any negated atoms in At. For $I \in S$, the one-step provability operator $T_{P,I}(M)$ consists of M together with the set of all $(a, J) \in At \times S$ such that **(1)** there exists a stationary-i rule $a \leftarrow B_1; B_i : H, O$ such that $(\mathbf{p}_1, \mathbf{p}_i) \in O \cap (\mathbb{GP}(M) \cup \{I\})^i$, $M \models (B_k, \mathbf{p}_k)$ for $k = 1, i$, and $H(\mathbf{p}_1, \mathbf{p}_i) = 1$, and $(a, J) = (a, \mathbf{p}_i)$ or **(2)** there exists an advancing rule $a \leftarrow B : A, O$. such that $\mathbf{p} \in O \cap (\mathbb{GP}(M) \cup \{I\})$ such that $J \in A(\mathbf{p})$ and $M \models (B, \mathbf{p})$.

An advancing rule is *inconsistent* with (M, I) if for all $\mathbf{p} \in O \cap (\mathbb{GP}(M) \cup \{I\})$ either $M \not\models (B, \mathbf{p})$ or $A(\mathbf{p}) \cap \mathbb{GP}(M) = \emptyset$. A stationary-$i$ rule is *inconsistent* with (M, I) if for all $(\mathbf{p}_1, \mathbf{p}_i) \in O \cap (\mathbb{GP}(M) \cup \{I\})^i$ there is a k such that $M_k \not\models (B_k, \mathbf{p}_k)$ or $H(\mathbf{p}_1, \mathbf{p}_i) = 0$.

We form the Gelfond-Lifschitz reduct of P over M and I, $P^{M,I}$ as follows. **(1)** Eliminate all rules that are inconsistent with (M, I). **(2)** If the advancing rule is not eliminated by (1), then replace it by $a \leftarrow B^+ : A^+, O^+$ where O^+ is the set of all $\mathbf{p}$ in $O \cap (\mathbb{GP}(M) \cup \{I\})$ such that $M \models (B^-, \mathbf{p})$ and $A(\mathbf{p}) \cap \mathbb{GP}(M) \neq \emptyset$, and $A^+(\mathbf{p}) = A(\mathbf{p}) \cap \mathbb{GP}(M)$. **(3)** If the stationary-i rule is not eliminated by (1), then replace it by $a \leftarrow B_1^+; B_i^+ : H|_{O^+}, O^+$ where O^+ is the set of all $(\mathbf{p}_1, \mathbf{p}_i)$ in $O \cap (\mathbb{GP}(M) \cup \{I\})^i$ such that $M \models (B_k^-, \mathbf{p}_k)$ for $k = 1, i$, and $H(\mathbf{p}_1, \mathbf{p}_i) = 1$.

Then M is a *stable model* of P with initial condition I if $\bigcup_{k=0}^{\infty} T_{P^{M,I},I}^k(\emptyset) = M$.

We will now introduce additional definitions which will be used later on in this paper. We say that an advancing algorithm A lets a parameter y be *free* if the domain of y is Y and for all generalized positions $\mathbf{p}$ and $\mathbf{q}$ and all $y' \in Y$, whenever $\mathbf{q} \in A(\mathbf{p})$, then there exist $\mathbf{q}' \in A(\mathbf{p})$ such that $y(\mathbf{q}') = y'$ and $\mathbf{q}$ and $\mathbf{q}'$ are identical in all the parameter values except possibly y. An advancing algorithm A *fixes* a parameter y if A does not let y be free.

We will use T to indicate a Boolean algorithm or a set constraint that always returns true. As a short hand notation, if we omit a Boolean algorithm or a set constraint from a rule, then by that we mean that T is used.

We say that a pair of generalized positions $(\mathbf{p}, \mathbf{q})$ is a **step** (with respect to a H-ASP program P) if there exists an advancing rule "$a \leftarrow B : A, O$" in P such that $\mathbf{p} \in O$ and $\mathbf{q} \in A(\mathbf{p})$. Then we will say that $\mathbf{p}$ is a **source** and $\mathbf{q}$ is a **destination**. We will assume that the underlying parameter space of P contains a parameter $Prev$. For a step $(\mathbf{p}, \mathbf{q})$, then we will have $Prev(\mathbf{q}) = (x_1(\mathbf{p}), ..., x_n(\mathbf{p}))$. We define the advancing algorithm $GeneratePrev$ as $GeneratePrev(\mathbf{p}) = \{\mathbf{q}|$ $q \in S$ and $Prev(\mathbf{q}) = (x_1(\mathbf{p}), ..., x_n(\mathbf{p}))\}$ for a generalized position $\mathbf{p}$. (This can be implemented efficiently using references, rather than copies of data). We can then define a Boolean algorithm $isStep(\mathbf{p}, \mathbf{q})$ equal true iff $Prev(\mathbf{q}) = (x_1(\mathbf{p}), ..., x_n(\mathbf{p}))$ and $[t(\mathbf{q}) = t(\mathbf{p}) + stepSize]$.

For two one-place Boolean algorithms A, B, the notation $A \vee B$, $A \wedge B$, or $\overline{A}$ means a Boolean algorithm that for every generalized position $\mathbf{q}$ returns $A(\mathbf{q}) \vee B(\mathbf{q})$, $A(\mathbf{q}) \wedge B(\mathbf{q})$ or *not* $A(\mathbf{q})$ respectively. The same holds for two place Boolean algorithms.

For a domain state at time t we will assume that subsequent action states are at time $t + 0.1$ and the subsequent domain states are at time $t + 1$. Thus, advancing algorithms executing in domain states will increment time by 0.1, and advancing algorithms executing in action states will increment time by 0.9.

Finally, for a set of atoms M, $rules(M)$ will denote a set of stationary-1 rules $\{m \leftarrow : T \mid m \in M\}$.

2   Action Language Hybrid $\mathcal{AL}$

In this section, we shall define Hybrid $\mathcal{AL}$. Our definitions mirror the presentation of $\mathcal{AL}$ given in [10].

Syntax. In Hybrid $\mathcal{AL}$, there are two types of atoms: **domain atoms** and **action atoms**. There are two sets of parameters: **domain parameters** and **time**. The domain atoms are partitioned into three sorts: **inertial**, **static** and **defined**. A **domain literal** is a fluent atom p or its negation $\neg p$. For a generalized position $\mathbf{q}$, let $\mathbf{q}|_{domain}$ denote a vector of domain parameters.

A **domain algorithm** is a Boolean algorithm P such that for all the generalized positions $\mathbf{q}$ and $\mathbf{r}$, if $\mathbf{q}|_{domain} = \mathbf{r}|_{domain}$, then $P(\mathbf{q}) = P(\mathbf{r})$. An **action algorithm** is an advancing algorithm A such that for all $\mathbf{q}$ and for all $\mathbf{r} \in A(\mathbf{q})$ $time(\mathbf{r}) = time(\mathbf{q}) + 0.9$. For an action algorithm A, the signature of A, $sig(A)$, is the vector of parameter indices $i_0, i_1, ..., i_k$ of domain parameters such that A fixes parameters $i_0, i_1, ..., i_k$.

Hybrid $\mathcal{AL}$ allows the following types of statements.

1. **Action association statements:** *associate a with A,*
2. **Signature statements:** *A has signature* $i_0, ..., i_k$,
3. **Causal laws:** *a causes* $\langle l_{nd}, L \rangle$ *if* $p_0, ..., p_m : P$,
4. **State constraints:** $\langle l, L \rangle$ *if* $p_0, ..., p_m : P$,
5. **Executability conditions:** *impossible* $a_0, ..., a_k$ *if* $p_0, ..., p_m : P$, and
6. **Compatibility conditions:** *compatible* a_0, a_1 *if* $p_0, ..., p_m : P$

where a is an action, A is an action algorithm, $i_0, ..., i_k$ are parameter indices, l_{nd} is a literal formed by an inertial or a static atom, L is a domain algorithm, $p_0, ..., p_m$ are domain literals, P is a domain algorithm, l is a domain literal, and $a_0, ..., a_k$ are actions $k \geq 0$ and $m \geq -1$. No negation of a defined fluent can occur in the heads of state constraints.

The following short-hand notation can be used for convenience.

(a) If L or P are omitted then the algorithm T is assumed to be used.
(b) For an action a, if the action association statement is omitted, then the action association statement *"associate a with* **0**" is implicitly used for a. Here, **0** is an action algorithm with an empty signature that for a generalized position **p**, produces the set of all generalized positions $\{\mathbf{q} : time(\mathbf{q}) = time(\mathbf{p}) + 0.9\}$.

Semantics. Similarly to $\mathcal{AL}$, a system description SD in Hybrid $\mathcal{AL}$ serves as a specification of the hybrid transition diagram $\mathcal{T}(SD)$ defining all possible trajectories of the corresponding dynamic system. Hence, to define the semantics of Hybrid $\mathcal{AL}$, we will define the states and the legal transitions of this diagram.

The H-ASP programs discussed below assume the parameter space consisting of parameters t (time), domain parameters and the parameter Prev. Such a parameter space will be called **the parameter space of** SD.

States. We let $\Pi_c(SD)$ be the logic program defined as follows.

1. For every state constraint $\langle l, L \rangle$ if $p_0, ..., p_m : P$, $\Pi_c(SD)$ contains the clause $l \leftarrow p_0, ..., p_m : P \vee \overline{L}$.
2. For every defined domain atom f, $\Pi_c(SD)$ contains the closed world assumption (CWA): $\neg f \leftarrow not\ f$.

For any set σ of domain literals, we let σ_{nd} denote the collection of all domain literals σ formed by inertial domain atoms and statics.

Definition 1. *Let* $(\sigma, \mathbf{q})$ *be a hybrid state. If* σ *is a complete and consistent set of domain literals, then* $(\sigma, \mathbf{q})$ *is a state of the hybrid transition diagram defined by a system description* SD *if* $(\sigma, \mathbf{q})$ *is the unique answer set of the program* $\Pi_c(SD) \cup rules(\sigma_{nd})$ *with the initial condition* **q**.

The definitions of the fluent dependency graph, weak acyclicity, sufficient conditions for well-foundedness are the same as those defined for $\mathcal{AL}$.

Transitions. To describe a transition $\langle (\sigma_0, \mathbf{p}), a, (\sigma_1, \mathbf{q}) \rangle$ we construct a program $\Pi(SD, (\sigma_0, \mathbf{p}), a)$, which as a similarly named program for $\mathcal{AL}$, consists of a logic program encoding of the system description SD, initial state $(\sigma_0, \mathbf{p})$ and an action a such that the answer sets of this program determine the states the system can move into after execution of a in $(\sigma_0, \mathbf{p})$.

The encoding $\Pi(SD)$ of the system description SD consists of the encoding of the signature of SD and rules obtained from the statements of SD. The encoding of the signature $sig(SD)$ into a set of stationary-1 rules is as follows:

(A) for each constant symbol c of sort *sort_name* other than fluent, static or action $sig(SD)$ contains: $sort\_name(c) \leftarrow$,

(B) for every defined fluent f of SD, $sig(SD)$ contains: $fluent(defined, f) \leftarrow$,

(C) for every inertial fluent f of SD, $sig(SD)$ contains: $fluent(inertial, f) \leftarrow$,

(D) for each static f of SD, $sig(SD)$ contains: $static(f) \leftarrow$, and

(E) for every action a of SD, $sig(SD)$ contains: $action(a) \leftarrow$

Next will specify *the encoding of statements of SD*.

(**1**) *For every action algorithm A*, we have an atom $\text{alg}(A)$. If A has signature $(i_0, ..., i_k)$, then we add the following rules for all $j \in \{0, ..., k\}$ that specify all the parameters fixed by A: $fix\_value(i_j) \leftarrow action\_state, exec(\text{alg}(A))$.

(**2**) *Inertia axioms for parameters*: For every domain parameter i, we have an advancing rule $discard \leftarrow action\_state, not\ fix\_value(i) : Default[i]$ where $Default[i](\mathbf{p}) = \{\mathbf{q}|\ \mathbf{p}_i = \mathbf{q}_i\}$. The inertia axioms for parameters will cause the values of parameters not fixed by one of the action algorithms to be copied to the consequent states. $discard$ here is a placeholder atom.

(**3**) *For every causal law: a causes $\langle l_{in}, L \rangle$ if $p_0, ..., p_m : P$*, $\Pi(SD)$ contains

(i) A stationary-1 rule generating atom $exec(\text{alg}(A))$ specifying that algorithm A associated with action a will be used:

$exec(\text{alg}(A)) \leftarrow action\_state, occurs(a), h(p_0), ..., h(p_m) : P$.

(ii) An advancing rule executing A to compute changes to domain parameters:

$h(l_{in}) \leftarrow action\_state, occurs(a), h(p_0), ..., h(p_m) : A, P$.

(iii) A stationary-2 rule to apply L to the successor states:

$h(l_{in}) \quad \leftarrow \quad action\_state, occurs(a), h(p_0), ..., h(p_m); : isStep \wedge [source(P) \vee dest(\overline{L})]$.

(**4**) *For a one-place boolean algorithm D*, $source(D)$ indicates a two-place Boolean algorithm $source(D)(\mathbf{p}, \mathbf{q}) = D(\mathbf{p})$, and $destination(D)(\mathbf{p}, \mathbf{q}) = D(\mathbf{q})$.

(**5**) *For every state constraint: $\langle l, L \rangle$ if $p_0, ..., p_m : P$*, $\Pi(SD)$ contains

$h(l) \leftarrow domain\_state, h(p_0), ..., h(p_m) : P \vee \overline{L}$.

(**6**) $\Pi(SD)$ *contains CWA for defined fluents*

$\neg holds(f) \leftarrow domain\_state, fluent(defined, f), not\ holds(f)$.

(**7**) *For every executability condition: impossible $a_0, ..., a_k$ if $p_0, ..., p_m : P$*, $\Pi(SD)$ contains

$\neg occurs(a_0) \vee ... \vee \neg occurs(a_k) \leftarrow action\_state, h(p_0), ..., h(p_m) : P$.

(**8**) $\Pi(SD)$ *contains inertia axioms for inertial fluents*. That is, for every inertial fluent f stationary-2 rules,

$holds(f) \leftarrow fluent(inertial, f), holds(f); not\ \neg holds(f) : isStep$ and

$\neg holds(f) \leftarrow fluent(inertial, f), \neg holds(f); not\ holds(f) : isStep$.

(9) $\Pi(SD)$ *contains propagation axioms for static and defined fluents.* These are used to copy statics and defined fluents from domain states to the successor action states. For every static or defined fluent f stationary-2 rules

$holds(f) \leftarrow domain\_state, \ static(f), \ holds(f); \ : isStep,$

$\neg holds(f) \leftarrow domain\_state, \ static(f), \ \neg holds(f); \ : isStep,$

$holds(f) \leftarrow domain\_state, \ fluent(defined, f), \ holds(f); \ : isStep,$ and

$\neg holds(f) \leftarrow domain\_state, \ fluent(defined, f), \ \neg holds(f); \ : isStep.$

(10) $\Pi(SD)$ *contains CWA for actions:* for every every action a, there is a clause
$\neg occurs(a) \leftarrow action\_state, \ not \ occurs(a).$

(11) *For every action algorithm* A, B *such that* $sig(A) \cap sig(B) \neq \emptyset$, we have a stationary-2 rule prohibiting executing the algorithms in the same state, unless they are explicitly marked as compatible

$fail \leftarrow action\_state, \ exec(\mathrm{alg}(A)), \ exec(\mathrm{alg}(B)),$

$not \ compatible(\mathrm{alg}(A), \ \mathrm{alg}(B)),$ and

$not \ compatible(\mathrm{alg}(B), \ \mathrm{alg}(A)); \ not \ fail : isStep.$

(12) *For every compatibility condition:* compatible a_0, a_1 if $p_0, ..., p_m$: P, $\Pi(SD)$, there is a clause

$compatible(\mathrm{alg}(A_0), \mathrm{alg}(A_1)) \leftarrow action\_state, h(p_0), ..., h(p_m),$

$occurs(a_0), occurs(a_1) : P$

where A_i is the action algorithm associated with a_i.

(13) $\Pi(SD)$ *has axioms for describing the interleaving of domain and action states.* These are a stationary-2 rule and an advancing rule:

$domain\_state \leftarrow action\_state; \ :$

$action\_state \leftarrow domain\_state : CreateActionState$

where for a generalized position $\mathbf{p}$,

$CreateActionState(\mathbf{p}) =$

$\{\mathbf{q}| \ \text{where} \ \mathbf{p}|_{domain} = \mathbf{q}|_{domain} \ \& \ time(\mathbf{q}) = time(\mathbf{p}) + 0.1\}.$

(14) $\Pi(SD)$ *contains rules making an action state with no actions invalid:* For every action a:

$valid\_action\_state \leftarrow action\_state, \ occurs(a)$

and a rule to invalidate the state without actions:

$fail \leftarrow action\_state, \ not \ valid\_action\_state, \ not \ fail.$

(15) $\Pi(SD)$ *contains the rule for generating the value of* Prev **parameter:**
$discard \leftarrow GeneratePrev, T.$

The inertia axioms and the propagation axioms guarantee that the set of fluents of a domain state and its successor action state are identical. The inertial axioms also guarantee that an action state and its successor domain state contain the same inertial fluents, if those fluents are not explicitly changed by causal laws.

A relation $holds(f, \mathbf{p})$ will indicate that a fluent f is true at a generalized position $\mathbf{p}$. $h(l, \mathbf{p})$ where l is a domain literal will denote $holds(f, \mathbf{p})$ if $l = f$ or $\neg holds(f, \mathbf{p})$ if $l = \neg f$. $occurs(a, \mathbf{p})$ will indicate that action a has occurred at $\mathbf{p}$. The encoding $h(\sigma_0, \mathbf{p})$ of the initial state is a set of stationary-1 rules:

$h(\sigma_0, \ \mathbf{p}) = \{h(l) \leftarrow: isDomainTime[t(\mathbf{p})] \mid l \in \sigma_0\}$

where $isDomainTime[x](\mathbf{q})$ returns true iff $t(\mathbf{q}) = x$.

Finally the encoding $occurs(a, \ \mathbf{p})$ of the action a is

$occurs(a, \mathbf{p}) = \{occurs(a_i) \leftarrow: isActionTime[t(\mathbf{p})] \mid a_i \in a\}$
where $isActionTime[x](\mathbf{q})$ returns true iff $t(\mathbf{q}) = x + 0.1$.

We then define $\Pi(SD, (\sigma_0, \mathbf{p}), a) = \Pi(SD) \cup h(\sigma_0, \mathbf{p}) \cup occurs(a, \mathbf{p})$.

Definition 2. *Let a be a nonempty collection of actions and $(\sigma_0, \mathbf{p})$ and $(\sigma_1, \mathbf{q})$ be two domain states of the hybrid transition diagram $\mathcal{T}(SD)$ defined by a system description $\mathcal{SD}$. A state-action-state triple $\langle(\sigma_0, \mathbf{p}), a, (\sigma_1, \mathbf{q})\rangle$ is a **transition** of $\mathcal{T}(SD)$ iff $\Pi(\mathcal{SD}, (\sigma_0, \mathbf{p}), a)$ has a stable model M with the initial condition $\mathbf{p}$, such that $(\sigma_1, \mathbf{q})$ is a hybrid state.*

Hybrid $\mathcal{AL}$ provides a superset of the functionality of $\mathcal{AL}$. We prove this by defining a translation of a description SD in $\mathcal{AL}$ into $E(SD)$, which is a description in Hybrid $\mathcal{AL}$. We will then show that there is a correspondence between the states and transitions of $\mathcal{T}(SD)$ with those of $\mathcal{T}(E(SD))$.

The signature of $E(SD)$ contains exactly the domain and action atoms of SD, and no domain parameters. For every action law: "a causes l_{in} if $p_0, ..., p_m$" of SD, $E(SD)$ contains an action law "a causes (l_{in}, T) if $p_0, ..., p_m : T$". For every state constraint: "l if $p_0, ..., p_m$" of SD, $E(SD)$ contains a state constraint "(l, T) if $p_0, ..., p_m : T$". Finally, for every executability condition: "impossible $a_0, ..., a_k$ if $p_0, ..., p_m$" of SD, $E(SD)$ contains an executability condition: "impossible $a_0, ..., a_k$ if $p_0, ..., p_m : T$". We then have the following two equivalence theorems.

Theorem 1. *Let SD be a system description in action language $\mathcal{AL}$, and let σ be a complete and consistent set of domain literals. Then σ is a state of a transition diagram $T(SD)$ iff for all generalized positions $\mathbf{q}$ from the parameter space of $E(SD)$ $(\sigma, \mathbf{q})$ is a state of the hybrid transition diagram $T(E(SD))$.*

Theorem 2. *Let SD be a system description in action language $\mathcal{AL}$. If a state-action-state triple (σ_0, a, σ_1) of $T(SD)$ is a transition of $T(SD)$, then for all generalized positions $\mathbf{q}_0$ and $\mathbf{q}_1$ such that $\mathbf{q}_0|_{domain} = \mathbf{q}_1|_{domain}$ and $t(\mathbf{q}_0) + 1 = t(\mathbf{q}_1)$ from the parameter space of $E(SD)$, $((\sigma_0, \mathbf{q}_0), a, (\sigma_1, \mathbf{q}_1))$ is a transition of $T(E(SD))$. Moreover, if a state-action-state triple $((\sigma_0, \mathbf{q}_0), a, (\sigma_1, \mathbf{q}_1))$ of $T(E(SD))$ is a transition of $T(E(SD))$ then (σ_0, a, σ_1) is a transition of $T(SD)$.*

Sketch of a proof. First we construct the translation $\Pi(SD, \sigma_0, a)$ of SD and a similar translation $\Pi(E(SD), (\sigma_0, \mathbf{q}_0), a)$ of $E(SD)$. There is an equivalence of the one step provability operators of the two Gelfond-Lifschitz transforms with respect to domain atoms. Using this, one can define a bijection between the set of stable models specifying transitions $\{(\sigma_0, a, \sigma_1)\}$ of $T(SD)$, and the set of stable models specifying transitions $\{((\sigma_0, \mathbf{q}_0), a, (\sigma_1, \mathbf{q}_1))\}$ of $\Pi(E(SD), (\sigma_0, \mathbf{q}_0), a)$.

3   Example

We will now revisit the example from Fig. 1 and describe it in Hybrid $\mathcal{AL}$. In our domain, the only action is *selectVideo*. The algorithm *selectVideoAlg* will produce a set of possible videos. Each such video will be stored in the parameter *video* of a possible state. The selected video quality will be checked by the domain algorithm *checkQuality*.

Our specification of actions has two statements:

associate selectVideo with selectVideoAlg and

selectVideoAlg has signature video.

In addition, there is a causal law,

selectVideo causes selected if not selected, and a state constraint,

malfunction if selected: -checkQualityAlg.

The translation of our specification into H-ASP is as follows.

```
% The encoding of the signature and action declaration:
fluent(inertial, selected):- . fluent(defined, malfunction):- . action(selectVideo):-
fix_value(video):- exec(alg(selectVideAlg))

% Inertia axioms for parameters
discard:- action_state, not fix_value(video): Default[video]

% Causal laws for: selectVideo causes selected if not selected
exec(alg(selectVideoAlg)):- action_state, occurs(selectVideo), -holds(selected)
holds(selected):- action_state, occurs(selectVideo), -holds(selected): selectVideoAlg, T
holds(selected):- action_state, occurs(selectVideo), -holds(selected);
     : isStep && [source(T) || dest(-T)]

% State constraint: malfunction if selected: -checkQualityAlg
holds(malfunction):- domain_state, holds(selected): -checkQualityAlg || -T

% CWA for the defined fluent:
-holds(malfunction):- domain_state, fluent(defined, malfunction), not holds(malfunction)

% Inertia axioms for the inertial fluent:
holds(selected):- fluent(inertial, selected), holds(selected); not -holds(selected);; isStep
-holds(selected):- fluent(inertial, selected), -holds(selected); not holds(selected);; isStep

% Propagation axioms: for defined fluents and CWA for the action:
holds(malfunction):- domain_state, fluent(defined, malfunction), holds(malfunction);; isStep
-holds(malfunction):- domain_state, fluent(defined, malfunction), -holds(malfunction);; isStep
-occurs(selectVideo):- action_state, not occurs(selectVideo)

% Interleaving of action and domain states, rules for generating action states:
domain_state:- action_state; : . action_state:- domain_state: CreateActionState

% Invalidate an action state with no actions; the rule for generating Prev parameter
valid_action_state:- action_state, occurs(selectVideo)
fail:- action_state, not valid_action_state, not fail
discard:- : GeneratePrev, T
```

We can now simulate our domain. A generalized positions will be written as a vector of 3 elements (*time, video, Prev*). The initial hybrid state has the generalized position $\mathbf{p}_0 = (0, \emptyset, ())$ (where $\emptyset$ is the initial value indicating that no data is available). The initial hybrid state can be encoded as:

```
domain_state:- : isDomainTime[0]
-holds(selected):- : isDomainTime[0] . -holds(malfunction):- : isDomainTime[0]
```

it is not difficult to see that it is a valid state according to our definitions. The action *selectVideo* executed at time 0 can be encoded as:

```
occurs(selectVideo):- : isActionTime[0]
```

We will assume that *selectVideoAlg* returns two videos: v1 and v2, and that checkQualityAlg succeeds on v1 and fails on v2. For brevity we will omit atoms specifying the signature, i.e. *fluent(inertial, selected)*, *fluent(defined, malfunction)* and *action(selectVideo)*, as these will be derived in every state. Our stable model will then consists of the following hybrid state encodings:

```
* Generalized position: (0, 0, ())
Atoms: -holds(selected). -holds(malfunction). domain_state

* Generalized position: (0.1, 0, Prev=(0))
Atoms: -holds(selected). -holds(malfunction). action_state. discard
   occurs(selectVideo). exec(alg(selectVideo)). valid_action_state. fix_value(video)

* Generalized position: (1, v1, Prev=(0))
Atoms: holds(selected). -holds(malfunction). domain_state. discard

* Generalized position: (1, v2, Prev=(0))
Atoms: holds(selected). holds(malfunction). domain_state. discard
```

4   Conclusions

In this paper we have introduced Hybrid $\mathcal{AL}$ - an extension of the action language $\mathcal{AL}$, that provides a mechanism for specifying both a transition diagram and associated computations for observing fluents and executing actions. This type of processing cannot be done easily with the existing action languages such as $\mathcal{AL}$ or $\mathcal{H}$ [6] without extending them. We think, however that this capability will be useful for improving computational efficiency in applications such as diagnosing malfunctions of large distributed software systems.

While the semantics of $\mathcal{AL}$ is defined using ASP, the semantics of Hybrid $\mathcal{AL}$ is defined using H-ASP - an extension of ASP that allows ASP type rules to control sequential processing of data by external algorithms. Hybrid $\mathcal{AL}$ and H-ASP can be viewed as part of the effort to expand the functionality of ASP to make it more useful along the lines of DLV^{DB} [13], VI [5], GRINGO [8] that allow interactions with the external data repositories and external algorithms. In [12], Redl notes that HEX programs [7] can be viewed as a generalization of these formalisms. Thus we will briefly compare H-ASP and HEX, as target formalisms for translating from a $\mathcal{AL}$-like action language. HEX programs are an extension of ASP programs that allow accessing external data sources and external algorithms via *external atoms*. The external atoms admit input and output variables, which after grounding, take predicate or constant values for the input variables, and constant values for the output variables. Through the external atoms and under the relaxed safety conditions, HEX programs can produce constants that don't appear in the original program.

There is a number of relevant differences between H-ASP and HEX. H-ASP has the ability to pass arbitrary binary information sequentially between external algorithm. While the same can be implemented in HEX, the restriction that the values of the output variables are constants means that a cache that uses constants as tokens, needs to be built around a HEX-based system to retrieve data. H-ASP also explicitly supports sequential processing required for $\mathcal{AL}$-like processing, whereas HEX does not. Such enhancements can also provide the potential for performance optimization. On the other hand, HEX allows constants returned by algorithms to be used as any other constants in the language. This gives HEX expressive advantage over H-ASP where the output of the algorithms cannot be easily used as regular constants.

In [4], the authors have described the use of H-ASP for diagnosing failures of Google's automatic whitelisting system for Dynamic Remarketing Ads, which is an example of a large distributed software system. The approach did not involve constructing a mathematical model of the diagnosed domain. The results of this paper can be viewed as a first step towards developing a solution to the problem of diagnosing malfunctions of a large distributed software system based on constructing a mathematical model of the diagnosed domain. The next steps in this development is to create a software system for Hybrid $\mathcal{AL}$.

Acknowledgments. We would like to thank Michael Gelfond for insightful comments that helped to enhance the paper.

References

1. Balduccini, M., Gelfond, M.: Diagnostic reasoning with a-prolog. TPLP **3**(4–5), 425–461 (2003)
2. Baral, C., Gelfond, M.: Reasoning agents in dynamic domains. In: Logic Based Artificial Intelligence, pp. 257–279. Kluwer Academic Publishers (2000)
3. Brik, A., Remmel, J.B.: Hybrid ASP. In: Gallagher, J.P., Gelfond, M. (eds.) ICLP (Technical Communications), vol. 11 of LIPIcs, pp. 40–50. Schloss Dagstuhl - Leibniz-Zentrum fuer Informatik (2011)
4. Brik, A., Remmel, J.: Diagnosing automatic whitelisting for dynamic remarketing ads using hybrid ASP. In: Calimeri, F., Ianni, G., Truszczynski, M. (eds.) LPNMR 2015. LNCS (LNAI), vol. 9345, pp. 173–185. Springer, Cham (2015). doi:10.1007/978-3-319-23264-5_16
5. Calimeri, F., Cozza, S., Ianni, G.: External sources of knowledge and value invention in logic programming. Ann. Math. Artif. Intell. **50**(3–4), 333–361 (2007)
6. Chintabathina, S., Gelfond, M., Watson, R.: Modeling hybrid domains using process description language. In: Vos, M.D., Provetti, A. (eds.) Answer Set Programming, Advances in Theory and Implementation, Proceedings of the 3rd International of ASP 2005 Workshop, Bath, UK, 27–29 September 2005, vol. 142 of CEUR Workshop Proceedings. CEUR-WS.org (2005)
7. Eiter, T., Ianni, G., Schindlauer, R., Tompits, H.: A uniform integration of higher-order reasoning and external evaluations in answer-set programming. In: Kaelbling, L.P., Saffiotti, A. (eds.) IJCAI 2005, Proceedings of the Nineteenth International Joint Conference on Artificial Intelligence, Edinburgh, Scotland, UK, 30 July–5 August 2005, pp. 90–96. Professional Book Center (2005)
8. Gebser, M., Kaufmann, B., Kaminski, R., Ostrowski, M., Schaub, T., Schneider, M.T.: Potassco: the Potsdam answer set solving collection. AI Commun. **24**(2), 107–124 (2011)
9. Gelfond, M., Inclezan, D.: Some properties of system descriptions of al$_d$. J. Appl. Non-class. Logics **23**(1–2), 105–120 (2013)

10. Gelfond, M., Kahl, Y.: Knowledge Representation, Reasoning, and the Design of Intelligent Agents: The Answer-Set Programming Approach. Cambridge University Press, New York (2014)
11. Gelfond, M., Lifschitz, V.: The stable model semantics for logic programming. In: ICLP/SLP, pp. 1070–1080 (1988)
12. Redl, C.: Answer set programming with external sources: algorithms and efficient evaluation. Ph.D. thesis, Vienna University of Technology (2015)
13. Terracina, G., Leone, N., Lio, V., Panetta, C.: Experimenting with recursive queries in database and logic programming systems. TPLP **8**(2), 129–165 (2008)
14. Verma, A., Pedrosa, L., Korupolu, M., Oppenheimer, D., Tune, E., Wilkes, J.: Large-scale cluster management at google with borg. In: Proceedings of the European Conference on Computer Systems (EuroSys), Bordeaux, France. ACM (2015)

moviola: Interpreting Dynamic Logic Programs via Multi-shot Answer Set Programming

Orkunt Sabuncu[1,2(✉)] and João Leite[2]

[1] TED University, Ankara, Turkey
orkunt.sabuncu@tedu.edu.tr
[2] NOVA LINCS, Universidade Nova de Lisboa, Caparica, Portugal
jleite@fct.unl.pt

Abstract. The causal rejection-based update semantics assign meanings to a Dynamic Logic Program (DLP), which is a sequence of logic programs each one updating the preceding ones. Although there are translations of DLPs under these update semantics to logic programs of Answer Set Programming (ASP), they have not led to efficient and easy to use implementations. This is mainly because such translations aim offline solving in a sense that the resulting logic program is given to an answer set solver to compute models of the current DLP and for any future updates the whole process has to be repeated from scratch. We aim to remedy this situation by utilizing multi-shot ASP, composed of iterative answer set computations of a changing program without restarting from scratch at every step. To this end, we developed a system called *moviola*, utilizing the multi-shot answer set solver *clingo*. Using the system, a user can interactively write a DLP, update it, compute its models according to various semantics on the fly.

1 Introduction

Dynamic knowledge bases incorporate new information that may not only augment the knowledge base, but also contradict with previous information. A DLP represents such knowledge base by a sequence of logic programs, each updates the preceding ones.

There are various causal rejection-based update semantics assign meanings to a DLP. They have been extensively studied and there are transformations of DLPs under these semantics to logic programs of ASP [1]. However, these transformations [5,9,10] have not led to efficient and easy to use implementations. The underlying reason for this is that these transformations foresee an offline solving process, i.e., the process ends after finding models of the input DLP. Hence, whenever a DLP is updated, the whole process of transforming and solving has to be repeated from scratch for the updated DLP. We aim to remedy this situation by utilizing multi-shot ASP, composed of iterative answer set computations of a changing program without killing the solver and restarting from scratch at every step. To this end, we developed a system called *moviola*[1],

---

[1] https://github.com/owizo/moviola.

© Springer International Publishing AG 2017
M. Balduccini and T. Janhunen (Eds.): LPNMR 2017, LNAI 10377, pp. 336–342, 2017.
DOI: 10.1007/978-3-319-61660-5_30

utilizing the multi-shot answer set solver *clingo*. Using the system, a user can interactively write a DLP, update it, compute its models according to various semantics on the fly.

In multi-shot ASP, it is not allowed to join two programs both having definitions of an atom (i.e., each program has a rule that has the same atom in the head). This is a problem since the solver fixes necessary conditions for an atom to be true considering the rules defining it in one program and these conditions cannot be altered when we want to join the other program. This issue becomes an obstacle when encoding update semantics of DLPs in multi-shot ASP. We overcome this obstacle by using a technique called *chaining*, which involves binding redefinitions of an atom by rule chains. Similar techniques have been used in other contexts [2,3].

The lack of modern implementations of update semantics of DLPs builds up a barrier to utilize DLP in various application domains. We hope *moviola* helps to bridge the gap between theory and practice of update semantics of DLPs.

2   Preliminaries

A DLP is a finite sequence of ground non-disjunctive logic programs (denoted by $\langle \mathcal{P}_i \rangle_{i<n}$), each one updating the preceding ones. Unlike normal logic programs, programs in a DLP may include rules having default negation in the heads. Although default negation in the head can be compiled away to form a normal logic program and it does not increase the expressive power of the program [4], it plays an important role in DLPs by facilitating updates with contradictory knowledge.

The causal rejection-based update semantics utilize the principle that a rule should be rejected when a more recent contradictory rule appears for assigning meanings to a DLP. Among this class of semantics are justified update (JU; [5]), update answer set (AS; [6]), dynamic stable models (DS; [7]), and refined dynamic stable models (RD; [8]) semantics. In this section we cover JU and RD semantics.

Two rules r and k are *in conflict*, denoted by $r \bowtie k$, when they have complementary head literals.[2] The set $all(\boldsymbol{P})$ is composed of all rules belonging to the programs in $\boldsymbol{P}$.

Definition 1 (JU-model). *Let $\boldsymbol{P} = \langle \mathcal{P}_i \rangle_{i<n}$ be a DLP over a set $\mathcal{A}$ of propositional atoms and $J \subseteq \mathcal{A}$ be an ASP interpretation. The set of rejected rules is defined as*

$$\mathsf{rej}_{\mathsf{JU}}(\boldsymbol{P}, J) = \{r \in \mathcal{P}_i \mid \exists j \exists k : i < j < n \ \wedge \ k \in \mathcal{P}_j \ \wedge \ r \bowtie k \ \wedge \ J \models B_k\},$$

where B_k is the body of k. J is a JU-model of $\boldsymbol{P}$ iff J is an answer set of the program $all(\boldsymbol{P}) \setminus \mathsf{rej}_{\mathsf{JU}}(\boldsymbol{P}, J)$.

---

[2] In this work, we assume that strong negation of atoms do not appear in logic programs of a DLP (refer to [9] for expansion of a DLP to make rule conflicts uniform).

Definition 2 (RD-model). *Let $P = \langle P_i \rangle_{i<n}$ be a DLP over a set $\mathcal{A}$ of propositional atoms and $J \subseteq \mathcal{A}$ be an ASP interpretation. The set of rejected rules is defined as*

$$\mathsf{rej}_{\mathsf{RD}}(P, J) = \{r \in P_i \mid \exists j \exists k : i \leq j < n \,\wedge\, k \in P_j \,\wedge\, r \bowtie k \,\wedge\, J \models B_k\},$$

where B_k is the body of k, and the set of default assumptions is defined as

$$\mathsf{def}(P, J) = \{\sim l \mid l \in \mathcal{A} \,\wedge\, \neg \exists r \in all(P) : (h = l \,\wedge\, J \models B_r)\},$$

where r is of the form $h \leftarrow B_r$. J is a RD-model of P iff $J' = least([all(P) \setminus \mathsf{rej}_{\mathsf{RD}}(P, J)] \cup \mathsf{def}(P, J))$ where $least(X)$ denotes the least model of program X with all literals treated as positive atoms and $J' = J \cup \sim(\mathcal{A} \setminus J)$.

Example 1. Let $P = \langle \{p.\}, \{\sim p \leftarrow \sim p.\} \rangle$ be a DLP. Observe that $M_1 = \{p\}$ is both a JU-model and a RD-model of P. Considering $M_2 = \{\}$, the only rule of the first program in P is in $\mathsf{rej}_{\mathsf{RD}}(P, M_2)$ and the default assumption $\sim p$ is not included in $\mathsf{def}(P, M_2)$. Thus, M_2 is not a RD-model given that the least model of the corresponding program does not satisfy the condition in RD semantics. However, M_2 is a JU-model of P although it is unintended (the second program P is a tautology and it is not expected to change models of the DLP before the update).

3   Encoding Dynamic Logic Programs via Answer Set Programming

We developed a translation that encodes RD semantics of DLPs via traditional (i.e., single-shot) ASP.[3] This translation establishes the foundation for the multi-shot ASP encoding used by *moviola*. It is in principle similar to the translation defined in [10], but it is developed in anticipation of its extension to a multi-shot encoding. Later in this section, we present the core of this multi-shot encoding.

Single-shot ASP encoding. Let $\mathcal{A}$ be a set of propositional atoms. We define $\mathcal{A}^n$ and $\mathcal{A}^c$ as the sets $\{a^n \mid a \in \mathcal{A}\}$ and $\{c_a \mid a \in \mathcal{A}\}$ of new propositional atoms, respectively. For a literal l, the transformed literal l^n is equal to p^n if $l = \sim p$ and to p if $l = p$ where $p \in \mathcal{A}$ and $p^n \in \mathcal{A}^n$. The transformation extends to sets of literals, rules, and programs, i.e., $B^n = \{l^n \mid l \in B\}$, r^n is $h^n \leftarrow B^n$. given a rule r of the form $h \leftarrow B$., and $P^n = \{r^n \mid r \in P\}$ given a program P. Let $P = \langle P_i \rangle_{i<n}$ be a DLP over a set $\mathcal{A}$ of propositional atoms. For each rule $r \in all(P)$, d_r is a new propositional atom (considering r as an id of the rule) and $\mathcal{R} = \{d_r \mid r \in all(P)\}$. For a rule $r \in all(P)$ of the form $h \leftarrow B$., the transformed rule r^d is $h \leftarrow B, \sim d_r$. where $d_r \in \mathcal{R}$. Given a program P, $P^d = \{r^d \mid r \in P\}$. Additionally, we define $\mathcal{A}^- = \{a \mid \sim a \text{ occurs in } P\}$.

The role of the transformation $(.)^n$ is to represent negative literals by a new positive atom. This is needed considering that an atom in $\mathcal{A}$ may have no

---

[3] Here, we explain the encoding for only RD semantics due to space constraints.

default assumption due to a rule with a satisfied body and the nature of *least* handling negative literals as positive ones. This technique is also applied in a similar translation [10] of RD semantics via ASP. Additionally, $c_p \in \mathcal{A}^c$ atom intuitively encodes conditions of when generation of default assumption for p must be avoided.

Next, we will define some program parts that will be utilized to assemble the transformed logic program encoding RD update semantics of DLPs.

Definition 3. *The base logic program* $\mathsf{B}(\boldsymbol{P})$*, rejection logic program* $\mathsf{R_{RD}}(\boldsymbol{P})$*, and defaults logic program* $\mathsf{D}(\boldsymbol{P})$ *are defined as:* $\mathsf{B}(\boldsymbol{P}) = \{(\mathcal{P}_i{}^n)^d | i < n\}$,

$$\mathsf{R_{RD}}(\boldsymbol{P}) = \{d_r \leftarrow B_k{}^n. \mid r \in \mathcal{P}_i, d_r \in \mathcal{R}, \exists j \exists k : i \leq j < n \wedge k \in \mathcal{P}_j \wedge r \bowtie k \wedge$$
$$B_k \text{ is the body of } k\},$$

$$\mathsf{D}(\boldsymbol{P}) = \{p^n \leftarrow \sim p, \sim c_p. \mid p \in \mathcal{A}^-\} \cup$$
$$\{\leftarrow p, p^n. \mid p \in \mathcal{A}^-\} \cup \{\leftarrow \sim p, \sim p^n. \mid p \in \mathcal{A}^-\} \cup$$
$$\{c_p \leftarrow B^n. \mid r \in all(\boldsymbol{P}) \text{ is a rule of the form } p \leftarrow B. \text{ and } p \in \mathcal{A}\}.$$

Lemma 1. *Let* $\boldsymbol{P} = \langle \mathcal{P}_i \rangle_{i<n}$ *be a DLP and* $R = \mathsf{B}(\boldsymbol{P}) \cup \mathsf{R_{RD}}(\boldsymbol{P}) \cup \mathsf{D}(\boldsymbol{P})$ *be the transformed program.* J *is a* RD*-model of* $\boldsymbol{P}$ *iff* J' *is an answer set of* R *s.t.* $J = J' \cap \mathcal{A}$.

Towards a multi-shot encoding. We explain the multi-shot ASP encoding used by *moviola* with an example DLP that is updated iteratively. Let $\mathcal{P}_1 = \{p \leftarrow \sim q.\}$ be the first logic program of a DLP $\langle \mathcal{P}_1 \rangle$ over a set $\mathcal{A}$ of propositional atoms. The following logic program is formed using the transformation defined in Definition 3 and captures the RD semantics of $\langle \mathcal{P}_1 \rangle$. Recall that $q^n \in \mathcal{A}^n$, c_p, $c_q \in \mathcal{A}^c$, $q \in \mathcal{A}^-$, and $d_1 \in \mathcal{R}$ (i.e., the identifier of the only rule of $\mathcal{P}_1$ is 1).

$$p \leftarrow q^n, \sim d_1. \text{ (a)} \qquad c_p \leftarrow q^n. \text{ (b)} \qquad q^n \leftarrow \sim q, \sim c_q. \text{ (c)} \qquad (1)$$
$$\leftarrow q, q^n. \qquad\qquad\qquad \leftarrow \sim q, \sim q^n. \qquad\qquad\qquad\qquad (2)$$

Its only answer set $\{p, q^n, c_p\}$ corresponds to the only RD-model $\{p\}$ of the DLP.

Considering $\langle \mathcal{P}_1 \rangle$, *moviola* does not use rules (1–2) naturally, but it generates a multi-shot program that is based on these rules. It uses a technique called *chaining*, which is an effective remedy for the problem of redefinitions in multishot ASP. During the transformation process of rules defining an atom, we need an additional auxiliary *chain rule* and a *chain atom* that anticipate a future update program having a rule that has the same atom in the head. The chain formed by the chain rule is at first *open*, since it will be *closed* later by a rule defining the corresponding chain atom in a transformed future update program. To this end, we define *tagged* versions of atoms used in the traditional ASP program. Regarding the atom p in rule (1.a), for instance, the tagged atom $^0 p$ is used in rule (3.a) to encode the only rule in $\mathcal{P}_1$ and $^1 p$ is used as a chain atom to encode the corresponding chain rule (3.b) for p.

$$^0 p \leftarrow q^n, \sim d_1. \text{ (a)} \qquad\quad ^0 p \leftarrow {}^1 p. \text{ (b)} \qquad\quad p \leftarrow {}^0 p. \text{ (c)} \qquad (3)$$

Formally, given a DLP $\langle \mathcal{P}_i \rangle_{i<n}$ and an atom a which is first defined in $\mathcal{P}_i$, the transformation uses $^{i-1}a^n$ for the rules defining a in $\mathcal{P}_i$ and $^ia^n$ for a's chain. Note that *moviola* uses non-tagged atoms in bodies of transformed rules corresponding to the rules of input DLP (for instance, q^n in rule (3.a)). Thus, whenever an atom is first defined in a DLP, a rule (for instance (3.c)) is added to define the non-tagged version of the atom.

Furthermore, in addition to atoms in $\mathcal{A}$ and $\mathcal{A}^n$, atoms in $\mathcal{A}^c$ and $\mathcal{R}$ also need chaining since future update programs may cause redefinitions of these atoms. Considering the example DLP $\langle \mathcal{P}_1 \rangle$, *moviola* generates the following rules in addition to rules (3).

$$^0q^n \leftarrow \sim q, \sim c_q. \qquad ^0q^n \leftarrow {}^1q^n. \qquad\qquad q^n \leftarrow {}^0q^n. \qquad (4)$$
$$^0c_p \leftarrow q^n. \qquad\qquad ^0c_p \leftarrow {}^1c_p. \qquad\qquad c_p \leftarrow {}^0c_p. \qquad (5)$$
$$\leftarrow q, q^n. \quad \text{(a)} \qquad \leftarrow \sim q, \sim q^n. \quad \text{(b)} \qquad d_1 \leftarrow {}^1d_1. \quad \text{(c)} \qquad (6)$$

The only answer set $S_1 = \{p, {}^0p, q^n, {}^0q^n, c_p, {}^0c_p\}$ of the multi-shot ASP program composed of rules (3–6) corresponds to the only RD-model of $\langle \mathcal{P}_1 \rangle$.

Next, let us update $\langle \mathcal{P}_1 \rangle$ with the program $\mathcal{P}_2 = \{p \leftarrow x.\}$. Considering the updated DLP $\langle \mathcal{P}_1, \mathcal{P}_2 \rangle$, *moviola* generates a program composed of the following rules and joins it with the previous program using *clingo*. Observe that redefinition of p via the only rule of $\mathcal{P}_2$ closes the open chain through atom ^1p in rule (7.a). Moreover, the new chain rule (7.b) encodes provision for a new chain that may be closed via future updates including a rule having p in the head. Similar chaining for c_p is achieved by rules (8).

$$^1p \leftarrow x, \sim d_2. \quad \text{(a)} \qquad ^1p \leftarrow {}^2p. \quad \text{(b)} \qquad d_2 \leftarrow {}^2d_2. \quad \text{(c)} \qquad (7)$$
$$^1c_p \leftarrow x. \qquad\qquad\qquad ^1c_p \leftarrow {}^2c_p. \qquad\qquad\qquad\qquad\qquad (8)$$

For the joined program, the multi-shot ASP solver of *moviola* computes S_1 again as the only answer set that corresponds to the only RD-model of $\langle \mathcal{P}_1, \mathcal{P}_2 \rangle$.

Recall that atoms in $\mathcal{R}$ may have redefinitions via future updates and for our running example, rules (6.c) and (7.c) already form new open chains. To illustrate encoding of rejecting rules in a multi-shot way, consider the update program $\mathcal{P}_3 = \{\sim p \leftarrow \sim p.\}$ and the resulting DLP $\langle \mathcal{P}_1, \mathcal{P}_2, \mathcal{P}_3 \rangle$. The only rule of $\mathcal{P}_3$ is in conflict with rules of $\mathcal{P}_1$ and $\mathcal{P}_2$. Hence, there must be rules that define d_1 and d_2, and may cause rules of $\mathcal{P}_1$ and $\mathcal{P}_2$ to be rejected. Considering the open chains via atoms $^1d_1$ and $^2d_2$, *moviola* generates the following rules.

$$^1d_1 \leftarrow p^n. \quad \text{(a)} \qquad ^2d_2 \leftarrow p^n. \quad \text{(b)} \qquad ^1d_1 \leftarrow {}^3d_1. \quad \text{(c)} \qquad ^2d_2 \leftarrow {}^3d_2. \quad \text{(d)} \qquad (9)$$

Since there may be future updates having rules that are in conflict with rules of $\mathcal{P}_1$ and $\mathcal{P}_2$, rules (9.c) and (9.d) generate new open chains.

In addition to rules (9), the following rules are generated when update $\mathcal{P}_3$ arrives.

$$^2p^n \leftarrow p^n, \sim d_3. \qquad ^2p^n \leftarrow {}^3p^n. \qquad ^2p^n \leftarrow \sim p, \sim c_p. \qquad d_3 \leftarrow {}^3d_3. \qquad (10)$$
$$p^n \leftarrow {}^2p^n. \qquad\qquad \leftarrow p, p^n. \qquad\qquad \leftarrow \sim p, \sim p^n. \qquad\qquad\qquad (11)$$

clingo computes the answer set S_1 again as the only answer set that corresponds to the only RD-model of $\langle \mathcal{P}_1, \mathcal{P}_2, \mathcal{P}_3 \rangle$.

4   Implementation

We implemented an interactive system called *moviola*, utilizing the multi-shot answer set solver *clingo* (version 5.1) and its Python scripting support. *moviola* is composed of a controller part that forms the interaction with the user and is written in Python, and an ASP meta-encoding capturing various update semantics of DLPs. The system is available online at a Github repository. (see footnote 3)

When a user enters an update logic program, *moviola* converts it to set of ASP facts. Later, these facts are fed into *clingo* with the meta-encoding [11] that practically implements the multi-shot ASP based transformation. The meta-encoding represents not just one (RD) semantics but all the causal rejection-based semantics. This is achieved by adding switch atoms that control which one of the semantics is active. This leads to a useful feature of *moviola* that the user may change the semantics anytime and investigate their differences. The reader may refer to its repository for demonstration of this feature and the full meta-encoding. (see footnote 3)

One aspect of the multi-shot ASP encoding is that definitions of chain atoms may come in a program joined later on. Hence, the grounder of *clingo* simplifies these chain atoms and their respective rules. To prohibit this, chain atoms are declared as external atoms [12]. Another aspect in multi-shot ASP is that the underlying module theory does not allow positive loops spanning over multiple programs that are joined [13]. For some DLPs, this situation may occur in the transformed encoding. To avoid unsound answers due to these undetected loops, we utilize the acyclicity theory feature of *clingo* [14].

5   Conclusion

We developed *moviola*, an interactive system that encodes various update semantics of DLPs via multi-shot ASP. It interprets DLPs using the multi-shot ASP solver *clingo*. Unlike previous implementations, *moviola* does not do redundant work by restarting computation from scratch at every step of update.

One important future work is to present the formalization of the transformation used by *moviola* and to provide its correctness. Although we have tested *moviola* with various small DLPs having theoretical importance, we have not conducted experiments involving large DLPs. Consequently, performing extensive empirical analysis is another important line of future research.

Acknowledgments. This work was partially supported by FCT under strategic project UID/CEC/04516/2013.

References

1. Baral, C.: Knowledge Representation, Reasoning and Declarative Problem Solving. Cambridge University Press, New York (2003)
2. De Cat, B., Denecker, M., Stuckey, P., Bruynooghe, M.: Lazy model expansion: interleaving grounding with search. J. Artif. Intell. Res. **52**, 235–286 (2015)
3. Gebser, M., Janhunen, T., Jost, H., Kaminski, R., Schaub, T.: ASP solving for expanding universes. In: Calimeri, F., Ianni, G., Truszczynski, M. (eds.) LPNMR 2015. LNCS (LNAI), vol. 9345, pp. 354–367. Springer, Cham (2015). doi:10.1007/978-3-319-23264-5_30
4. Janhunen, T.: On the effect of default negation on the expressiveness of disjunctive rules. In: Eiter, T., Faber, W., Truszczyński, M. (eds.) LPNMR 2001. LNCS (LNAI), vol. 2173, pp. 93–106. Springer, Heidelberg (2001). doi:10.1007/3-540-45402-0_7
5. Leite, J.A., Pereira, L.M.: Generalizing updates: from models to programs. In: Dix, J., Pereira, L.M., Przymusinski, T.C. (eds.) LPKR 1997. LNCS, vol. 1471, pp. 224–246. Springer, Heidelberg (1998). doi:10.1007/BFb0054796
6. Eiter, T., Fink, M., Sabbatini, G., Tompits, H.: On properties of update sequences based on causal rejection. Theor. Pract. Logic Program. **2**(6), 711–767 (2002)
7. Alferes, J.J., Leite, J., Pereira, L., Przymusinska, H., Przymusinski, T.: Dynamic updates of non-monotonic knowledge bases. J. Logic Program. **45**(1–3), 43–70 (2000)
8. Alferes, J.J., Banti, F., Brogi, A., Leite, J.A.: The refined extension principle for semantics of dynamic logic programming. Stud. Logica **79**(1), 7–32 (2005)
9. Slota, M., Leite, J.: On condensing a sequence of updates in answer-set programming. In: Proceedings of the 23rd International Joint Conference on Artificial Intelligence (2013)
10. Banti, F., Alferes, J.J., Brogi, A.: Operational semantics for DyLPs. In: Bento, C., Cardoso, A., Dias, G. (eds.) EPIA 2005. LNCS, vol. 3808, pp. 43–54. Springer, Heidelberg (2005). doi:10.1007/11595014_5
11. Gebser, M., Kaminski, R., Schaub, T.: Complex optimization in answer set programming. Theor. Pract. Logic Program. **11**(4–5), 821–839 (2011)
12. Gebser, M., Kaminski, R., Kaufmann, B., Schaub, T.: *Clingo* = ASP + control: preliminary report. In: Technical Communication of the 30th International Conference on Logic Programming (2014)
13. Oikarinen, E.: Modular answer set programming. In: Dahl, V., Niemelä, I. (eds.) ICLP 2007. LNCS, vol. 4670, pp. 462–463. Springer, Heidelberg (2007). doi:10.1007/978-3-540-74610-2_46
14. Bomanson, J., Gebser, M., Janhunen, T., Kaufmann, B., Schaub, T.: Answer set programming modulo acyclicity. Fund. Inform. **147**(1), 63–91 (2016)

Adjudication of Coreference Annotations via Answer Set Optimization

Peter Schüller[⊠]

Computer Engineering Department, Faculty of Engineering,
Marmara University, Istanbul, Turkey
peter.schuller@marmara.edu.tr

Abstract. We describe the first automatic approach for merging coreference annotations obtained from multiple annotators into a single gold standard. Merging is subject to hard constraints (consistency) and optimization criteria (minimal divergence from annotators) and involves an equivalence relation over a large number of elements. We describe two representations of the problem in Answer Set Programming and four objective functions suitable for the task. We provide two structurally different real-world benchmark datasets based on the METU-Sabanci Turkish Treebank, and we report our experiences in using the Gringo, Clasp, and Wasp tools for computing optimal adjudication results on these datasets.

Keywords: Coreference resolution · Adjudication · Answer set programming

1 Introduction

Coreference Resolution [10,19,20,23,25] is the task of finding phrases in a text that refer to the same real-world entity. Coreference is commonly annotated by marking subsequences of tokens in the input text as *mentions* and putting sets of mentions into *chains* such that all mentions in a chain refer to the same, clearly identifiable entity in the world. For example in the text *"John is a musician. He played a new song. A girl was listening to the song. 'It is my favorite,' John said to her."* [21] we can identify the following mentions.

[John] $^{(i)}$ is [a musician] $^{(ii)}$. [He] $^{(iii)}$ played [a new song] $^{(iv)}$.
[A girl] $^{(v)}$ was listening to [the song] $^{(vi)}$.
"[It] $^{(vii)}$ is [[my] $^{(ix)}$ favorite] $^{(viii)}$," [John] $^{(x)}$ said to [her] $^{(xi)}$.

Roman superscripts denote mention IDs, chains in this text are as follows: {(i), (iii), (ix), (x)} (John, He, my, John); {(iv), (vi), (vii)} (a new song, the song, It); and {(v), (xi)} (A girl, her), where roman numbers again refer to mention IDs.[1]

This work has been supported by The Scientific and Technological Research Council of Turkey (TUBITAK) under grants 114E430 and 114E777.

[1] Mention pairs $(i)/(ii)$ and $(vii)/(viii)$ are in a predicative relationship and therefore (per convention) not considered coreferent.

© Springer International Publishing AG 2017
M. Balduccini and T. Janhunen (Eds.): LPNMR 2017, LNAI 10377, pp. 343–357, 2017.
DOI: 10.1007/978-3-319-61660-5_31

For building and testing automatic coreference resolution methods, annotated corpora, i.e., texts with mention and chain annotations, are an important resource.

Adjudication is the task of combining mention and chain information from several human annotators into one single "gold standard" corpus. These annotations are often mutually conflicting, and resolving these conflicts is a task that is *global* on the document level, i.e., it is not possible to decide the truth of the annotation of one token, mention, or chain, without considering other tokens, mentions, and chains in the same document.

We here present results and experiences obtained in a two-year project for creating a Turkish coreference corpus based on the METU-Sabanci Turkish Treebank [26]. For creating this coreference corpus, we collected a total of 475 annotations for 33 distinct documents over two separate annotation cycles. Merging such annotations manually to create a gold standard is a tedious task. With sometimes more than 10 distinct annotations per document in our datasets, tool support is necessary for the task, however only manual adjudication tools exist. Hence we developed a (semi)automatic solution based on Answer Set Programming (ASP) [6,7,16,22], a logic programming and knowledge representation paradigm that allows for a declarative specification of problems and is suitable for solving large-scale combinatorial optimization problems.

Our contributions are as follows.

- We formalize the *problem* of coreference adjudication, introduce four *objective functions* that have practical relevance for both our datasets, and describe *input and solution representations* of the problem in our application in Sect. 3.
- We provide two *ASP encodings* in Sect. 4: MM (mention-mention) explicitly represents the transitive closure of the chain equivalence relation, while CM (chain-mention) avoids this explicit representation.
- We describe and provide two real-life *datasets*,[2] outline their properties and differences, and report on *experiments* with unsatisfiable-core optimization and stratification using the tools Gringo [17], Clasp [18], and Wasp [3] in Sect. 5.
- Finally, in Sect. 6, we reflect on *insights* about developing ASP applications, analyzing bottlenecks in such applications, and specific issues with optimization.

Our tool is the first automatic tool for coreference adjudication, and our datasets are the first published datasets for automatic adjudication, because usually only the final result of adjudication (i.e., the gold standard corpus) gets published. The approach is not specific to Turkish, works well in practice, and we have used it as a basis for the publicly available CaspR tool[3] for performing (semi-) automatic adjudication of coreference annotations in CoNLL format.

The computational problem of finding minimal repairs for inconsistent annotations is related to finding minimal repairs for databases [9] or ontologies [13], and to managing inconsistency in multi-context systems [12]. These related works

---

[2] https://bitbucket.org/knowlp/asp-coreference-benchmark.
[3] https://bitbucket.org/knowlp/caspr-coreference-tool.

aim to find a minimal change in the system that makes it globally consistent, which is similar to merging mutually inconsistent coreference annotations. Our work and these applications have in common, that a change that fixes one inconsistency, might introduce another one.

ASP encodings for transitivity are present in many applications. In particular ASP encodings for acyclicity properties which make use of transitive closure have been studied by Gebser et al. [15] who report effects of tightness similar as in our experiments. Different from Gebser et al. we consider optimization problems and compare different ASP solvers and several optimization algorithms.

Note that we perform adjudication of coreference annotations and not coreference resolution, i.e., we do not aim to predict coreference annotations on a given text. Denis and Baldridge described a system based on Integer Linear Programming [11] which performs coreference resolution and classification of named entities, and includes transitivity of mention-mention links, similar to our ASP encodings.

2   Preliminaries

2.1   Coreference Resolution

Coreference resolution is the task of finding phrases in a text that refer to the same entity [10,19,20,23,25]. We call such phrases *mentions*, and we call a group of mentions that refers to one entity *chains*. Formally we can describe mention detection and coreference resolution as follows. Given a document D which is a sequence of tokens $w_1, \ldots, w_n$, mention detection is the task of finding a set $M = \{(f_1, t_1), \ldots, (f_m, t_m)\}$ of mentions, where a mention (f, t) is a pair of indexes, $1 \leq f \leq t \leq n$, such that the span of the mention goes from token index f to token index t in D. Given a set M of mentions, coreference resolution is the task of partitioning M into a set of chains P such that all sequences of tokens $w_{f_i}, \ldots, w_{t_i}$ in all mentions (f_i, t_i) in one chain refer to the same entity.

Example 1 (ctd.). We have 31 tokens (including punctuation). Some of the tokens are $w_1 = $ 'John', $w_2 = $ 'is', ..., $w_{30} = $ 'her', $w_{31} = $ '.', the set of mentions is $M = \{(1, 1), (3, 4), (6, 6), \ldots, (30, 30)\}$, the correct chains are $P = \{\{(1, 1), (6, 6), (23, 23), (27, 27)\}, \ldots, \{(12, 13), (30, 30)\}\}$, where $(12, 13)$ represents 'A girl' and $(30, 30)$ represents 'her'. □

Mentions can be part of other mentions, but in that case they (usually) cannot be coreferent. Moreover mentions are (usually) phrases, therefore if mentions m and m' are overlapping, then m is either properly contained in m' or vice versa.

Given a set of mentions, there are exponentially many potential solutions to the coreference resolution problem, and finding a globally optimal solution is NP-hard according to most measures of coreference optimality [29].

2.2 Answer Set Programming

ASP is a logic programming paradigm which is suitable for knowledge representation and for finding solutions for computationally (NP-)hard problems [6,7,16,22]. An ASP program contains *rules* of the form

```
Head :- Body.
```

`Head` is a first-order atom, a *choice* construction, or empty, and `Body` is a conjunction of first-order atoms. Intuitively, a rule makes the head logically true if the body is satisfied. Rules without body are *facts* (the head is always true) and rules without head are *constraints* (if the body is satisfied, the candidate solution is discarded). Choices of the form `L { A1 ; ... ; AN } U` generate all solution candidates where between L and U atoms from the set $\{A1, \ldots, AN\}$ are true. *Weak constraints* of the form

```
:~ Body. [Cost@1,Tuple]
```

define objectives for combinatorial optimization: a weak constraint incurs cost `Cost` for each unique tuple `Tuple` where `Body` is satisfied in the answer set.

For details of syntax and semantics we refer to the ASP-Core-2 standard [8].

3 Automatic Coreference Adjudication

Coreference adjudication can be formalized as follows.

> Given a document D of tokens and $u \geq 2$ partitions $P_1, \ldots, P_u$ of mentions in D, we search for a partitioning P of M that minimally differs from $P_1, \ldots, P_u$.

Clearly, annotations might be contradictory and we need to ensure certain structural constraints in the solution, moreover it is not immediately apparent what 'minimally' means, and if it is defined then there could be multiple solutions of the same quality.

A solution might also contain equivalences between mentions that are not present in any annotation. This is because chains are equivalence relations, and if we merge equivalence relations that are not subsets of each other, the new equivalence relation is the reflexive, symmetric, and transitive closure of the original relations.

Example 2 (ctd.). Assume that in parallel to chain $\{(i), (iii), (ix), (x)\}$ in the Introduction, we obtain (from another annotator) a chain $\{(i), (ii), (x)\}$. If we merge these chains naively by merging the sets, we obtain a single chain $\{(i), (ii), (iii), (ix), (x)\}$ although no annotator indicated that (ii) and (iii) belong to the same chain.   □

In the following we consider input and output chains as links between mentions (mention-mention links) and describe objective functions for selecting preferred solutions.

```
1    % Links implicitly given by annotators.
2    link(Ann,M1,M2) :- cm(Ann,C,M1), cm(Ann,C,M2), M1 < M2.

3    % Guess which links to use.
4    { uselink(Ann,M1,M2) } :- link(Ann,M1,M2).

5    % Represent canonical mentions and links between them.
6    clink(MID1,MID2) :- uselink(Ann,M1,M2),
7      mention(Ann,M1,From1,To1), mention(Ann,M2,From2,To2),
8      MID1=mid(From1,To1), MID2=mid(From2,To2), MID1 < MID2.
9    clink(MID1,MID2) :- uselink(Ann,M2,M1),
10     mention(Ann,M1,From1,To1), mention(Ann,M2,From2,To2),
11     MID1=mid(From1,To1), MID2=mid(From2,To2), MID1 < MID2.

12   % Reflexive symmetric transitive closure of clink/2.
13   cc(X,Y) :- clink(X,Y).
14   cc(X,Y) :- cc(Y,X).
15   cc(X,Z) :- cc(X,Y), cc(Y,Z).

16   % Smallest mention in one SCC of cc/2 becomes representative of the chain.
17   notrepresentative(Y) :- cc(X,Y), X < Y.
18   resultcm(X,Y) :- cc(X,Y), not notrepresentative(X).
19   resultchain(C) :- resultcm(C,_).

20   % Require more than one mention in a chain.
21   :- resultchain(C), #count { Y : resultcm(C,Y) } <= 1.
22   % Forbid a mention within another mention in the same chain.
23   :- resultcm(C,mid(Fr1,To1)), resultcm(C,mid(Fr2,To2)),
24      Fr1 <= Fr2, To2 <= To1, (Fr1,To1) != (Fr2,To2).
```

```
25   % [V, U, VA, UA] Cost for not using a link, corresponding to annotations.
26   :~ not clink(M1,M2), cmomitcost(M1,M2,Cost). [Cost@1,M1,M2,omit]
27   % [V,    VA   ] Cost for using a link, corresponding to non-annotations.
28   :~ clink(M1,M2), cmusecost(M1,M2,Cost). [Cost@1,M1,M2,use]
29   % [V, U       ] Forbid creating an implicit link that was not annotated.
30   :- cc(X,Y), X < Y, not clink(X,Y).
31   % [    VA, UA] Minimize number of implicit links that were not annotated.
32   :~ cc(X,Y), X < Y, not clink(X,Y). [1@1,X,Y]
```

Fig. 1. MM encoding and objective function variations.

3.1   Objective Functions

We build a preference relation where we incur cost under the following conditions.

(i) Omitting a mention-mention link provided by an annotator.
(ii) Using a mention-mention link of an annotator, where a number of annotators did not provide the same link.
(iii) Putting two mentions into a chain where no annotator gave any evidence for this.

We incur separate cost for each annotator who provided a link in (i), and for each annotator who did not provide a link that was used in (ii).

Concretely, we use the following objective functions in our application:

[V] Cost 2 for (i), cost 1 for (ii), and (iii) is a hard constraint (cost ∞).
[U] Cost 2 for (i), no cost for (ii), and (iii) is a hard constraint (cost ∞).
[VA] Cost 2 for (i), cost 1 for (ii), and cost 1 for (iii).
[UA] Cost 2 for (i), no cost for (ii), and cost 1 for (iii).

Note that cost for (ii) is higher than for (i) because we observed that annotators are more likely to miss links than to add spurious links. Moreover, using cost 1 in [U] would yield the same preference, but using cost 2 permits a more uniform ASP encoding.

Intuitively, V indicates "voting" (a link given by only one annotator can be dismissed if many other annotators do not give the same link) while U indicates "use as many mentions as possible" (no cost for (ii)), and A indicates that additional links are allowed at a cost. The main idea of these objectives is to use given information optimally to produce an overall consistent solution.

Our objectives are motivated by properties of our datasets: if mentions are given, annotators only disagree on assignment of mentions to chains, and in this case objectives [V] and [VA] make sure that the result reflects the opinion of the majority of annotators. Contrarily, if mentions are not given, annotators often disagree on mentions, hence [V] and [VA] would eliminate most mentions completely; in this case [U] and [UA] are useful. The BLANC [24] coreference resolution evaluation measure is based on counting existing and non-existing mention-mention links, similar as we do in [V] and [VA].

3.2 Input and Output ASP Representation

Given annotator inputs $P_1, \ldots, P_u$ where each chain $C_c \in P_a$ is a set of mentions of form (f_j, t_j), we represent each mention j in chain d of annotator a as a fact $cm(a, c, j)$, and each mention (f_j, t_j) of annotator a as a fact $\mathtt{mention}(a, j, f_j, t_j)$.

Example 3 (ctd.). Tokens in our running example are numbered with integers.

$$\text{John}^1 \text{ is}^2 \text{ a}^3 \text{ musician}^4 .^5 \text{ He}^6 \text{ played}^7 \text{ a}^8 \text{ new}^9 \text{ song}^{10} .^{11}$$

Consider that annotator a1 created chain $c1 = \{a, b, c\}$ containing mentions $a = (1,1)$ for 'John', $b = (3,4)$ for 'a musician', and $c = (6,6)$ for 'He'. Annotator a2 annotated chain $c2 = \{d, e\}$ containing mentions $d = (4,4)$ for 'musician' and $e = (6,6)$ for 'He'. Then these annotations are represented as follows:

```
1   mention(a1,a,1,1).    mention(a1,b,3,4).    mention(a1,c,6,6).
2   mention(a2,d,4,4).    mention(a2,e,6,6).
3   cm(a1,c1,a). cm(a1,c1,b). cm(a1,c1,c). cm(a2,c2,d). cm(a2,c2,e).
```

where c1 and c2 represent the first and second chain, respectively.   □

```
1    % We reuse lines 1-11 of MM encoding for guessing used links.

2    % Obtain canonical mentions from clink/2.
3    cmention(M) :- clink(M,_).  cmention(M) :- clink(_,M).

4    % Number of chains from each annotator and highest number.
5    countchain(Ann,N) :- cm(Ann,_,_), N = #count { C : cm(Ann,C,_) }.
6    maxchain(N) :- N = #max { C : countchain(_,C) }.
7    % Assume we will not need more than this amount of chains.
8    chainlimit(N*6/5) :- maxchain(N).

9    % Guess which chain IDs to use in result.
10   { resultchain(X) : X = 1..Max } :- chainlimit(Max).
11   resultchain(Y) :- resultchain(X), Y = 1..(X-1). % Symmetry breaking.

12   % Guess which canonical mention becomes part of which chain.
13   1 { resultcm(C,M) : resultchain(C) } 1 :- cmention(M).

14   % Synchronize guessed result with clink/2.
15   :- clink(X,Y), resultcm(C,X), not resultcm(C,Y).
16   :- clink(X,Y), not resultcm(C,X), resultcm(C,Y).

17   % We reuse lines 20-24 of MM encoding for structural constraints.
```

```
18   % [V, U, VA, UA] Cost for not using a link, corresponding to annotations.
19   :~ not clink(M1,M2), cmomitcost(M1,M2,Cost). [Cost@1,M1,M2,omit]
20   % [V,    VA  ] Cost for using a link, corresponding to non-annotations.
21   :~ clink(M1,M2), cmusecost(M1,M2,Cost). [Cost@1,M1,M2,use]
22   % [V, U     ] Forbid creating an implicit link that was not annotated.
23   :- resultcm(C,X), resultcm(C,Y), X < Y, not clink(X,Y).
24   % [      VA, UA] Minimize number of implicit links that were not annotated.
25   :~ resultcm(C,X), resultcm(C,Y), X < Y, not clink(X,Y). [1@1,X,Y]
```

Fig. 2. CM encoding and objective function variations.

The answer sets of our logic program represent a set of chains without annotator information, represented as atoms of the form `resultcm(Chain,mid(Fr,To))`
which indicates that in chain `Chain` there is a mention from token `Fr` to token `To`.

Example 4 (ctd.). Assume we merged the annotations of the previous example
into two chains $v = \{(1,1),(3,4)\}$ and $w = \{(4,4),(6,6)\}$, this is represented as
follows.

```
1    resultcm(v,mid(1,1))  resultcm(v,mid(3,4))
2    resultcm(w,mid(4,4))  resultcm(w,mid(6,6))
```

□

4   ASP Encodings

We next provide two ASP encodings, MM and CM, which model coreference
adjudication and objective functions from Sect. 3.1. MM explicitly represents the
transitive closure of mention-mention links, while CM avoids this by assuming
an upper limit on the number of chains and guessing which mention belongs to
which chain.

4.1   MM: Mention-Mention Encoding

Figure 1 shows the MM encoding. The rule in line 2 represents annotated mention-mention links, line 4 guesses whether to use each link, lines 6–11 canonicalize links by removing their annotator information and representing canonical mentions (*cmentions*) by the term mid(F,T) which represents that the mention spans from token F to token T in the document.[4] In lines 13–15 we represent the reflexive, symmetric, and transitive closure of clink/2 in cc/2, which therefore represents result chains as strongly connected components (SCCs) of cmentions. In lines 17–18 we represent each chain by its lexicographically smallest cmention and in line 19 we define that such cmentions represent chains. Constraints in lines 21–24 require that each chain contains at least 2 mentions, and that no chain contains two mentions where one is contained in the other.

The bottom part of Fig. 1 shows constraint variations for realizing objectives using auxiliary predicates cmomitcost/3 and cmusecost/3 that are discussed in Sect. 4.3. Line 26 incurs cost for not using annotators' input ([U], [UA], [V], and [VA]), line 28 incurs cost for using annotators' information ([V] and [VA]), line 30 forbids to put two mentions into a chain if there is no evidence for that ([V] and [U]), and line 32 alternatively incurs a cost for such mention pairs ([VA] and [UA]).

```
1   % Number of annotators and mentions vs canonical mentions.
2   nf(N) :- N = #count { Ann : mention(Ann,_,_,_) }.
3   cfmention(Ann,M,mid(From,To)) :- mention(Ann,M,From,To).

4   % Which noncanonical and canonical mentions are in the same chain.
5   fsamechain(Ann,FM1,FM2) :- cm(Ann,C,FM1), cm(Ann,C,FM2), FM1 != FM2.
6   cmsamechain(M1,M2) :- fsamechain(Ann,FM1,FM2),
7     cfmention(Ann,FM1,M1), cfmention(Ann,FM2,M2), M1 < M2.

8   % Count direct evidence from annotators for M1 and M2 being in same chain.
9   evidence(M1,M2,Ev) :- cmsamechain(M1,M2), Ev > 0, Ev = #count { Ann :
10    fsamechain(Ann,F1,F2), cfmention(Ann,F1,M1), cfmention(Ann,F2,M2),F1<F2 }.

11  % Usage cost 1 for each annotators who did not put M1/M2 into same chain.
12  cmusecost(M1,M2,N-K) :- cmsamechain(M1,M2), evidence(M1,M2,K), N-K>0, nf(N).
13  % Omission cost 2 for each annotator who put M1/M2 into the same chain.
14  cmomitcost(M1,M2,2*K) :- cmsamechain(M1,M2), evidence(M1,M2,K).
```

Fig. 3. Common rules for representing cost of using/omitting annotated mention-mention links.

4.2   CM: Chain-Mention Encoding

In the CM encoding shown in Fig. 2 we directly guess resultcm/2 and do not derive it from clink/2 as in MM. CM reuses lines 1–11 and lines 20–24 from MM for guessing which annotator information to use, and for structural constraints. Rules in line 3 represent cmentions that appear in the canonical links

---

[4] We use two rules because mention and cmention IDs can have different lexicographic order.

obtained from these guesses, in lines 5–8 we find the maximum number of chains from a single annotator and use $\frac{6}{5}$ times that value as an assumption for the maximum amount of chains in the result.[5] In lines 10–13 we guess chain IDs $1, \ldots, n$ that exist in the result, and we guess which cmention belongs to which chain (recall, that in MM we defined this relation as being the lexicographically smallest representative of the SCCs of the transitive closure of clink/2). Constraints in lines 15–16 ensure, that links in clink/2 are fully represented by resultcm/2.

The lower part of Fig. 2 shows objective functions for CM. Lines 18–21 are the same as in MM, however transitivity (weak) constraints in lines 22–25 are different: we have no transitive closure cc/2 at our disposal, so we need to make an explicit join over resultcm/2, to rule out links that are not founded by annotations in uselink/3.

4.3   Common Rules for Adjudication Evidence

Both MM and CM encodings use predicates cmomitcost/3 and cmusecost/3, which are defined deterministically from input facts as shown in Fig. 3. Line 2 gets the number of annotator inputs for relating the weight of one annotator's input compared with all annotators, line 3 joins mentions with cmentions, line 5 represents which mentions are in the same chain, and line 6 does the same for cmentions. Based on this, line 9 accumulates for each pair (M1, M2) of cmentions that are potentially in the same chain, how many annotators actually provided evidence for putting them into the same chain. Finally, lines 12 and 14 accumulate 'usage cost' of 1 for each annotator who did not put both cmentions in such a pair into the same chain, 'omission cost' of 2 for each annotator who put them into the same chain. Omission cost is used in objectives [U], [UA], [V], and [VA], while usage cost is applied only in [V] and [VA].

5   Evaluation

We first describe our datasets and then experimental results.

5.1   Datasets

The datasets that prompted development of this application are based on the METU-Sabanci Turkish Treebank [26]. Table 1 shows the properties of both datasets. Documents were annotated in two distinct annotation cycles: for DS1, annotators had to produce mentions and chains, while for DS2, mentions were given and could be assigned to chains or removed by annotators. This yielded a large amount of cmentions for DS1 (on average 316.7 per document), while DS2 contains fewer cmentions (159.8). DS1 is also smaller, it is based on 21 documents from the corpus while DS2 covers the whole corpus and has more annotations

---

[5] We found that this assumption can safely be used in practice.

Table 1. Dataset properties (real-world dataset based on METU-Sabanci Turkish Treebank).

Dataset	DS1 (**21** instances)			DS2 (**33** instances)		
	min	avg	max	min	avg	max
# Annotators	6	6.5	8	9	10.3	11
# Chains	78	132.7	197	34	294.6	599
# Mentions	247	**596.6**	1289	117	**1561.4**	3897
# Canonical mentions	169	**316.7**	702	17	**159.8**	358
Longest chain	7	33.1	66	6	28.4	70

(10.3) per document, more annotated chains (294.6), and more mentions (1561.4) than DS1.

In practice we observe the following: due to disagreement on cmentions in DS1, adjudication with [V] or [VA] eliminates most data, hence using [U] or [UA] is more useful. On the other hand the 'voting' of [V] and [VA] can utilize the larger amount of annotations per cmention. This yields reasonable results for DS2. DS1 and DS2 are structurally quite different: DS1 has several instances where nearly all mentions are (transitively) connected to all other mentions, while this does not occur in DS2.

5.2   Experiments

Experiments were performed on a computer with 48 GB RAM and two Intel E5-2630 CPUs (total 16 cores) using Debian 8 and at most 7 concurrently running jobs, each job using at most two cores (we did not use tools in multi-threaded mode), and results are averaged over 5 repeated runs. As systems we used Gringo and Clasp from Clingo 5 (git hash 933fce) [17,18] and Wasp 2 (git hash ec8857) [3]. We use Clasp with parameter `--opt-strategy=usc,9` and Wasp with `--enable-disjcores`, i.e., both systems use unsat-core based optimization with stratification and disjoint core preprocessing.

Table 2 shows experimental results. We limited memory usage to 5 GB and time usage to 300 sec, and columns MO (memory out), resp. TO (time out), show the percentage of runs that aborted because of exceeding the memory limit, resp. the time limit. Columns SAT and OPT show the percentage of runs that found some solution and the first optimal solution, respectively, and T and M give average time and memory usage. Columns T_{grd}, Opt, Chc, and Cnf give average instantiation time, optimality of the solution ($\frac{UB-LB}{LB}$), number of choices and number of conflicts for those 44 instances where Gringo+Clasp never exceeded memory or time. Overall we performed runs for 3 systems, 2 encodings, 4 objectives, and both datasets (54 instances), which yields 1296 distinct configurations. If memory or time limits were exceeded, this happened during solving. Note that, due to rounding, some percentages do not add up to 100.

Table 2. Experimental results accumulated on a high level and on the level of concrete use cases.

Accumulation		MO %	TO %	SAT %	OPT %	T sec	M MB	T_{grd} sec	Opt avg	Opt max	Chc #	Cnf #
System	Clingo	5	0	**26**	**70**	96	738	5.7				
	Gringo+Clasp	5	0	**26**	**70**	96	**695**	2.7				
	Gringo+Wasp	8	19	21	53	145	1075	2.6				
Encoding	CM	**2**	0	40	58	137	567	3.7	6.8	368.9	1M	331K
	MM	8	0	11	**82**	54	822	**1.6**	**1.0**	**84.7**	132K	5K
Objective	[U]	2	0	35	63	125	**398**	1.8	1.2	15.2	2M	538K
	[V]	2	0	3	95	**24**	397	1.8	0.0	0.0	177K	22K
	[UA]	7	0	54	39	177	994	3.5	14.4	368.9	606K	99K
	[VA]	7	0	11	81	**58**	990	3.5	0.0	0.6	132K	13K
Scenario	**DS1 / [U]/ CM**	0	0	**100**	0	300	384	2.8	**5.8**	**15.2**	5M	2M
	DS1/ [U] / MM	10	0	5	86	36	806	1.0	0	0	8K	314
	DS1 / [UA] / CM	10	0	90	0	282	1474	8.3	37.1	368.9	1M	338K
	DS1 / [UA] / MM	19	0	38	43	136	1206	1.0	5.9	84.7	458K	38K
	DS2 / [V] / CM	0	0	6	94	35	166	1.6	0.0	0.0	465K	69K
	DS2 / [V] / MM	0	0	0	**100**	**13**	376	1.9	0	0	70K	35
	DS2 / [VA] / CM	0	0	21	79	87	506	3.7	0.0	0.6	323K	39K
	DS2 / [VA] / MM	6	0	15	79	64	1034	2.0	0.0	0.0	84K	134

System Accumulation shows a comparison between Clingo, Gringo+Clasp, and Gringo+Wasp: Wasp clearly has worse performance in this application with respect to memory as well as time. Columns T_{grd}, Opt, Chc, Cnf are not measurable for all runs of Wasp, therefore we omit them for System comparison. Interestingly, Clingo requires more memory than running Gringo+Clasp in a pipe. T_{grd} of Clingo includes preprocessing time (we discuss this issue in Sect. 6). As many Wasp runs exceeded memory, we present only results for Gringo+Clasp in the remaining table.

Encoding Accumulation shows that choosing between MM and CM means a trade-off between time and memory: CM exceeds memory in fewer runs, while MM finds more optimal solutions, moreover solutions of CM are further away from the optimum (on average 6.8) in comparison with MM (on average 1.0).

Objective Accumulation shows that voting-based objectives ([V] and [VA]) are easier to optimize than other objectives, and even suboptimal solutions are often close to optimal. Moreover strict constraints for transitive links ([V] and [U]) require significantly less memory and instantiation time than weak constraints.

Scenario Accumulation shows practically relevant scenarios, accumulated over single datasets: for DS1 non-voting-objectives are practically relevant, while for DS2 the opposite holds. DS2 can be automatically adjudicated with encoding MM and [V] with all optimal solutions in the given time and memory. For DS1 the most feasible configuration is [U] with encoding CM: no run exceeds memory but unfortunately also no optimal solutions are found. In practice, any solution is better than exceeding memory, and by increasing resource limits to 8 GB and 1200 sec we can obtain optimal solutions for [U] with MM and suboptimal solutions for [UA] with CM for all instances of DS1.

To analyze the bottleneck in instantiating both encodings, we have modified Gringo version 4.5 to print the number of instantiations of each non-ground rule.[6] For an instance of average difficulty and the UA objective function, the main instantiation effort of MM encoding is the transitive closure (Fig. 1 line 15, 725K instantiations), while for CM it is the weak constraint for transitivity (Fig. 2 line 25, 5M instantiations). These rules clearly dominate over the next most frequently instantiated rules (12K instances for MM, 112K instances for CM). Although CM instantiates significantly more rules than MM, CM requires less memory and exceeds memory limits less often than MM. A significant difference in the structure of CM and MM encodings is, that CM is tight [14], while encoding MM is not (due to the transitive closure in the guessing part).

6   Conclusion

We have developed an ASP application for automatic adjudication of coreference resolution annotations along with two structurally distinct real-world benchmark datasets. We consider this problem solved (at least for these datasets) and integrated our encodings in a tool for automatic adjudication of CoNLL-format coreference data.

We have learned the following lessons in this project.

6.1   Approximation, Modularity, and Declarativity

The abstract task we solve is quite straightforward. However, to make its computation feasible, we need to resort to approximations (as in assuming a maximum number of chains in the CM encoding), and we have the possibility to 'trade time for space', just as in classical algorithm development (MM vs. CM is such a trade-off). Careful tuning of the encoding is necessary (e.g., realizing lines 15 and 16 in Fig. 2 using an auxiliary atom and a single constraint makes the encoding perform significantly worse) and the need for such tuning makes ASP less declarative than we would expect (or want) it to be. Preliminary encodings [27] used different objective function formulations, however these encodings were not usable in practice without resorting to aggressive approximations that degraded results. Still, the modularity of ASP also facilitates tuning and finding better formulations: our encodings share many rules although their essential way of representing the search space is very different.

---

[6] https://github.com/peschue/clingo/tree/grounder-stats.

6.2   ASP Optimization

Both unsat-core (USC) optimization [4] and stratification [1,5] are essential to the applicability of ASP in this application (obtaining some solution is always more important than optimality). Branch-and-bound (BB) optimization is performing so much worse than USC in this application that we omitted any numbers. In connection with USC optimization we noticed that additional symmetry breaking for the CM encoding (preventing solutions with permutated chain IDs) reduces performance of USC optimization and improves performance of BB (though not enough to make it competitive with USC). Experiments with unsat-core-shrinking of WASP [2] and lazy constraints [28] have not yielded better results than using a purely rule-based encoding.

6.3   Instantiation Issues

Analyzing the number of rules instantiated by non-ground rules can be useful, but it can also be misleading: in this application the encoding instantiating more constraints (CM) requires less memory in search (probably because of tightness of CM). A separate issue is, that using strings for filenames (instead of integers) significantly slows down grounding and also increases memory used during grounding (infeasible results in [27] were partially caused by this). We conclude that using strings in ASP can be a big (and non-obvious) performance issue while at the same time it can be imagined that the grounder transparently converts strings to integers if no string processing operations are required. Moreover, we note that measuring instantiation time with Clingo is impossible, as Clingo does not report preprocessing and instantiation times separately. This makes comparisons with other systems difficult, hence we opted to compare mainly with Gringo+Clasp, which, surprisingly, also uses slightly less memory than Clingo.

References

1. Alviano, M., Dodaro, C., Marques-Silva, J., Ricca, F.: Optimum stable model search: algorithms and implementation. J. Logic Comput. (2015). doi:10.1093/logcom/exv061
2. Alviano, M., Dodaro, C.: Anytime answer set optimization via unsatisfiable core shrinking. Theor. Pract. Logic Program. 16(5–6), 533–551 (2016)
3. Alviano, M., Dodaro, C., Leone, N., Ricca, F.: Advances in WASP. In: Calimeri, F., Ianni, G., Truszczynski, M. (eds.) LPNMR 2015. LNCS (LNAI), vol. 9345, pp. 40–54. Springer, Cham (2015). doi:10.1007/978-3-319-23264-5_5
4. Andres, B., Kaufmann, B., Matheis, O., Schaub, T.: Unsatisfiability-based optimization in clasp. In: International Conference on Logic Programming (ICLP), Technical Communications, pp. 212–221 (2012)
5. Ansótegui, C., Bonet, M.L., Levy, J.: SAT-based MaxSAT algorithms. Artif. Intell. 196, 77–105 (2013)
6. Baral, C.: Knowledge Representation, Reasoning, and Declarative Problem Solving. Cambridge University Press, New York (2004)

7. Brewka, G., Eiter, T., Truszczynski, M.: Answer set programming at a glance. Commun. ACM **54**(12), 92–103 (2011)
8. Calimeri, F., Faber, W., Gebser, M., Ianni, G., Kaminski, R., Krennwallner, T., Leone, N., Ricca, F., Schaub, T.: ASP-Core-2 Input language format. Technical report, ASP Standardization Working Group (2012)
9. Chomicki, J., Marcinkowski, J.: Minimal-change integrity maintenance using tuple deletions. Inf. Comput. **197**(1), 90–121 (2005)
10. Clark, J.H., González-Brenes, J.P.: Coreference resolution: current trends and future directions. Lang. Stat. II Lit. Rev. (2008). http://www.cs.cmu.edu/~jhclark/
11. Denis, P., Baldridge, J.: Global joint models for coreference resolution and named entity classification. Procesamiento del Lenguaje Nat. **42**, 87–96 (2009)
12. Eiter, T., Fink, M., Schüller, P., Weinzierl, A.: Finding explanations of inconsistency in multi-context systems. Artif. Intell. **216**, 233–274 (2014)
13. Eiter, T., Fink, M., Stepanova, D.: Towards practical deletion repair of inconsistent DL-programs. In: European Conference on Artificial Intelligence, pp. 285–290 (2014)
14. Erdem, E., Lifschitz, V.: Tight logic programs. Theor. Pract. Logic Program. **3**(4–5), 499–518 (2003)
15. Gebser, M., Janhunen, T., Rintanen, J.: ASP encodings of acyclicity properties. J. Logic Comput. **25**, 613–638 (2015)
16. Gebser, M., Kaminski, R., Kaufmann, B., Schaub, T.: Answer Set Solving in Practice. Morgan Claypool (2012)
17. Gebser, M., Kaminski, R., König, A., Schaub, T.: Advances in *gringo* series 3. In: Delgrande, J.P., Faber, W. (eds.) LPNMR 2011. LNCS (LNAI), vol. 6645, pp. 345–351. Springer, Heidelberg (2011). doi:10.1007/978-3-642-20895-9_39
18. Gebser, M., Kaufmann, B., Schaub, T.: Conflict-driven answer set solving: from theory to practice. Artif. Intell. **187–188**, 52–89 (2012)
19. Hirst, G.: Anaphora in Natural Language Understanding: A Survey. Springer, New York (1981)
20. Kehler, A., Kertz, L., Rohde, H., Elman, J.L.: Coherence and coreference revisited. J. Semant. **25**(1), 1–44 (2008)
21. Lee, H., Chang, A., Peirsman, Y., Chambers, N., Surdeanu, M., Jurafsky, D.: Deterministic coreference resolution based on entity-centric, precision-ranked rules. Comput. Linguist. **39**(4), 885–916 (2013)
22. Lifschitz, V.: What is answer set programming? In: AAAI Conference on Artificial Intelligence pp. 1594–1597 (2008)
23. Ng, V.: Supervised noun phrase coreference research: the first fifteen years. In: Association for Computational Linguistics (ACL), pp. 1396–1411 (2010)
24. Recasens, M., Hovy, E.: BLANC: implementing the rand index for coreference evaluation. Nat. Lang. Eng. **17**(4), 485–510 (2010)
25. Sapena, E., Padró, L., Turmo, J.: Coreference Resolution. Technical report, TALP Research Center, UPC (2008)
26. Say, B., Zeyrek, D., Oflazer, K., Özge, U.: Development of a corpus and a treebank for present day written Turkish. In: Current Research in Turkish Linguistics (Proceedings of the Eleventh International Conference of Turkish Linguistics, 2002), pp. 183–192. Eastern Mediterranean University Press (2004)
27. Schüller, P.: Adjudication of coreference annotations via finding optimal repairs of equivalence relations. In: International Workshop on Experimental Evaluation of Algorithms for Solving Problems with Combinatorial Explosion (RCRA), pp. 57–71 (2016)

28. Schüller, P.: Modeling variations of first-order horn abduction in answer set pro-gramming. Fund. Inform. **149**, 159–207 (2016)
29. Stoyanov, V., Eisner, J.: Easy-first coreference resolution. In: International Con-ference on Computational Linguistics (COLING), pp. 2519–2534 (2012)

Author Index

Printed in the United States
By Bookmasters